Introduction to Clinical Aspects of the Autonomic Nervous System

Introduction to Clinical Aspects of the Autonomic Nervous System

Volume 2

Sixth Edition

OTTO APPENZELLER
University of New Mexico, Albuquerque, NM, United States; New Mexico Health Enhancement and Marathon Clinics, Research Foundation, Albuquerque, NM, United States

GUILLAUME J. LAMOTTE
Department of Neurology, Movement Disorders, and Autonomic Disorders, The University of Utah, Salt Lake City, UT, United States

ELIZABETH A. COON
Department of Neurology, Autonomic Disorders, Mayo Clinic, Rochester, MN, United States

ELSEVIER

ACADEMIC PRESS
An imprint of Elsevier

Academic Press is an imprint of Elsevier
125 London Wall, London EC2Y 5AS, United Kingdom
525 B Street, Suite 1650, San Diego, CA 92101, United States
50 Hampshire Street, 5th Floor, Cambridge, MA 02139, United States
The Boulevard, Langford Lane, Kidlington, Oxford OX5 1GB, United Kingdom

ISBN: 978-0-323-96011-3

For Information on all Academic Press publications
visit our website at https://www.elsevier.com/books-and-journals

Publisher: Nikki P. Levy
Acquisitions Editor: Natalie Farra
Editorial Project Manager: Sam Young
Production Project Manager: Omer Mukthar
Cover Designer: Vicky Pearson Esser

Typeset by MPS Limited, Chennai, India

Contents

Foreword

Otto Appenzeller, Emeritus Professor of Neurology at the University of New Mexico, is a pioneer in the autonomic field, converting this "Cinderella of medicine" from a research curiosity to medical practice. His initial book enabled some of that thrust. The foreword by Ray Adams to the fifth edition captures well the situation with clinicians of the day when a young investigator from Sydney, Australia, arrived at the Massachusetts General Hospital. He reveled in the academic environment and the influence of neurologic giants like Ray Adams and C. Miller Fisher. In turn, he provides novel insights into autonomic dysfunction in disorders like Guillain–Barré syndrome and acute pandysautonomia. The sustainability of the effort is evident in this sixth edition, now grown into two volumes with 25 chapters. The volumes reflect Otto's interest in autonomic physiology and the historical evolution of some autonomic disorders. The book also provides historical aspects of some autonomic disorders.

The sixth edition greatly benefited from the addition of two new authors, Elizabeth Coon and Guillaume Lamotte, who bring into the field cutting-edge autonomic neurology and autonomic neuroscience. Their involvement has resulted in a revision of each of the chapters. Additionally, they have added their experience derived from the clinical autonomic laboratory, adding the quantitative dimension, describing the severity and distribution of autonomic failure. Their contributions include standardized autonomic function tests, modern biomarkers, novel chapters on postural tachycardia syndrome (POTS), and the dys-autonomias related to Covid-19 infection. This blend of the unique aspects of Otto's contributions with evolving new approaches promises to give the book a new life.

Phillip A. Low

Acknowledgments

Nearly half a century since the appearance of the first edition should have increased the number of individuals who importantly contributed to the success of this book. Surprisingly, however, modern technology and easy access to databases have allowed many helpful labors for the sixth edition to remain anonymous. This is, for the most part, regrettable because human interactions added significant dimensions to the readability, relevance, and clinical immediacy of the book. We would like to thank Dr. Negin Badihian for her assistance with the figures and the cover of the book. We must also not forget those who contributed generously to previous editions: Drs. T. K. Von Storch, G. B. Marcus, and D. Scott (first edition), Dr. G. Ogin (second edition), Dr. E. Collins and Prof. Yen Tsai (third edition), Drs. M. Appenzeller, P. Appenzeller, S. Wood, and R. Greene, and my son Tim Appenzeller, now on the editorial staff of Science (fourth edition). Previous editions were typed, proofread, and shepherded through the publishing process by many devoted helpers: Mrs. Grace Wilson, Mrs. Polly Gauthier, and Mrs. Vi Farmer (first edition), Mrs. Connie Sokolowski (second edition), Ms. Katherine Miller (third edition), and Ms. Pamela Livingston (fourth edition). The authors themselves used computer technology to produce the fifth and sixth editions.

Throughout the quarter of the century of the development of the book, and more so recently, our families have supported our endeavors with great understanding and sacrifice and we acknowledge with gratitude the only human interactions for the sixth edition.

Many new illustrations were taken from other sources, and references to the generous permissions for reproduction are found in the legends. Previously published materials are cited in the list of references at the end of the book.

Elsevier Science has shown a remarkably sustained interest in the Autonomic Nervous System.

Introduction

Since the appearance of the fifth edition a number of important advances in knowledge needed to be made available to clinicians. These include the effects of the coronavirus that emerged in December 2019 (Covid-19) on the autonomic nervous system and updates in other clinical autonomic disorders. In this volume, we discuss important clinical autonomic disorders such as syndromes associated with chronic autonomic failure, peripheral autonomic neuropathies, autonomic dysfunction in patients with spinal cord injuries, or disorders associated with chronic orthostatic intolerance. We also discuss autonomic testing in the laboratory and autonomic function in circadian rhythms, sleep, pupillary function, aging, exercise, and exposure to an extreme environment (high altitude and space). Similar to Volume 1 of the sixth edition, the reader will find key references to historical studies with foundational work that still influences the field of autonomic medicine today as well as up-to-date references.

Otto Appenzeller
Guillaume J. Lamotte
Elizabeth A. Coon

CHAPTER 1

Circadian rhythms

1.1 Introduction

The designation, circadian, was introduced by Halberg in 1959 and is now generally used to denote a 24-hour rhythm in body function (Halberg, 1959). Cycle is used for the repeating part of the rhythm. The time occupied by a cycle is called a period. The German word Zeitgeber denotes environmental influences that set the phase of biological rhythms. In this section the circadian variations in an environment that may act as Zeitgeber will not be considered, even though 40 have been recognized (Aschoff, 1955). How much of this rhythmicity in human body function is of endogenous origin, that is, biologic or due to an internal clock or produced by Zeitgeber is uncertain (Brown, 1959; Logan & McClung, 2019).

In fasting and resting subjects, body temperature shows circadian variations and it is lowest in the early morning (Bornstein & Völker, 1926). The rhythmicity in metabolic rate which accompanies the change in temperature is thought to be caused by the circadian temperature oscillation which is about 1°C. This circadian rhythm probably depends on variation in the activity of temperature-regulating mechanisms rather than changes in any one component of this mechanism (Mills, 1964). When changes in light-darkness or working time are made, it takes anything from 3 days to 3 weeks for the new rhythmicity to appear. There is, therefore, some evidence for endogenous rhythmicity of temperature regulation which is initiated by the hypothalamus, but it is not known if this rhythm resides in the hypothalamus or if it occurs in response to the activity of other structures.

The rhythm of the pulse rate seems to adapt immediately to 18- or 28-hour days and may under certain circumstances be completely dissociated from temperature periodicity (Kleitman & Kleitman, 1953). This suggests that pulse rate periodicity is largely determined by rest and activity, rather than by endogenous factors.

Sleep and wakefulness show circadian oscillations. Kleitman studied two subjects in a cave, living on a 28-hour day schedule. One of the subjects

Introduction to Clinical Aspects of the Autonomic Nervous System
DOI: https://doi.org/10.1016/B978-0-323-96011-3.00006-1

adapted his sleep-wakefulness and temperature cycle but the other did not. At the end of 1 week, they were exactly out of phase. The subject who did not adapt maintained his 24-hour cycle throughout the 32-day study period and the other persisted for some time in the 28-hour cycle after emerging from the cave (Kleitman, 1949). These observations and others (Mills, 1964; Von Aschoff & Wever, 1962) show that the sleep—wakefulness rhythm is probably endogenous, but whether it arises in the central nervous system or is secondary to the changes in body temperature or adrenal activity is not known.

The circadian rhythms, though sensitive to light in almost all mammals, were hitherto thought to be insensitive to light in humans. The synchronization of circadian rhythms to the 24-hour period was attributed in humans to social contacts and the sleep-waking schedule.

1.2 Effect of light on the circadian system

Human neuroanatomic structures necessary for light entrainment of circadian rhythms are the same as those found in other mammals, where the rhythmicity is related to photic entrainment. The discovery of intensity-dependent neuroendocrine response to bright light in animals led researchers to study the response in humans. In one such study, using only a single subject, the subject's neuroendocrine levels were measured over 7 consecutive evenings before and after exposure to bright light. The sleep—wake cycle and social contacts remained unaltered during the experiment. When the subject was exposed to bright light in the evening, a 6-h delay in the shift of the circadian pacemaker, evidenced by recordings of body temperature and cortisol secretion, was found. The shift was rapid and large and remained as stable as the circadian rhythms recorded before the experiment. These results change previous ideas, mentioned elsewhere, about circadian phase resetting capacity in humans, and imply that exposure to very bright light can reset the human circadian pacemaker, which controls a large number of daily physiologic, behavioral, and cognitive rhythms (Czeisler et al., 1986, 1989). Indeed, exposure to bright light can desynchronize circadian rhythms from their normal periodicity (external desynchronization). For instance, a circadian rhythm can be entrained to periods slightly longer or shorter than 24 hours by manipulating the normal light-dark cycle. Research has indicated that the response of the circadian pacemaker to light depends on the timing, intensity, duration, and the number of consecutive daily exposures to light. The brighter the light, the greater is its entrainment effect. A properly timed and intensity light exposure may even halt the circadian clock (Czeisler, 1995).

Air travel across time zones, by producing an abrupt phase shift of the normal light-dark cycle, can result in a transient external desynchronization of the sleep–wake cycle. Changes in the sleep–wake cycle are usually accompanied by alteration of other rhythms and the phase relationships that they normally keep with each other (internal desynchronization). The typical manifestation of this type of desynchronization is known as jet lag (Winget, DeRoshia, Markley, & Holley, 1984). The severity and duration of jet lag depend on the distance traveled across time zones (i.e., the degree of shift of the dark-light cycle). Because free-running rhythms are slightly greater than 24 hours, the severity of jet lag also depends on the direction of travel. Westbound travel, which results in a subjectively longer day, is, therefore, better tolerated than eastbound travel. As bright light entrains circadian rhythms, exposure to light can be used as a treatment to normalize the phase mismatches between circadian rhythms and the light-dark cycle and between forced sleep–wake periods.

While dramatically altered light-dark cycles such as those encountered by space flight crews during terrestrial orbits (the light-dark cycle lasts 80–140 minutes with 30%–40% corresponding to darkness) can be very disruptive of sleep–wake rhythms, satisfactory sleep and alertness have been maintained by keeping a constant sleep–wake cycle relation with the normal terrestrial rhythm (Stampi, 1994).

Endogenous oscillators have major roles in anticipating the fluctuations in the environment associated with the Earth's rotations on its axis every 24 hours. However, there are also lunar-related cycles which include semilunar (~15 days) and lunar cycles (29.5 days) which have important implications for the reproductive cycles of organisms. These cycles, circadian and noncircadian, entrain to environmental "Zeitgebers" (time-givers). The most important of is the light cycle, but temperature cycles, social stimuli, and seasonal photoperiodic changes can also be effective in "entraining" a biological clock to its optimal phase. It is entrainment that determines the time of day or chronotype, yet this property of circadian clocks is less well understood than free-running rhythm.

Extreme environmental conditions such as constant light or long or short photoperiods that are prevalent in polar regions also have effects as yet not fully understood in humans (Costa & Kyriacou, 2021).

1.3 Effects of melatonin on the circadian system

Galen (129 AD–216) named the pineal gland after the nuts found in the cones of the stone pine but the firm establishment of the pineal gland as an endocrine organ emerged later following the isolation of melatonin in 1958 (Lerner, Case, & Takahashi, 1960). Melatonin (N-acetyl-5-methoxytryptamine), produced by the pineal gland during darkness at night and suppressed by exposure to bright light, has recently been shown to play a major role in the regulation of circadian rhythms. Melatonin stimulates the suprachiasmatic nucleus providing an additional mechanism in which the light-dark cycle can entrain the circadian pacemaker. The use of the melatonin rhythm as a marker for the biological clock, and the use of melatonin levels as a practical method of following the different phases of the circadian rhythm (Lewy, Wehr, Goodwin, Newsome, & Markey, 1980), has permitted important advances in the study of circadian rhythms. Levels assayed in blood (or saliva) are most significant when collected in dim light (dim light melatonin onset). Loss of rhythms in melatonin has been observed in several neurodegenerative diseases and melatonin supplementation may have mild beneficial effects in dementia (Riemersma-van der Lek et al., 2008). The pineal gland is photosensitive in lower vertebrates, whereas in humans information about the light-dark cycle reaches the pineal gland via neuronal networks emerging from retinal ganglion cells containing melanopsin. The pineal gland is largely sympathetically innervated and pinealocytes are thought to be of two distinct subtypes: type 1 "light" cells secreting serotonin and type 2 "dark" cells secreting melatonin.

In humans, a physiological (0.5 mg) dose of melatonin shifts the circadian pacemaker by about 12 hours out of phase with the cycle entrained by normal light (Czeisler et al., 1989; Lewy, Ahmed, Jackson, & Sack, 1992). The property of melatonin to shift circadian rhythms has been used to treat jet lag and selected sleep disturbances. Melatonin appears to act as a dark pulse: when taken in the late afternoon, it acts as if darkness was extended into the afternoon and sleep onset is advanced. When taken in the morning, its effects are similar to extending darkness into the morning and sleep termination is delayed. For westward travel, melatonin has been taken at the destination's local bedtime for 4 or 5 days after arrival. For eastbound travel, preflight treatment in the late afternoon for 1 or 2 days is followed by local bedtime doses after arrival.

There is evidence for wider roles for melatonin. For example, melatonin may play an important role in the reduction of neurovascular oxidative stress (Karolczak & Watala, 2021) and in cancer therapy as an inhibitor of hypoxia-induced tumor growth (Bastani, Akbarzadeh, & Rastgar Rezaei, 2021). Finally, pinealocytes are not the sole source of melatonin as indicated by the lack of specific sleep impairment and change in sleep—wake cycle after pinealectomy (Krieg, Slawik, & Meyer, 2012).

1.4 Other circadian rhythms in human

There is a circadian periodicity of corticosteroid secretion (Halberg, 1959). Peak values in normal subjects are found just before awakening (Bartter & Delea, 1962). This circadian rhythm of adrenal activity is apparently not induced by physical environmental factors. Isolated subjects in a room without daylight who reverse their sleeping habits, activity, and mealtimes also reverse plasma corticosteroid periodicity (Migeon et al., 1956). This rhythm is said to be absent in patients with impaired consciousness (Eik-Nes & Clark, 1958). Studies of circadian rhythm of 17-hydroxycorticosteroid content in the plasma in patients with circumscribed pretectum, temporal lobe, or hypothalamic lesions have shown marked abnormalities. It is postulated that this is due to interference with regulatory pathways which affect ACTH release (Krieger & Krieger, 1966).

Circadian rhythmic changes of 5-hydroxytryptamine have been found in the serum of healthy males and patients with mental retardation. 5-Hydroxytryptamine was evaluated in the serum after clotting which was presumably released from platelets. The crest of this circadian rhythmicity in subjects living on a normal routine daily activity occurred between 4 and 14 hours. The cause of the rhythmicity is at present not clear (Halberg, Anderson, Ertel, & Berendes, 1967).

Distinct diurnal variations in serotonin levels have been demonstrated both in whole brain and in specific regions of the brain. The highest levels occur during the period of light. The possibility that the changing levels in serotonin are related to sleep and to some episodic disorders with a tendency to make their appearance at certain times of the light-dark cycle must be entertained. For example, serotonin is a vasoactive substance that might have an important role in the genesis of headaches such as migraine or cluster headaches. Whether the tendency for cluster headaches to appear at night can be related to the diurnal variability of serotonin in specific regions of the brain or perhaps to the varying levels in the blood is uncertain. Measurements of serotonin and 5-hydroxyindoleactic acid (5HIAA) in patients with cluster headache suggest an increase in serotonin metabolism and involvement in central serotoninergic networks contributing to cluster headache pathogenesis (D'Andrea, Granella, Alecci, & Manzoni, 1998).

The excretion of urine and its various components also has a circadian rhythm, but this does not have any direct relation to the nervous system because it is found in denervated transplanted kidneys. The nervous system may, however, indirectly affect urine flow and its composition through its control of cortisol release.

It has been known for some time that the eosinophilic and lymphocytic counts in the peripheral blood fall and the neutrophil count rises shortly after waking (Bartter & Delea, 1962; Sharp, 1960; Von Domarus, 1931). There is some evidence that this leukocytic rhythm depends on adrenocortical activity (Doe, Flink, & Goodsell, 1956), since the morning leukopenia does not occur in Addison's disease or after bilateral adrenalectomy (Halberg, Visscher, Flink, Berge, & Bock, 1951). A circadian rhythm has been found in the erythrocyte production in patients with active blood regeneration by counting reticulocytes in the peripheral blood (Goldeck, 1948).

Periodicity in different body functions emerges in infancy. A slow pulse rate at night is first seen at 6 weeks of age and the difference from the pulse rate observed during the day increases after the first year. The appearance of pulse rate periodicity coincides with that of body temperature which is lowest at 2300 hours, and it is thought that the temperature is associated with the pulse rate rhythm in this age group (Kleitman & Ramsaroop, 1948).

Evidence for an effect of vitamin B_{12} on the circadian clock has come from human experiments which show that the injection of the vitamin, but not placebo, causes an elevation of the rectal temperature toward the evening hours and this elevation coincides with increased alertness as assessed by a visual analog scale (Uchiyama, Mayer, Okawa, & Meier-Ewert, 1995).

The distribution of sleep between day and night becomes unequal at about 3–6 weeks of age (Kleitman & Engelmann, 1953). Periodicity in the flow of urine is first noticed at 4 weeks of age. The height of urine flow occurs from 0600 to 1400 hours. Sodium and potassium excretion become periodic at 15 weeks. It is not known if the appearance of these rhythms is due to the maturation of an endogenous clock or if maturation itself permits the infant to respond to Zeitgeber (Mills, 1966).

1.5 Aging and circadian rhythms

Experiments in rats have shown that aging alters the circadian rhythms for glucose utilization and alpha1-adrenoceptor levels in the suprachiasmatic nucleus and alters the circadian preovulatory release of luteinizing hormone (Wise, Cohen, Weiland, & London, 1988). While the mechanisms underlying the age-related circadian rhythm changes are unknown, some evidence suggests a role for decreased catecholamine function (Meites, 1990). Circadian rhythms present in young animals can be aged by depleting brain monoamines and circadian rhythms in older animals can be rejuvenated by transplantation of fetal suprachiasmatic tissue (Turek et al., 1995).

As melatonin rhythms are gradually lost with aging, melatonin and the pineal gland have been implicated in the processes of both aging and age-related diseases (Reiter, 1995). Some theories have relied on the progressive loss of the circadian rhythmicity of melatonin to explain the weakening and desynchronization of other circadian rhythms that are believed to contribute to aging and to increased susceptibility to age-related diseases. Other theories propose that the progressive loss of the melatonin cycle provides a switch for genetically programmed aging at the cellular level and that the waning of the melatonin rhythm determines the rate of aging. In addition, as melatonin is the most potent hydroxyl radical scavenger known and, thus, may have a role in protecting macromolecules, especially DNA, against free radical attack, an age-related decrease in melatonin may contribute to aging and the onset of age-related diseases.

Some pathological conditions have circadian rhythmicity and this may be contributed to by physiological changes. An example is the occurrence of hemoptysis commonly seen between 0600 and 0900 hours and 1800 and 2100 hours. This coincides with circadian changes in the blood content of the lungs (Dissmann, 1950). There is also an increase in the incidence of deaths during the early morning hours which may depend on a similar mechanism (Halberg, 1960). Numerous studies have shown a variety of circadian rhythms, but whether these are consequent on the setting of an internal clock or an unrecognized Zeitgeber is not clear. Moreover, this rhythmicity may merely be the outward manifestation of an imposed change of habit or environment. The circadian clock, operative in some rhythmic changes, is tentatively sited in the hypothalamus, which is acted upon by the cortex and it, in turn, influences many organs through numerous neural and endocrine effector systems.

In animals, ablation of the suprachiasmatic nuclei eliminates some circadian rhythms. These nuclei make a substantial contribution to the organization of the rhythms. The female estrous, a longer period rhythm, is also regulated in part by circadian mechanisms and is eliminated by suprachiasmatic ablation. The necessity of the suprachiasmatic nuclei for proper organization of circadian rhythm suggests that the neurons in this area of the hypothalamus generate their own rhythms but that this is probably just an orchestrating function and the mammalian pacemaker activities are more widespread. Experiments with hamsters suggest that nitric oxide (NO) production is involved in the light-induced phase delay of the hamster's circadian system because pretreatment with N-nitro-L-arginine methyl ester (L-NAME), an inhibitor of NO production, significantly attenuated light-induced phase delays of hamster wheel running activity at night. The effect of L-NAME was reversed by the coadministration of L-arginine. These effects of NO manipulation were attributed to the production of NO by the suprachiasmatic nucleus (Watanabe, Ono, Shibata, & Watanabe, 1995).

Some circadian rhythms survive suprachiasmatic ablation and some tissues can continue circadian rhythmicity even if isolated from the animal and they continue responsive to hormonal influences which are known to modify the properties of the pacemaker system.

Several elements have been identified in neurons that are necessary for adequate modeling of circadian pacemakers. These include single endogenous oscillators or populations of oscillator mechanisms in the suprachiasmatic nuclei, which integrate oscillations into a circadian frame. Visual projections are capable of transmitting the effect of light on circadian rhythms, and endocrine glands produce hormones that affect the activity of circadian systems. Not all mechanisms have been fully identified, and it remains a rich field for research to unravel the complicated interactions between environmental entrainment stimuli and endogenous physiologic rhythm mechanisms (Rusak, 1979).

One approach for the study of photoentrainment is pharmacologic. The use of drugs could shed light on neurotransmitters important in photoentrainment and the mechanism by which they influence circadian pacemakers. The cholinergic agonist carbachol mimics the effect of light on rat pineal serotonin-*N*-acetyltransferase. Therefore, it is probable that acetylcholine is involved in photoentrainment of mammalian circadian rhythms (Zatz, 1979). Light acts via the retinohypothalamic projection to the suprachiasmatic nucleus of the hypothalamus. Therefore, acetylcholine seems to be the responsible neurotransmitter in this tract. But the true mechanism of carbachol action and that of light in photoentrainment are as yet unknown.

Circadian rhythms in humans have been shown to persist in isolated subjects without environmental cues. These free-running rhythms usually last longer than 24 hours. If body temperature is measured over several days, it is found that the rhythm usually lasts 24 hours. In male subjects who exercised on a bicycle ergometer, there was no difference in temperature between periods of exercise carried out seven times per 24 hours and nonexercising periods. Circadian rhythms are usually independent of physical workload and external perturbations are compensated for by as yet unknown mechanisms. The deviation of free-running periods in humans from 24-hour cycles observed in isolated individuals is not due, as previously suggested, to lack of exercise during the isolation periods (Wever, 1979).

Behavioral studies suggest that the cumulative effects of daily living cause fatigue, which, figuratively speaking, can be seen as a vessel made empty by the outflow of energy. Energy flows out of the full vessel in response to need: tasks must be performed and obstacles overcome. A person placed in an environment with poor illumination, excessive heat or cold, loud noise, and poor ventilation will soon be fatigued by the struggle of coping with these pressures. A person suffering from anxiety, disease, or poor diet will also experience fatigue. So, too, will a person burdened by responsibilities or overworked either physically or mentally.

Physiologic measurement of fatigue of muscles (Edwards, 1986) suggests that fatigue may originate in the brain. This type of fatigue is characterized by less force generated by voluntary effort than the force produced by direct electrical stimulation of the same muscles. This central type of fatigue is thought to result from failure to activate voluntarily the same number of motor units or with the same stimulation frequency of motor units, as is possible with direct electrical stimulation of muscles. Another type of fatigue may originate in the periphery. In this type, force is lost both after repeated voluntary contraction or stimulated contraction of a muscle. This peripheral type of fatigue is thought to be due to impaired neuromuscular transmission or impairment of propagation of muscle-action potentials. Peripheral fatigue may be of low-frequency type when selective loss of force occurs at low-stimulation frequencies with a decrement of surface-recorded action potentials. This type may be of low frequency also when selective loss of force at low-stimulation frequencies is observed without decrement of surface-recorded surface action potentials. Peripheral types of fatigue are thought to result from impaired excitation-contraction coupling in muscles.

Another measure of fatigability is the magnitude of the perceived exertion of the exercising individual. A behavioral scale from no fatigue (6) to unbearable (20) is widely employed to measure the effect of exercise on fatigability. The concept of perceived exertion was first proposed by Borg in 1962 (Borg, 1962). The correlation between the rating of perceived exertion (RPE) and exercise-induced cardiac acceleration has been found to be very high. A modification of the rather cumbersome RPE to a symptom scale has been published. The two scales have been used and favorably compared in patients with coronary artery disease (Hare, Hamid, & Hakki, 1985).

Fatigability is a leading symptom in some patients with multiple sclerosis. Amantadine hydrochloride, given in the usual doses (100 mg twice daily), may be beneficial in about 60%–70% of patients in alleviating fatigability (Taus, Giuliani, Pucci, D'Amico, & Solari, 2003). The mechanism by which fatigability is improved with this drug remains, however, obscure.

In a double-blind sequential study using amantadine and placebo in 10 patients with multiple sclerosis, the effectiveness of the drug was confirmed in 60% of patients. Responders to amantadine, when compared with nonresponders, had significant increases in b-endorphin-b-lipotropin in their circulation. A significant decrease in blood lactate was found in responders to the drug. This study confirms the usefulness of the drug in treating the fatigue that accompanies multiple sclerosis and shows that a measurable effect on neuropeptides in the circulation can accompany the improvement in fatigability (Rosenberg & Appenzeller, 1988). Hypothetically, the drug may act by releasing dopamine and noradrenaline from central stores. These catecholamines are known to stimulate opioid receptors in the brain. Consequently, a secondary release of opioids from the pituitary into the circulation may occur, accounting for the well-being and decrease in fatigue.

A similar effect of amantadine is thought to underlie its usefulness in the treatment of withdrawal symptoms from cocaine in former addicts, where fatigability forms part of the withdrawal syndrome (Tennant & Sagherian, 1987). This effect on fatigability in multiple sclerosis may not be specific to the disease and the drug was also effective in some patients with other brain diseases (previous brain abscess).

1.6 Circadian disturbances and the COVID-19 pandemic

The COVID-19 pandemic has changed lifestyle and well-being. Sleep and circadian rhythms have important effects on physiology, behavior, emotion, and cognition. A "circadian misalignment" has been demonstarted in the entire population due to the pandemic. Home confinement ("lockdown") is a major factor in driving circadian misalignment and poor sleep quality. The effects on recovery from infection and on emotional well-being remain to be evaluated (Salehinejad, Azarkolah, Ghanavati, & Nitsche, 2021).

1.7 Molecular and circuit-based aspects of the circadian system

At a cellular level, there is a set of conserved clock proteins with a transcriptional-translational feedback loop that mediates daily oscillations in gene expression. The basic helix-loop-helix PER-ARNT Sim transcriptional factor BMAL1, which heterodimerizes with CLOCK (or NPAS2), drives transcription of a wide array of clock-controlled genes. These clock-controlled genes regulate key processes involved in neurodegeneration such as redox homeostasis, inflammation, proteostasis, and metabolism. The circadian clock oscillates in a cell-autonomous manner and is tuned to a 24-hour period by multiple layers of posttranslational regulation. The suprachiasmatic nucleus of the hypothalamus, which receives light:dark input from the retina, is responsible for the synchronicity between oscillators at the cellular level and between the different organs in the body. Sleep influences the cellular circadian clock, the peripheral circadian clock, and the activity of the suprachiasmatic nucleus in a bidirectional relationship. Therefore the interplay between the circadian clock and sleep is complex. Mutations in circadian clock genes manifest as sleep disturbances and sleep deprivation can alter the expression of clock genes (Musiek & Holtzman, 2016).

1.8 Circadian system and neurodegeneration

The circadian system regulates hippocampal-dependent learning and has potent effects on cognition (Smarr, Jennings, Driscoll, & Kriegsfeld, 2014). Impairment of the circadian system has been found in neurodegenerative disorders. In Alzheimer's disease, there is asynchronous clock gene expression between different brain regions and abnormal expression of clock genes in peripheral blood cells (Cermakian, Lamont, Boudreau, & Boivin, 2011). Beta-amyloid, a protein that is associated with the pathogenesis of Alzheimer's disease, facilitates BMAL1 degradation in neuronal cells (Song et al., 2015).

Disrupted or abnormal activity of the suprachiasmatic nucleus is also evident in other neurodegenerative conditions such as Parkinson's disease or Huntington's disease. Epidemiological studies suggest that less robust circadian rhythms or more fragmented activity patterns are risk factors for future dementia (Tranah et al., 2011). Loss of circadian regulation of glucose and lipid metabolism, immune system function, and hormone secretion could predispose the brain to degeneration through increased inflammation and impaired neurogenesis. Disruption of the suprachiasmatic nucleus-mediated rhythms occurs during jet lag when an individual flies across several time zones. Intercontinental flight attendants subjected to frequent jet lag exhibited hippocampal atrophy when compared to nonjetlagged colleagues (Cho, 2001).

The circadian clock and sleep are interconnected systems that impact brain function and neurodegeneration by multiple mechanisms. The exact function of the core circadian clock in different brain regions and cell types is poorly understood, and future studies should explore how specific clock-controlled pathways in the brain might influence neurological and psychiatric disorders.

References

Aschoff, J. (1955). Exogene und endogene Komponente der 24-Stunden-Periodik bei Tier und Mensch. *Die Naturwissenschaften*, *42*, 569–575.

Bartter, F. C., & Delea, C. S. (1962). A map of blood and urinary changes related to circadian variations in adrenal cortical function in normal subjects. *Annals of the New York Academy of Sciences*, *98*, 969–983.

Bastani, S., Akbarzadeh, M., Rastgar Rezaei, Y., et al. (2021). Melatonin as a therapeutic agent for the inhibition of hypoxia-induced tumor progression: a description of possible mechanisms involved. *International Journal of Molecular Sciences*, *22*(19), 10874.

Borg, G. (1962). *Physical Performance and Perceived Exertion*. Lund: Gleerup.

Bornstein, A., & Völker, H. (1926). Über die Schwankungen des Grundumsatzes. *Zeitschrift für die gesamte experimentelle Medizin*, *53*, 439–450.

Brown, F. A., Jr. (1959). Living clocks. *Science (New York, N.Y.)*, *130*(3388), 1535–1544.

Cermakian, N., Lamont, E. W., Boudreau, P., & Boivin, D. B. (2011). Circadian clock gene expression in brain regions of Alzheimer's disease patients and control subjects. *Journal of Biological Rhythms*, *26*(2), 160–170.

Cho, K. (2001). Chronic 'jet lag' produces temporal lobe atrophy and spatial cognitive deficits. *Nature Neuroscience*, *4*(6), 567–568.

Costa, R., & Kyriacou, C. (2021). Editorial: entrainment of biological rhythms. *Frontiers in Physiology*, *12*, 757000.

Czeisler, C. A. (1995). The effect of light on the human circadian pacemaker. *Ciba Foundation Symposium*, *183*, 254–290, discussion 290–302.

Czeisler, C. A., Allan, J. S., Strogatz, S. H., Ronda, J. M., Sánchez, R., Ríos, C. D., ... Kronauer, R. E. (1986). Bright light resets the human circadian pacemaker independent of the timing of the sleep-wake cycle. *Science (New York, N.Y.)*, *233*(4764), 667–671.

Czeisler, C. A., Kronauer, R. E., Allan, J. S., Duffy, J. F., Jewett, M. E., Brown, E. N., & Ronda, J. M. (1989). Bright light induction of strong (type 0) resetting of the human circadian pacemaker. *Science (New York, N.Y.)*, *244*(4910), 1328–1333.

D'Andrea, G., Granella, F., Alecci, M., & Manzoni, G. C. (1998). Serotonin metabolism in cluster headache. *Cephalalgia: an International Journal of Headache*, *18*(2), 94–96.

Dissmann, E. (1950). Lungenblutung and 24 Stundenrhythmus (Studien zur klinischen Rhythmologie). *Acta Tubera Scandinav.*, *24*, 277–292.

Doe, R. P., Flink, E. B., & Goodsell, M. G. (1956). Relationship of diurnal variation in 17-hydroxycorticosteroid levels in blood and urine to eosinophils and electrolyte excretion. *The Journal of Clinical Endocrinology and Metabolism*, *16*(2), 196–206.

Edwards, R. H. T. (1986). Central vs. peripheral mechanisms of fatigue in exercise. *Muscle and Nerve*, *9*(Suppl. 39), 5.

Eik-Nes, K., & Clark, L. D. (1958). Diurnal variation of plasma 17-hydroxycorticosteroids in subjects suffering from severe brain damage. *The Journal of Clinical Endocrinology and Metabolism*, *18*(7), 764–768.

Goldeck, H. (1948). Der 24 studen rhythmus der erythropoese. *Arzneimittel Forsch*, *2*, 22–27.

Halberg, F. (1959). [Physiologic 24-hour periodicity; general and procedural considerations with reference to the adrenal cycle]. *Internationale Zeitschrift fur Vitaminforschung. Beiheft*, *10*, 225–296.

Halberg, F. (1960). Temporal coordination of physiologic function. *Cold Spring Harbor Symposia on Quantitative Biology*, *25*, 289–310.

Halberg, F., Anderson, J. A., Ertel, R., & Berendes, H. (1967). Circadian rhythm in serum 5-hydroxytryptamine of healthy men and male patients with mental retardation. *International Journal of Neuropsychiatry*, *3*(4), 379–386.

Halberg, F., Visscher, M. B., Flink, E. B., Berge, K., & Bock, F. (1951). Diurnal rhythmic changes in blood eosinophil levels in health and in certain diseases. *The Journal-Lancet*, *71*(8), 312–319.

Hare, T. W., Hakki, D., & Lowenthal, D. T. (1985). Simplified scale for rating perceived exertion in patients with coronary artery disease. *Annals of Sports Medicine and Research*, *2*, 64–68.

Karolczak, K., & Watala, C. (2021). Melatonin as a reducer of neuro- and Vasculotoxic oxidative stress induced by homocysteine. *Antioxidants.*, *10*(8), 1178.

Kleitman, N. (1949). Biological rhythms and cycles. *Physiological Reviews*, *29*(1), 1–30.

Kleitman, N., & Engelmann, T. G. (1953). Sleep characteristics of infants. *Journal of Applied Physiology (Bethesda, Md.: 1985)*, *6*(5), 269–282.

Kleitman, N., & Kleitman, E. (1953). Effect of non-twenty-four-hour routines of living on oral temperature and heart rate. *Journal of Applied Physiology (Bethesda, Md.: 1985)*, *6*(5), 283–291.

Kleitman, N., & Ramsaroop, A. (1948). Periodicity in body temperature and heart rate. *Endocrinology*, *43*(1), 1–20.

Krieg, S. M., Slawik, H., Meyer, B., et al. (2012). Sleep disturbance after pinealectomy in patients with pineocytoma WHO°I. *Acta Neurochirurgica*, *154*, 1399–1405.

Krieger, D. T., & Krieger, H. P. (1966). Circadian variation of the plasma 17-hydroxycorticosteroids in central nervous system disease. *The Journal of Clinical Endocrinology and Metabolism*, *26*(9), 929–940.

Lerner, A. B., Case, J. D., & Takahashi, Y. (1960). Isolation of melatonin and 5-methoxyindole-3-acetic acid from bovine pineal glands. *The Journal of Biological Chemistry*, *235*, 1992–1997.

Lewy, A. J., Ahmed, S., Jackson, J. M., & Sack, R. L. (1992). Melatonin shifts human circadian rhythms according to a phase-response curve. *Chronobiology International*, *9*(5), 380–392.

Lewy, A. J., Wehr, T. A., Goodwin, F. K., Newsome, D. A., & Markey, S. P. (1980). Light suppresses melatonin secretion in humans. *Science (New York, N.Y.)*, *210*(4475), 1267–1269.

Logan, R. W., & McClung, C. A. (2019). Rhythms of life: circadian disruption and brain disorders across the lifespan. *Nature Reviews. Neuroscience*, *20*(1), 49–65.

Meites, J. (1990). Aging: hypothalamic catecholamines, neuroendocrine-immune interactions, and dietary restriction. *Proceedings of the Society for Experimental Biology and Medicine. Society for Experimental Biology and Medicine (New York, N.Y.)*, *195*(3), 304–311.

Migeon, C. J., Tyler, F. H., Mahoney, J. P., Florentin, A. A., Castle, H., Bliss, E. L., & Samuels, L. T. (1956). The diurnal variation of plasma levels and urinary excretion on 17-hydroxycorticosteroids in normal subjects, night workers and blind subjects. *The Journal of Clinical Endocrinology and Metabolism*, *16*(5), 622–633.

Mills, J. N. (1964). Circadian rhythms during and after three months in solitude underground. *The Journal of Physiology*, *174*(2), 217–231.

Mills, J. N. (1966). Human circadian rhythms. *Physiological Reviews*, *46*(1), 128–171.

Musiek, E. S., & Holtzman, D. M. (2016). Mechanisms linking circadian clocks, sleep, and neurodegeneration. *Science (New York, N.Y.)*, *354*(6315), 1004–1008.

Reiter, R. J. (1995). The pineal gland and melatonin in relation to aging: a summary of the theories and of the data. *Experimental Gerontology*, *30*(3–4), 199–212.

Riemersma-van der Lek, R. F., Swaab, D. F., Twisk, J., Hol, E. M., Hoogendijk, W. J., & Van Someren, E. J. (2008). Effect of bright light and melatonin on cognitive and noncognitive function in elderly residents of group care facilities: a randomized controlled trial. *JAMA: the Journal of the American Medical Association*, *299*(22), 2642–2655.

Rosenberg, G. A., & Appenzeller, O. (1988). Amantadine, fatigue, and multiple sclerosis. *Archives of Neurology*, *45*(10), 1104–1106.

Rusak, B. (1979). Neural mechanisms for entrainment and generation of mammalian circadian rhythms. *Federation Proceedings*, *38*(12), 2589–2595.

Salehinejad, M. A., Azarkolah, A., Ghanavati, E., & Nitsche, M. A. (2021). Circadian disturbances, sleep difficulties and the COVID-19 pandemic. *Sleep Medicine*, *91*, 246–252.

Sharp, G. W. G. (1960). Reversal of diurnal leukocyte variations in man. *The Journal of Endocrinology*, *21*, 107–114.

Smarr, B. L., Jennings, K. J., Driscoll, J. R., & Kriegsfeld, L. J. (2014). A time to remember: the role of circadian clocks in learning and memory. *Behavioral Neuroscience*, *128* (3), 283–303.

Song, H., Moon, M., Choe, H. K., Han, D. H., Jang, C., Kim, A., ... Mook-Jung, I. (2015). Aβ-induced degradation of BMAL1 and CBP leads to circadian rhythm disruption in Alzheimer's disease. *Molecular Neurodegeneration*, *10*, 13.

Stampi, C. (1994). Sleep and circadian rhythms in space. *Journal of Clinical Pharmacology*, *34* (5), 518–534.

Taus, C., Giuliani, G., Pucci, E., D'Amico, R., & Solari, A. (2003). Amantadine for fatigue in multiple sclerosis. *Cochrane Database of Systematic Reviews (Online)* (2), Cd002818.

Tennant, F. S., Jr., & Sagherian, A. A. (1987). Double-blind comparison of amantadine and bromocriptine for ambulatory withdrawal from cocaine dependence. *Archives of Internal Medicine*, *147*(1), 109–112.

Tranah, G. J., Blackwell, T., Stone, K. L., Ancoli-Israel, S., Paudel, M. L., Ensrud, K. E., ... Yaffe, K. (2011). Circadian activity rhythms and risk of incident dementia and mild cognitive impairment in older women. *Annals of Neurology*, *70*(5), 722–732.

Turek, F. W., Penev, P., Zhang, Y., Van Reeth, O., Takahashi, J. S., & Zee, P. (1995). Alterations in the circadian system in advanced age. *Ciba Foundation Symposium*, *183*, 212–226, discussion 226–234.

Uchiyama, M., Mayer, G., Okawa, M., & Meier-Ewert, K. (1995). Effects of vitamin B12 on human circadian body temperature rhythm. *Neuroscience Letters*, *192*(1), 1–4.

Von Aschoff, J., & Wever, R. (1962). Spontanperiodik des Menschen bei Ausschluss aller Zeitgeber. *Die Naturwissenschaften*, *49*, 337–342.

Von Domarus, A. (1931). Die Bedeutung der Kammerzählung der Eosinophilen für die Klinik. *Deutsches Archiv fur Klinische Medizin*, *171*, 333–358.

Watanabe, A., Ono, M., Shibata, S., & Watanabe, S. (1995). Effect of a nitric oxide synthase inhibitor, N-nitro-L-arginine methylester, on light-induced phase delay of circadian rhythm of wheel-running activity in golden hamsters. *Neuroscience Letters*, *192*(1), 25–28.

Wever, R. A. (1979). Influence of physical workload on freerunning circadian rhythms of man. *Pflugers Archiv: European Journal of Physiology*, *381*(2), 119–126.

Winget, C. M., DeRoshia, C. W., Markley, C. L., & Holley, D. C. (1984). A review of human physiological and performance changes associated with desynchronosis of biological rhythms. *Aviation, Space, and Environmental Medicine*, *55*(12), 1085–1096.

Wise, P. M., Cohen, I. R., Weiland, N. G., & London, E. D. (1988). Aging alters the circadian rhythm of glucose utilization in the suprachiasmatic nucleus. *Proceedings of the National Academy of Sciences of the United States of America*, *85*(14), 5305–5309.

Zatz, M. (1979). Photoentrainment, pharmacology, and phase shifts of the circadian rhythm in the rat pineal. *Federation Proceedings*, *38*(12), 2596–2601.

CHAPTER 2

Sleep and wakefulness

2.1 Introduction

The definition of sleep remains descriptive since the precise biological substrate of the sleep-wakefulness cycle is still not fully understood. Sleep is a reversible recurring state in which the organism remains for some time with reduced muscular activity and is less perceptive and reactive to the environment. The true biological function of sleep is poorly understood. Two main brain states of sleep exist—sleep with and without rapid eye movement (REM)—and each state plays an important role in maintaining homeostasis. There are sleep-inducing stimuli in the surroundings and tiredness is the subjective manifestation of these, as well as internal factors which culminate in sleep. The onset and termination of sleep seem to be regulated by a biological clock that regulates the alternating cycles of sleep and wakefulness and that is entrained by external cyclical stimuli (mainly the 24-hour light-dark cycle). Less powerful stimuli include meal timing and social interactions. The distinction between wakefulness and sleep seems an obvious one and yet there is evidence that many of the characteristics of sleep, with the exception of certain electroencephalogram (EEG) patterns and the Babinski response, can be obtained in recumbent relaxed subjects. On the other hand, the change from the waking state to sleep is accompanied by physiological phenomena found only in sleep and the sleeping state can be induced in animals by electrical stimulation of certain areas of the brain suggesting that sleep is an active process.

2.2 Normal sleep physiology

The most extensively used test of sleep is the EEG, which is part of polysomnography. The EEG has been most helpful in assessing various stages of sleep. Arousal thresholds and sleep EEG were used by Loomis, Harvey, and Hobart (1937) to divide the sleep EEG into five stages: A–E. Numerous variations in this classification have been made by either merging some stages or subdividing others. At the onset of sleep, a number of easily recognizable changes in the restful waking EEG occur. These are

Introduction to Clinical Aspects of the Autonomic Nervous System
DOI: https://doi.org/10.1016/B978-0-323-96011-3.00003-6

characterized by general slowing of brain waves and the appearance of spindles. The subdivision into further unique patterns, although useful, has definite limitations since there are rapid shifts from one stage to another which make identification of a given sleep stage sometimes difficult. Dement and Kleitman (1957) observed that the EEG may suddenly change from a slow-wave and spindle pattern to a fast-wave low-voltage trace, resembling the arousal pattern of an awakened subject. Since the subject shows all outward signs of sleep, this phase has been called paradoxical, desynchronized, fast-wave, low-voltage, activated, or REM sleep. During this phase, characteristic REMs are seen and it usually appears after an initial slow-wave phase of sleep lasting about 60 minutes (Fig. 2.1). On the basis of these observations, the normal adult sleep EEG has been divided into REM and non-REM (NREM) sleep. During the descending stage 1 there is initially fragmentation and disappearance of alpha activity, the frequency diminished but the amplitude of waves is higher. Fourteen cycles per second (sleep spindles) appear in stage 2. High-amplitude 1–2 cycles per second delta waves appear in stage 3 and are the dominant rhythm in stage 4. All these four stages are considered phases of NREM sleep. Just before the appearance of the first REM, there is a return of relatively low-voltage fast nonspindling activity resembling stage 1 and these persist until REM ceases. Coinciding with REM are short periods of 2–3 cycles per second saw-tooth waves. These periods recur at about 80-minute intervals throughout the night and constitute about 20% of the total night sleep in a young adult.

Factors that tend to alter the individual diurnal proportions of the three sleep states (wakefulness, NREM, and REM sleep) include anxiety (Dement, 1965), hypnosis (Stoyva (Stoyva, 1965), catching up with REM

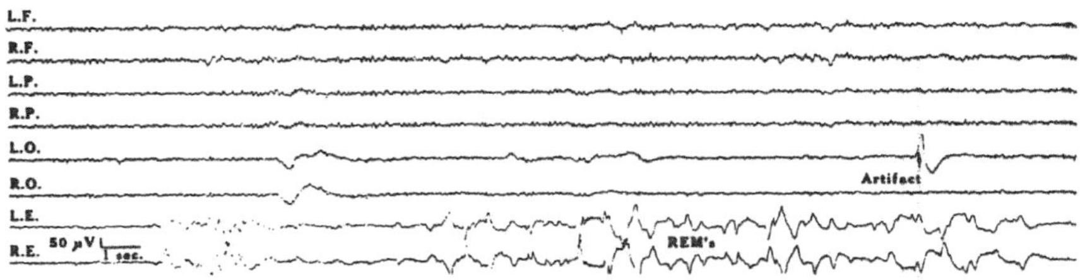

Figure 2.1 EEG recording during REM sleep from a single night recording. Eight channels in monopolar arrangement against ear lobes. *L*, left; *R*, right; *F*, frontal; *P*, parietal; *O*, occipital; *E*, outer canthus of eye. *From Appenzeller, O., & Fischer, A.P. (1968). Disturbances of rapid eye movements during sleep in patients with lesions of the nervous system. Electroencephalography and Clinical Neurophysiology, 25(1):29–35.*

sleep previously experimentally interrupted (Dement, 1965), and psychotic states (Oswald et al., 1963). The effect of some drugs and their withdrawal and the effects of age are also important.

Dreaming appears to be confined to REM sleep, although some imagery and other ideational material may persist through NREM sleep (Foulkes, 1964; Rechtschaffen, Verdone, & Wheaton, 1963). This has greatly helped in the study of dreams and their physiological correlates. Frequent changes of gaze or staring during a dream are correlated with the rapidity or relative paucity of REM (Wolpert & Trosman, 1958). Respiratory irregularities are also linked to dream experiences and apnea may occur when the subject dreams that he or she is laughing or talking (Roffwarg, Muzio, & Dement, 1966). During REM sleep, there seems to be an elevation of gastric hydrochloric acid in patients with peptic ulcers. Adrenal cortical steroids and free fatty acids in the plasma of normal subjects may be elevated and vary with the dream content (Roffwarg et al., 1966). The study of a variety of body functions during dreams has shown surprising alterations, which appear to be related to the dream content and dreaming has become more an expression of alterations of these functions rather than a subjective phenomenon removed from scientific scrutiny.

There are several other known physiological characteristics of the REM state (Peever & Fuller, 2017). During NREM sleep, the respiration and heart rate are relatively slow; blood pressure is low. On the other hand, respiration decreases in frequency and can become irregular, and heart rate and blood pressure increase to levels close to waking values during the REM sleep (Irfan, Schenck, & Howell, 2017). During REM sleep, muscle tone is practically abolished except for limb, trunk, face twitches, and, of course, it is preserved in extraocular muscles. Tendon jerks are profoundly diminished during REM states whereas they are only slightly decreased during NREM sleep. The pupils are miotic during both REM and NREM sleep. Penile erections occur during REM sleep and detumescence appears with a return of NREM sleep.

The study of animals during sleep has shown that all species so far examined have alternating REM and NREM sleep, and this has facilitated the recognition of some changes in structures not ordinarily accessible in humans. Awakening by noise during sleep causes an increase in blood flow in the posterior hypothalamus (Birzis & Tachibana, 1964) but undisturbed sleep is associated with a 0.5°C fall in hypothalamic temperature and wide but significant oscillations in hypothalamic temperature occur which are not seen during wakefulness (Adams, 1963). Specific hypothalamic circuitry exists to cool the brain and simultaneously induce NREM sleep (Harding et al., 2018). The lower temperature during NREM is important for the expression of the cold-inducible RNA-binding protein and RNA-binding motif protein 3 genes, which play a role in structural remodeling and synaptic function. The lower temperature during sleep slows inhibitory postsynaptic currents. During sleep, there is no drop in oxygen consumption of the brain, whereas in coma or other pathological states a striking drop in oxygen uptake occurs (Mangold et al., 1955). A remarkable increase in blood flow in various structures of the brain and subcortical nuclei during REM sleep has been found in animals.

The evolution of sleep patterns has been studied in humans and animals (Valatx, Jouvet, & Jouvet, 1964). During maturation, there is a gradual reduction in the proportion of REM sleep. In infancy, wakefulness does not persist for very long. Soon after the onset of sleep, REM periods occur and these are not limited to certain times of sleep and are not of fixed duration. After motor development has been achieved, the total amount, as well as the percentage of REM sleep, diminishes. The pattern approaches the adult proportion of REM and NREM sleep in mid-adolescence. In the newborn, contrary to the adult, there is an immediate onset of REM at the beginning of sleep and thereafter, there is a rhythm to the periodic occurrence of REM sleep in babies (Roffwarg et al., 1966). The NREM sleep EEG in the newborn resembles the adult stage 2 EEG (Dement-Kleitman classification) and is distinguished by slow high-amplitude waves and bursts of 13–15 cycles per second fast waves. Low-voltage fast activity, like that of adults, accompanies neonatal REM sleep and the REMs are predominantly vertical in this age group. The percentage of REM sleep in neonates is extremely variable but is about 50% of total sleep compared with about 14%–16% in adults over age fifty (Lairy, 1962). Some other physiological phenomena are related to REM sleep in infants. At the end of NREM sleep EEG, there is an increase in gross body, facial, and sucking movements until the EEG changes to a low-voltage fast-wave activity. The onset of REMs is also characterized by the appearance of mouth movements resembling sucking, although this might precede REMs by some minutes. This sucking persists throughout REM sleep.

Minor episodes of respiratory irregularities during the REM period occur but marked differences amongst individuals in the frequencies of these irregularities are found. Periodic breathing can appear in normal subjects during NREM sleep also but pauses in respiration are unusual and probably never exceed 10 seconds in duration. The presence of hypercapnia during normal sleep has been reported but not substantiated by subsequent investigation in normal subjects.

Significant changes in arterial and pulmonary artery pressure occur during normal sleep. Blood pressure decreases during the first hour of sleep and then increases again in the second part of the night. Direct relationships between sleep stages and arterial pressure changes have been found. The lowest systolic and diastolic pressures occur in stages 3 and 4 in NREM sleep. Low blood pressures occur in REM sleep also but higher arterial pressures tend to occur in REM sleep than in stages 3 and 4 NREM sleep. Values may exceed the highest level recorded in these subjects during wakefulness. In general, however, it is agreed that wide swings in arterial pressure occur during REM sleep, which are not connected with changes in heart rate or arterial tone.

Variations in pulmonary arterial pressures were recorded with a periodicity of 20–30 seconds occurring during states 1 and 2 of NREM sleep. An increase of 2–4 mm Hg above levels recorded during wakefulness was found. No significant pulmonary arterial pressure fluctuations occurred during stages 3 or 4 of NREM sleep.

During NREM sleep, parasympathetic activity increases, whereas there is a reduction in cardiac sympathetic activity (Jafari, 2017). Muscle sympathetic nerve activity (MSNA) falls with the transition from wake to NREM sleep, whereas MSNA is higher in REM sleep than during wakefulness. During the transition from NREM to REM, bursts of vagus nerve activity may result in cardiac pauses. The decrease in sympathetic activity during sleep explains the dipping in nocturnal blood pressure in healthy individuals. Autonomic instability with fluctuations in blood pressure and respiratory rate have been reported during REM sleep, possibly due to increased activity in the pedunculopontine nucleus, via its connections with sympathetic/respiratory control centers (nucleus tractus solitarius and rostral ventrolateral medulla). The irregular breathing patterns associated with REM sleep may be attributed to the instability of chemoreflex and baroreflex control during REM sleep.

Whether the moon cycle can modulate human nocturnal activity and sleep has been a matter of controversy. To examine the hypothesis that the lunar phase affects sleep, Casiraghi and colleagues conducted a study in communities living in environments that range from a rural setting with and without access to electricity in indigenous Toba/Qom communities in Argentina to a highly urbanized postindustrial setting in the United States. The authors reported that all three communities of Toba/Qom, including those in the urbanized setting, showed a strong association of sleep timing with the moon cycle. Furthermore, sleep recordings obtained from 464 University of Washington undergraduate students revealed a similar pattern, suggesting that sleep changes across the moon cycle may still be present in urbanized environments. This study raises the possibility that moon gravity could have a time-of-day specific effect on sleep (Casiraghi et al., 2021).

2.3 The neurobiology of wakefulness and sleep

2.3.1 Wakefulness

There are two major pathways that contribute to the promotion of alertness and awakening: a monoaminergic and glutamatergic ventral pathway and a dorsally oriented cholinergic pathway (Fig. 2.2). The first pathway originates

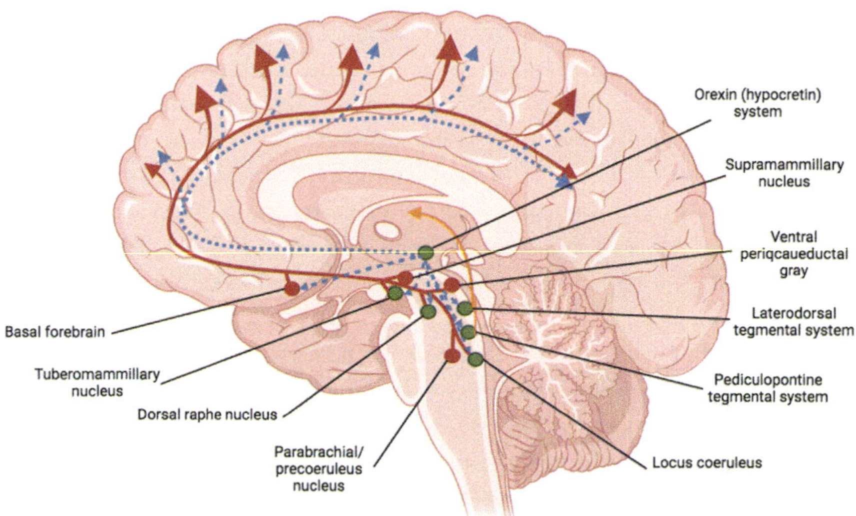

Figure 2.2 Brain structures responsible for wakefulness. The glutamatergic parabrachial/preceruleus and supramammillary nuclei and the γ-aminobutyric acid–mediated (GABAergic), the dopaminergic ventral periaqueductal gray, and the cholinergic basal forebrain project to the cerebral cortex (red). These structures are responsible for wakefulness. The modulatory monoaminergic (noradrenergic locus coeruleus, histaminergic tuberomammillary nucleus, serotonergic dorsal raphe nucleus) and cholinergic (laterodorsal tegmental and pedunculopontine tegmental) systems are represented in green with a dorsal pathway feeding into the thalamus (*orange arrow*). The orexin (hypocretin) system stabilizes the wake state and its primary projections are indicated in blue (*dashed line*). *Adapted from Schneider, L. (2020). Neurobiology and neuroprotective benefits of sleep.* Continuum 26*(4), 848–870.*

in the upper brainstem in cells that contain monoamine neurotransmitters (norepinephrine, serotonin, dopamine, and histamine) and projects to the hypothalamus. Lateral hypothalamic/perifornical areas containing orexin neurons influence the projections of the ventral monoaminergic pathway to the basal forebrain. The ultimate result is cortical arousal. The exact role of each monoamine neurotransmitter in promoting wakefulness is not fully understood; however, these neuronal populations exhibit their highest activity during wakefulness, diminishing activity during NREM sleep, while they remain quiet during REM sleep. The involvement of multiple neurotransmitters might indicate redundancy of the system to avoid impairment. It also explains the effects of several medications whose actions increase or decrease the levels of these neurotransmitters (Table 2.1). The locus coeruleus exert potent wake-promoting actions via activation of noradrenergic β- and α1-receptors located within multiple subcortical structures, including the general regions of the medial septal area, the medial preoptic area, and the lateral hypothalamus. Changes to the glutamatergic neuronal populations can induce profound changes in wakefulness, whereas none of the monoaminergic systems appear to be necessary for the maintenance of wakefulness to a significant degree. Therefore it is possible that monoamine neurotransmitters modulate a more potent glutamatergic wake-promoting system from the parabrachial/precoeruleus and supramammillary nuclei (Saper & Fuller, 2017).

Table 2.1 Medication classes that affect sleep and wakefulness.

Medications	Mechanism
Impairment in wakefulness	
α1 Antagonists	Inhibit noradrenergic signaling
α2 Agonists	Inhibit noradrenergic signaling
Anticholinergics	Inhibit acetylcholine signaling
Antihistamines	Inhibit histamine signaling
Benzodiazepines	GABA signaling
Narcotics	Central κ-opioid agonism
Tricyclic antidepressants, Selective serotonin reuptake inhibitors, Antipsychotics, Monoamine oxidase inhibitors	Serotonin autoinhibition, anticholinergic activity, and antihistaminergic activity
Dopamine agonist	DA2 inhibitory receptor activation
Impairment in sleep	
Cholinesterase inhibitors	Augmentation of acetylcholinergic tone
Antiparkinsonian agents (e.g., levodopa, amantadine)	Increase dopaminergic signaling

The second pathway involved in the promotion of wakefulness takes origin from cholinergic neurons in the upper pons (laterodorsal tegmental and pedunculopontine tegmental nuclei) and activates parts of the thalamus that are responsible for maintaining the transmission of sensory information to the cerebral cortex. These cholinergic neurons are most active during both wake and REM sleep (Fig. 2.2).

Another system that is crucial in maintaining a given sleep—wake state is the orexin/hypocretin system located in the lateral hypothalamus. These neurons have extensive excitatory projections throughout the arousal circuitry but also directly excite the cerebral cortex and basal forebrain (Fig. 2.2).

2.3.2 Induction and maintenance of sleep

The main sources of sleep-promoting GABAergic activity are the ventrolateral and median preoptic areas. These neurons are most active during NREM sleep and REM sleep, whereas they are relatively quiescent during wake. GABAergic neurons in the parafacial zone have a potent sleep-inducing effect (Venner et al., 2019). Neurons of the ventrolateral preoptic nucleus inhibit the discharge of wake-promoting monoaminergic cells. Acetylcholine neurons from the laterodorsal tegmental and pedunculopontine tegmental nuclei are crucial for the physiological change from NREM to REM sleep. Melanin-concentrating neurons of the lateral hypothalamus promote REM sleep through the facilitation of NREM to REM transitions and REM sleep prolongation (Komagata et al., 2019).

The ascending activating reticular system is inhibited by the hypnogenic centers and thus closely related to sleep onset and maintenance. Electrophysiological evidence supports the contention of a tonic inhibitory interaction between the arousal system of the reticular activating system in the brainstem and the preoptic hypnogenic center. This inhibitory relationship facilitates sleep onset by a positive feedback action. Physiological evidence exists to show that spontaneous sleep can result from disinhibition or inhibition, both of which achieve deactivation of the brainstem arousal systems in the reticular formation.

Serotonin was first believed to be a true neuromodulator of sleep because the destruction of serotonin neurons of the raphe system or the inhibition of serotonin synthesis with parachlorophenylalanine induced severe insomnia, which could be reversed by injection of 5-hydroxytryptophan. Nevertheless, the demonstration that the electrical activity of serotonin perikarya and the release of serotonin are increased during waking and decreased during sleep was in direct contradiction to this hypothesis. Subsequent experiments have suggested that the release of serotonin during waking may initiate a cascade of genomic events in some hypnogenic neurons located in the preoptic area. Thus when serotonin is released during waking, it leads to a homeostatic regulation of slow-wave sleep (Jouvet, 1999).

Cortical structures also actively contribute to sleep homeostasis and the global control of vigilance states. Specifically, projections from the prefrontal cortex—a part of the brain sensitive to homeostatic sleep pressure and a hub for the generation of slow waves during NREM sleep—to the lateral hypothalamus might represent a direct pathway through which the neocortex could modulate the control of vigilance states. Layer 5 pyramidal neurons also project to thalamic nuclei and could influence sleep regulation through corticothalamic loops. Layer 5 pyramidal neurons also connect to brainstem structures and other parts of the cortex. Therefore layer 5 pyramidal neurons may play a crucial role in initiating slow-wave sleep and regulating the subcortical circuitry responsible for sleep–wake switching (Krone et al., 2021).

2.3.3 Sleep–wake transition

Saper and colleagues proposed the flip-flop switch model of sleep–wake regulation to allow for rapid sleep–wake transition (Saper, Fuller, Pedersen, Lu, & Scammell, 2010). In this model, the ventrolateral and median preoptic nuclei inhibit the components of the ascending monoaminergic and cholinergic arousal pathways in both the hypothalamus and the brainstem and promote sleep. The ascending arousal systems are also capable of inhibiting the ventrolateral preoptic nucleus. This mutually inhibitory relationship of the arousal- and sleep-promoting pathways produces the conditions for a "flip-flop" switch. This model can generate rapid and complete transitions between waking and sleeping states.

In 1982 Borbély proposed the two-process model of sleep (Borbély, 1982). In this model, a homeostatic drive to sleep (Process S) depends on the duration of wakefulness and level of physical and cognitive activity. Another system (Process C) functions to integrate environmental and organismal cues into circadian rhythms that orient alertness to the time most suited for waking activities. Among state regulating substances, adenosine appears to play an important role in sleep promotion. It accumulates during wakefulness to increase the physiological pressure to sleep.

2.3.4 Body temperature

Body temperature has a profound effect on paradoxical sleep cycle. In a cat, the normal body temperature is about 39.5°C. If the temperature falls to around 30°C, the duration of each paradoxical sleep period is increased from 6 to about 20 minutes. The intervals between these periods are, however, widened so that the total time spent in paradoxical sleep is not markedly altered. The intravenous administration of hypertonic saline to such animals results in a 25% increase in paradoxical sleep time. This increase is most striking in animals left with a functional hypothalamo-pituitary island. It is believed, therefore, that blood osmolality in turn affects that of the brain and is responsible for the increase in paradoxical sleep time.

In humans the duration of sleep has been related to circadian changes in body temperature. In a temporal isolation experiment, subjects slept longer when they went to sleep at or shortly after the peak of the circadian temperature oscillation. Those who went to sleep during the temperature trough slept less (Czeisler, Weitzman, Moore-Ede, Zimmerman, & Knauer, 1980). All subjects awoke during the rising phase of the temperature cycle and went to bed most frequently shortly after the circadian temperature trough (when alertness was decreased). Thus the onset and duration of sleep in humans appear to depend in part on the interaction between different biological clocks.

2.4 Rapid eye movement sleep and rapid eye movement sleep behavior disorder

2.4.1 Physiology of rapid eye movement sleep

In animals, the mechanism underlying REM sleep has been extensively studied. In the cat, ablation in the rostral pons causes abolition of REM and absence of periods of low-voltage fast-activity sleep in the EEG (Jouvet, 1962; Peever & Fuller, 2017). Only wakefulness and NREM sleep occur until the animal dies. Decortication, on the other hand, abolishes NREM sleep. Low-voltage fast activity is seen in the mesencephalic EEG, REMs, respiratory irregularities, and a decrease in muscle tone appears at regular intervals. The suggestion is therefore made that the cortex is necessary for NREM and the rostral pons for REM sleep. Similar to wakefulness, REM sleep is characterized by intense cholinergic activity from the basal forebrain and pontine nuclei, which activate cortical activity. The combination of an "active" cortex and muscle paralysis allows the brain to perform the consolidation of emotionally laden memories.

In humans, evidence suggests that the glutamatergic neurons of the subcoeruleus nucleus (also called the sublaterodorsal nucleus) located in the dorsolateral pons are crucial to generate REM sleep (Boeve et al., 2007). Lesions in the subcoeruleus nucleus have been linked to REM sleep behavior disorder in humans (Limousin et al., 2009; Xi & Luning, 2009). The subcoeruleus nucleus induces muscle atonia via descending glutamatergic projections to the ventromedial medulla, which inhibits the output of spinal motor neurons via glycinergic and GABAergic premotor neurons. The ventrolateral periaqueductal gray and lateral pontine tegmentum plays a role in REM sleep by regulating the firing of the subcoeruleus nucleus. Larger network of brainstem and hypothalamic regions are involved in the regulation of REM sleep with various neurotransmitters, including GABA (preoptic region, ventromedial medulla), orexin, and melanin-concentrating hormone (lateral hypothalamus), serotonin (dorsal raphe), and norepinephrine (locus coeruleus). REM-off–related neurons whose activation suppresses REM sleep are located in the locus coeruleus (Boeve et al., 2007) (Fig. 2.3).

It has long been a puzzle how dreams, which predominantly occur during REM sleep, might be related to drug-induced hallucinations and psychosis. The manifestations of these three states have similarities and are to some extent functional modifications of brain structures connected to sleep and autonomic function. The functions of specific neurons may be grossly disturbed during hallucinations, psychosis, and dreaming. Hallucinogenic drug action, dreams, and amphetamine psychosis share effects that are at least in part attributable to a decrease in central serotonergic neurotransmission. The function of brain dopamine-containing structures is also modified, but its role is less certain, particularly during REM sleep and dreaming. The serotonergic and dopaminergic systems in the brain seem to be modulators of sensory and affective processes, but their influence is not limited to these states and extends to the motor and other functions as well.

2.4.2 Rapid eye movement sleep behavior disorder

In 1959 a feline model was used to identify the key structures of the brainstem involved in REM sleep. Lesions of the pontine tegmentum in cats resulted in the absence of skeletal muscle paralysis despite the preservation of several other features of REM sleep. At night, the cats would intermittently and without provocation lash out against imagined threats

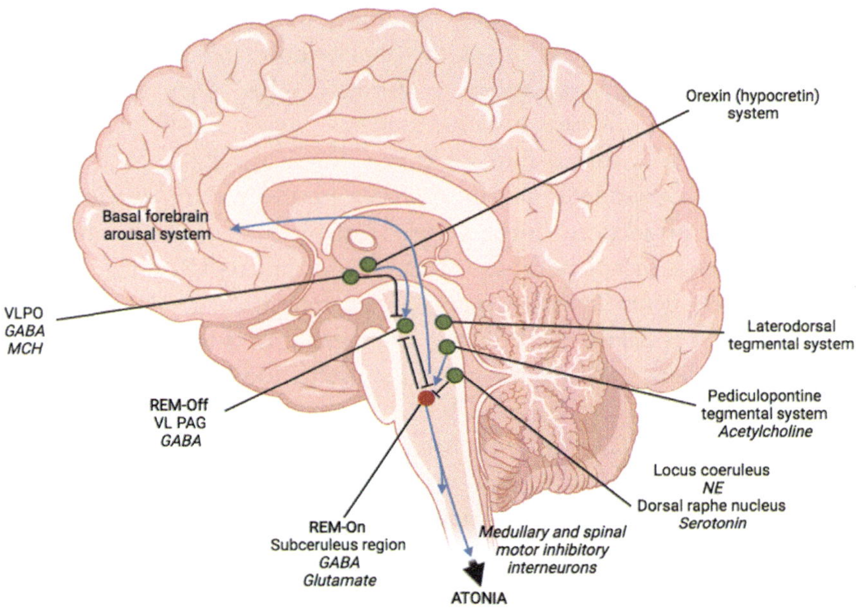

Figure 2.3 Structures responsible for REM sleep. The REM-off region comprises the ventrolateral periaqueductal gray (VL PAG) and the lateral pontine tegmentum. GABAergic neurons of the ventrolateral periaqueductal gray inhibit the REM-on neurons. The REM-on region is represented by the locus subceruleus in the rostral pons. The REM-on neurons in turn inhibit the neurons of the VL PAG (flip-flop switch). Glutamatergic projections from the locus subceruleus to other parts of the brainstem initiate the motor and autonomic manifestations of REM sleep, while projections to the basal forebrain mediate behavioral and electroencephalographic components of REM sleep. The locus subceruleus projects to spinal interneurons (*direct route*) that causes active inhibition of skeletal muscle activity in REM sleep. The locus subceruleus also projects to the magnocellular reticular formation (*indirect route*), which inhibits spinal motor neurons via glycinergic and GABAergic projections causing active inhibition of skeletal muscle activity in REM sleep. The REM-on neurons are under the control of forebrain neurons, including neurons that release the excitatory neuropeptide orexin, neurons in the ventrolateral preoptic nucleus (VLPO) that release the inhibitory neurotransmitter gamma-aminobutyric acid (GABA), and hypothalamic neurons that release the inhibitory neuropeptide melanin-concentrating hormone (MCH). Neurons in the locus ceruleus and dorsal raphe nucleus inhibit the REM generator, whereas cholinergic neurons in the pediculopontine and laterodorsal tegmental nuclei promote REM sleep. During REM sleep, serotoninergic neurons of the median raphe nucleus are virtually silent.

(Jouvet, 1965). REM sleep behavior disorder in humans was well documented by Schenck in 1986 when numerous bursts of motor activity that superficially resembled arousals were noted during REM sleep in a patient's overnight polysomnographic study (Schenck, Bundlie, Ettinger, & Mahowald, 1986). Over the years, it has become clear that REM sleep behavior disorder, characterized by periods of REM sleep without atonia and dream enactment behaviors, was not rare. While most cases of RBD are caused by α-synuclein neurodegeneration (80% of patients with isolated REM sleep behavior disorder will be later diagnosed with Parkinson's disease, dementia with Lewy bodies, or multiple system atrophy), brainstem lesions from an ischemic stroke, demyelinating lesion, or neoplasm are capable of causing REM sleep behavior disorder. Cases of REM sleep behavior disorder have also been described in narcolepsy. Medications inhibiting the reuptake of serotonin and norepinephrine have also been associated with REM sleep behavior disorder. It is debated if these medications directly cause the disorder or unmask an underlying predisposition for REM sleep behavior disorder. Loss of signal in the locus subcoeruleus can sometimes be seen with high-resolution MRI in patients with REM sleep behavior disorder compared to controls (Ehrminger et al., 2016). The source of the motor manifestations during REM sleep in patients with REM sleep behavior disorder is unknown. Nevertheless, evidence suggests that the basal ganglia networks are bypassed during REM sleep. Indeed, the movements observed during periods of REM in patients with Parkinson's disease or multiple system atrophy are usually fast with no bradykinesia. Functional imaging studies support this observation with an ictal-SPECT study revealing activation of premotor areas but no involvement of the basal ganglia during REM sleep behavior disorder (Mayer, Bitterlich, Kuwert, Ritt, & Stefan, 2015). Finally, there is evidence that different motor networks are active during movements in REM sleep behavior disorder compared with motor control in the waking state in patients with deep brain stimulation of the subthalamic nucleus (Hackius, Werth, Sürücü, Baumann, & Imbach, 2016).

2.5 Sleep deprivation

Striking effects of experimental deprivation of REM sleep are observed. There is a gradual development of tachycardia to 60%–80% above normal, muscular weakness, photophobia, and increases in food and water intake occur, and some animals die after about 3 weeks of REM sleep deprivation. Throughout this period, NREM sleep occurs normally. The recovery of REM sleep at the end of experimental deprivation takes about half the deprivation time (Jouvet, 1965). The effect of sleep rather than REM sleep deprivation has also been studied in humans. Numerous observers have noted lapses in performance. These are particularly noticeable in vigilance experiments in which they can be seen to last 1–2 seconds. During such a lapse in performance, the subjects become glassy and the head drops forward as the neck muscles relax. If aroused during the lapse in the early stages of sleep deprivation, transient visual experience may occur, but later in the experiment dream-like states are reported on arousal. These lapses are particularly frequent in the early morning hours and seem to be related to the fall in body temperature at that time. The onset and termination of lapses can be predicted from the EEG because a sudden change from the waking to low-voltage 5–6 cycles per second dominant frequency occurs and this changes back again to a waking pattern at the end of the lapse (West, Janszen, Lester, Lester, & Cornelisoon, 1962). The finger blood vessels show a transient constriction followed by dilatation as the lapse begins and they constrict again at the end of the lapse. From these and other studies, it is clear that sleep deprivation impairs performance and is also associated with a number of other phenomena.

2.6 Excessive sleep

There are four clinically recognizable sleep disorders: (1) when patients complain that they sleep too much (hypersomnia); (2) when the sleep is disturbed (insomnia); (3) where relatives complain about their sleep is being disturbed by the patient; these include sleepwalking, bedwetting, nightmares, sleep talking, and REM sleep behavior disorder; and (4) when other illness affects sleep secondarily; this occurs in endocrine disorders, nocturnal asthma, nocturia, and in epilepsy, for example. Numerous difficulties with the study of sleep continue to bedevil clinical evaluations of patients with sleep disorders. Thus it is not clear whether the investigation of sleep should be carried out in the patient's own home or in sleep laboratories, but even worse is the general lack of interest of physicians in the diagnosis and management of patients with sleep problems. Table 2.1 summarizes medication classes that affect sleep and wakefulness.

2.6.1 Can healthy young individuals sleep too much?

"Too much" sleep would imply sleeping longer than is biologically necessary. Klerman and colleagues demonstrated that the average healthy young human cannot chronically sleep over 10 hours per day (Klerman, Barbato, Czeisler, & Wehr, 2021). The authors analyzed sleep data from two human studies with sleep opportunities much longer than people usually stay in bed (14–16 hours per day for 3–28 days). The values of total sleep duration, REM sleep duration, and non-REM sleep durations increased in the first few days suggesting rebound recovery from sleep loss. This was followed by decreasing values of total sleep duration, REM sleep duration, and non-REM sleep durations and increasing sleep latency for the next few days. This pattern is consistent with a homeostatic process, likely reflecting recovery from prior behaviorally induced insufficient sleep (or sleep debt) (Klerman et al., 2021).

2.6.2 Narcolepsy

Narcolepsy has been labeled hypersomnia but evaluation of the duration of sleep in narcoleptics shows that sleep rarely exceeds the normal amount in a 24-hour period. Narcolepsy is characterized by attacks of sleep associated with cataplexy, sleep paralysis, hypnagogic hallucinations, and a general disruption of nocturnal sleep. During the day, there are short naps that are irresistible; they are frequent and persistent and usually are the symptom bringing the patient to the attention of a physician. Monotonous environments are conducive to more frequent attacks but the patient is able to resist attacks to some degree by maintaining alertness during interesting tasks. Sometimes patients are unable to appropriately respond to the environment without visibly falling asleep and may get injured because of these lapses. Almost all patients have warning signals of impending sleep by their irresistible urge to do so. By recognizing the imminence of the attack, they can adjust their situation and avoid embarrassment or injury. After a short nap, they are fully alert again and greatly refreshed. When the patient is drowsy, headache may be a prominent symptom and visual disturbances such as blurred vision or burning eyes occur. These discomforts increase with voluntary resistance to sleep.

When hypnagogic hallucinations are the first symptom in young individuals, they may be of a terrifying nature, particularly when the associated sleep disorder has not yet been recognized. Soon, however, the hallucinatory nature of the images is recognized and their association with dreams is established. With hypnagogic hallucinations, there is sometimes a feeling of inability to move, and this may add to the terror of the hallucinatory experiences.

During the transition period between wakefulness and sleep, paralysis with inability to perform voluntary movements may appear.

Complete recovery of motor function occurs if the patient is touched but sleep paralysis may last for 20 or more minutes.

A sudden loss of muscle tone characterizes cataplexy. This is associated with an inability to perform any voluntary movement. The severity of the attack may vary from total loss of motor function to impairment of voluntary activity of certain muscle groups only. Extraocular muscles are not involved in the process. Most commonly there is sagging of the jaw, a forward drop of the head, and buckling of the knees. In severe attacks, however, a sudden loss of muscle tone may lead to falls. Cataplexy is usually elicited in response to emotional stress, fatigue, and sometimes heavy meals.

Though most patients with narcolepsy suffer from the classic tetrad of sleep attacks, cataplexy, hypnagogic hallucinations, and sleep paralysis, some also have disturbed nocturnal sleep. In such patients, sleep is interrupted by frequent wakefulness. Though total sleep time is no greater than that found in subjects of similar age, patients with narcolepsy seem to sleep at inappropriate times.

The assumption is made that there is a close relationship between the mental activity of the patient during wakefulness immediately preceding a REM period and hypnagogic hallucinations and sleep paralysis. Cataplectic attacks do not occur during sleep but are regarded as dissociated manifestations of REM processes. The only condition in which there is spontaneous nonreciprocal motor inhibition in humans is during REM sleep. During cataplectic attacks, the same phenomena are observed with total simultaneous inhibition of EMG action potentials in different muscle groups. This is incidentally also associated with the abolition of tendon reflexes. The sparing of motor inhibition of the extraocular muscles parallels the lack of inhibition of electromyographic activity in these muscles in normal subjects during REM periods. This explanation of cataplexy does not account for the attacks induced by strong emotions. How emotions can inhibit tonic muscle discharges is not known at present.

Narcolepsy is concentrated in families, and it has been shown that relatives of a narcoleptic patient run a 200-fold greater risk for narcolepsy than other subjects in the general population. Narcolepsy associated with cataplexy forms a specific syndrome but daytime sleepiness may be caused by other conditions (Table 2.2). There are also a number of patients who have daytime sleep attacks shown to be associated with REM periods without the other components of narcolepsy. They may continue to have just sleep attacks for many years before sleep paralysis hypnagogic hallucinations or cataplectic attacks appear.

Table 2.2 Clinical features of disorders of hypersomnolence.

	Narcolepsy type 1	Narcolepsy type 2	Kleine–Levin syndrome	Idiopathic hypersomnia
Excessive daytime sleepiness	Yes	Yes	Yes	Yes
Cataplexy	Common	No	No	No
Sleep hallucinations	Common	Less common	Not typical	Not typical
Sleep paralysis	Common	Less common	Not typical	Not typical
Disturbed nocturnal sleep	Common	Less common	No	Not typical
Long sleep time	No	May be present	Very prolonged during episodes	Common
Prolonged sleep inertia	Rare	May be present	Not typical between episodes	Common

Narcolepsy type 1 is a disorder of orexin deficiency and may be diagnosed by testing orexin levels in the CSF. Orexin levels are low or absent in people with narcolepsy type 1 but normal in people with other forms of hypersomnia. Testing for human leukocyte antigen-DQB1*06:02 may be considered as this allele is present in 98% of people with orexin-deficiency narcolepsy type 1; however, it can be present in up to 30% of controls. The cause of narcolepsy type 2 is unknown but approximately 20% of people initially diagnosed with narcolepsy type 2 have or will develop orexin deficiency.

Table 2.2 summarizes the clinical features of four major central disorders of hypersomnolence: narcolepsy type 1, narcolepsy type 2, idiopathic hypersomnia, and Kleine–Levin syndrome.

2.6.3 Other syndromes associated with hypersomnia

Another clinical syndrome consists of hypersomnia with sleep drunkenness. These patients have daily sleep spells, which resemble slow-wave sleep narcolepsy but they also appear confused, disoriented, and react slowly and inappropriately to the environment. The confusion disappears as the day progresses but there might be short recurrences in association with sleep periods. It has been postulated that hypersomnia with sleep drunkenness involves an excessive increase of stage 3 and 4 NREM sleep.

Lesions of the dorsomedial and anteroventral thalamic nuclei, thought to interrupt pathways connecting the amygdala and prefrontal cortex with the hypothalamus, are found in fatal familial insomnia. Fatal familial insomnia is an autosomal dominant inherited prion disease caused by D178N mutation in the prion protein gene (PRNP D178N) accompanied by the presence of a methionine at the codon 129 polymorphic site on the mutated allele. Fatal familial insomnia is characterized by a severe sleep disorder, dysautonomia, motor signs, and abnormal behavior together with primary atrophy of selected thalamic nuclei and inferior olives, and expansion to other brain regions with disease progression (Cortelli et al., 1991). Fatal familial insomnia is characterized by dysfunction of the sleep and autonomic regulating systems of the thalamus, hypothalamus, parabrachial nucleus, periaqueductal gray, and nucleus tractus solitarius. There is severe sympathetic hyperactivity with tachycardia, hypertension, and hyperthermia, which is hypothesized to result from overexcitation of the rostral ventrolateral medulla (Benarroch & Stotz-Potter, 1998). This disorder presents usually in the 3rd or 4th decade of life with progressive fatal insomnia, sympathetic overactivity, and endocrine and motor disturbances (ataxic gait and myoclonus).

2.7 Abnormal apneic periods during sleep

Several types of apneic episodes during sleep have been recognized. In some patients, the diaphragm may cease movement for an inordinate length of time. No muscle activity in respiratory muscles, intercostal muscles, or diaphragm can be recorded. This type of apnea has been termed central apnea. A different type of pathologic sleep disturbance is obstructive apnea. In this condition, in spite of persistent contractions of the diaphragm of increasing strength, there is no airflow through the upper airways. A tremendous rise of intrathoracic pressure occurs during these episodes. Increasing muscle activity is recorded in the inspiratory intercostal muscles throughout the period of apnea. When suddenly the obstruction is overcome and the apneic period is interrupted, there is a great snort or snore. Some patients appear to have a mixed variety of apneic episodes during sleep consisting of central and obstructive syndromes. Clinically, however, patients who have predominantly diaphragmatic central apnea complain of insomnia whereas those with mainly obstructive apnea usually complain of hypersomnia and sleepiness during the day.

The examination of patients with predominant obstructive sleep apnea might show clear evidence of the cause of the disturbance. Partial airway obstruction due to enlargement of tonsils or adenoids may be present. Cardiomegaly and cor pulmonale, in addition to an alveolar hypoventilation syndrome and CO_2 retention, can be seen. Patients with obstructed nasopharynx due to hypertrophic adenoids and resulting pulmonary hypertension and cardiac failure are usually children. In adults, however, significant obesity and daytime somnolence can be found and some of these patients have hypertrophic adenoids; others though were found to have thyroid enlargement. Removal of the obstruction improves the irresistible urge to sleep during the day and improves cardiovascular abnormalities.

Certain anatomic abnormalities can lead to sleep apnea. Periods of sleep apnea due to obstruction have been reported in patients with abnormalities of the craniovertebral region involving the first and second cervical vertebrae often associated with platybasia and basilar impression.

In some patients who complain of hypersomnia and have diurnal somnolence with sleep apnea documented by extensive studies, the causative anatomical lesions cannot be found. Perhaps in some of them, the central nervous system may be important in causing some obstruction of the upper airways during sleep.

The syndrome of obesity, cardiorespiratory abnormalities, and sleep apneic episodes has been extensively studied but it is not a homogenous disorder. This syndrome encompasses patients who present with obesity, micrognathia, secondary changes in the cardiovascular system, and hypersomnolence. It includes children with upper airway obstruction; those who have platybasia and basilar impression; some who have only obesity with secondary cardiovascular changes with or without abnormal responses of respiration to CO_2. Progressive alveolar hypoventilation develops in all these patients. In addition, there are a number of patients with similar abnormalities but without obesity. Some suffer from Ondine syndrome, now called congenital central hypoventilation syndrome (CCHS). Neonatal-onset CCHS is characterized by apparent hypoventilation with monotonous respiratory rates and shallow breathing either during sleep only or while awake as well as asleep. Signs of autonomic dysregulation include decreased heart rate beat-to-beat variability, sinus pauses, altered temperature regulation, and altered pupillary response to light. Some children have altered development of neural crest-derived structures (i.e., Hirschsprung disease) and/or tumors of neural crest origin (neuroblastoma, ganglioneuroma, and ganglioneuroblastoma). Neurocognitive delay is variable and possibly influenced by breath-holding periods, prolonged sinus pauses, need for artificial ventilation, and seizures. Later-onset CCHS is characterized by alveolar hypoventilation during sleep and attenuated manifestations of autonomic dysfunction. The diagnosis of CCHS is established by suggestive findings and a heterozygous PHOX2B pathogenic variant identified on molecular genetic testing. Rapid-onset obesity with hypothalamic dysfunction, hypoventilation, and autonomic dysregulation (ROHHAD) is a rare disorder characterized by dramatic weight gain over a 3- to a 12-month period between ages 1.5 and 10 years followed by hypothalamic dysfunction, central alveolar hypoventilation, and autonomic dysfunction (Overeem, van Litsenburg, & Reading, 2021).There is frequent co-occurrence of neuroblastic tumors and ROHHAD may present as a paraneoplastic syndrome with ZSCAN1 autoantibodies (Mandel-Brehm et al., 2022).

Sleep apnea plays an important role in the development of cardiovascular abnormalities. Many patients with this syndrome have been shown to have extreme changes in pulmonary artery pressure, which increases during apneic episodes and does not always return to baseline levels at the end of apnea. This is particularly found in patients with predominant obstructive apneic periods. High systemic blood pressures occur in such patients also. During sleep, blood pressure rises progressively with successive REM stages and further increases during REM sleep. Levels of 210/110 mm Hg have been recorded during night sessions in such patients. The heart rate also increases and parallels the rise in blood pressure. As mentioned earlier, adrenergic hyperactivity and abnormalities in cardiac function are manifestations of sleep disorders particularly associated with sleep apnea. Studies to determine autonomic instability have shown that urine and plasma catecholamine concentrations were increased in those with sleep apnea. Such patients also showed excessive increases in heart rate during isoproterenol infusions consistent with adrenergic hyperresponsiveness. What is even of greater interest, however, is that a large proportion of patients with sleep apnea had mitral valve prolapse. This suggests that the fatigue and lassitude commonly found in those with this cardiac abnormality might be neurogenically determined. Patients with panic attacks and excessive cardiac responses to isoproterenol have previously been reported (Frohlich, Tarazi, & Dustan, 1969), but the frequent association of these attacks with mitral valve prolapse has also been recognized. The hyperresponsiveness of these patients to isoproterenol suggests that a peripheral rather than central abnormality might be pathogenetically important. Thus autonomic instability may be a feature of excessive daytime drowsiness, and this includes mitral valve prolapse. Patients with mitral valve prolapse are known to show changes in sinus arrhythmia and abnormal increases in heart rate during the upright posture. The assessment of excessive daytime drowsiness and lethargy should include careful sleep history and clinical, electrocardiographic, and other necessary examinations of the heart for the presence of mitral valve prolapse (Clark, Boudoulas, Schaal, & Schmidt, 1980).

Evidence of autonomic disturbances in narcolepsy has also been found. Studies of heart rate and blood flow in the forearm in response to contralateral isometric grip, respiratory sinus arrhythmia, and heart rate responses to Valsalva maneuver have been examined in six male patients with narcolepsy. These examinations were made while the patients were on and off amphetamine. Surprisingly, slow heart rates, reduced forearm blood flow, which increased on the contralateral handgrip, and less than normal sinus arrhythmia were found. Amphetamine, however, increased sinus arrhythmia to normal levels. It was concluded that narcolepsy can be accompanied by a decrease in autonomic reactivity of the cardiovascular system. The functional abnormalities involved both sympathetic and parasympathetic systems and were therefore attributed to a central disturbance of autonomic reflex function (Sachs & Kaijser, 1980). From these studies, it cannot be determined whether autonomic abnormalities in sleep apnea or narcolepsy are of primary pathogenic significance or merely epiphenomena. Patients with narcolepsy also show excess hippus, which is independent of the degree of ambient illumination. This has been attributed to an imbalance of the autonomic nervous system. It can be abolished by small doses of amphetamine.

The study of cellular receptors and behavior of cells may have relevance to neurologic disease. In patients with fatal sleep disorders, autonomic dysfunction, and degeneration of thalami, the abnormality may be at ligand recognition sites of certain cells. In healthy individuals, specific sleep receptors may recognize one or more ligands, and sleep may be induced when the ligands bind to specific receptors. In genetically determined sleep failure, a specific defective protein may be found in affected family members. This protein may be part of a receptor, an enzyme involved in the synthesis, or an effector ligand, or another ligand involved in the appropriate processing of chemical information. Another possibility is that the defective gene is implicated in a regulatory rather than structural role.

The phenomenon of sleep is still far from clearly understood even though conventional metabolic, enzymatic, and neurophysiological studies have been carried out on sleep for decades. More mechanistic, rather than descriptive, aspects of sleep have emerged.

2.8 Neurodegeneration and sleep

Accumulating evidence indicates that circadian and sleep disturbances may play an important role in the pathogenesis of neurodegenerative diseases by promoting processes such as inflammation and synaptic damage. In neurodegenerative diseases such as Parkinson's disease, Alzheimer's disease, or Huntington's disease, sleep disturbances may precede the onset of more typical symptoms, in some cases by decades. Sleep deprivation increases cerebrospinal fluid markers of neuronal injury and alters plasma markers of inflammation in humans (Benedict et al., 2014). In the case of amyloid-beta, a protein that is thought to be involved in the pathogenesis of Alzheimer's disease, amyloid-beta concentrations in the extracellular space in the brain exhibit clear circadian oscillations, rising during the active period and falling during rest (Huang et al., 2012; Kang et al., 2009). In transgenic mouse models of Alzheimer's disease, sleep deprivation greatly accelerates amyloid plaque deposition (Kang et al., 2009). Genetic deletion of orexin in mice modestly increases sleep time but strongly suppresses the formation of amyloid plaques. Fragmented sleep in cognitively normal individuals is a risk factor for the future development of symptomatic Alzheimer's disease (Lim, Kowgier, Yu, Buchman, & Bennett, 2013). Tau protein is also correlated with concentrations of orexin in the cerebrospinal fluid and a higher concentration of orexin in the CSF is associated with increased amounts of phosphorylated tau. Increased wakefulness promoted by orexin may cause increased neuronal activity, leading to elevated Aβ production and aggregation. Wakefulness also increases sympathetic output, suppressing glymphatic system function, which could result in decreased clearance of pathogenic proteins (such as amyloid-beta, tau, or synuclein). Sleep loss also promotes oxidative stress, inflammation, and a loss of synaptic homeostasis. There is evidence that sleep regulates the removal of proteins and other molecules from the brain through the glymphatic system and slow-wave sleep is associated with increased glymphatic flow (Iliff et al., 2012). Noradrenergic neurons may regulate the glymphatic system and increased sympathetic noradrenergic output, as is seen during waking, suppresses glymphatic clearance. Further studies are needed to investigate the effect of chronic sleep disturbances on glymphatic function. It is recommended that adults sleep 7 or more hours per night on a regular basis to promote optimal health. Nevertheless, sleeping more than 9 hours per night might be associated with increased health risks. In a cross-sectional study, both short and long sleep durations were associated with worse outcomes for older adults, such as greater amyloid-beta burden, greater depressive symptoms, higher body mass index, and cognitive decline (Winer et al., 2021). Treatment of insomnia with cognitive behavioral therapy has been shown to have an overall benefit in the prevention of incident and recurrent major depression in older adults with insomnia disorder (Irwin et al., 2021).

Sleep disturbances were also reported in patients with progressive supranuclear palsy (Gilman, 2005). All patients in this study had hyposomnia and absent or markedly decreased REM sleep. Abnormal non-REM sleep patterns were also recorded. These abnormalities were attributed to the involvement of pontine structures by the disease.

The absence of REM and altered non-REM sleep has been found in some patients with spinocerebellar degeneration who had slow saccades. This emphasizes the importance of pontine tegmental structures in the generation of REM sleep. Saccadic eye movements are initiated by neurons in the pontine tegmentum; it is not surprising, therefore to find a close relationship between clinical abnormalities in saccades and REM sleep (Osorio & Daroff, 1980).

2.9 Treatment of sleep disorders

Pharmacologic treatment of sleep disorders includes medications from a variety of different classes. Table 2.3 summarizes the treatment of the most common sleep disorders.

2.10 Some nocturnal disorders associated with various stages of sleep

A relationship between sleep and migraine has long been recognized but until the advent of proper electroencephalographic staging of sleep, this association was purely anecdotal. Data have been presented (Dexter & Riley, 1975), which suggest that arousals with nocturnal migraine and nocturnal cluster headaches have a time-locked relationship to REM sleep. It has been suggested that REM periods occur at intervals

Table 2.3 Treatments of common sleep disorders.

Disorder	Recommended treatment
Narcolepsy	Modafinil/armodafinil (enhance dopaminergic neurotransmission)
Idiopathic hypersomnia *(Off label)*	Traditional psychostimulants (e.g., methylphenidate, etc.)
	Sodium oxybate: approved for both sleepiness AND cataplexy
	Solriamfetol (dopamine- and norepinephrine-reuptake inhibitor)
Kleine–Levin syndrome *(Off label)*	Pitolisant (histamine H3 inverse agonist/antagonist)
	REM suppressing antidepressant for cataplexy
Obstructive sleep apnea	Weight loss for overweight patients
	Positional therapy
	Positive airway pressure therapy
	Mandibular-adjusting oral appliance therapy
	Surgery (e.g., uvulopalatopharyngoplasty)
	Stimulation of the hypoglossal nerve
REM sleep behavior disorder	Clonazepam
	Melatonin
Circadian delayed sleep–wake phase disorder	Timed exposure to light and melatonin
	Bright light therapy
Advanced sleep–wake phase disorder	Evening bright light to delay circadian rhythms
Irregular sleep–wake rhythm disorder	Melatonin around bedtime
Non–24-hour sleep–wake rhythm disorder	Melatonin
Insomnia	Cognitive behavioral therapy
	Benzodiazepine receptor agonists: eszopiclone, zaleplon, zolpidem
	Orexin (hypocretin) receptor antagonist: suvorexant
	Benzodiazepines: triazolam, temazepam
	Melatonin agonist: ramelteon
	Heterocyclics: doxepin, trazodone

throughout a 24-hour period even not always associated with behavioral correlates of sleep. If this is true, the possibility exists that even daytime migraine or cluster headache attacks are somehow induced by a physiological stage, which is akin to the REM stage of sleep despite the manifestly awake subject. It is possible that during REM sleep, the lowest pain arousal threshold occurs and that this accounts for the waking of patients with migraine or cluster headaches. In keeping with this interpretation is the temporal association of nocturnal angina with REM sleep stages. Whether these relationships of pain syndromes with REM sleep imply a connection with impaired central inhibition of raphe neurons, which may result in hyperalgesia, also remains at present entirely hypothetical (Appenzeller, 1975).

Nocturnal ulcer pain may be related to slow-wave sleep. Growth hormone release in the early morning in association with slow-wave sleep is attributed to inhibition of somatostatin secretion from the hypothalamus. This in turn is related to inhibition of gastrointestinal somatostatin release causing an outpouring of gastrin and insulin. At the same time, the levels of cortical steroids are low and the insulin effect is not opposed. This increases gut motility through an increase in vagal tone, which results in an increase of acid secretion and delivery to the duodenum. In patients who have duodenal ulcers and a reduced capacity to buffer acid because of a deficiency in secretin response, early morning pain occurs (Tormey, Davis, & Darragh, 1978).

In a rare mucopolysaccharidosis (Scheie syndrome), sleep apnea has also been found. In two brothers with the syndrome, significant cerebral hypoxia and ventricular extrasystoles were present. A tracheostomy performed in one brother gave relief and possibly prevented a myocardial infarction, which had occurred in the other sibling. In this condition where obstructive features of the airways could be demonstrated during wakefulness, hypoxia occurred only during sleep, suggesting that central nervous dysfunction and peripheral mechanical factors may be operative together to produce significant sleep apnea (Perks et al., 1980).

References

Adams, T. (1963). Hypothalamic temperature in the cat during feeding and sleep. *Science, 139*(3555), 609–610.

Appenzeller, O. (1975). Hypothesis-pathogenesis of vascular headache of the migrainous type: the role of impaired central inhibition. *Headache, 15*, 177–179.

Benarroch, E. E., & Stotz-Potter, E. H. (1998). Dysautonomia in fatal familial insomnia as an indicator of the potential role of the thalamus in autonomic control. *Brain Pathology, 8*(3), 527–530.

Benedict, C., Cedernaes, J., Giedraitis, V., Nilsson, E. K., Hogenkamp, P. S., Vågesjö, E., ... Schiöth, H. B. (2014). Acute sleep deprivation increases serum levels of neuron-specific enolase (NSE) and S100 calcium binding protein B (S-100B) in healthy young men. *Sleep, 37*(1), 195–198.

Birzis, L., & Tachibana, S. (1964). Local cerebral impedance and blood flow during sleep and arousal. *Experimental Neurology, 9*, 269–285.

Boeve, B. F., Silber, M. H., Saper, C. B., Ferman, T. J., Dickson, D. W., Parisi, J. E., . . . Braak, H. (2007). Pathophysiology of REM sleep behaviour disorder and relevance to neurodegenerative disease. *Brain, 130*(Pt 11), 2770–2788.

Borbély, A. A. (1982). A two process model of sleep regulation. *Human Neurobiology, 1*(3), 195–204.

Casiraghi, L., Spiousas, I., Dunster, G. P., McGlothlen, K., Fernández-Duque, E., Valeggia, C., & de la Iglesia, H. O. (2021). Moonstruck sleep: synchronization of human sleep with the moon cycle under field conditions. *Science Advances*, 7(5).

Clark, R. W., Boudoulas, H., Schaal, S. F., & Schmidt, H. S. (1980). Adrenergic hyperactivity and cardiac abnormality in primary disorders of sleep. *Neurology, 30*(2), 113–119.

Cortelli, P., Parchi, P., Contin, M., Pierangeli, G., Avoni, P., Tinuper, P., . . . Lugaresi, E. (1991). Cardiovascular dysautonomia in fatal familial insomnia. *Clinical Autonomic Research, 1*(1), 15–21.

Czeisler, C. A., Weitzman, E., Moore-Ede, M. C., Zimmerman, J. C., & Knauer, R. S. (1980). Human sleep: its duration and organization depend on its circadian phase. *Science, 210*(4475), 1264–1267.

Dement, W., & Kleitman, N. (1957). The relation of eye movements during sleep to dream activity: an objective method for the study of dreaming. *Journal of Experimental Psychology, 53*(5), 339.

Dement, W. C. (1965). Recent studies on the biological role of rapid eye movement sleep. *American Journal of Psychiatry, 122*(4), 404–408.

Dexter, J. D., & Riley, T. L. (1975). Studies in nocturnal migraine. *Headache, 15*(1), 51–62.

Ehrminger, M., Latimier, A., Pyatigorskaya, N., Garcia-Lorenzo, D., Leu-Semenescu, S., Vidailhet, M., . . . Arnulf, I. (2016). The coeruleus/subcoeruleus complex in idiopathic rapid eye movement sleep behaviour disorder. *Brain, 139*(Pt 4), 1180–1188.

Foulkes, D. (1964). Theories of dream formation and recent studies of sleep consciousness. *Psychological Bulletin, 62*, 236–247.

Frohlich, E. D., Tarazi, R. C., & Dustan, H. P. (1969). Hyperdynamic ß-adrenergic circulatory state: increased ß-receptor responsiveness. *Archives of Internal Medicine, 123*(1), 1–7.

Gilman, S. (2005). Functional imaging with positron emission tomography in multiple system atrophy. *Journal of Neural Transmission, 112*(12), 1647–1655.

Hackius, M., Werth, E., Sürücü, O., Baumann, C. R., & Imbach, L. L. (2016). Electrophysiological evidence for alternative motor networks in REM sleep behavior disorder. *The Journal of Neuroscience, 36*(46), 11795–11800.

Harding, E. C., Yu, X., Miao, A., Andrews, N., Ma, Y., Ye, Z., . . . Franks, N. P. (2018). A neuronal hub binding sleep initiation and body cooling in response to a warm external stimulus. *Current Biology, 28*(14), 2263–2273, e2264.

Huang, Y., Potter, R., Sigurdson, W., Santacruz, A., Shih, S., Ju, Y. E., . . . Bateman, R. J. (2012). Effects of age and amyloid deposition on Aβ dynamics in the human central nervous system. *Archives of Neurology, 69*(1), 51–58.

Iliff, J. J., Wang, M., Liao, Y., Plogg, B. A., Peng, W., Gundersen, G. A., . . . Nedergaard, M. (2012). A paravascular pathway facilitates CSF flow through the brain parenchyma and the clearance of interstitial solutes, including amyloid β. *Science Translational Medicine, 4*(147), 147ra111.

Irfan, M., Schenck, C. H., & Howell, M. J. (2017). Non-rapid eye movement sleep and overlap parasomnias. *Continuum (Minneap Minn), 23*(4, Sleep Neurology), 1035–1050.

Irwin, M. R., Carrillo, C., Sadeghi, N., Bjurstrom, M. F., Breen, E. C., & Olmstead, R. (2021). Prevention of incident and recurrent major depression in older adults with insomnia: a randomized clinical trial. *JAMA Psychiatry, 79*(1), 33–41.

Jafari, B. (2017). Sleep architecture and blood pressure. *Sleep Medicine Clinics*, *12*(2), 161–166.

Jouvet, M. (1962). Research on the neural structures and responsible mechanisms in different phases of physiological sleep. *Archives Italiennes de Biologie*, *100*, 125–206.

Jouvet, M. (1965). Paradoxical sleep–A study of its nature and mechanisms. *Progress in Brain Research*, *18*, 20–62.

Jouvet, M. (1999). Sleep and serotonin: an unfinished story. *Neuropsychopharmacology*, *21*(2 Suppl), 24s–27s.

Kang, J. E., Lim, M. M., Bateman, R. J., Lee, J. J., Smyth, L. P., Cirrito, J. R., … Holtzman, D. M. (2009). Amyloid-beta dynamics are regulated by orexin and the sleep-wake cycle. *Science*, *326*(5955), 1005–1007.

Klerman, E. B., Barbato, G., Czeisler, C. A., & Wehr, T. A. (2021). Can people sleep too much? Effects of extended sleep opportunity on sleep duration and timing. *Frontiers in Physiology*, *12*, 2204.

Komagata, N., Latifi, B., Rusterholz, T., Bassetti, C. L. A., Adamantidis, A., & Schmidt, M. H. (2019). Dynamic REM sleep modulation by ambient temperature and the critical role of the melanin-concentrating hormone system. *Current Biology*, *29*(12), 1976–1987, e1974.

Krone, L. B., Yamagata, T., Blanco-Duque, C., Guillaumin, M. C. C., Kahn, M. C., van der Vinne, V., … Vyazovskiy, V. V. (2021). A role for the cortex in sleep-wake regulation. *Nature Neuroscience*, *24*(9), 1210–1215.

Lairy, G. C. (1962). Action du Librium sur 1'electroencephalogramme. *Revue Neurologique.*, *106*, 152–166.

Lim, A. S., Kowgier, M., Yu, L., Buchman, A. S., & Bennett, D. A. (2013). Sleep fragmentation and the risk of incident 'Alzheimer's disease and cognitive decline in older persons. *Sleep*, *36*(7), 1027–1032.

Limousin, N., Dehais, C., Gout, O., Héran, F., Oudiette, D., & Arnulf, I. (2009). A brainstem inflammatory lesion causing REM sleep behavior disorder and sleepwalking (parasomnia overlap disorder). *Sleep Medicine*, *10*(9), 1059–1062.

Loomis, A. L., Harvey, E. N., & Hobart, G. A. (1937). Cerebral states during sleep, as studied by human brain potentials. *Journal of Experimental Psychology*, *21*(2), 127–144.

Mandel-Brehm, C., Benson, L. A., Tran, B., Kung, A. F., Mann, S. A., Vazquez, S. E., et al. (2022). ZSCAN1 autoantibodies are associated with pediatric paraneoplastic. April 25, Online ahead of print. ROHHAD. *Annals of Neurology*. Available from http://doi.org/10.1002/ana.26380.

Mangold, R., Sokoloff, L., Conner, E., Kleinerman, J., Therman, P. O., & Kety, S. S. (1955). The effects of sleep and lack of sleep on the cerebral circulation and metabolism of normal young men. *Journal of Clinical Investigation*, *34*(7, Part 1), 1092–1100.

Mayer, G., Bitterlich, M., Kuwert, T., Ritt, P., & Stefan, H. (2015). Ictal SPECT in patients with rapid eye movement sleep behaviour disorder. *Brain*, *138*(Pt 5), 1263–1270.

Osorio, I., & Daroff, R. B. (1980). Absence of REM and altered NREM sleep in patients with spinocerebellar degeneration and slow saccades. *Annals of Neurology*, 7(3), 277–280.

Oswald, I., Berger, R. J., Jaramillo, R. A., Keddie, K. M., Olley, P. C., & Plunkett, G. B. (1963). Melancholia and barbiturates: a controlled EEG, body and eye movement study of sleep. *British Journal of Psychiatry*, *109*, 66–78.

Overeem, S., van Litsenburg, R. R. L., & Reading, P. J. (2021). Sleep disorders and the hypothalamus. *The Handbook of Clinical Neurology*, *182*, 369–385.

Peever, J., & Fuller, P. M. (2017). The biology of REM sleep. *Current Biology*, *27*(22), R1237–R1248.

Perks, W. H., Cooper, R. A., Bradbury, S., Horrocks, P., Baldock, N., Allen, A., ... Prowse, K. (1980). Sleep apnoea in Scheie's syndrome. *Thorax*, *35*(2), 85–91.

Rechtschaffen, A., Verdone, P., & Wheaton, J. (1963). Reports of mental activity during sleep. *Canadian Psychiatric Association Journal*, *8*, 409–414.

Roffwarg, H. P., Muzio, J. N., & Dement, W. C. (1966). Ontogenetic development of the human sleep-dream cycle. *Science*, *152*(3722), 604–619.

Sachs, C., & Kaijser, L. (1980). Autonomic control of cardiovascular reflexes in narcolepsy. *Journal of Neurology, Neurosurgery, and Psychiatry*, *43*(6), 535–539.

Saper, C. B., & Fuller, P. M. (2017). Wake-sleep circuitry: an overview. *Current Opinion in Neurobiology*, *44*, 186–192.

Saper, C. B., Fuller, P. M., Pedersen, N. P., Lu, J., & Scammell, T. E. (2010). Sleep state switching. *Neuron*, *68*(6), 1023–1042.

Schenck, C. H., Bundlie, S. R., Ettinger, M. G., & Mahowald, M. W. (1986). Chronic behavioral disorders of human REM sleep: a new category of parasomnia. *Sleep*, *9*(2), 293–308.

Stoyva, J. M. (1965). Posthypnotically suggested dreams and the sleep cycle. *Archives of General Psychiatry*, *12*, 287–294.

Tormey, W. P., Davis, W. G., & Darragh, A. S. (1978). Nocturnal ulcer pain associated with slow-wave sleep. *Lancet*, *1*(8078), 1340–1341.

Valatx, J. L., Jouvet, D., & Jouvet, M. (1964). Évolution électroencéphalographie des différents états de sommeil chez le chaton. *Electroencephalography and Clinical Neurophysiology.*, *17*, 218–233.

Venner, A., Todd, W. D., Fraigne, J., Bowrey, H., Eban-Rothschild, A., Kaur, S., & Anaclet, C. (2019). Newly identified sleep-wake and circadian circuits as potential therapeutic targets. *Sleep*, *42*(5).

West, L. J., Janszen, H. H., Lester, B. K., Lester, B. K., & Cornelisoon, F. S., Jr. (1962). The psychosis of sleep deprivation. *Annals of the New York Academy of Sciences*, *96*, 66–70.

Winer, J. R., Deters, K. D., Kennedy, G., Jin, M., Goldstein-Piekarski, A., Poston, K. L., & Mormino, E. C. (2021). Association of short and long sleep duration with amyloid-β burden and cognition in aging. *JAMA Neurology*, *78*(10), 1187–1196.

Wolpert, E. A., & Trosman, H. (1958). Studies in psychophysiology of dreams. I. Experimental evocation of sequential dream episodes. *AMA Archives of Neurology & Psychiatry*, *79*(5), 603–606.

Xi, Z., & Luning, W. (2009). REM sleep behavior disorder in a patient with pontine stroke. *Sleep Medicine*, *10*(1), 143–146.

CHAPTER 3
The pupil

3.1 Anatomy and physiologic function of the pupil

The pupil functions by regulating the amount of light transmitted into the retina and thereby helps in focusing the image on it. The size of the pupil is regulated by the musculature of the iris, and the dilator and sphincter pupillae. These muscles are made up of smooth muscle fibers, which are derived from the neural ectoderm, whereas the rest of the iris is derived from the mesoderm. Because the color of the iris depends on pigment, which collects in chromatophores, and this pigmentation is dependent on normal innervation, a delay in innervation might be a cause of heterochromia iridis, a condition in which the iris of one eye has a different color from that of the other. The anterior surface of the iris is divided into two concentric zones: the peripheral ciliary zone and the central pupillary zone. These zones are divided by a ridge called the collarette.

The sphincter pupillae is a circular band of fibers surrounding the pupillary aperture. It is attached on one side to the pigment epithelium of the iris and blends on the other side with fibers of the dilator muscle. The sphincter pupillae is, at most, 1mm wide but it can shorten some 87% of its length, apparently more than any other muscle in the body (Adler, 1950). The sphincter pupillae is supplied by crossed and uncrossed parasympathetic fibers which reach it through the third nerve and take origin in the paired nuclei of Edinger—Westphal (Crouch, 1936). The dilator pupillae muscle is a layer of muscle fibers situated in the posterior portion of the stroma of the iris supplied by sympathetic fibers. The dilator is attached peripherally to the ciliary body and centrally to the constrictor pupillae. The dilator and constrictor pupillae muscles are thought to be antagonistic in activity, but the sphincter seems more powerful than the dilator because histamine instillation into the conjunctival sac, which stimulates both constrictor and dilator fibers, results in marked constriction of the pupil. The pupil may also dilate due to relaxation of the constrictor pupillae without dilator muscle activity (Walsh, 1957) (Fig. 3.1).

The normal pupil is 2.5—5.5 mm in diameter. The pupil is smaller in infancy and widens progressively until adolescence when it attains its largest

Introduction to Clinical Aspects of the Autonomic Nervous System
DOI: https://doi.org/10.1016/B978-0-323-96011-3.00010-3

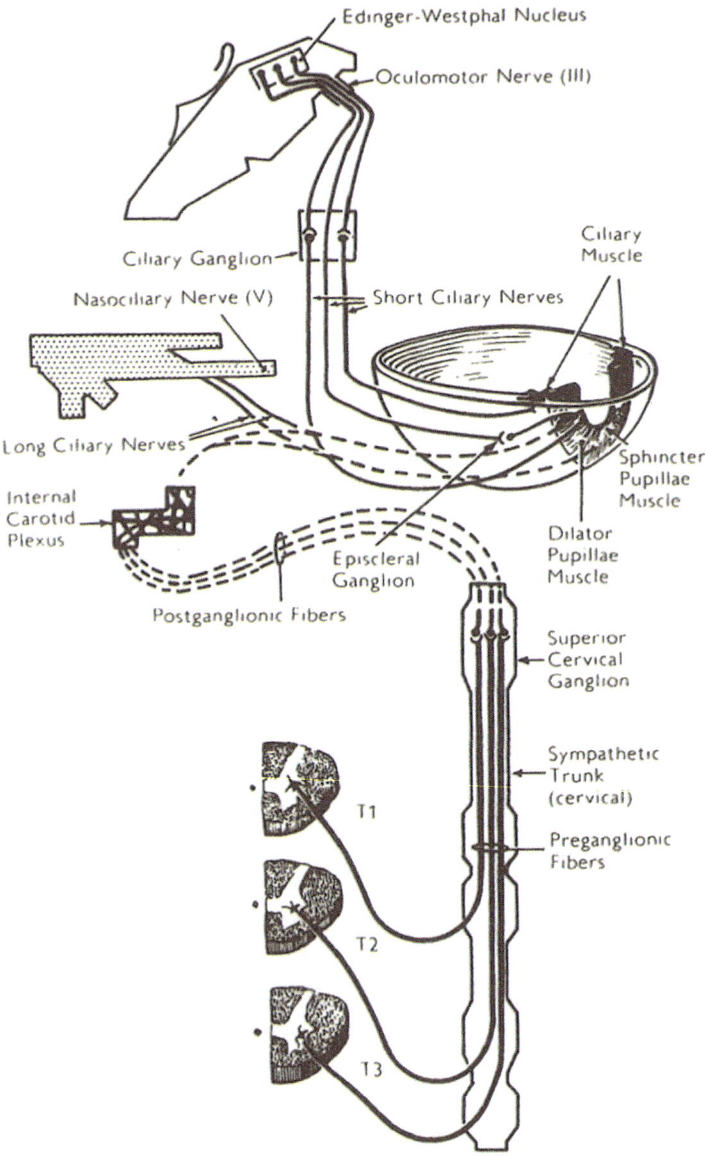

Figure 3.1 Diagram to show the autonomic innervation of the dilator and sphincter pupillae muscles. Dotted lines: postganglionic sympathetic fibers. *Drawn by M. Norviel.*

diameter and becomes narrower again in old age. The magnitude of the pupillary responses parallels the size of the pupil and most active pupillary reactions are obtained in young people (Larsson & Osterlind, 1944). Small pupils (miosis) are usually 2.5 mm or less in diameter, large pupils (mydriasis) are more than 6 mm in diameter. Unequal pupils (anisocoria) may be due to different illumination of the eye or a physiological variant, apart from indicating disease.

The pupil dilates and contracts continuously in response to changes in light, convergence-accommodation, and sensory input from the entire surface of the body, and with respiration. These continuous variations in size are dependent on the age of the individual.

The term hippus describes continuous rhythmic variations in pupil size which can be observed under constant illumination in some subjects. The etiology of these rhythmic changes in pupillary size is not clear. A working definition of hippus has been proposed (Yoss, Moyer, & Hollenhorst, 1970). This definition restricts the term to bilateral synchronous rhythmic changes in pupillary size. The large amplitude of the change is visible to the unaided eye and is independent of illumination or convergence. Hippus has to be distinguished from pupillary unrest characterized by strong contractions and dilatations of the pupil of small amplitude and which occur in response to a strong stimulus of light or convergence. Generally, hippus is considered a variant of normal pupillary unrest, and it has been compared to physiologic tremor or attributed to an imbalance between sympathetic and parasympathetic activity of the autonomic nervous system. It occurs in association with respiration and has been attributed to a subcortical extrapyramidal irritation, but it is regarded by most clinicians as of little diagnostic value because it occurs in normal subjects. Nevertheless, hippus has been reported in neurosyphilis, tumors of the brain, epilepsy, multiple sclerosis, chorea, posterior fossa lesions other than tumors, schizophrenia, meningitis, coma, and with the use of barbiturate and paraldehyde and other drug intoxication. Decreased hippus may occur in myasthenia gravis. Normally, hippus varies in frequency from 8 to 14 per minute and the change in pupillary diameter usually exceeds 1 mm. The area of the pupil may approximately double when it changes from constriction to dilatation. Using a paired-eye control study design with three cohorts receiving either 1.0% tropicamide (an antagonist of the parasympathetic nervous system) in light, 1.0% tropicamide in dark, or 10% phenylephrine (an agonist of the sympathetic nervous system), Turnbull and colleagues demonstrated that pupillary hippus could be extinguished by antagonizing the parasympathetic nervous system, whereas agonizing the sympathetic nervous system led to pupillary dilation without affecting hippus. This study suggests that hippus originates from central parasympathetic activity, and not from increased sympathetic activity, or oscillations in the balance between parasympathetic and sympathetic nervous systems at the pupil (Turnbull, Irani, Lim, & Phillips, 2017).

While it has been widely accepted that the continuous change in pupil size depends on the reciprocal activity of parasympathetic and sympathetic fibers, there is some experimental evidence in animals to suggest that the sympathetic might not be important in pupillary dilatation. In rabbits, for example, it could not be demonstrated that there were dilator pupillae fibers but parasympathetic fibers to the constrictor muscle could be easily seen (Langworthy & Ortega, 1943). Moreover, the vasculature of the iris also seemed important in determining the size of the pupil for if the intravascular pressure was artificially raised, the pupil became smaller and enlarged again when the pressure fell. Vasoconstriction without pressure changes also caused enlargement of the pupil whereas vasodilatation had the opposite effect. Under physiological conditions, however, vascular influences on pupil size have not been clearly demonstrated. In animals, the pupil dilated reflexively in response to painful stimuli and this dilatation was due to inhibition of parasympathetic discharges rather than sympathetic activity (Ury & Gellhorn, 1939). In a study of the ciliospinal reflex in patients with clinically complete lesions of the spinal cord, it was found that unilateral faradic stimulation of the chest produced pupillary widening most marked on the side opposite the stimulus whereas stimulation on the face had no effect (Arieff & Pyzik, 1953). From this, it was concluded that sympathetic activity rather than parasympathetic inhibition was responsible for pupillary dilatation in response to noxious stimuli in man. It was argued that the cord lesions prevented inhibitory impulses descending from the face and painful impulses ascending through the cord from the chest. Faradic stimulation of the chest must have therefore activated the sympathetic outflow to produce pupillary dilatation and this response was similar to the pupillary dilatation in response to pain (ciliospinal reflex).

3.1.1 Pupillary constriction

The pupil reacts to light by constriction due to a reflex whose first neuron is in the rods and cones of the retina. The majority of fibers related to pupillary constriction in response to light originate near the macula (Hess, 1908). Other retinal areas can also initiate this reflex but require more light (Walsh, 1957). A third photosensitive cell type has been discovered within the ganglion cell layer. A minority (<5%) of retinal ganglion cells are intrinsically photosensitive. These intrinsically photosensitive retinal ganglion cells employ melanopsin to mediate direct depolarization by light in a manner completely independent of rod- and cone-mediated photoactivation (Provencio et al., 2000). Rods and cones are responsible for rapid modulation in pupillary size with changes in ambient light sensitivity, whereas intrinsically photosensitive retinal ganglion cells mediate a slower but more sustained pupillary response to light (Gooley et al., 2012). Intrinsically photosensitive retinal ganglion cells do not generate vision-forming signals as rods and cones do, but they do project to the suprachiasmatic nucleus, serving as the sole retinal regulator of the circadian rhythm. The impulses pass backward in the optic nerve and tract of the same and opposite sides decussating in part in the chiasm. The pupillomotor fibers leave the optic tract and continue in the superior brachium to reach the pretectal region where they synapse. It is thought that the pupillomotor fibers synapse in the pretectal region and do not enter the superior colliculus. The axons originating from each olivary pretectal nucleus innervate both the ipsilateral nucleus of Edinger–Westphal through intercalated neurons and the constrictor center on the contralateral nucleus of Edinger–Westphal through fibers in the posterior commissure (Fulton, 1938). The occipital lobes are also connected to the constrictor centers because electrical stimulation of the occipital cortex produces pupillary constriction (Karplus & Kreidl, 1909). This output from the visual cortex down to the dorsal midbrain helps govern the accommodation reflex. From the accessory nucleus of Edinger–Westphal, which is the constrictor center, fibers pass in the third nerve to the ciliary ganglion where they synapse. Postganglionic fibers in short ciliary nerves go to the choroidal, ciliary, and iridic plexuses. Fibers from the nucleus of Edinger–Westphal to the iridic plexus are both crossed and uncrossed (Crouch, 1936) and this explains the simultaneous constriction of both pupils when only one eye is exposed to light. Another pathway for pupillary constriction not passing through the ciliary ganglion has also been postulated (Walsh, 1957) (Figs. 3.2 and 3.3).

The reaction of the pupils to accommodation and convergence of the eyes is constriction. This reaction is synergistic in both eyes and probably initiated by the cortex. Animal studies have demonstrated the role of the peristriate visual cortex (areas 19 and 22) and the lateral suprasylvian areas (Ohtsuka & Sato, 1996). Accommodation does not seem to be completely dependent on convergence, and pupillary constriction on accommodation can occur without convergence. In blind subjects, voluntary convergence can give rise to pupillary constriction. Afferent impulses for the pupillary

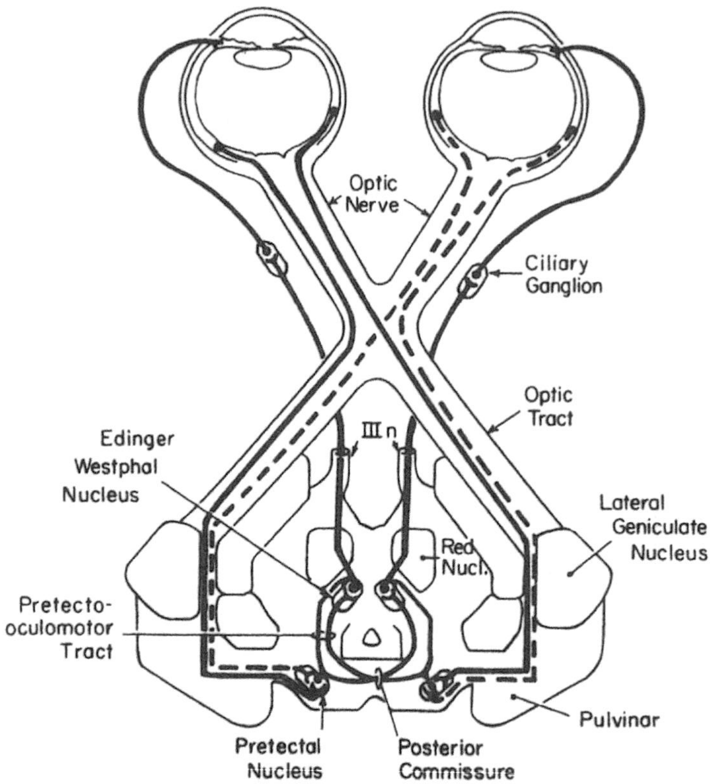

Figure 3.2 Diagram illustrating pathways and connections in the visual system related to the light reflex (*solid line*). *Drawn by M. Norviel.*

accommodation-convergence reaction are visual and proprioceptive but either alone is effective in initiating this response. The final efferent pathway for pupillary constriction in response to light is separate from the one induced by accommodation-convergence. This was demonstrated in a case of peripheral injury to the eye, which abolished the direct and consensual light reflex in the right eye with associated paralysis of the inferior rectus and inferior oblique muscles and bilaterally normal accommodation-convergence (Nathan & Turner, 1942). In a study of ten similar cases, the authors suggested that convincing patients in whom such dissociation can be demonstrated must fulfill several criteria: (1) the pupil on the normal side contracts when the affected eye is exposed to light; (2) there is no response of the affected pupil to light and no dilatation in the dark either on direct or consensual stimulation; (3) there is a normal reaction to accommodation-convergence bilaterally and symmetrically and both pupils

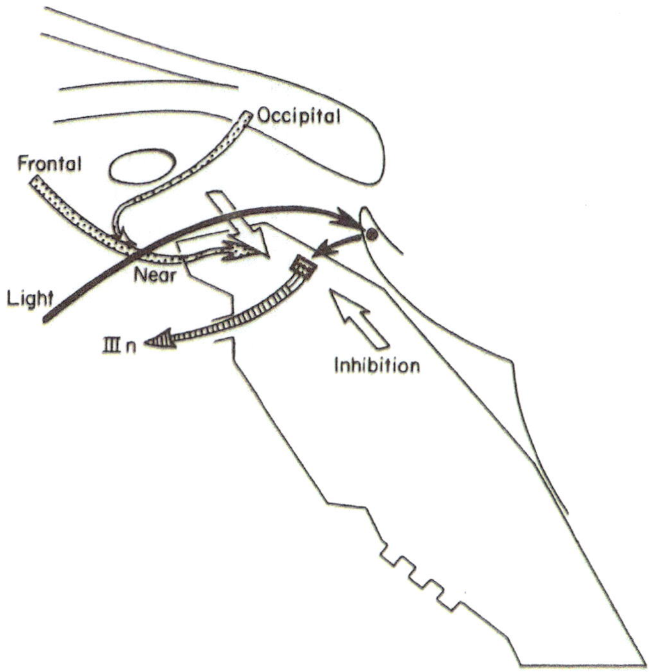

Figure 3.3 Diagram illustrating brain structures which affect the pupillary responses to light and accommodation. *Drawn by M. Norviel.*

dilate again afterward; and (4) the lesion is peripheral. With such strict criteria, a dissociation of the light reflex from the accommodation-convergence has only rarely been demonstrated.

3.1.2 Pupillary dilation

Pupillary dilatation can occur after stimulation of the sensory, motor, and frontal cortices and this can be obtained also after interruption of the sympathetic pathways so that the dilatation is presumably due to inhibition of the activity of the sphincter pupillae. Several claims for a pupillary dilator center in the brain have been made but the position of this center is not known. The sympathetic pathway that innervates the pupillary dilator is a three-neuron system. The majority of the first-order neurons are located in the paraventricular and arcuate nuclei of the ipsilateral hypothalamus. These neurons send axons caudally through the lateral brainstem and cervical spinal cord to synapse on the second-order neurons that form the ciliospinal nucleus of Budge from the C8 to T2 spinal cord levels. The axons of these cell bodies synapse on the third-order neurons that form the superior cervical (stellate) ganglion. The postsynaptic fibers are not myelinated and go on to innervate three distinct targets: the superior and inferior Müller tarsal muscles, the pupillary dilator, and the sweat glands of the face. These fibers destined to innervate the pupillary dilator and superior and inferior tarsal muscles join the internal carotid nerve, which divides into the internal carotid and cavernous plexuses after it enters the skull. Fibers from the cavernous sympathetic plexus pass to the nasociliary division of the trigeminal nerve and then to the dilator pupillae muscle through the long ciliary nerve (from the ophthalmic branch of the trigeminal nerve) and through the short ciliary nerves, which arise from the ciliary ganglion. Epinephrine and norepinephrine stimulate β1-adrenergic receptors on the target muscles. On the afferent side, the connections for pupillary dilatation are not well known. They probably pass from the retina to midbrain structures and hence into the ciliospinal center of Budge (Walsh, 1957).

3.1.3 Pharmacology

The pharmacology of the iris depends on its dual sympathetic and parasympathetic innervation. Numerous blockers and stimulators can be used to unravel the cause of pupillary dysfunction. These pharmacologic agents and their site of action are illustrated in Figs. 3.4 and 3.5.

3.1.4 Other pupillary reflexes

Several reflexes leading to changes in pupil size have been described (Belliveau, Somani, & Dossani, 2021). The orbicularis or lid reflex is unilateral and induces constriction of the pupil on forceful closure or opening of the eye. Although this reflex is frequently absent in normal individuals it can be seen in eyes that do not react to light or convergence (Boshes & Mayer, 1935) and in Argyll Robertson pupils. Because cocainization of the facial nerve abolishes the reflex it has been suggested that pupillary constriction is a consequence of an overflow of stimuli from the facial to the oculomotor nerve but the anatomical structures responsible for the orbicularis reflex and their location are not known.

The trigeminus, oculopupillary, or oculosensory reflex is elicited by stimulation of the cornea or conjunctiva and consists of bilateral brief dilatation after which sustained constriction of the pupil occurs. The structures subserving this response are not known but it has been suggested that impulses reach the trigeminal sensory nucleus and hence somehow the medial longitudinal fasciculus and third nerve nuclei. This reflex

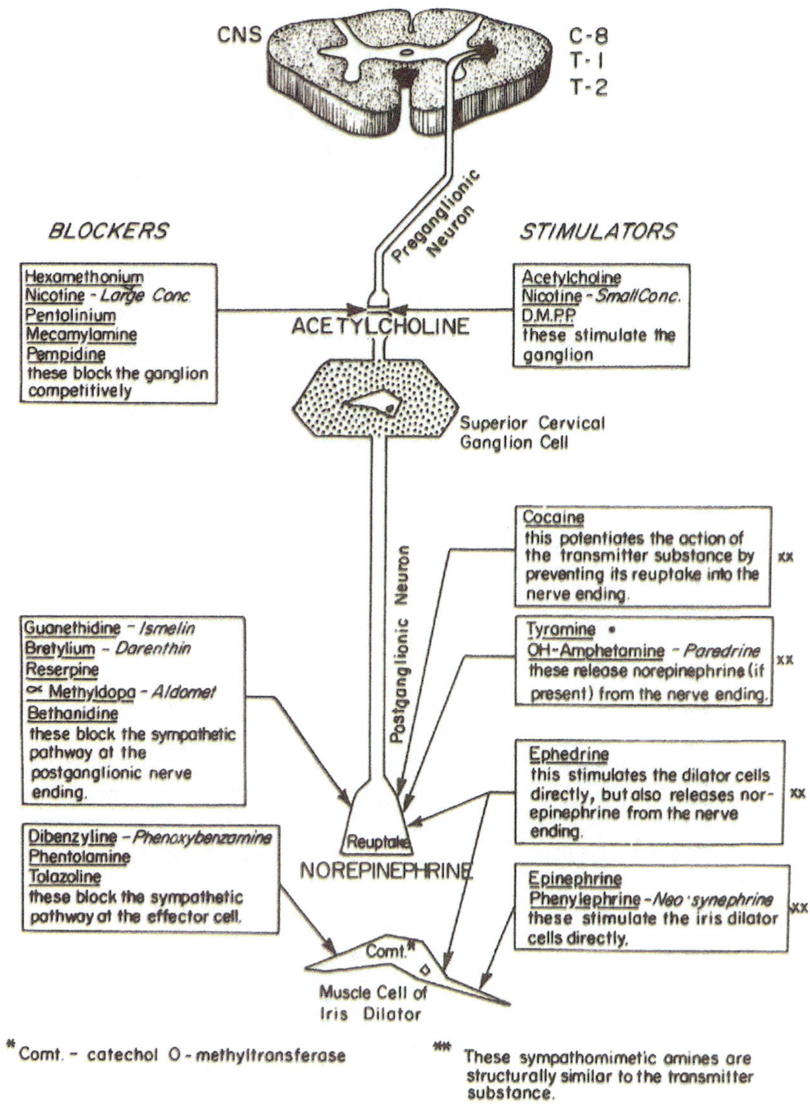

Figure 3.4 Diagrammatic representation of the pharmacology of the sympathetic innervation of the iris. *Drawn by M. Norviel.*

cannot be elicited in some otherwise normal eyes and is absent in the Argyll Robertson pupil.

The pupils dilate as a result of emotional stress and after peripheral stimulation (psychosensory reflex). This response is preserved in animals after interruption of the sympathetic fibers and pupillary constriction occurred after

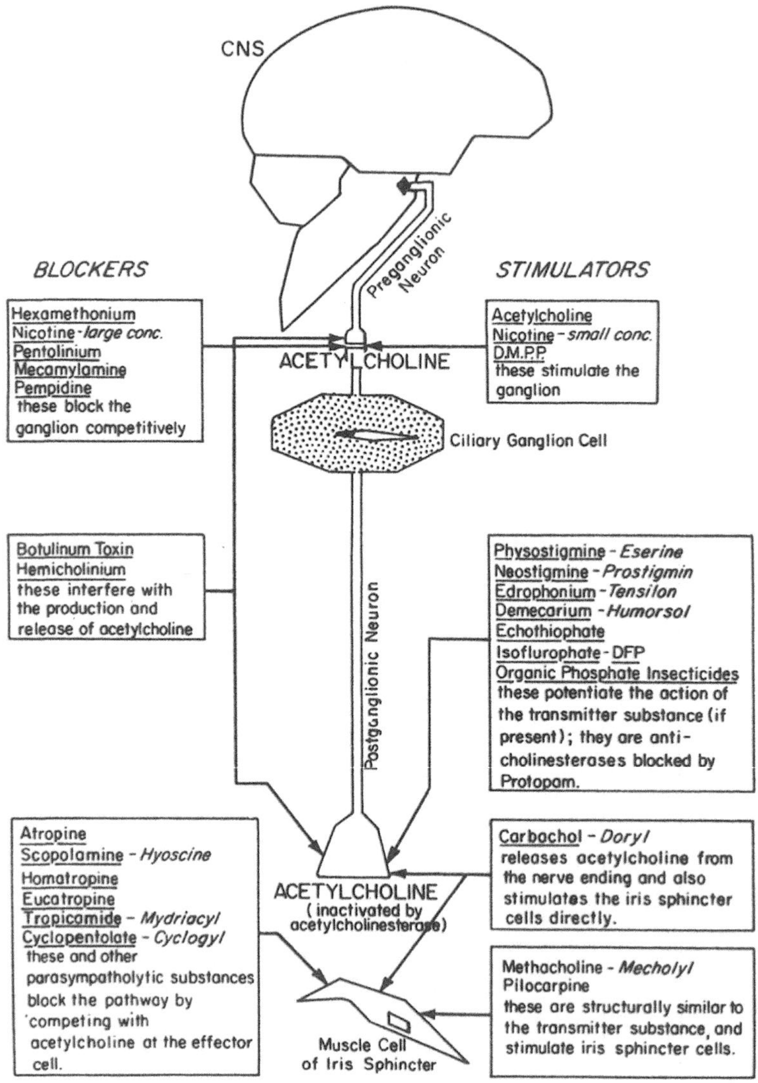

Figure 3.5 Diagrammatic representation of the pharmacology of the parasympathetic innervation of the iris. *Drawn by M. Norviel.*

decortication. In animals with an intact cortex and sympathetic, but sectioned third nerve, this reflex remained normal (Walsh, 1957). From these experiments, it was concluded that the pupillary dilatation of the psychosensory reflex was the result of both parasympathetic inhibition and sympathetic stimulation (Adler, 1950; Duke-Elder, 1932–1938). In the cat, stimulation of the sciatic nerve led to pupillary dilatation (Weinstein & Bender, 1941). This reflex was not affected by section of the sympathetic trunk in the neck or by hypothalamic lesions but was abolished by section of the third nerve. In monkeys, however, oculomotor nerve section was without effect but section of the cervical sympathetic trunk abolished the reflex. The pupillary dilatation in response to sciatic nerve stimulation is therefore due both to parasympathetic inhibition and sympathetic activity, and the predominant mechanism for this reflex varies with the species studied.

During induction of anesthesia, the pupil dilates and this is attributed to cortical stimulation and consequent inhibition of parasympathetic activity. In the second stage of anesthesia, the pupil constricts due to loss of cortical control or cessation of tonic sympathetic activity which arises in the hypothalamus (Ury & Gellhorn, 1939). During deep anesthesia, the pupil dilates again because of decreased constrictor activity (Adler, 1950).

The vestibular pupillary reflex can be elicited by caloric or rotational stimuli which induce pupillary dilatation succeeded by hippus. It is assumed that stimuli pass through the medial longitudinal fasciculus and affect the pupilloconstrictor center in the brainstem. Transient bilateral pupillary contraction giving way to dilatation can be induced by stimulation of the cochlea with a tuning fork (cochlear pupillary reflex) (Nishimura, 1927) and pupillary dilatation on deep inspiration and contraction during deep expiration (vagotonic pupillary reflex). While these pupillary responses can be easily observed, their mechanism is poorly understood and the explanations proffered have so far little experimental support. A summary of pupillary reflex responses is given in Table 3.1.

The introduction of visual-evoked responses for diagnostic purposes has led to an interest in establishing the reliability of this clinical test. The flash-evoked visual response is variable and this decreases its clinical usefulness. On the other hand, pattern-evoked responses are thought to be more stable between recording sessions in individuals; nevertheless, they are influenced by the width of the field of vision, by square size and luminance, and by visual acuity. In further studies of the stability of pattern-evoked visual responses, it became clear that pupil size also affected the pattern-evoked visual response. If the pupil size is artificially varied in healthy subjects and in patients with multiple sclerosis who already are known to have abnormal latencies, a slight depression of amplitude was seen irrespective of whether the pupil was small or large but the most obvious effect was a decrease in latencies with large pupils which has been attributed to an increased amount of light entering the eye and not to a

Table 3.1 Some reflex responses.

Reflex	Stimulus	Afferents	Central structures	Efferents	Effects	Remarks
Light	Light	Outer limbs of rods and cones of retina and intrinsically photosensitive retinal ganglion cells; optic nerve and tracts of same and opposite side; superior brachium	Pretectal region; ipsilateral and contralateral accessory nuclei of Edinger–Westphal	Branch of 3rd nerve to the inferior oblique; ciliary ganglion, short ciliary nerves; choroid, ciliary and iridic plexuses; pupilloconstrictor muscle	Constriction to light	Simultaneous contraction of both pupils when only one is exposed to light
	Darkness	Retina; midbrain	Adjacent to Edinger–Westphal nucleus; hypothalamus lateral to infundibulum; close to oculomotor nerve; medulla; 4th–6th thoracic cord segments	1st-3rd thoracic nerves, paravertebral sympathetic chain and chain ganglia in neck; internal carotid nerve; cavernous sympathetic plexus; nasociliary division of 5th nerve; long ciliary nerves; dilator pupillae	Dilatation in the dark	
Ciliospinal	Scratching of skin of neck	Cutaneous afferents from stimulated area			Dilatation on side of painful stimulus to neck	4th–6th thoracic cord segment also known as ciliospinal center of Budge
Convergence-accommodation	Focusing on near object; pro prioceptive impulses; blurring of image	Retina -> lateral geniculate nucleus via the optic tract -> Visual cortex	Striate cortex (area 18 and 19) association pathways; visual association cortex; superior colliculus; pretectal areas; Perlia's nucleus	1. Pupillary constric tion: fibers from the Edinger–Westphal nucleus project via the 3rd nerve -> ciliary ganglion -> short ciliary nerves -> iris sphincter	Contraction of ciliary muscles; constric tion of pupils; convergence of eyes	In blind persons voluntary convergence produces pupillary constriction

				2. Lens accommodation: fibers from the Edinger–Westphal nucleus project via the 3rd nerve -> ciliary ganglion -> short ciliary nerves -> ciliary muscle 3. Convergence: fibers from the medial rectus subnucleus innervate the bilateral medial rectus muscles		
Orbicularis or lid	Forceful opening of closed eyelids	In 7th nerve	Oculomotor nucleus	3rd nerve	Unilateral constriction of pupil	May be absent in normal individuals. May be preserved in eye without light or convergence reflexes or in Argyll Robertson pupil

(*Continued*)

Table 3.1 (Continued)

Reflex	Stimulus	Afferents	Central structures	Efferents	Effects	Remarks
Trigeminus or oculopupillary or oculosensory	Pinching or pricking skin of eyelid	5th nerve	5th nerve nucleus; median longitudinal fasciculus; 3rd nerve nucleus	3rd nerve	Immediate dilatation soon giving way to constriction of pupil	May be absent in normal individuals. Not found in Argyll Robertson pupil
Psychosensory	Stimulation of sensory nerves; emotional stimuli		Sensory-motor cortex	Sympathetic; parasympathetic (inhibition); 3rd nerve in some species	Dilatation of pupils	May play a role in the changes in pupil size observed during various stages of anesthesia and sleep
Vestibular pupillary	Caloric or rotational stimuli to labyrinth	8th nerve	Medial longitudinal fasciculus; constrictor centers	3rd nerve	Dilatation of pupils soon followed by hippus	
Cochlear pupillary	Tuning fork	8th nerve	Medial longitudinal fasciculus; constrictor centers	3rd nerve	Dilatation of pupil followed by contraction	
Vagotonic pupillary	Deep inspiration; deep expiration				Dilatation with inspiration; Contraction with expiration	Relation to pupillary changes during Cheyne–Stokes respiration

change in the functional capacity of the visual system. These differences were seen in both patients and controls, and it might be necessary to take pupil size into account when comparing individual groups and assessing latencies statistically in experimental situations (Hawkes & Stow, 1981).

3.1.5 Pupillometry

Simple tests of autonomic function usually available in practice depend on the measurement of cardiovascular reflexes, and these have been extensively employed in determining the integrity of the autonomic nervous system. The smooth muscle of the iris is entirely innervated by the autonomic nervous system, and it is natural that such an easily accessible structure should arouse interest as a possible test site for autonomic function.

Pupil cycle time (PCT) is a simple way of measuring autonomic reflex activity (Martyn & Ewing, 1986) (Fig. 3.6). It is easy to induce persistent regular oscillations of the pupil using a slit lamp. The period of these oscillations can be measured with a stopwatch, and it is prolonged in patients with evidence of autonomic neuropathy or central autonomic failure.

The test was first described by Stern (1944) and was modified little for the evaluation of a larger number of normal subjects, patients with diabetic neuropathy, and six patients with progressive autonomic failure. The normal PCT was found to be 946 ± 120 milliseconds when the average time for 100 cycles of dilatation and constriction was determined. The upper limit of normal is 1150 milliseconds. There was no correlation with age in normal subjects (19–65 years, mean, 34 years). Various pharmacologic tests using parasympathetic blockade and sympathomimetic drugs suggest that PCT is most sensitive to disturbances in the efferent parasympathetic supply to the iris, and while there was no good correlation with tests of cardiovascular autonomic function, the test, nevertheless, provides a simple and novel way of quantifying yet another autonomic function and should be further explored in patients with autonomic disturbances from other causes. Using edge-light PCT as a probe of the pupillary light circuit in migraine, Cortez and colleagues demonstrated significantly increased PCT in patients with probable, episodic, and chronic migraine, compared to controls. Additionally, increased PCT correlated with the presence of craniofacial autonomic symptoms, linking pupillary circuit dysfunction to peripheral trigeminal sensitization (Cortez, Rae, Millsap, McKean, & Brennan, 2019).

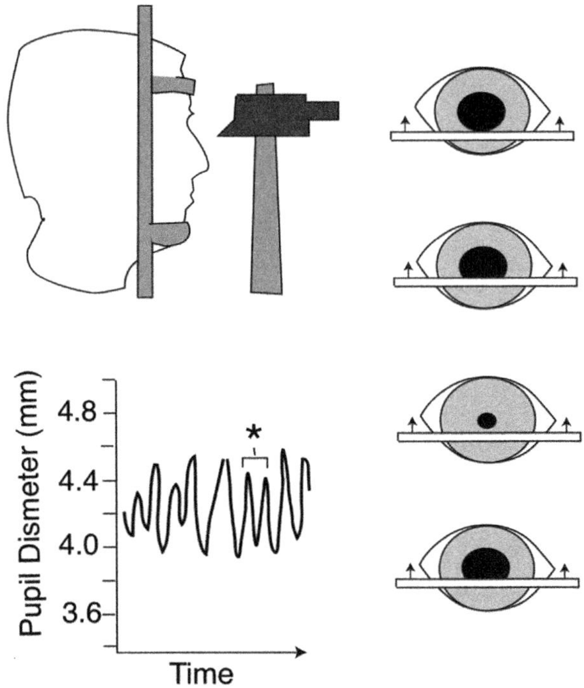

Figure 3.6 Examination technique for edge-light pupil cycle time (PCT). The subject is seated in front of a slit lamp. The slit lamp beam moves over the iris until contact is made with the pupillary margin, initiating pupillary constriction. Once out of the light, the pupil then spontaneously redilates, eventually returning the edge of the pupil to contact the light beam. This method produces oscillations between dilated and constricted states. One cycle is the time it takes for a complete loop of the pupil reflex arc to be completed. *Drawn by M. Norviel.*

A portable pupilcyclemeter (Fig. 3.7) is available. With this machine, a significant shortening of PCT at altitude was found, and advanced age and altitude had additional significant effects on PCT at high altitude only (Appenzeller, 1989) (Figs. 3.8 and 3.9). PCT has also been used to study autonomic function in patients with headaches and other neurologic diseases (Appenzeller, 1989). In cluster headache, a latent Horner syndrome causes a significant shortening of PCT (ipsilateral to the headache), whereas in peripheral neuropathies (of various etiologies) PCT is significantly prolonged (Fig. 3.10).

The psychological significance of pupillary movements has also been studied (Goldwater, 1972), and the reactivity of the iris muscle as a measure of autonomic dysfunction in schizophrenic patients has been examined (Rubin & Barry, 1972a, 1972b). Pupillary dilatation and constriction

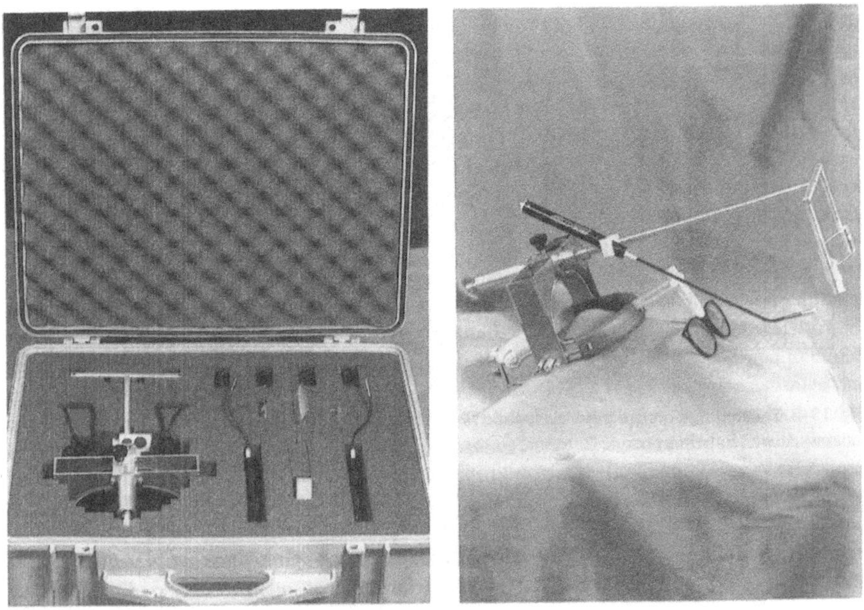

Figure 3.7 A portable pupilcycle-meter (right). The specially designed (patent pending) pupilcycle-meter is mounted on a headband worn by the operator. Subjects rest their forehead on the acrylic plastic band. The distance between operator and subject can be changed for focusing. A modified flashlight provides illumination. The carrying case contains different strength magnifying glasses, spare flashlights, and electronic stopwatches (left).

were taken as expressions of sympathetic adrenergic and parasympathetic cholinergic activity. A reciprocal relationship between sympathetic and parasympathetic system activity on the pupil which characterizes normal subjects during stress and after recovery was not found in schizophrenic patients. In patients with schizophrenia with psychosis and those with schizophrenia in remission, disorganized and unintegrated autonomic responses to stress and after recovery from stress were found. This study suggested a possible irreversible defect in central autonomic control which seems to be independent of the clinical state of the patient or the cause of the illness. Alteration of pupillary light responses has also been reported in several neurodegenerative disorders such as Alzheimer's disease or Parkinson's disease. Abnormalities of pupillary movements reported in Parkinson's disease include reduced amplitude of constriction, increased latency, and decreased velocity of constriction, while the baseline pupil diameter may be increased or not significantly different compared to healthy individuals (Fotiou et al., 2009; Micieli et al., 1991).

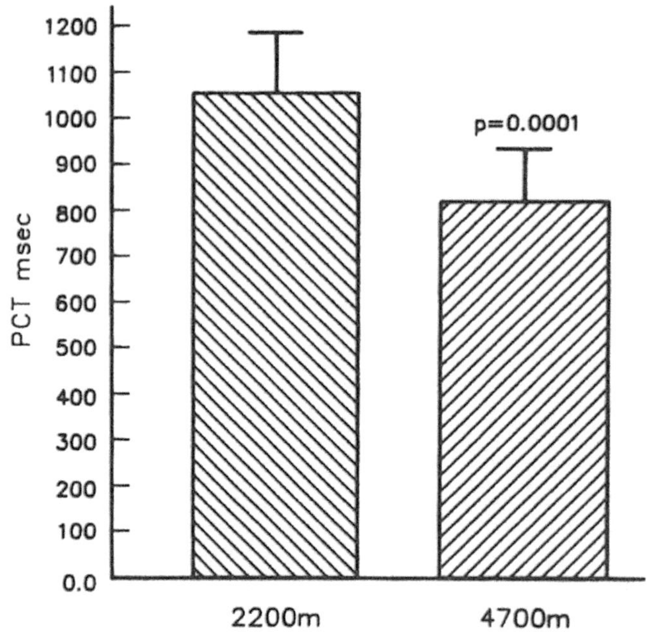

Figure 3.8 Shortening of mean pupil cycle time (PCT) at altitude. This is tentatively attributed to the altitude-associated hyperadrenergic state.

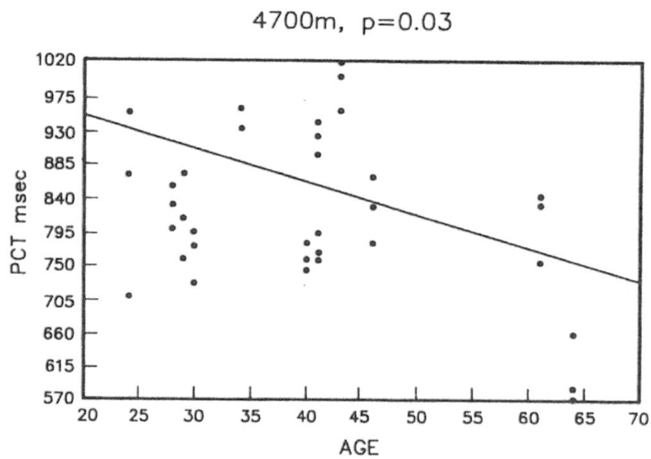

Figure 3.9 A significant shortening, with advancing age, of pupil cycle time, not seen at lower altitude.

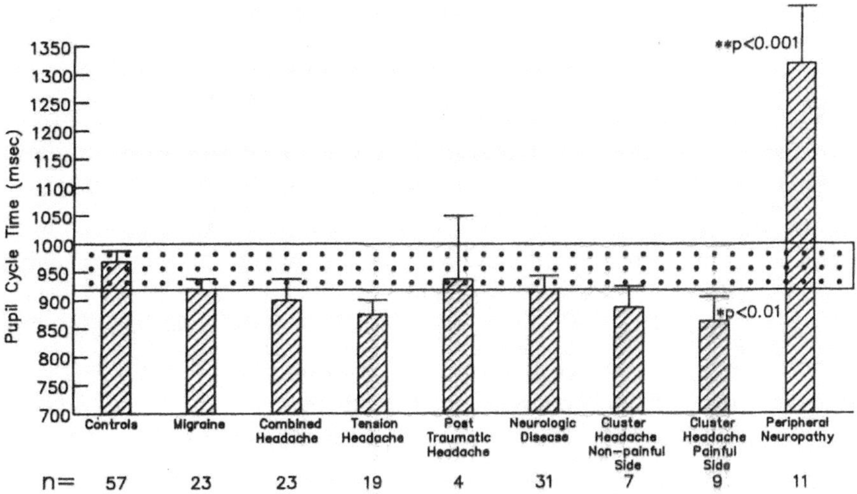

Figure 3.10 Histogram showing pupil cycle times (PCT). The area between the horizontal bar represents PCT values 2 standard deviations (SD) above and below the mean value for controls. Vertical T-bar = 1 SD. Source: *From Appenzeller, O. (1989). Pupil cycle time (PCT) in headache and other neurologic diseases. In: F.C. Rose (Ed.),* New advances in headache research: The 7th Migraine Trust international symposium *(pp. 25–29). Smith-Gordon, London.*

Future studies will benefit from technological advances such as low-cost hardware, continuous home monitoring, wearable devices, and the potential use of artificial intelligence that may allow a more accurate screening, follow-up, and management of patients with neurodegenerative disorders.

3.2 Disordered pupillary function

3.2.1 Horner syndrome

This symptom complex was first delineated by Bernard (1853). The fully developed syndrome was extensively studied in patients who had upper thoracic and lower cervical sympathectomies. It consists of pupillary constriction, ptosis, enophthalmos, a rise in temperature, and a lack of sweating on the ipsilateral side of the face.

The pupillary constriction is most marked when the 7th and 8th cervical and 1st thoracic roots have been interrupted or when the inferior cervical sympathetic ganglion has been removed. The pupil continues to react to light, accommodation-convergence, and the lid and psychosensory reflexes are retained. Cocaine instillation into the conjunctiva does not produce the normally seen dilatation of the pupil. The smallest diameter of the pupil occurs initially after the interruption of the sympathetic supply but with time there is progressive enlargement. The affected pupil usually remaining smaller compared the unaffected pupil. The relative difference in margin-to-reflex distance between eyes usually does not exceed 1 mm to 2 mm in a patient with Horner syndrome. The narrowing of the palpebral fissure results from interruption of the innervation of smooth muscle fibers in the eyelids, causing ptosis and elevation of the lower lid, both of which improve with time. Although enophthalmos is often described in Horner syndrome, this might be only apparent because of the droopiness of the upper lid and elevation of the lower lid ("pseudo enophthalmos"). In some cases of Horner syndrome associated with trauma, however, enophthalmos is very marked and this has been attributed to the protrusion of orbital fat through fractures of the orbit. In other cases, the enophthalmos has been ascribed to the denervation of smooth muscle fibers within the orbit but at histological examination of orbital contents, smooth muscle fibers are not always found (Whitnall, 1932). Another feature of the Horner syndrome is occasionally increased intraocular pressure which has been attributed to engorgement of vessels and excessive ciliary secretion (Magitot, 1931). On the other hand, a feature that is not part of the triad but can be a helpful clue to identify cases of partial Horner syndrome is conjunctival injection due to vasodilation resulting from decreased sympathetic tone and decreased intraocular pressure relative to the fellow eye resulting from a decrease in aqueous humor production by the ciliary body, a process that is stimulated by the sympathetic nervous system. It has been also assumed that depigmentation can result after sympathetic denervation. This is rarely if ever seen in the adult but lack of pigmentation rather than loss of pigment occurs and is seen in infants with perinatal injuries to the brachial plexus and involvement of sympathetic fibers. Injuries to sympathetic fibers after two years of age do not result in a change in eye color because the pigmentation of the iris is by then complete (Walsh, 1957). With the onset of Horner syndrome, there is an initial rise in skin temperature on the affected side of the face due to interruption of vasomotor fibers and consequent vasodilatation, but the loss of sweating may contribute to this rise in skin temperature also. Changes in the quantity of tear secretion have been reported but this is not usually a prominent feature and the amount of tears may be either increased or decreased.

There are numerous causes of Horner syndrome that are related to lesions affecting the sympathetic pathways in the medulla, cord, or periphery, and these include stroke, traumatic or surgical lesions of the cord, syringomyelia, or surgical interventions in or near the thoracic inlet. The clinical presentation of Horner syndrome depends on the location of the lesion. The first-order neurons originate in the hypothalamus, which is rarely the site of injury. The most common location for first-order neuron injury is in the dorsolateral medulla (e.g., Wallenberg syndrome) but may occur more rostrally. Spinal cord lesions rostral to the ciliospinal nucleus of Budge are encountered, particularly in patients with Brown–Séquard syndrome. First-order Horner syndrome is accompanied by ipsilateral ptosis and anhidrosis of the ipsilateral face and body. The second-order neuron is often injured by thoracic neoplastic disease whereas the common carotid artery is a less frequent site of injury. Second-order Horner syndrome is accompanied by ipsilateral ptosis and ipsilateral facial anhidrosis. The third-order neuron ascending the carotid is prone to injury in the setting of carotid dissection. Third-order Horner syndrome is accompanied by ipsilateral ptosis of the upper and lower eyelids without anhidrosis. A combination of cocaine and hydroxyamphetamine eye drops has been used to detect sympathetic denervation of the pupil. The instillation of hydroxyamphetamine causes third-order neurons to secrete catecholamines. Therefore, a miotic pupil from a third-order Horner syndrome will not dilate in response to hydroxyamphetamine as the damaged neuron cannot respond to the secretagogue. In cases of first- or second-order neuron lesions, a mydriatic response will be observed. In clinical practice, the two-step test process of using cocaine and hydroxyamphetamine has been replaced by apraclonidine testing. Apraclonidine is a weak α-2 agonist and dilates the supersensitive pupil. Two drops of apraclonidine 0.5% will not change the pupillary size in the unaffected eye, whereas it will cause pupillary dilation in the affected eye. Denervation supersensitivity usually occurs within a few days of injury but may take up to a week. A positive test occurs when reversal of anisocoria is seen after 45–60 minutes (Bouffard, 2019).

Rare causes of Horner syndrome have been reported as an extension of suppurative otitis media (Metzner & Wölfflin, 1915) and after tonsillectomy (Hammer, 1933). A Horner syndrome is also part of a syndrome that includes anomalies of the sternum, abnormal pigmentation of the areola, facial asymmetry, heterochromia iridis, trophic changes in the hands with malformed fingers, disturbed sensory perception, and pyramidal tract signs. This syndrome is attributed to abnormal development of the medulla and cervical spinal cord (Passow, 1933). Lesions in the thalamus have been reported to give rise to an ipsilateral Horner syndrome and experimentally this was reproduced in dogs by lesions placed in the optic thalamus (Walsh, 1957). After coronary bypass graft surgery, peripheral nerve damage is not uncommon, and brachial plexus dysfunction has been reported in 7% of patients. Other peripheral nerve complications include sympathetic damage resulting in Horner syndrome. Mydriasis from sympathetic overactivity has been produced experimentally by anastomosis of the phrenic nerve and cut stump of the sympathetic chain (Cannon, Binger, & Fitz, 1915). Unilateral mydriasis has appeared in bronchiectasis, apical pulmonary tuberculosis, and tumors of the lung. This was associated with ipsilateral flushing of the cheek and attributed to irritation of the sympathetic trunk (Walsh, 1957).

Since the advent of levodopa therapy for Parkinson's disease, several effects of this substance on the size of pupils have been noted. Pupillary dilatation, a benign transient side effect of levodopa treatment for Parkinson's disease occurs and this has been attributed to a stimulatory effect of dopamine or of its metabolites on the sympathetic nervous system. Pupillary dilatation produced by levodopa is bilateral but in patients with subclinical Horner syndrome, the dilatation is confined to the uninvolved eye. Thus, levodopa brings out a latent Horner syndrome not apparent before treatment (Weintraub, Gaasterland, & Van Woert, 1970). Further investigation of a number of patients with Parkinson's disease and controls showed that topical dopamine and the chronic oral administration of levodopa produced dilatation of the pupil only in eyes with an intact sympathetic innervation. Single doses of levodopa do not affect pupil size, but chronic administration of this drug and topical instillation of dopamine into the conjunctival sac produce pupillary dilatation through a peripheral increase in the activity of intact sympathetic pathways and not through alpha-receptor stimulation.

Cervical spine disorders can be associated with Horner syndrome. This may be seen in spinal concussion and vertebral dislocation. It may be a transient complication after anterior cervical fusion and is occasionally seen after severe cervical trauma. In one patient, however, a previously healthy man, a Horner syndrome occurred ipsilateral to burning chest pain which extended to the arm. Extensive investigation showed a large osteophyte that appeared to encroach upon the neural foramen between the first and second thoracic vertebrae. Denervation was found on electromyography of muscles innervated by the lower cervical and first thoracic cord segments, and the chest pain was attributed to causalgia. Three years later, the pain had disappeared, but the Horner syndrome persisted (Bernard & Perlo, 1980).

In congenital Horner syndrome, the fully developed form of the syndrome is rarely seen but ptosis, miosis, facial anhidrosis, and hypochromia of the affected iris are usually found. This is often associated with congenital brachial palsy (Klumpke palsy). The pathogenesis of unilateral Horner syndrome, which appears in the first year of life has now been clarified by the use of different drugs affecting pupil size. Clinically, patients seem to have preganglionic lesions but some had a partial mydriatic failure when hydroxyamphetamine hydrobromide was instilled in the conjunctival sac. In addition, there was supersensitivity to phenylephrine hydrochloride. This paradoxical response has been interpreted to show orthograde transsynaptic dysgenesis of the postganglionic neuron. This condition has previously been demonstrated in the sympathetic nervous system in newborn animals. The failure of hydroxyamphetamine to induce mydriasis must mean failure of postganglionic sympathetic innervation, but in the setting of a congenital Horner syndrome, this may be a secondary phenomenon and is consistent with preganglionic lesions as well (Weinstein, Zweitel, & Thompson, 1980).

Spasm of the iris sphincter has also been reported. This may be intermittent and is attributed to contractions of the pupillary sphincter and ciliary muscle. On examination, contractions of the intraocular muscles several times per minute are found and this can induce spontaneous darkening of vision in the affected eye. The symptoms continue for days and may be preceded by an ipsilateral severe headache. This phenomenon is, after extensive investigation, attributed to the excitability of the ciliary ganglion because an intermittent segmental spasm of the iris sphincter with a similar time course had previously been described in a patient with Adie's syndrome (Czarnecki & Thompson, 1976). The frequency of the spasms and speed of contractions of the muscles suggests parasympathetic stimulation (Thompson & Corbett, 1980).

The significance of lipid deposition in the corneal stroma (arcus senilis) for the prediction of vascular disease or hypercholesterolemia is not clear. A Horner syndrome, however, can affect the deposition of lipid in the cornea and presumably also in other parts of the eye such as Drusen formation in the choroid, or asteroid hyalosis. Therefore, lipid deposition is probably related to blood flow. Experimentally, a change in the velocity of flow to the eye associated with hypercholesterolemia in rabbits can cause more severe changes mainly leading to an abnormally heavy deposition of cholesterol (Zauberman & Livni, 1981). We have observed a marked decrease or total absence of arcus senilis ipsilateral to chronic Horner syndrome of many years duration. Three patients were encountered, two 15 or more years after pneumonectomy and one after a traumatic lesion to the upper spinal cord with traumatic syringomyelia. Arcus senilis was only observed on the unaffected side. Conversely, sympathetic overactivity in another patient with Raeder's syndrome who had a severe ipsilateral headache after extensive sinus surgery responded to sympathetic blockade with local anesthetic for prolonged periods. This patient had only arcus senilis ipsilateral to the headache and the hypertonic sympathetic nervous system (Fig. 3.11). The mechanism by which sympathetic paralysis or overactivity affects the deposition of lipid in the corneal stroma is hypothetical. One proposal is that chronic sympathetic denervation is associated with denervation supersensitivity and consequent vasoconstriction in response to circulating catecholamines. The decrease in blood flow leads to a decrease or absent deposition of lipid. In support of this suggestion is the known impairment of lipid deposition (absence of arcus senilis) in some patients with complete internal carotid occlusion ipsilateral to the occluded blood vessel. Many such patients, however, also have an associated Horner syndrome and both denervation supersensitivity and pure mechanical obstruction of blood flow might be important in the genesis of this asymmetric arcus senilis. The increase in the arcus senilis in the one patient with sympathetic overactivity must then be attributed to β-adrenergic stimulation with vasodilatation and increased flow ipsilateral to the lesion (Raeder's syndrome). Further support for this interpretation is the patient with classic Argyll Robertson pupil shown in Fig. 3.12 (right eye) where the pupil is also small and ptosis is

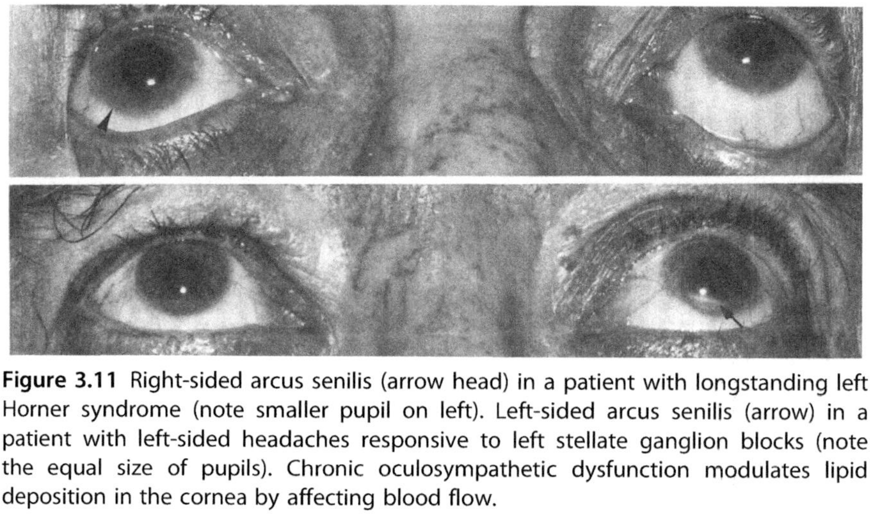

Figure 3.11 Right-sided arcus senilis (arrow head) in a patient with longstanding left Horner syndrome (note smaller pupil on left). Left-sided arcus senilis (arrow) in a patient with left-sided headaches responsive to left stellate ganglion blocks (note the equal size of pupils). Chronic oculosympathetic dysfunction modulates lipid deposition in the cornea by affecting blood flow.

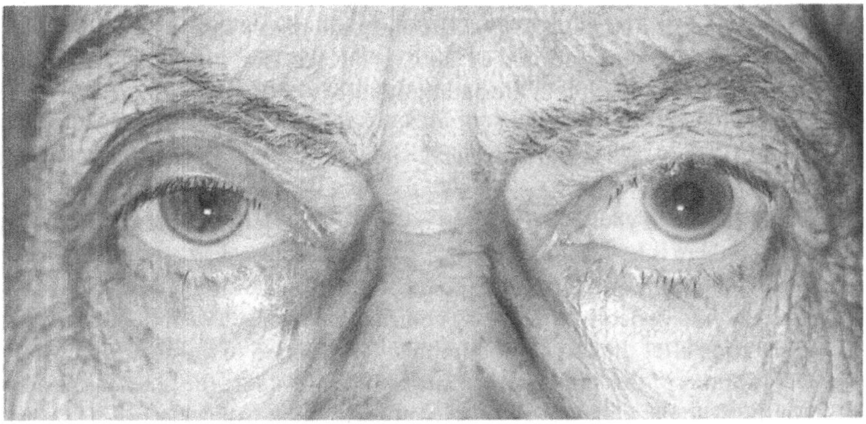

Figure 3.12 Eyes of a patient with tabes dorsalis. The right pupil had all the classical features of the Argyll Robertson pupil and ptosis can also be seen. The left pupil is large and not responsive to light or accommodation. The left retina remained sensitive to light and extraocular movements were normal.

frequently present. Thus, a postganglionic lesion or perhaps excessive tonic discharges primarily arising from postganglionic sympathetic structures affect blood flow and thus the deposition of cholesterol in the corneal stroma causing asymmetric arcus senilis (Dagen & Appenzeller, 1985).

3.2.2 The Argyll Robertson pupil

The characteristics of this pupillary abnormality were first described by Robertson and Argyll (1869a, 1869b). The diagnosis depends on five main characteristics: (1) the retina is light-sensitive; (2) the pupil does not contract to light; (3) the contraction of the pupil is preserved on accommodation for near vision; (4) the pupil contracts with physostigmine but atropine produces poor dilatation; (5) the pupil must be small. The connection between this syndrome and syphilis was established by Babinski and Charpentier (1901) who recognized that when the Argyll Robertson pupil is permanent and unchanged under different intensities of light or previous exposure to darkness, and both the direct and consensual light reflexes are absent with preservation of the accommodation reflex, the patient almost always suffers from syphilis. When all the features of the Argyll Robertson pupil are present except for miosis, the diagnosis of syphilis is less certain and a large pupil unresponsive to light but responsive to accommodation can also be seen in midbrain lesions. In one patient with proven syphilis of the nervous system, manifested as classical tabes dorsalis, one pupil had all the features of the classical Argyll Robertson variety, the other was large and unresponsive to both light and accommodation, although vision and extraocular movements remained unimpaired (Fig. 3.12). Many other pupillary changes have been described in association with the Argyll Robertson pupil such as irregularity of the pupil and iris atrophy, which has been attributed to sympathetic denervation. These features, however, are less constant. The Argyll Robertson pupil does not dilate on painful stimulation of the skin of the neck (ciliospinal reflex) but it has been reported that dilatation occurs within fifteen minutes after death. However, the opportunity to observe such a change must be very rare (McGrath, 1932).

The site of the lesion responsible for the Argyll Robertson pupil is disputed and experimentally such a pupillary abnormality has not been produced. It has been suggested that the lesion is peripheral in the third nerve or ciliary ganglion, however, some reports have not documented sectoral hypokinesis, and involvement of the dorsal midbrain might contribute to the Argyll Robertson pupil in some patients. Syphilis of the nervous system is initiated by syphilitic meningitis so that the third nerve might be involved by subsequent arachnoiditis affecting predominantly superficially situated pupillomotor fibers. In keeping with this interpretation is the observation that the Argyll Robertson pupil has rarely occurred in association with herpes zoster (Naquin, 1954). It has also been suggested that the pupillomotor fibers might be vulnerable to syphilitic meningitis as they enter the midbrain and the lesion might therefore be on the afferent side (Ingvar, 1928). Because the Argyll Robertson pupil is frequently associated with ptosis, the sympathetic is likely involved in its production, and one site where a single lesion could explain the usual bilaterality of the pupillary abnormalities is in the midbrain just lateral to the aqueduct of Sylvius. Here both the afferent fibers for the light reflex and the pupillodilator fibers are close together and crossed and uncrossed fibers passing to the oculomotor nuclei could be affected by a lesion (Merritt & Moore, 1933). It has also been postulated that lesions in the gray matter around the aqueduct could account for the loss of the light reflex (Redlich, 1922; Wilson, 1921) but the failure to produce pupillary constriction on stimulation of this area makes this an unlikely site for the cause of the Argyll Robertson pupils (Magoun & Ranson, 1935).

Argyll Robertson pupil can occur in pseudotabetic syndromes which include ataxia, areflexia, and, in the eye, miosis with light-near dissociation bilaterally. In one case with malignant lymphoma, negative serologic tests for syphilis, and autopsy-proven involvement of the ciliary ganglion were reported (Lapresle & Man, 1979). The midbrain, in this case, was entirely normal and the findings support the proposition that the Argyll Robertson pupil may arise from peripheral lesions but does not fully explain the pathophysiology of this condition.

3.2.3 The tonic pupil

This abnormality is more common in young females and is characteristically unilateral. It usually consists of mydriasis and there is no response or a very poor reaction to direct and consensual light stimulation. Slow dilatation on prolonged exposure to the dark and slow constriction in light is observed. There is a long delay in pupillary contraction during accommodation-convergence and slow dilatation after the end of the stimulus. The pupil is often irregularly shaped with sectoral hypokinesis or vermiform writhing movements. The reaction of the pupil to mydriatics is preserved. Descriptions of this pupil, in patients who also had lost the tendon jerks, by Adie (1931, 1932) have led to the eponym of Adie's syndrome. When areflexia is not present the term tonic pupil is used for this abnormality. A tonic pupil characteristically changes its size when examined on different occasions. Although the majority of tonic pupils are benign and idiopathic, cases have been reported with giant cell arteritis, orbital inflammation due to sarcoid, neoplasm, and Guillain–Barré syndrome among others (Bowie & Givre, 2003; Foroozan, Buono, Savino, & Sergott, 2003; Müller, Prass, & Zschenderlein, 2005; Toth & Fletcher, 2005). Similar pupillary abnormalities with areflexia have been described in diabetics and in patients with influenza, encephalitis, alcoholic polyneuropathy, and hypothyroidism. A transient tonic pupil may also be seen in rheumatoid arthritis (Victor et al., 1977). The tonic pupil contracts strongly when a weak solution (2.5%) of mecholyl is instilled into the conjunctival sac, whereas a normal pupil is not affected by a solution of this strength. This has led to the hypothesis that the tonic pupil results from partial parasympathetic denervation due to a disturbance in the ciliary ganglion or structures peripheral to it (Scheie, 1940). The response of the tonic pupil to mecholyl or other similar drugs has been repeatedly confirmed (Sprofkin, 1953; Walsh, 1957). Similarly, the tonic pupil should constrict to 0.125% pilocarpine after approximately 45 minutes, whereas the fellow eye is unaffected. A positive pilocarpine test is characterized by reversal of the anisocoria. Mydriatic pupils due to a third nerve palsy may also constrict to 0.125% pilocarpine.

The importance of the tonic pupil, which on cursory examination is fixed to light and may occur in association with areflexia, lies in its frequent confusion with the Argyll Robertson pupil occurring in a setting of tabes dorsalis, and it has been stressed that a nonreactive pupil which is changeable in size is not likely to be due to syphilis (Walsh, 1957). While Adie's tonic pupil is dilated initially, it constricts over time because of the aberrant reinnervation by accommodative fibers ("little old Adie's pupil") (Sarao, Elnahry, & Sharma, 2021). In cases of chronic bilateral miotic Adie's pupils, the slow tonic redilation response following near reaction may be challenging to see at the bedside.

Ross syndrome is characterized by Adie's tonic pupil, hyporeflexia, and segmental anhidrosis. Unexplained injury to postganglionic cholinergic fibers accounts for the usual clinical deficits. Tonic pupil and reduced sweating are thought to reflect selective dysfunction of cholinergic parasympathetic and sympathetic fibers, whereas the pathogenesis of hypo-/areflexia may reflect a decrease in synaptic efficacy between the spindle afferents and the alpha motoneurons (Macefield, 2012). A patient who also had a complete postganglionic Horner syndrome has been described (Wolfe, Galetta, Teener, Katz, & Bird, 1995). In these patients, in addition to the postganglionic cholinergic damage, complete focal injury to sympathetic postganglionic neurons or axons was also documented. In a case series of 26 patients, Lamotte and colleagues demonstrated that the pattern of autonomic dysfunction in Ross syndrome was suggestive of a limited and selective autonomic ganglioneuropathy with clinical stability or slow progression over time and it was rarely associated with markers of autoimmunity. Sudomotor impairment was prominent, whereas cardiovagal and adrenergic functions are normal or only mildly impaired in a minority of patients (Lamotte et al., 2021).

A small pupil may be due to a spasm of the sphincter pupillae. It can occur in meningitis or be the result of the instillation of pilocarpine or eserine. It is also seen after morphine or histamine injection. A small (2 mm) nonreactive pupil together with downward deviation of the eyes is indicative of a thalamic hemorrhage. Small (pinpoint) pupils are the hallmark of pontine hemorrhage or massive infarction. This small pupil has been attributed to interruption of the descending sympathetic pupillodilator fibers and preservation of the parasympathetic innervation. Hitherto it has been stated that such pupils fail to react to light but the mechanism of this failure remained unknown. Careful clinical observation with a hand lens and bright light in four cases with massive pontine lesions has shown that the pupil, although 1 mm in diameter, still reacted to light. The previously overlooked reaction to light has been attributed to the extreme miosis (Fisher, 1967). Lesions in the subthalamus produce a pupil 2–3 mm in diameter and this helps differentiate such pupils from those due to extensive lesions in both the pons and midbrain where they are about 4–6 mm in diameter and presumably result from both sympathetic and parasympathetic paralysis. The absence of pinpoint pupils in the Horner syndrome or in complete cervical cord destruction, where one would expect uninhibited parasympathetic constrictor activity, has not been satisfactorily explained. Cocaine does not usually dilate miotic pupils due to sympathetic paralysis, but dilatation does occur after instillation of atropine, and the reaction to light and convergence-accommodation is preserved in such pupils. Senile miosis has been attributed to the deposition of a hyaline material at the pupillary margin which is usually preceded by arteriosclerotic changes in the vessels of the iris and iris muscle (Larsson & Osterlind, 1944). In rare instances, such pupils are not reactive to light or convergence-accommodation (Walsh, 1957).

While hippus has little diagnostic localizing value if it occurs unilaterally, it may point to some abnormalities in the innervation of the pupil. Hippus has been observed in many disorders including the recovery phase from a third nerve paralysis, meningitis, cerebral tumor, and during Cheyne–Stokes respiration.

Segmental paralysis of the iris sphincter is a common occurrence in Adie's pupils. Magnification is needed to appreciate its presence, and it was obvious in all Adie's pupils who had the slightest remnant of a reaction to light. In a series of 122 patients carefully studied, 109 showed this phenomenon, and follow-up examination over a period of years showed progressive denervation and paralysis around the circumference of the iris. Although segmental paralysis of the sphincter of the iris is characteristic of Adie's syndrome, it is not confined to this condition and can be found after blunt injuries to the globe, from chickenpox, peripheral neuropathies, in diabetic neuropathy, and in the Argyll Robertson pupil. Partial failure of the iris to react to light can also be induced by preganglionic third nerve lesions and segmental contractions may be found in patients with aberrant regeneration of the ocular motor nerve (Thompson, 1978).

3.2.4 The large pupil

Monocular mydriasis results from disruption of parasympathetic input to the pupillary sphincter. Mydriasis may be due to damage to the oculomotor nerve, damage to the ciliary ganglion or its axons on their route to the pupillary sphincter, pharmacologic dilation of the pupil, or damage to the sphincter pupillae itself. In clinical practice, the patient should be queried about a history of iris trauma. Clinicians should evaluate for a third nerve palsy and evolving third nerve palsies from compression may present with pupillary abnormalities before oculomotor abnormalities. The two most common causes of third nerve palsies include compression and microvascular ischemia. As indicated above, tonic pupils should constrict to 0.125% pilocarpine. If there is no evidence of third nerve palsy or tonic pupil, the absence of pupillary constriction following the instillation of two drops of 2% pilocarpine indicates that the mydriasis is pharmacologic (e.g., tropicamide eye drops, ipratropium nebulizers). A unilaterally large pupil, which is fixed to light and does not react to convergence-accommodation, may occur after injury to the eye. Usually, this disturbance is only transient. The cause of this traumatic iridioplegia is not fully understood but it has been attributed to a depression of parasympathetic impulses and additional sympathetic overactivity. Dilated and fixed pupils are also encountered in syphilitic general paresis and juvenile paresis. In such patients, there is also widespread iris atrophy. The Parinauds syndrome which consists of pupillary dilatation usually preceding the paralysis of upward gaze is seen in tumors or other lesions involving the quadrigeminal plate. The enlargement of the pupil with unilateral brain swelling has been attributed to hippocampal herniation through the tentorium with consequent compression of the ipsilateral third nerve. In such patients, there is also frequently contralateral hemianopsia due to ischemia of the occipital cortex because of compression of the posterior cerebral artery. If the herniation has persisted for some time, the pupillary dilatation is usually bilateral and multiple cranial nerve palsies occur due to crowding, and damage to the brainstem, papilledema, and hydrocephalus may be present (Lyle, 1954).

Uni- or bilateral pupillary dilatation may occur in temporal arteritis. In such cases, it is often associated with ptosis and diplopia and involvement of the 6th nerve on one or both sides. The affection of the third nerve or its branches by the pathological process does not always imply that blindness will occur, but in most cases the third nerve involvement with pupillary dilatation preceded blindness. In patients with temporal arteritis the reaction of the pupil to convergence-accommodation is usually preserved and the oculomotor paralysis, if it occurs, is often transient and responds to appropriate treatment. A survey showed an incidence of 15% of oculomotor paralysis in this disorder (Fisher, 1959). The third nerve palsy often improves rapidly after steroid treatment is started.

Unilateral pupillary dilatation may occur during and immediately after seizures and this anisocoria is transient. The physiologic mechanism or anatomical pathways responsible for this phenomenon are not known (Pant, Benton, & Dodge, 1966). While in animals cortical stimulation in various areas induces pupillary changes, these are usually bilateral. In humans, however, unilateral dilatation of the pupil is seen on the side of a stroke involving the deep and superficial territory of the middle cerebral artery. It is often difficult in such patients to decide whether the large or small pupil is the abnormal one but the poor or absent reaction to light or the presence of hippus may help in the identification of the abnormal side. In seizures, as in strokes, a large abnormal pupil may occur on the side of a lesion or the contralateral side. The practical importance of anisocoria in patients with seizures lies in the suspicion that this pupillary abnormality might signify third nerve compression due to temporal lobe herniation through the tentorium. When the patient is unconscious and no history is available or some evidence of trauma is found, the decision to embark on further diagnostic procedures or to attribute the enlarged pupil to seizures is a difficult one and may sometimes lead to unnecessary investigations when anticonvulsants would have been indicated. In practice, however, it appears safer to further study the patient than wait until irreversible brainstem compression due to tentorial herniation has occurred (Pant et al., 1966).

In occlusion of the internal carotid artery, ipsilateral dilatation of the pupil has been observed and attributed to probable ischemia of the iris (Fisher, 1967). While it is likely that in some such cases, particularly if complicated by diabetes and rubeosis iridis, this may be the cause, mydriasis was also seen in some other patients with stroke who had no radiological evidence of internal carotid occlusion, and in whom the middle cerebral artery territory was affected clinically. Extensive midbrain damage may produce dilatation of the pupils, which is, however, less marked than in midbrain compression due to tentorial herniation (Fisher, 1967). Tuberculous meningitis may cause dilated and poorly reactive or fixed pupils. The correct diagnosis of the etiology of the meningitis was suggested in one case after the pupils became large and unreactive. The assumption was made that the basal exudate in tuberculous meningitis might have involved the third nerve with consequent parasympathetic paralysis resulting in pupillary dilatation (Fisher, 1967).

A well-known complication of diabetes is autonomic neuropathy but this usually occurs late in the disorder. Parasympathetic denervation in diabetes is common and 81% of patients with the disease were found to react to the instillation of methacholine in a manner suggestive of denervation supersensitivity of the pupil (Sigsbee, Torkelson, Kadis, Wright, & Reeves, 1974). The incidence of this abnormality was even higher in patients with a clinical course of longer than 2 years duration. Though other evidence of autonomic neuropathy was sometimes present, particularly in patients who have had the disease for some time, this was only a small proportion of those who showed parasympathetic denervation supersensitivity of the pupil.

Unilateral dilatation of the pupil in isolation, without evidence of other neurologic deficits, sometimes occurs. In most such patients, the cause remains enigmatic. Nevertheless, unilateral mydriasis has been described in association with known disorders and the benign idiopathic variety must be distinguished from others. Episodic dilatation of a pupil must be separated from mydriasis in cranial nerve palsies, Horner syndrome in which the unaffected side is larger, pupillary paresis, tonic pupils, and seizures. Unilateral dilatation of the pupil might be a clinical manifestation of a variety of disorders amongst which ophthalmoplegic migraine must be considered. In most patients with unilateral episodic idiopathic dilatation of the pupil, head pain or eye pain is usually present. Nevertheless, in most patients, the etiology remains obscure but the absence of other neurologic signs, the occasional distortion of pupil contour, and on follow-up the lack of serious neurologic disease suggests that this might be an entity of its own and that the mydriasis is probably of peripheral origin (Hallett & Cogan, 1970) (Table 3.2).

Posttraumatic unilateral mydriasis has also been described (Vijayan & Dreyfus, 1975). This is a condition that occurs after soft tissue injury to the neck usually around the carotid artery sheath. Headache after this injury occurs without a history of headache before the event. The headaches are ipsilateral to the side of injury, are throbbing and severe, and suggest migraine. There is striking ipsilateral mydriasis and sweating during headaches. Clinical examination is otherwise normal but the examination of the pupil with pharmacologic agents suggests partial sympathetic denervation either immediately after a headache or even during headache-free intervals. The administration of propranolol, a β-blocker abolishes the headache and improves mydriasis and excessive sweating. The condition is attributed to direct injury of the carotid perivascular sympathetic plexus with resultant overactivity of β receptors.

Table 3.2 Normal pupil and pupillary abnormalities.

	Normal pupil	Afferent defect	Argyll Robertson syndrome	Third nerve defect	Adie's syndrome	Horner syndrome
Size in dim light	Large size varies with age and excitability of subject	Normal	Small	Normal	May be smaller in longstanding cases	Small
Size in bright light	Small	Larger than normal	Same as in dim light	Larger than normal	Larger than normal	Smaller than normal
Unilateral or bilateral	-	Either	Predominantly bilateral	Either	Usually unilateral (80%–90%)	Usually unilateral
Direct light reflex	Brisk, magnitude varies with age and excitability of subject	Reduced; laterality and degree of dysfunction related to field loss	Reduced or absent	Reduced on side of lesion when either eye is stimulated	Absent or reduced when either eye is stimulated. If present, contraction is delayed slow, smooth and prolonged, or segmental	Normal
Consensual light reflex	Same as direct light reflex	Same as direct light reflex	Reduced or absent			Normal
Anisocoria in dim light	No	No	Common	No	Usually present	Moderate, with the affected pupil smaller
Anisocoria in bilateral bright light	No	No	Common	Pupil larger on side of lesion	Tonic pupil larger than normal due to poor light reaction	Less than in dim light

(*Continued*)

Table 3.2 (Continued)

	Normal pupil	Afferent defect	Argyll Robertson syndrome	Third nerve defect	Adie's syndrome	Horner syndrome
Contraction to near vision	Varies with degree of accommodation convergence and effort, but does not exceed maximum light reaction	Normal, with normal vision. If blind voluntary accommodation or convergence may be accompanied by normal pupillary constriction	Extent may be normal or reduced	Reduced or absent on side of lesion	More marked than light reflex, and may be even larger than the near contraction of the normal pupil. The contraction is delayed, slow, smooth and prolonged	Normal
Dilatation to far vision	Quick	Normal	Quick	Quick	Delayed and slower than normal (tonic)	Quick
Psychosensory reflex dilatation	Quick, bilateral dilatation. On arousal extensive dilatation	Normal	Present or absent or masked by spastic miosis	Normal	Normal	Reduced (the anisocoria increases)
Dilatation to sudden dimming of light	Quick	Reduced	Reduced or absent	Quick	Reduced, delayed and slower than normal (tonic)	Present but reduced
Dilatation to atropine	Sphincter paralysis; pupil dilates to larger than diameter in darkness; loss of light reflex	Normal but dilatation reduced when pupil larger to begin with	May dilate normally; final diameter may be smaller	Normal but range reduced because pupil large to begin with	Normal but range of dilatation reduced because pupil large to begin with	Range of dilatation normal but anisocoria persists because pupil smaller to begin with

Dilatation to cocaine	Extensive mydriasis; reduction but no loss of light reflex	Normal	Normal or reduced	Normal	Normal	Post- and pregang lionic neuron—none. Central slight or none
Dilatation to epinephrine 1: 1000 (3 drops)	No	No	No	No	No	Postganglionic neuron—larger than normal. Pregang lionic neuron—normal. Central normal
Constriction to mecholyl 2.5%	No	No	No	No	Constriction	No
Pupil shape	Round	Round	Round or distorted	Usually, round	Round or distorted	Round
Other signs		Impaired visual acuity or loss of visual field	Patchy atrophy of the iris, sometimes	1. Ptosis, accommodative or extraocular muscle pareses 2. Other signs	1. Asymmetric areflexia or hyporeflexia 2. Accommodation is reduced, absent or tonic	Ipsilateral anhidrosis, vasodila tation, slight ptosis and elevation of lower lid

(*Continued*)

Table 3.2 (Continued)

	Normal pupil	Afferent defect	Argyll Robertson syndrome	Third nerve defect	Adie's syndrome	Horner syndrome
Location of defect		Retina, optic nerve, chiasm, tract	Midbrain, interrupting light reflex arc near oculomotor nuclei, sparing near vision pupillary pathway and Edinger–Westphal nuclei	Edinger–Westphal nucleus, oculomotor nerve	Ciliary ganglion or short ciliary nerves	Along sympathetic pathways

Arguments about autonomic dysfunction in migraine have continued unabated since the early sixties. Examining ocular autonomic function in migraineurs with a strictly unilateral headache, it was found that the pupil on the symptomatic side dilated more slowly and less extensively in darkness. This was taken to indicate a sympathetic impairment and correlated with increased facial blood flow on the symptomatic side (Drummond, 1990). It remains unclear if this finding represents an epiphenomenon secondary to trigeminal activation and tests for sympathetic activation in other areas of autonomic control were not performed in this study.

Though it has been assumed that headache is not a sequel to surgery on the carotid system, in fact, several patients with postcarotid endarterectomy hemicrania have been described (Leviton, Caplan, & Salzman, 1975). Whether this might represent a similar syndrome without mydriasis remains to be determined by further observation both of posttraumatic and postendarterectomy cases.

In hypothermia, pupils are usually constricted. Once bilateral pupillary dilatation occurs (usually nonreactive to light) serious brainstem cooling with dysfunction is present. In this instance, brain herniation (due to unsuspected subdural hematoma) would not clinically be signaled by anisocoria until rewarming when it may be too late to avoid progressive relentless brain edema.

References

Adie, W. J. (1931). Argyll Robertson pupils true and false. *Brit. Med. J., 2*, 136–138.

Adie, W. J. (1932). Complete and incomplete forms of the benign disorder characterized by tonic pupils and absent tendon reflexes. *Brit. J. Ophthal., 16*, 449–461.

Adler, F. H. (1950). *Physiology of the Eye, Clinical Application*. St. Louis: C.V. Mosby.

Appenzeller, O. (1989). Pupil cycle time (PCT) in headache and other neurologic diseases. In F. C. Rose (Ed.), *New advances in headache research: The 7th Migraine Trust international symposium* (pp. 25–29). London: Smith-Gordon.

Arieff, A. J., & Pyzik, S. W. (1953). The ciliospinal reflex in injuries of the cervical spinal cord in man. *Arch. Neurol. Psychiat. (Chicago), 70*, 621–629.

Babinski, G., & Charpentier, C. (1901). De l'abolition des réflexes pupillaires dans ses relations avec la syphilis. *Bull. Soc. Méd. Paris, 18*, 502–505.

Belliveau, A. P., Somani, A. N., & Dossani, R. H. (2021). Pupillary Light Reflex. *StatPearls. Treasure Island (FL): StatPearls Publishing.*

Bernard, C. (1853). Recherches expérimentales sur le grand sympathique et spécialement sur l'influence que le section de ce nerf exerce sur la chaleur animal. *C.R. Soc. Biol., 5* (pt. 2), 277.

Bernard, P. G., & Perlo, V. P. (1980). Homer syndrome with causalgia. *Neurology, 30*, 534–535.

Boshes, B., & Mayer, L. L. (1935). The lid-closure reflex of the pupil. *Amer. J. Ophthal, 18*, 1048–1051.

Bouffard, M. A. (2019). The pupil. *Continuum (Minneap Minn)*, *25*(5), 1194–1214.

Bowie, E. M., & Givre, S. J. (2003). Tonic pupil and sarcoidosis. *Am. J. Ophthalmol.*, *135* (3), 417–419.

Cannon, W. B., Binger, C. A. L., & Fitz, R. (1915). Experimental hyperthyroidism. *Amer. J. Physiol.*, *36*, 363–364.

Cortez, M. M., Rae, N., Millsap, L., McKean, N., & Brennan, K. C. (2019). Pupil cycle time distinguishes migraineurs from subjects without headache. *Front Neurol*, *10*, 478.

Crouch, R. L. (1936). The efferent fibers of the Edinger–Westphal nucleus. *J. Compar. Neurol.*, *64*, 365–373.

Czarnecki, J. S. C., & Thompson, H. S. (1976). Spontaneous cyclic segmental sphincter spasms in Adies tonic pupil. *Am. J. Ophthalmol.*, *82*, 636–637.

Dagen, D., & Appenzeller, O. (1985). Oculosympathetic dysfunction. Effects on arcus senilis. *Ann. Neurol*, *18*, 150.

Drummond, P. D. (1990). Disturbances in ocular sympathetic function and facial blood flow in unilateral migraine headache. *J. Neurol. Neurosurg. Psychiat.*, *53*, 121–125.

Duke-Elder, W. S. (1932–1938). *Textbook of Ophthalmology*. St. Louis: C.V. Mosby.

Fisher, C. M. (1959). Ocular palsy in temporal arteritis. *Minnes. Med.*, *42*, 1258–1268, 1430–1437; 1617–1630.

Fisher, C. M. (1967). Some neuro-ophthalmological observations. *J. Neurol. Neurosurg. Psychiat.*, *38*, 383–392, 30.

Foroozan, R., Buono, L. M., Savino, P. J., & Sergott, R. C. (2003). Tonic pupils from giant cell arteritis. *Br. J. Ophthalmol.*, *87*(4), 510–512.

Fotiou, D. F., Stergiou, V., Tsiptsios, D., Lithari, C., Nakou, M., & Karlovasitou, A. (2009). Cholinergic deficiency in Alzheimer's and Parkinson's disease: evaluation with pupillometry. *Int. J. Psychophysiol.*, *73*(2), 143–149.

Fulton, J. F. (1938). *Physiology of the Nervous System*. London: Oxford University Press.

Goldwater, B. C. (1972). Psychological significance of pupillary movements. *Psychol. Bull.*, *77*, 340–355.

Gooley, J. J., Ho Mien, I., St Hilaire, M. A., Yeo, S. C., Chua, E. C., van Reen, E., ... Lockley, S. W. (2012). Melanopsin and rod-cone photoreceptors play different roles in mediating pupillary light responses during exposure to continuous light in humans. *J. Neurosci.*, *32*(41), 14242–14253.

Hallett, M., & Cogan, D. G. (1970). Episodic unilateral mydriasis in otherwise normal patients. *Arch. Ophthalmol.*, *84*, 130–136.

Hammer, C. (1933). Einseitiger Hornerscher Symptomenkomplex nach beiderseitiger Gaumenmandelausschälung. *Klin. Mbl. Augenheilk*, *90*, 79–80.

Hawkes, C. H., & Stow, B. (1981). Pupil size and the pattern evoked visual response. *J. Neurol. Neurosurg. Psychiat.*, *44*, 90–91.

Hess, C. (1908). Untersuchungen zur physiologie und pathologie des pupillenspieles. *Arch. Augenheilk*, *60*, 327–389.

Ingvar, S. (1928). The pathogenesis of the Argyll Robertson phenomenon. *Bull. Johns Hopk. Hosp.*, *43*, 363–396.

Karplus, J. P., & Kreidl, A. (1909). Gehirn und Sympathicus. I. Zwischenhirnbasis and Halssympathicus. *Arch. Ges. Physiol. (Bonn)*, *129*, 138–144.

Lamotte, G., Sandroni, P., Cutsforth-Gregory, J. K., Berini, S. E., Benarroch, E. E., Shouman, K., ... Coon, E. A. (2021). Clinical presentation and autonomic profile in Ross syndrome. *J. Neurol.*, *268*(10), 3852–3860.

Langworthy, O. R., & Ortega, L. (1943). The iris: Innervation of the iris of the albino rabbit as related to its function. Theoretical discussion of abnormalities of the pupils observed in man. *Medicine*, *22*, 287–361.

Lapresle, J., & Man, H. X. (1979). Signe d'Argyll Robertson dans un cas de lymphome malin avec lésions ciliaires. *Revue Neurologique*, *135*, 515–525.

Larsson, S., & Osterlind, G. (1944). Studies on the cause of senile miosis and rigidity of the pupil. *Acta Ophthal*, *211*, 25.

Leviton, A., Caplan, L., & Salzman, E. (1975). Severe headache after carotid endarterectomy. *Headache*, *15*, 207–210.

Lyle, D. (1954). Eye symptoms produced by tentorial herniation from increased intracranial pressure. *Congress Ophthal. Acta*, *3*, 1845–1854.

Macefield, V. G. (2012). Selective autonomic failure: Ross syndrome. *Clin. Neurophysiol.*, *123*(8), 1479–1480.

Magitot, A. (1931). *L'Iris*. Paris: Octave Doin Editeur.

Magoun, H. W., & Ranson, S. W. (1935). The central path of the light reflex: a study of the effects of lesions. *Arch. Ophthalmol.*, *13*, 791–811.

Martyn, C. N., & Ewing, D. J. (1986). Pupil cycle time: a simple way of measuring an autonomic reflex. *J. Neurol. Neurosurg. Psychiat.*, *49*, 771–774.

McGrath, W. M. (1932). Observations upon abnormalities of the pupils and iris in tabes dorsalis, general paralysis and taboparesis with considerations of their bearing upon pathogenesis of Argyll Robertson pupil. *J. Mental Sci.*, *78*, 362–373.

Merritt, H. H., & Moore, M. (1933). The Argyll Robertson pupil; anatomic-physiologic explanation of phenomenon, with survey of its occurrence in neurosyphilis. *Arch. Neurol. Psychiat*, *30*, 357–373.

Metzner, R., & Wölfflin, E. (1915). Klinische und experimentelle Untersuchungen über Halssympathicuslahmung. *Arch. Ophthalmol.*, *89*, 308–322.

Micieli, G., Tassorelli, C., Martignoni, E., Pacchetti, C., Bruggi, P., Magri, M., & Nappi, G. (1991). Disordered pupil reactivity in Parkinson's disease. *Clin. Autonom. Res.*, *1* (1), 55–58.

Müller, N. G., Prass, K., & Zschenderlein, R. (2005). Anti-Hu antibodies, sensory neuropathy, and Holmes-Adie syndrome in a patient with seminoma. *Neurology*, *64*(1), 164–165.

Naquin, H. A. (1954). Argyll Robertson Pupil. Following herpes zoster ophthalmitis with remarks on the efferent pupillary pathways. *Amer. J. Ophthal.*, *38*, 23–33.

Nathan, P. W., & Turner, J. W. A. (1942). Efferent pathway for pupillary contraction. *Brain*, *65*, 343–351.

Nishimura, H. (1927). Variation en diamètre pupillaire par l'excitation directe du nerf acoustique. *C.R. Soc. Biol. (Paris)*, *97*, 158–160.

Ohtsuka, K., & Sato, A. (1996). Descending projections from the cortical accommodation area in the cat. *Invest. Ophthalmol. Vis. Sci.*, *37*(7), 1429–1436.

Pant, S. S., Benton, J. W., & Dodge, P. R. (1966). Unilateral pupillary dilatation during and immediately following seizures. *Neurology (Minneap.)*, *16*, 837–840.

Passow, A. (1933). Horner syndrome, heterochemie und status dysraphicus, ein symptomencomplex. *Arch. Augenheilk*, *107*, 1–51.

Provencio, I., Rodriguez, I. R., Jiang, G., Hayes, W. P., Moreira, E. F., & Rollag, M. D. (2000). A novel human opsin in the inner retina. *J. Neurosci.*, *20*(2), 600–605.

Redlich, E. (1922). Zur Pathologie der reflectorischen Pupillenstarre. *Wien. Klin. Wschr*, *35*, 756–761.

Robertson., & Argyll, D. (1869a). Four cases of spinal miosis: with remarks on the action of light on the pupil. *Edinb. Med. J.*, *15*, 487–493.

Robertson., & Argyll, D. (1869b). On an interesting series of eye symptoms in a case of spinal disease with remarks on the action of Belladonna on the iris. *Edinb. Med. J.*, *14*, 696–708.

Rubin, L. S., & Barry, T. J. (1972a). The effect of the cold pressor test on pupillary reactivity of schizophrenics in remission. *Biol. Psychiat.*, *5*, 181–197.

Rubin, L. S., & Barry, T. J. (1972b). The reactivity of the iris muscles as an index of autonomic dysfunction in schizophrenic remission. *J. Nerv. Mental Dis.*, *155*, 265–276.

Sarao, M. S., Elnahry, A. G., & Sharma, S. (2021). Adie Syndrome. Treasure Island, FL: StatPearls Publishing.
Scheie, H. G. (1940). Site of disturbance in Adies syndrome. *Arch. Ophthalmol.*, *24*, 225–237.
Sigsbee, B., Torkelson, B., Kadis, G., Wright, J. W., & Reeves, A. G. (1974). Parasympathetic denervation of the iris in diabetes mellitus. *J. Neurol. Neurosurg. Psychiat.*, *37*, 1031–1035.
Sprofkin, B. E. (1953). Adies syndrome and the Adler-Scheie mecholyl test. *Neurol*, *3*, 910–915.
Stern, H. J. (1944). A simple method for the early diagnosis of abnormalities of the pupillary reaction. *Br. J. Ophthal.*, *28*, 275–276.
Thompson, H. S., & Corbett, J. J. (1980). Spasms of the iris sphincter. *Ann. Neurol.*, *8*, 547–549.
Thompson, H. S. (1978). Segmental palsy of the iris sphincter in Adies syndrome. *Arch. Ophthalmol.*, *96*, 1615–1620.
Toth, C., & Fletcher, W. A. (2005). Autonomic disorders and the eye. *J. Neuro-ophthalmol.*, *25*(1), 1–4.
Turnbull, P. R., Irani, N., Lim, N., & Phillips, J. R. (2017). Origins of pupillary hippus in the autonomic nervous system. *Invest. Ophthalmol. Vis. Sci.*, *58*(1), 197–203.
Ury, B., & Gellhorn, E. (1939). Role of sympathetic system in reflex dilatation of pupil. *J. Neurophysiol.*, *2*, 268–275.
Victor, D. I., Green, W. R., et al. (1977). A non-permanent tonic pupil in rheumatoid arteritis. *Canad. J. Neurol. Sci.*, *4*, 209–212.
Vijayan, N., & Dreyfus, P. M. (1975). Posttraumatic dysautonomic cephalalgia. *Arch. Neurol.*, *32*, 649–652.
Walsh, F. B. (1957). *Clinical Neuro-Ophthalmology*. Baltimore, MD: Williams and Wilkins.
Weinstein, E. A., & Bender, M. B. (1941). Pupillodilator reactions to sciatic and diencephalic stimulation. Comparative study of cat and monkey. *J. Neurophysiol*, *4*, 44–50.
Weinstein, J. M., Zweitel, T. J., & Thompson, H. S. (1980). Congenital Horners syndrome. *Arch. Ophthalmol.*, *98*, 1074–1078.
Weintraub, M. I., Gaasterland, D., & Van Woert, M. H. (1970). Pupillary effects of Levodopa therapy. Development of anisocoria in latent Horners syndrome. *New Engl. J. Med.*, *283*, 120–123.
Whitnall, S. E. (1932). *Anatomy of the Human Orbit and Accessory Organs of Vision*. London: Oxford University Press.
Wilson, S. A. K. (1921). Some problems in neurology: 1. The Argyll Robertson pupil. *J. Neurol. Psychopath.*, *2*, 1–25.
Wolfe, G. I., Galetta, S. L., Teener, J. W., Katz, J. S., & Bird, S. J. (1995). Site of autonomic dysfunction in a patient with Ross' syndrome and postganglionic Horner's syndrome. *Neurology*, *45*(11), 2094–2096.
Yoss, R. E., Moyer, N. J., & Hollenhorst, R. W. (1970). Hippus and other spontaneous rhythmic pupillary waves. *Amer. J. Ophthal*, *70*, 935–941.
Zauberman, H., & Livni, N. (1981). Experimental vascular occlusion in hypercholesterolemic rabbits. *Invest. Ophthalmol. Vis. Sci.*, *21*(2), 248–255.

CHAPTER 4

Trophic disorders

Trophic disorders may be defined as changes or lesions in limbs and occasionally other structures which are not directly related to diseases of these organs but associated with disturbances of their innervation. This definition is purely descriptive since the mechanism by which trophic disturbances are induced remains unknown. Trophic disturbances can be seen in association with the central nervous system and peripheral nerve lesions and may manifest at birth. More commonly, however, they are seen in association with diseases affecting adults. Disorders that, in part may represent trophic disturbances and are evident at birth may be classified on clinical grounds into 1) congenital absence of muscles, 2) congenital neuromuscular disorders with localized weakness, 3) congenital neuromuscular disorders associated with contractures and deformities about joints (Dodge, 1961) and 4) disorders affecting the skin and subcutaneous tissues. This classification is arbitrary and it is likely that many disturbances which are included in this section will be reclassified when their pathogenesis is understood.

4.1 Congenital absence of muscles

Le Double (1897) extensively reviewed muscular variations in humans based on dissections. The congenital absence of muscles, changes in their structure, size, and origin, as well as variations in sex and racial incidence of these abnormalities are recorded.

Although the congenital absence of muscles is asymptomatic at birth, when certain muscles, such as the diaphragm or abdominal wall are involved, there may be serious functional disturbances, due not only to the absent muscles but because of other associated congenital abnormalities. The absence of the diaphragm is sometimes associated with respiratory distress and in infants with deficiency of the abdominal muscles, difficulties in respiration and defecation.

Associated defects in urogenital organs have also been found (Garrod & Davies, 1905; Osler, 1901; Silverman & Huang, 1950). Other associated abnormalities with absent pectoral muscles are defects in mammary glands,

DOI: https://doi.org/10.1016/B978-0-323-96011-3.00009-7

syndactyly and microdactyly (Brown & McDowell, 1940; Krabbé, 1949; Morley, 1923), and skeletal deformities (Dodge, 1961).

Absent muscles may not be obvious on clinical examinations since the space they normally occupy may be filled with fat and connective tissue, which has the outline of the muscle it replaces but is usually of smaller bulk. Occasional muscle fibers and nerve twigs can be seen on histological examination of this tissue (Schlesinger, 1900).

The pathogenesis of this disorder is obscure. Clinicopathological descriptions are incomplete and it is difficult to decide whether the disorder is a denervation atrophy or myopathy. Changes in the spinal cord are found in some cases but a decreased number of anterior horn cells does not mean they are the cause of the disorder since changes in these cells can be induced by intrauterine removal of a developing muscle (Dunnebacke, 1953; Lichtenstein, 1939; Schlesinger, 1900). An extensively examined case with multiple congenital abnormalities, in addition to absence of several muscles, showed that the latter was probably related to abnormalities of the spinal cord which was destroyed over two or three segments (Dodge, 1961).

4.2 Congenital neuromuscular disorders with localized weakness

The presence of bilateral facial and external rectus palsies was thought to represent an entity and has been given the eponym of Möbius (1888). The cases were extensively reviewed by Evans (1955) and Sprofkin and Hillman (1956). It is now clear that in such patients, in addition to weakness of facial and extraocular muscles, there are a number of congenital defects in other organs. The clinical signs are recognized early because of the difficulty in sucking and the immobile facies when crying. The disturbance is attributed to a paucity of neurons in the facial nucleus, but there is at least one case on record in which the facial nerve nucleus was normal, although there were abnormalities of the olive and cerebellum (Pitner, Edwards, & McCormick, 1965). On the other hand, biopsies in the region of the frontalis muscle in three cases showed no recognizable muscle fibers. The nervous system of these patients was not, however, examined (Richards, 1953). Some cases appear to have a destructive process rather than hypoplasia of cranial nerve nuclei (Fenyes, 1937). It is possible, of course, that what is generally considered as Möbius syndrome may be caused by a variety of conditions, some of which may result in weakness of facial and extraocular muscles due to a loss of innervation of the musculature and this may be a manifestation of a trophic disorder.

In the newborn guinea pig soleus muscle, an approximately equal number of Type I and Type II fibers are found. The number of Type II muscle fibers gradually diminishes and by 6 weeks of age, only Type I extrafusal fibers are found. This change can be prevented by sciatic denervation of the muscle in the newborn and delayed by sciatic nerve crush. It is therefore attributed to regulatory trophic function of the maturing motor neuron, which seems to exert a controlling influence on the histochemical appearance of the developing muscle (Karpati & Engel, 1967).

4.3 Congenital neuromuscular disorders associated with contractures and deformity about joints

Diseases of the nervous system are sometimes manifested first by contractures or deformities. Neurological dysfunction in such infants may be obvious but in many, no definite disease of the nervous system is apparent. The most common disorder in this group is congenital club foot. In the majority of such cases, no abnormalities of the nervous system have been found but spinal cord lesions were found in others, and these were held responsible for the talipes deformity (Courtillier, 1897; Michaud, 1870). There was a failure of development of anterior horn cells at the junction of the thoracic and lumbar cord segment, together with deficiencies in Clarke's column, and a reduction in the number of medullated fibers in the anterior roots. It is also suggested that some cases are due to traction damage in utero to peripheral nerves in the legs (Dittrich, 1930). The muscles in the leg in talipes may show uniform small fibers. In other cases, there was marked variation in muscle fiber size (Adams, Denny-Brown, & Pearson, 1954). From these studies, it seems likely that congenital club foot may be caused by a variety of different disorders and that probably in a minority the disturbance is a trophic one.

4.3.1 Congenital elevation of the shoulder (Sprengel's deformity)

In this condition, which is usually unilateral, there is a change in the position and shape of the scapula. The lower angle of the scapula is nearer the spine and the shoulder higher than that of the unaffected side. Scoliosis is usually also present. The movements of the arms are normal but no abduction of the shoulder beyond 90 degrees is possible (Greig, 1924). In a pathologically examined case, there were abnormalities in the trapezius muscle which was missing below the scapular spine and it was replaced by connective tissue (Fairbank, 1913–1914). This disorder is probably related to defective innervation of the trapezius or serratus anterior muscles or, if associated with the Klippel–Feil deformity, the atrophy of the muscles may be secondary to spinal cord lesions (Adams et al., 1954).

4.3.2 Deformities of several joints (arthrogryposis multiplex congenita and amyoplasia congenita)

This condition is characterized by a deformity of multiple joints with contracture of muscle. Numerous names have been attached to this disorder. The term arthrogryposis multiplex congenita was introduced by Stern (1923) and the name amyoplasia congenita was given to it by Sheldon (1932). A typical case is obvious at birth with curved joints and rigidity of limbs; the latter may interfere with delivery and they may be fixed in a variety of positions. In some cases, the condition is restricted to part of one limb. Many cases in which the contractures were associated with amyotonia congenita have been recognized. Affected limbs are small but the joints appear unusually large and hypotonia and weakness are found. In addition, the skin and subcutaneous tissues are thickened and rather loose. In one case, the most important lesions were in the spinal cord where anterior horn cells had completely disappeared from several cord segments. This was associated with a reduction in anterior root fibers. In another case, there was, in addition, incomplete fissuration, the smallness of the brain, and enlargement of the ventricles. In the muscles, the most conspicuous histological change was the smallness of muscle fibers; some muscles could not be found at all, and others were replaced by connective tissue (Adams et al., 1954). This condition, like many others in this section, can probably be caused by a variety of different disorders but at present, the most widely held view is that it is due to defective development or intrauterine degeneration of the anterior horn cells. This, however, cannot entirely explain the clinical picture since the widespread contractures are probably related to some other disturbance than the loss of innervation of muscles alone.

4.4 Disorders affecting the skin and subcutaneous tissue

Chronic hereditary trophedema or Milroys disease is characterized by persistent edema of the lower limbs. It is usually found in several members of the family and often in several generations (Milroy, 1928). The edema may appear shortly after birth but may be delayed until puberty. This, once established, remains permanently, and there is no pain or evidence of obstruction of the arterial or venous systems. This disorder has been attributed to disturbances in vasomotor control, blood vessels, or changes in autonomic neurons in the spinal cord (McGuire & Zeek, 1932; Ormsby & Montgomery, 1954). No convincing physiological study of vasomotor control or pathological examination of the nervous system in such a case has hitherto been reported.

4.4.1 Hypermelanosis (increased brown melanin pigmentation)

Hypermelanosis is a feature of neurofibromatosis and polyostotic fibrous dysplasia. It results from the increased activity of epidermal melanocytes. In neurofibromatosis, the café-au-lait spots often have a dermatomal distribution, but the role of the nervous system in the genesis of this disorder of pigmentation is not yet elucidated.

4.5 Trophic disorders appearing after birth

Disorders of the central nervous system, which might give rise to trophic disorders and are not usually evident at birth, have been recognized. Of these, parietal wasting has been well described and numerous cases are on record (Critchley, 1966; Winkelman & Silverstein, 1932, 1935). This may also be evident in patients with injuries to the parietal lobe sustained in the perinatal period or childhood and is then referred to as infantile hemiplegia with contralateral dysgenesis. Clinically, wasting of muscles occurs on the side opposite the parietal lobe lesion, but occasionally minor wasting may be noted on the ipsilateral side. The atrophy is seen in association with vascular accidents, after which it may develop within a day or two, although usually taking about a week. It is also seen with slowly progressive lesions such as tumors or due to degeneration. The earliest signs of wasting are usually found in the small muscles of the hand and this, at first, may only be detectable on palpation. Wasting spreads to the entire upper limb and may affect predominantly the shoulder girdle. The lower limb is less affected, if at all, and rarely half the tongue, palate, or face may be wasted. The wasting is usually associated with a progressive diminution in tone and a decrease in tendon jerks. Commonly, patients show only mild wasting and the entire limb may appear smaller if it is associated with a nonprogressive parietal lesion. The skin becomes less coarse and shiny and the hand is usually warm and slightly moist or, on occasion, colder. The fingers are tapered and nails brittle and the hair may be lost. Deformities are not common. On occasion, there is puffiness and edema of the limbs so that wasting is not easily appreciated (Gordon, 1927). X-ray examinations of the bones of the affected limb are usually not remarkable except in long-standing cases, particularly those dating from infancy in which hypoplasia and sometimes demineralization may be evident. Wasting of parietal origin is usually associated with hemiparesis. There are, however, cases with wasting without concomitant motor weakness, and it is recognized that unilateral wasting with astereognosis is a parietal lobe syndrome (Critchley, 1966).

Parietal wasting may occasionally be reversible, particularly after early removal of a noninvasive cerebral tumor. Histological examination of muscle fibers in affected muscles shows them to be smaller and edema is often prominent (Winkelman & Silverstein, 1932).

There is no doubt about the occurrence of wasting in limbs opposite to parietal lobe disease but the mechanism of this phenomenon is not understood. Numerous theories have been advanced to explain these findings; none to date have proved fully satisfactory. Charcot (1879) believed that the parietal lesion leads to changes in the anterior horn of the spinal cord and that these are the cause of wasting. His views were based on a pathologically examined case in which a hemorrhage in the contralateral hemisphere was associated with spinal sclerosis. Similar findings were described by Brissaud (1879), and others have shared Charcot's views (Schaffer, 1897; Steinert, 1903). In a case of cerebral muscle atrophy, Dejerine (1889) found changes in the peripheral nerve but none in the anterior horn cells, and he believed that the parietal lesion was responsible for the peripheral nerve changes. Primary affection of joints was thought to be the cause of parietal wasting by De La Tourette (1897) and he called this arthritic muscular atrophy. Vasomotor changes induced by the vasomotor center in the brain were also thought to be the cause of parietal wasting or angiomyopathy (Luzzato, 1903; Marinesco, 1898). Still another center was proposed by Quincke (1893). He thought that the cortex close to the sensory pathways controls the nutrition of the muscles and, if this is destroyed, wasting ensues.

Disturbances in vasomotor control of the affected limb have been noted by Van Bogaert and Ley (1933) in a patient with a lesion in the parietoparacentral area, manifested by sympathetic paralysis, which lasted several weeks. The patient had no changes in deep sensibility or stereognosis but had ataxia of parietal origin as well.

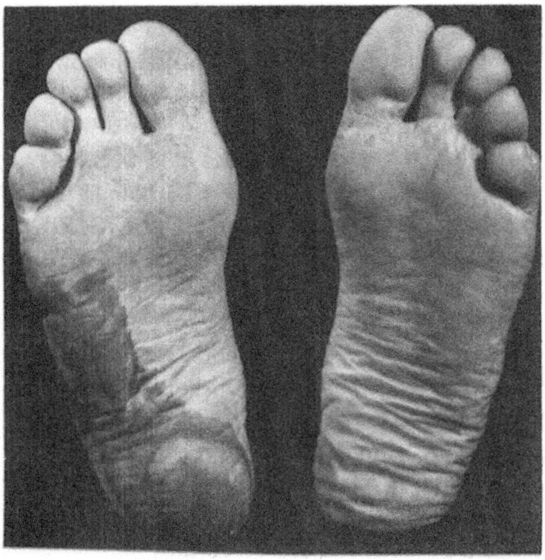

Figure 4.1 Feet of a patient with left thalamotomy. The blisters on the right sole appeared without trauma 12 h after the operation and could represent a trophic change related to the thalamic lesion.

Thalamotomies for Parkinson's disease have also resulted in blistering of the skin on the side opposite the lesion (Fig. 4.1). In such patients, it is possible that this represents a trophic change and thus another manifestation of autonomic dysfunction.

4.6 The influence of the nervous system on the triple response of Lewis

When the skin is injured by a variety of agencies, the normal vascular reaction is threefold: the vessels dilate locally, the skin blisters or wheals locally and the surrounding skin shows a bright red flare, which is due to spreading dilatation of small vessels. This reaction has been called the triple response of Lewis. Lewis, Harris, and Grant (1927) demonstrated that the flare component is absent in denervated skin, but can be elicited for a few days after section until the nerve degenerates. Thereafter, only the wheal and central red area, without the spreading flare, is seen. It is also known that transection of dorsal roots has no effect on the flare, but it is abolished seven or more days after dorsal root ganglionectomy (Chapman & Goodell, 1949). This is direct evidence in humans that the integrity of afferent fibers, which arise in dorsal root ganglion cells, is essential for the diffuse cutaneous vasodilatation (flare), which forms part of the normal response to injury of the

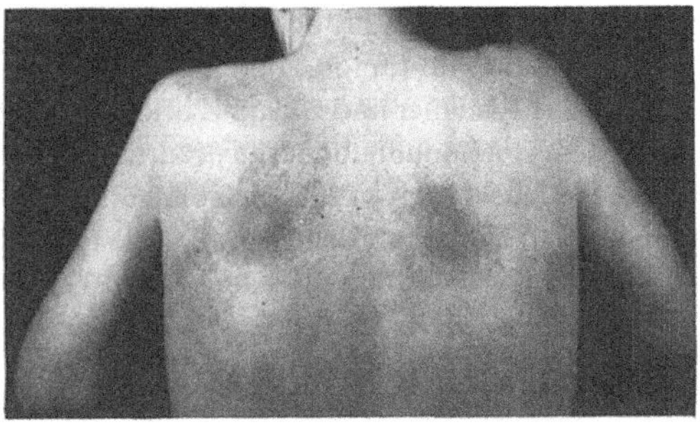

Figure 4.2 Larger flare on the left-back is associated with right deep and superficial middle cerebral artery territory infarction. The triple response of Lewis was produced by the intracutaneous injection of 0.1 mL of 1:1,000 solution of histamine phosphate. Note the equal size of the wheal and surrounding red area and the dermatomal distribution of the flare on the left side. *From Appenzell and McAndrews (1966).*

skin. While the inflammatory reaction is not abolished by peripheral denervation, it is clear that it is modified in denervated skin. Moreover, it can be shown that in subjects with intact peripheral nerves, the central nervous system modulates the extent of the flare and thus in turn influences the response to injury. In patients with lesions in the deep and superficial territory of the middle cerebral artery, the flare response is enhanced on the contralateral side of the body (Fig. 4.2), whereas lesions of the spinothalamic tract or thalamus decrease the extent of the flare on the contralateral side of the body. In cord lesions, a decrease of the flare in dermatomes below the lesion has also been demonstrated (Reed, Pidgeon, & Becker, 1961). The mechanism by which the central nervous system affects this peripheral axon reflex is not clear but trophic changes and painful syndromes, occasionally seen after central lesions, may be related to the modulation by the central nervous system of the flare after injury to the skin.

Ulceration of the skin is seen in spinal cord disease. Most commonly this is due to repeated injury or unrelieved pressure but impaired axonal reflexes, deficits in vasomotor control, and nutrition of the skin may also be important. Pressure sores in denervated dermatomes after spinal cord lesions heal, however, with the relief of pressure, improvement in nutrition, control of infection, and correction of anemia. The repeated injuries to hands of patients with syringomyelia often result in ulcers, which heal when subsequent injuries are prevented. All these trophic ulcers, if they heal at all, take much longer to do so than ulcers in normally innervated skin, and this may perhaps be due to the altered inflammatory responses in denervated areas.

The axon reflex is traditionally discussed in textbooks in the context of the triple response, popularized by Sir Thomas Lewis. Although the evidence for the existence of the axon reflex has always been controversial, those strongly committed to belief in its existence have gained ascendancy. The response consists of both a flare and plasma extravasation. The evidence that both phenomena are subserved by the same mechanism is incomplete. The flare and plasma extravasation are components of neurogenic inflammation, but it is suggested that these manifestations are separate phenomena. For example, very low-frequency stimulation of peripheral nerve of about one per second is sufficient to produce detectable changes in skin blood flow, but it requires much higher frequencies of stimulation to induce plasma extravasation (Magerl, Szolczanyi, & Westerman, 1987; Martin, Basbaum, & Kwiat, 1987). It has been suggested that this difference results from a different response mode of polymodal nociceptor-afferent axons, one being activated at low-impulse frequencies, causing vasodilatation but no other sensations, and unlikely to activate spinal reflexes. The second mode of higher impulse frequencies causes further vasodilatation, including activation of spinal reflexes and sensation, and produces plasma extravasation (Szolcsanyi, 1988).

All explanations of the axon reflex require that some primary sensory neurons release a chemical agent in association with impulse activity, and this can occur either directly or via mast cells, causing both local vasodilatation and/or plasma extravasation. Much experimental work, coming particularly from the rat, has implicated substance-P as a candidate for the chemical agent released in concert with the activation of the axon reflex. But other peptides, for example, calcitonin gene-related peptide (CGRP), vasoactive intestinal polypeptide (VIP), somatostatin, and neurokinin-A are also found in small sensory neurons; and the latter two have properties that make them candidates for chemical agents involved in the axon reflex.

The anatomic substrate of the axon reflex has continued to elude demonstration, and the branching of the main sensory axon, postulated by Lewis and his predecessors, has never been found. This may have been related to experiments carried out in the rat, and it seems more likely that human skin does indeed contain branching anatomic substrate for the axon reflex, as originally envisaged by Lewis. In the rat, however, there is a very close match between the receptive field and the area of extravasation, so that no branching of the primary sensory afferent axon is necessary to explain the response in this animal. In the human, the flare following a scratch to the skin spreads many millimeters from the site of injury, and it is highly improbable that a neurochemical agent that is released by the injury can diffuse such distances. Preliminary work on pigskin is promising and shows similar responses to those found in humans, where the neuroanatomic substrate of the axon reflex might be demonstrable in this experimental animal.

A great deal of effort is expended on identifying and characterizing the neuropeptides and peripheral neurons and substances that might act as chemical intermediaries in the flare component of the triple response of Lewis and in the extravasation of plasma that occurs simultaneously. The growing case made for substance-P involvement in this response is being supported by the use of laser Doppler flowmeters applied to the human skin. The difficulty with the neuroanatomical evidence of afferent branching in the skin should soon be overcome by careful studies in animals and perhaps also in human skin (Lisney & Bharali, 1989).

After injury of the skin by chemical stimuli, the excitation of noxious afferents causes vasoconstriction, which has been found to be sustained and almost always more pronounced on the stimulated side when compared to the opposite hand (Magerl, Koltzenburg, & Schmitz, 1996). The sympathetic skin responses to noxious stimuli, mostly vasoconstriction, have in the past been shown to be generalized and difficult to separate from those that occur with startle and general arousal response. Moreover, they show progressive decline in responsiveness and habituation on repeated application, characteristics common to autonomic reflexes. These characteristics are found with brief stimuli, however, when relatively longer-lasting chemical noxious stimuli are applied to the human skin, which produce either itch (histamine) or burning pain (mustard oil), the vasoconstriction is sustained and asymmetric (Magerl et al., 1996). Thus, reflex vasoconstriction in humans in response to noxious stimulation is like reflex vasodilatation in response to heating (Appenzeller & Scott, 1971): asymmetric and homotopic to the area of noxious stimulation.

4.7 The influence of the nervous system on myoedema

Contraction after percussion of skeletal muscles was first described by Graves and Stokes (1830) and Schiff (1858). Muscle contraction after percussion normally gives rise to small waves that leave a knot at the margin of percussion. This lasts only a few seconds and electromyographic studies have shown that action potentials, which are irregular and persist throughout the visible contraction and perhaps beyond, accompany the small waves. Two responses have been recognized: an immediate electrical response, which is triggered by mechanical stimulation of the muscle, and a delayed electrical response, which is a spinal reflex. Myoedema (Tait, 1871) is characterized however by a local contraction (mounding) in response to percussion of the muscle which is not associated with a recordable electrical discharge (Gelfan & Bishop, 1933). Moreover, this mounding is fatiguable.

Myoedema has been recognized in patients with wasting disorders and at one time, it was considered important in the recognition of cavity tuberculosis when atrophy of pectoral muscles was usually found overlying an apical cavity (Tait, 1871). In addition to syndromes associated with malnutrition such as sprue, the cachexia of cancer, acute infections, gastritis, and the debility of chronic alcoholism, myoedema has also been recognized in myxedema. The exact mechanism of the production of myoedema is not clear, but because of its prominence in debilitating disorders and the persistence of the phenomenon after death (Fig. 4.3), it has been attributed to an abnormal but inherent muscular excitability predominantly related to the contents of the muscle fiber rather than to depolarization of the muscle membrane (Denny-Brown & Pennybacker, 1938). There is no reasonable doubt now that myoedema is a physiologic phenomenon. Its duration and ease of demonstration at the bedside are largely dependent upon skin and muscle thickness (Jones & Parkes, 1955). The possibility that the intensity of myoedema is modulated by nervous system lesions has also been explored (Appenzeller, 1975). Contrary to previous reports, which suggested that myoedema can

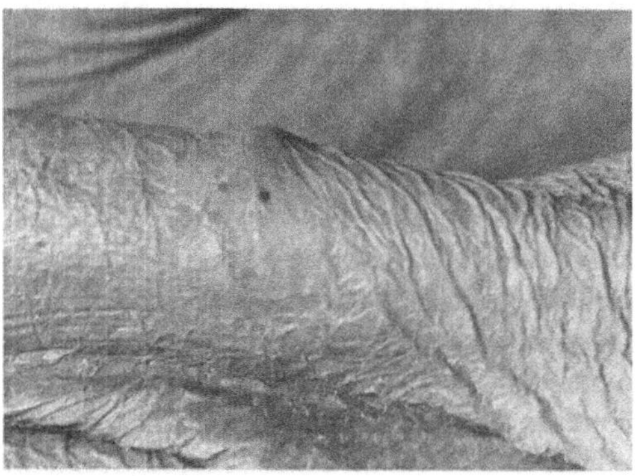

Figure 4.3 Myoedema of biceps muscle elicited 1 h after death. *From Appenzeller, O. Percussion contraction of muscles without action potentials (myoedema)—Effects of nervous system lesions. Trans. Amer. Neurol.-Ass. 100 (1975) 159–161.*

most easily be elicited in the pectoral muscles, the examination of 27 patients showed myoedema on percussion of the brachioradialis muscle in most patients, whereas other muscles including the pectoral muscles did not show myoedema. The phenomenon was always symmetric in all but two patients. It was most commonly found in patients with peripheral neuropathies (19 patients), 2 had incomplete mid and lower thoracic cord lesions and both had myoedema in the brachioradialis, an area that was not involved by the nervous system lesion. One patient with myxedema and right hemiparesis and another without myxedema but infarction in the deep and superficial supply territory of the left middle cerebral artery had myoedema on both sides. A patient with a pure sensory stroke and another with a lesion involving the spinothalamic pathways had myoedema on the uninvolved side of the body only. From this study, it was concluded that myoedema is a widespread phenomenon in patients with neurological disorders and is not confined to those with clinical evidence of malnutrition or endocrine disturbances. Myoedema persists in the face of corticospinal tract lesions and also in patients with such lesions and myxedema. The suppression of this phenomenon by a pure sensory stroke or spinothalamic tract lesion suggests that the central nervous system can modulate the response even though it has been attributed entirely to an independent myogenic mechanism. Whether this is the result of a direct effect of the central nervous system on muscle or due to removal of trophic function by the lesion has not been decided, but this is yet another example of central nervous system modulation of what appears to be a purely peripheral phenomenon.

4.8 Lesions of the peripheral nervous system

Lesions of the peripheral nervous system are commonly associated with trophic disorders. The characteristic atrophy of muscles is well known and is attributed to the lack of trophic influence of motor fibers since only minor changes are observed in muscles that remain innervated from experimentally isolated cord segments. The trophic changes of the skin are obvious and usually most marked in hands and feet, particularly when pain is a feature of the nerve lesion or the affected parts are subjected to continuous trauma or applications of excessive heat or cold, which may not be appreciated as such because of sensory loss (Haymaker & Woodhall, 1953). The skin becomes tight and smooth, later shiny and transparent and the subcutaneous tissue may also be atrophic. Pigmentation, common in chronically denervated areas, may occur and eczema-like changes may be seen sharply localized to denervated dermatomes. Minor mechanical or thermal trauma leads to indolent ulcers, particularly in fingers and toes, which may be painful even in the presence of extensive sensory loss in the affected parts. Fibrosis of subcutaneous tissue has been found, particularly in partial nerve injury, and the overlying skin is elevated by the contraction of the fibrous tissue into heavy folds. Occasionally clubbing of fingers is observed. The fingernails, in areas of severe sensory loss, become transversely striped, thickened, ridged, brittle, and often claw-like (Figs. 4.4 and 4.5). Nail growth, however, is not always slowed, except when the arterial supply to the limb has also become impaired (Sunderland & Ray, 1952). The hair is thinned or disappears from denervated areas of limbs but occasionally hypertrichosis is seen, particularly on the forearms. Most marked trophic changes usually occur in traumatic nerve lesions and chronic sensory neuropathies. Trophic ulcerations of the feet are very common in diabetic neuropathy, with marked sensory loss (Figs. 4.6–4.8).

The regeneration of extremities in some vertebrates depends upon the nervous system and because of this, the fine structure of nerves in the regenerating forelimb of adult newts has been studied. There was the sprouting of nerve fibers, which invaded the blastema and epidermis. An increased number of smooth-surfaced channels of endoplasmic reticulum and microtubules as well as a number of membrane-bound dense granules were seen. These granules were not normally found in nerves, and it was postulated that they contained a trophic substance or hormone that was transported down the axon and that was related to the regeneration of the limb.

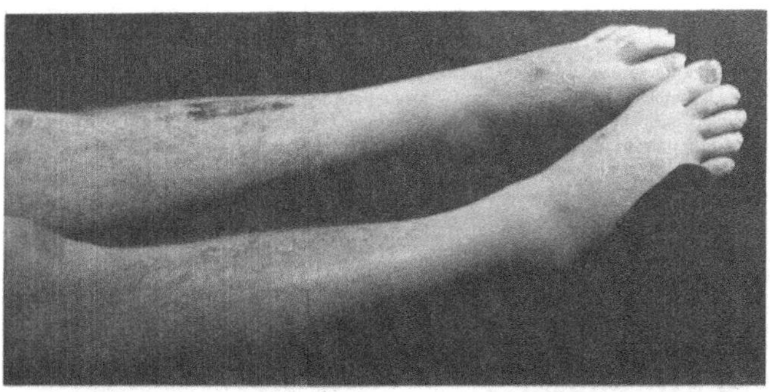

Figure 4.4 Lower limbs of a patient with severe peripheral neuropathy showing trophic changes. The skin is tight and shiny; transverse skin folds can be seen in the region of the ankles.

Figure 4.5 Toenails of the same patient as in Fig. 14.4, showing transverse ridging of 2nd and 3rd toenails.

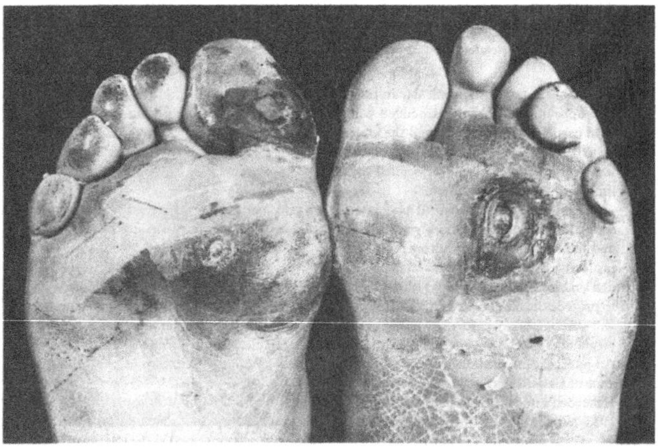

Figure 4.6 Trophic ulcers in diabetic polyneuropathy.

The fine structure of sensory ganglion cells during limb regeneration was also examined. There was an increase in the number of lipofuscin bodies and mitochondria after amputation. Newly formed membrane-bound dense granules similar to those seen in the nerves were also found in the sensory ganglion cells. These structural alterations might be a morphological basis for trophic changes in peripheral nerves and sensory ganglia contributing to the understanding but not explaining the trophic function of the nervous system.

4.9 Neuroarthropathies

Affections of joints as a result of denervation are common. The first description is usually attributed to Mitchell (1831). However, Charcot (1868) gave a

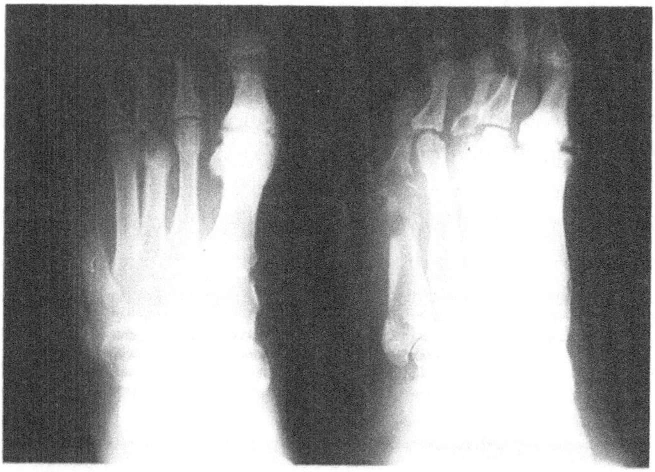

Figure 4.7 X-rays of the left foot show marked destruction of joint and bone (compared with Figs. 4.4–4.6).

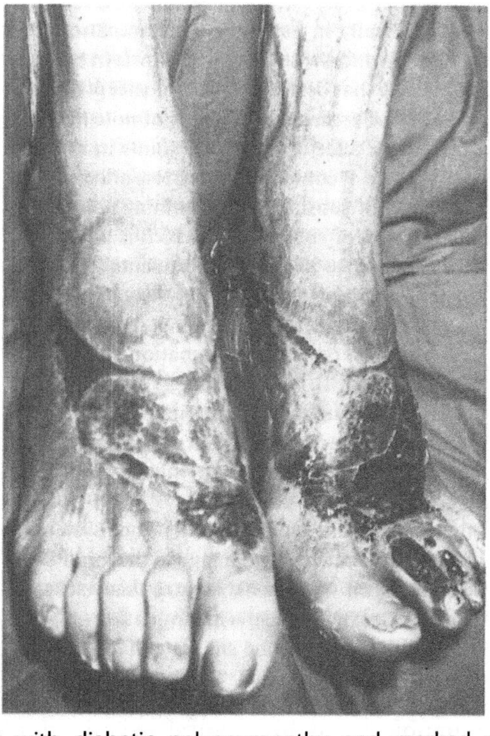

Figure 4.8 Patient with diabetic polyneuropathy and marked sensory loss. After stepping into a hot bath, the level of the water is indicated by the upper level of the ulcers.

detailed account of this condition and noted its frequent association with tabes dorsalis. Similar arthropathies are found in syringomyelia and in patients with peripheral nerve lesions. The commonest association of neuroarthropathies is with tabes dorsalis but it is also seen in syringomyelia, peripheral nerve lesions due to injury, neuropathies of sensory type (Fig. 4.7), and leprous neuritis. Rarely, neuroarthropathies are seen after spinal cord lesions due to injury and congenital malformations. Experimentally, it can be shown that joint denervation in association with trauma produces the characteristic changes of neuroarthropathy. Denervation alone, however, is not effective (Eloesser, 1917). It seems that continuous function of a denervated joint is necessary before the characteristic clinical features of excessive mobility, subluxation, and finally dislocation appears. Later, this is associated with heterotrophic cartilage and bone formation. The joint cartilage is worn away and the underlying bone may adapt to its new role of serving as an articular surface. Tabetic neuroarthropathies usually involve lower limbs and spinal column joints, whereas those associated with syringomyelia are most often found in the upper limbs or upper spinal column.

After hemiplegia, several ill-understood but well-recognized phenomena have been described, which affect the paralyzed limbs and include swelling, cyanosis, excessive sweating, and the shoulder-hand syndrome (Friedmann, Schwartz, Trubek, & Steinbrocker, 1953). Less commonly recognized is the protective effect of hemiplegia on the development of rheumatoid or osteoarthritis with Heberden's nodes (Bland & Eddy, 1968). It seems from the reported cases that the hemiplegia has to precede the onset of arthritis and that no evidence has so far been produced that hemiplegia will affect an established arthritic change. In the cases reported, the interval between the hemiplegia and the clinical onset of arthritis, which was confined to the normal side, was variable. An interval of 9 and 10 years elapsed in two cases after the hemiplegia before the appearance of Heberden's nodes on the uninvolved side (Forestier, 1935). In another case, Heberden's nodes appeared on the normal hand only 8 months after a stroke (Winter, 1952). Thirty-nine years after the onset of hemiplegia, rheumatoid arthritis developed on the normal side only, even though the patient was employed in the interim and fully ambulant (Jacqueline, 1953). The clinical impression of sparing of the paralyzed side in rheumatoid arthritis was confirmed by the absence of characteristic rheumatoid changes on radiographical and histological examination of the synovium of these joints (Thompson & Bywaters, 1962). A similar sparing effect has been reported for tophaceous gout (Glynn & Clayton, 1976).

Peripheral nerve lesions preceding the onset of arthritis appear to afford some protection to the involved parts also. Heberden's nodes were not found in the denervated joints in some cases of osteoarthritis (McEwen, 1940; Stecher & Karnosh, 1947). In one patient with complete ulnar and incomplete radial nerve palsies on the right, rheumatoid arthritis developed 6 years later, sparing the right hand in spite of widespread involvement of other joints (Kamermann, 1966). The mechanism of this protective effect of nervous system lesions on arthritis is not understood but it is clearly not related to immobilization alone because many protected joints remain quite useful. It might, however, be due to the alterations in the inflammatory response, which is modulated by lesions of the nervous system, or relative disuse since the involved parts would naturally be used less frequently. Caution in the interpretation of sparing effects of neural lesions on paralyzed joints needs to be exercised and this sparing effect may be nonspecific simply due to less exposure to trauma rather than to any specific trophic manifestations.

The influence of the nervous system on manifestations of joint disease has long been recognized. In rheumatoid arthritis, the pain is thought to be more than just a consequence of the disease. In animal models, mobility and significant benefits on joint destruction can be observed by alleviating pain. The interaction of pain and inflammation with the nervous system has been suspected for some time. In experimental arthritis, the section of relevant ascending spinal pathways that transmit sympathetic and sensory information to the central nervous system does alleviate symptoms. If animals with experimental arthritis are given a choice, they prefer solutions that contain nonsteroidal antiinflammatory agents or opiate analgesics to plain water.

The behavioral accompaniments of pain in animals are found to be expressed in neurophysiological and neurochemical changes in the peripheral and central nervous systems. Afferents from the joints, for example, have a resting discharge that is not seen in normal controls, and joints in the affected limb become sensitive to movement and slight degrees of pressure that do not cause responses in control animals. There is also a high level of background activity in neurons of the spinal cord that receive afferents from the inflamed joints. And similarly, inappropriate response patterns of neurons in the thalamus and somatosensory cortex, with a disproportionate number of cells responding to joint stimulation in these areas, when compared to normal rats, have been found. This abnormal response pattern has been interpreted to show long-term changes in central nervous processing mechanisms and in the primary afferent responses.

Primary afferent nociceptors have been known to be important contributory factors in inflammation, specifically in neurogenic inflammation. Peripheral nerve section inhibits neurogenic inflammation and the severity of arthritis in humans, and experimental application of capsaicin, a substance that destroys C-fiber nociceptive afferents, causes a reduction of swelling and hyperalgesia of arthritic joints. It is of note also, that ankle joints, which are more severely affected than knee joints in arthritis, have lower nociceptive thresholds and higher substance-P content than the less-affected knee joints. This suggests that there is an increased density of substance-P containing nerve fibers in ankle joints. It has also been shown that the infusion of substance-P, which is a proposed mediator of C-fiber neurogenic inflammation into the knee joints of experimentally affected rats, increases the severity of the inflammation and joint destruction.

Evidence that the peripheral sympathetic nervous system is involved in arthritis has also been found. Sympathectomy, for example, reduces inflammation and joint injury in experimental arthritic models and animals treated with β-adrenergic receptor blockers; specifically, β-2 antagonists have a significant reduction in joint injury. The β-2 blockade is effective even if given after the clinical onset of arthritis, and in patients with rheumatoid arthritis, propranolol and regional sympathetic blocks with guanethidine seem to reduce pain (Levine, Fye, Heller, Basbaum, & Whiting-O'Keefe, 1986).

A proposed model for the importance of the sensory and sympathetic nervous systems in arthritis includes the impairment of C-fiber nociceptors induced by damage to the joint. Damaged nociceptors transmit information to the spinal cord while axon reflexes that release inflammatory mediators are activated, which then, in turn, exacerbate the inflammation on a neurogenic basis and in turn influence C-fiber excitability. This scheme suggests that the pain is part of the disease rather than the result of joint destruction alone, and this proposal is in line with clinical and experimental observations.

Somatosympathetic reflexes activated by C-fiber activity are thought to excite sympathetic postganglionic neurons, causing generalized or excess autonomic discharges. These reflexes could explain the stress-associated exacerbation of rheumatoid arthritis. Sympathetic axons may also release mediators of neurogenic inflammation that activate C-fibers and thus perpetuate the vicious cycle of neurogenic and arthropathic interactions. Similar mechanisms are perhaps also operative in reflex sympathetic dystrophy, which may be associated with inflammation of synovial joints and is always accompanied by severe pain.

This proposal cannot, however, entirely explain the pathogenesis of joint destruction in rheumatoid arthritis. The etiology of the disorder is not fully understood, and the disease extends to other tissues as well as joints. A possible role for autoantibodies and infective factors must also be considered. The involvement of the cardiovascular system and of peripheral nerves and respiratory complications in rheumatoid arthritis is not easily explained by purely autonomic factors. Nevertheless, the significant role of the sympathetic nervous system in perpetrating and modulating joint dysfunction in rheumatoid arthritis is clearly not confined to this disease, but it has also been suspected in other types of degenerative joint disease (Fitzgerald, 1989).

4.10 Hypertrophic osteoarthropathy (Bamberger–Marie syndrome)

This condition is characterized by clubbing of the fingers and toes, periostitis with new bone formation, and arthritis. Most cases are seen in association with disease in other parts of the body, particularly lungs, heart, liver, gastrointestinal tract, and occasionally in thyrotoxicosis after thyroidectomy. Rarely, it is seen in syringomyelia. A hereditary form is associated with thickened skin (Touraine–Solente–Gole syndrome). The lesion is an inflammatory periostitis and arthritis and involves the adjacent subcutaneous tissue. Pain may be severe. The commonest association nowadays is with carcinoma of the lung, although previously pulmonary suppurative processes were more common. There is no satisfactory explanation for this disorder, except that occasionally it regresses within days after section of the vagi. It is thought that the interruption of an abnormal pulmonary vascular reflex by this procedure may account for the decrease in the peripheral circulation which accompanies the improvement of hypertrophic osteoarthropathy.

4.11 Interactions between central and peripheral neurons and their target tissues

Trophic phenomena are related to the formation of connections between a nerve and its target. The maintenance of the integrity of target cells is exemplified by the close dependence of muscles on its nerve supply, and the regulation of some properties of target cells by innervating neurons that adapt the target to an intended task but which can be changed by changing its innervation (an example is the regulation of fast-slow, white-red muscle fiber properties). Clearly, until the mechanism of trophic interactions has been clarified, this definition remains tentative.

In the peripheral nervous system, supporting cells play an important role in maintaining neurons; an example is the effect of nerve growth factor that is essential in the maintenance of sympathetic neurons in vivo and in vitro. The nerve growth factor is produced by supporting cells or glial cells, but it can also be produced by glioma cells in vitro and a number of nonneural tissues. Enhanced development of peripheral neurons in tissue cultures can also be achieved by a number of unidentified other factors, which have been shown to reside in cultured glial cells and regenerating peripheral neurites.

The role of the nervous system in the maintenance of muscle is obvious and well documented. Whether acetylcholine, which is the established neurotransmitter at the neuromuscular junctions, is necessary for trophic functions in the maintenance of muscle has not been established. It is, for example, not clear if the acetylcholine-mediated neuromuscular transmission is alone sufficient for trophic function in vitro or in vivo. Considerable evidence has, however, accumulated to show that cholinergic blockade of neuromuscular transmission produces the same effects as seen after denervation. Cholinergic neuromuscular transmission is at least necessary, if not entirely sufficient, for trophic neuromuscular interactions. There are other components whose importance in nerve–muscle relationships has not been fully documented. These include disuse and spontaneous nonaction potential-induced release of acetylcholine. In a number of clinical situations and animal experiments, use does seem necessary for the normal trophic influence of nerves on muscles. Disuse results in clinical and morphological signs of partial denervation. The role of spontaneous nonaction potential release of acetylcholine in trophic activity has not been determined.

There are other noncholinergic factors that play a role in trophic activities. These include axoplasmic flow and adequate activity of muscles. Some nerve trophic activity is closely related to sufficient muscle contractions because stimulation experiments show that usage alone can almost be sufficient to maintain the muscle. Spontaneous release of acetylcholine on the other hand is not alone sufficient for normal trophic function but the possibility that it may contribute cannot be excluded.

Axonal transport of some as yet unidentified substances, spontaneous release of acetylcholine, and usage all interact to a varying degree in the control of the sensitivity of muscles to acetylcholine and establish the trophic function of the nerve. The sum of these actions accounts for normality and trophic regulation becomes a measure of the degree of activity of these three or other as yet unidentified influences of nerve on muscle. Moreover, acetylcholine plays a very important role in neuromuscular interactions and maintenance of trophic function but its role is not established for other organs (Drachman, 1974).

The wide scope of neurotrophic relations, which includes maintenance of connections and the development and regeneration of parts of the nervous system and other tissues of the body, has led to the suggestion that the term neurotrophic relations be restricted to functional manifestations necessary for the long-term maintenance and regulations of structures by their innervation but which are independent of nerve impulses. Since a number of different components are involved in intercellular relationships such as cell-to-cell contact, in addition to neuron and nonneural cell contract, it cannot be expected that all trophic relations are based on the same mechanisms.

Hippocampal cell cultures by themselves develop neurites during the first day of culture and resemble other central or peripheral neurons. However, the survival of such cells is limited and only a few neurons with intact processes usually remain after a few days. When astrocytes are grown on coverslips, which are added to the culture medium, the survival of hippocampal neurons is considerably increased. Further studies of astrocyte-neuronal relations in the culture medium, which contained astroglial cells, is important because when added to growing neurons, it has similar effects on the survival of hippocampal neurons. Moreover, the influence of such a conditioned growth medium is reversible indicating that growth-promoting agents released by astroglial cells must continuously be present and that insignificant amounts permanently bind to neurons to enhance neuronal survival. This demonstration in vitro of the release of trophic substances other than nerve growth factor by glial cells supports their trophic function on neurons. But, the molecular mechanisms of this trophic effect remain to be explored (Banker, 1980).

References

Adams, R. D., Denny-Brown, D., & Pearson, C. M. (1954). *Diseases of muscle, a study in pathology*. New York: Paul B. Hoeber.

Appenzeller, O. (1975). Percussion contraction of muscles without action potentials (myoedema)—Effects of nervous system lesions. *Trans. Amer. Neurol.-Ass.*

Appenzeller, O. & D. Scott (1971). Migraine vasomotor function and basophilic leukocyte. In *Proceedings of the international headache symposium, Elsinore, Denmark.* 16–18.

Banker, G. A. (1980). Trophic interactions between astroglial cells and hippocampal neurons in culture. *Science (New York, N.Y.)*, *209*, 809–810.

Bland, J. H., & Eddy, W. M. (1968). Hemiplegia and rheumatoid hemiarthritis. *Arthr. Rheum. (N.Y.)*, *11*, 72–80.

Brissaud, E. (1879). De l'atrophie musculaire dans l'hémiplégie. *Rev. Mens. Méd. Chir.*, *3*, 616–642.

Brown, J. B., & McDowell, F. (1940). Syndactylism with absence of the pectoralis major. *Surgery*, 7, 599–601.

Chapman, L. F., & Goodell, H. (1949). The participation of the nervous system in the inflammatory reaction. *Ann. N.Y. Acad. Sci.*, *116*, 990–1017.

Charcot, J. M. (1868). Sur quelques arthropathies qui paraissent dépendre d'une lésion du cerveau ou de la moelle épinière. *Arch. Physiol. Norm. Path. I*, 161–178.

Charcot, J.M. (1879). Lectures on diseases of the nervous system. Translated by G. Sigerson. H.C. Lea, Philadelphia.

Courtillier, L. (1897). Contribution à l'étiologie et à pathogénie du pied-bot congénitale. *Arch. Gen. Méd.*, *1*, 536–563.

Critchley, M. (1966). *The parietal lobes*. New York: Hafner Publishing Co.

Dejerine, J. (1889). De la névrite périphérique dans l'atrophie musculaire des hémiplégiques. *C.R. Soc. Biol. (Paris)*, *1*, 523–530.

De La Tourette, G. (1897). Pathogénie et prophylaxie de l'atrophie musculaire et des douleurs des hémiplégiques. *Bull. Soc. méd. Hôp. Paris*, *14*, 499–513.

Denny-Brown, D., & Pennybacker, J. B. (1938). Fibrillation and fasciculations in voluntary muscle. *Brain*, *61*, 311.

Dittrich, R. J. (1930). Pathogenesis of congenital clubfoot (pes equinovarus); anatomical study. *J. Bone Jt. Surg.*, *12*, 373–399.

Dodge, P. R. (1961). Congenital neuromuscular disorders. *Res. Publ. Assoc. Nerv. Ment. Dis.*, *38*, 479–533.

Drachman, D. B. (1974). The role of acetylcholine as a neurotrophic transmitter. *Ann. N. Y. Acad. Sci.*, *228*, 160–176.

Dunnebacke, T. H. (1953). The effects of the extirpation of the superior oblique muscle on the trochlear nucleus in the chick embryo. *J. Compar. Neurol.*, *98*, 155–177.

Eloesser, L. (1917). On the nature of neuropathic affections of the joints. *Ann. Surg.*, *66*, 201–207.

Evans, P. R. (1955). Nuclear agenesis—Mobius syndrome. The congenital facial diplegia syndrome. *Arch. Dis. Childh.*, *30*, 237–243.

Fairbank, H. A. T. (1913–1914). Congenital elevation of the scapula; a series of eighteen cases, with a detailed description of a dissected specimen. *Brit. J. Surg.*, *1*, 553–572.

Fenyes, I. (1937). Zur Frage der Entstehung von angeborenen Beweglichkeitsstörungen im Gehirnnervenbereich; ein klinisch-anatomischer Beitrag. *Arch. Psychiat.*, *106*, 296–311.

Fitzgerald, M. (1989). Arthritis and the nervous system. *Trends in Neurosci.*, *12*, 86–87.

Forestier, J. (1935). Rhumatisme d'Heberden chez une hémiplégique: lésions articulaires limitées au côté sane. *Revista Neurol.*, *63*, 442–448.

Friedmann, H. H., Schwartz, S., Trubek, M., & Steinbrocker, O. (1953). The pararheumatic arthropathies. *Ann. Int. Med.*, *38*, 732–758.

Garrod, A. E., & Davies, L. W. (1905). On a group of associated congenital malformations, including almost complete absence of the muscles of the abdominal wall, and abnormalities of the genito-urinary apparatus. *Med. Chir. Trans.*, *88*, 363–382.

Gelfan, S., & Bishop, G. H. (1933). Conducted contractures without action potentials in single muscle fibers. *Amer. J. Physiol.*, *103*, 237.

Glynn, J. J., & Clayton, M. L. (1976). Sparing effect of hemiplegia on tophaceous gout. *Ann. Rheum. Dis.*, *35*, 534, 535.

Gordon, A. (1927). Segmental trophic edema of cerebral origin. *J. Nerv. Mental Dis.*, *66*, 381–389.

Graves, R. J., & Stokes, W. (1830). Clinical report of cases in the medical wards of the Meath Hospital during the session 1828 and 1829. *Dublin Hosp. Rep.*, *5*, 70.

Greig, D. M. (1924). Congenital high scapula, with which is included considerations of brevicollis. *Edinb. Med. J.*, *31*, 22–24.

Haymaker, W., & Woodhall, B. (1953). *Peripheral nerve injuries. Principles of diagnosis.* Philadelphia: W.B. Saunders Co.

Jacqueline, F. (1953). A case of evoluted polyarthritis with localization contralateral to a hemiplegia. *Rev. Rheum.*, *20*, 323–324.

Jones, M. P., & Parkes, W. E. (1955). Myoidema. *Clin. Sci.*, 97–100.

Kamermann, J. S. (1966). Protective effect of traumatic lesions on rheumatoid arthritis. *Ann. Rheum. Dis.*, *25*, 361–363.

Karpati, G., & Engel, W. K. (1967). Neuronal trophic function, a new aspect demonstrated histochemically in developing soleus muscle. *Arch. Neurol. Psychiat. (Chicago)*, *17*, 542–545.

Krabbé, K. (1949). Les lésions embryonnaires la lumière des défectuosités mammaire et pectorale, de la syndactylie et de la microdactylie. *Acta Psychiat. Scand.*, *24*, 539–546.

Le Double, A.F. (1897). Traité des Variations du Système Musculaire de l'Homme et de leur Signification au Point de Vue de l'Anthropologie Zoologique. Schleicher Frères, Paris.

Levine, J. D., Fye, K., Heller, P., Basbaum, A. I., & Whiting-O'Keefe, Q. (1986). Clinical response to regional intravenous guanethidine in patients with rheumatoid arthritis. *J. Rheumatol.*, *13*(6), 1040–1043.

Lewis, T., Harris, K. E., & Grant, R. T. (1927). Influence of the cutaneous nerves on various reactions of the cutaneous vessels. *Heart (Br. Card. Soc.)*, *14*, 1–17.

Lichtenstein, B. W. (1939). Congenital absence of the abdominal musculature; associated changes in the genitourinary tract and in the spinal cord. *Amer. J. Dis. Child.*, *58*, 339–348.

Lisney, S. J. W., & Bharali, L. A. M. (1989). The axon reflex: an outdated idea or a valid hypothesis? *News Physiol. Sci.*, *4*, 45–48.

Luzzato, A. M. (1903). Über vasomotorische Muskelatrophie. *Dtsch. Z. Nervenheilk*, *23*, 482–498.

Magerl, W., Koltzenburg, M., Schmitz, J. M., et al. (1996). Asymmetry and time-course of cutaneous sympathetic reflex responses following sustained excitation of chemosensitive nociceptors in humans. *J. Autonom. Nerv. Sys.*, *57*, 63–72.

Magerl, W., Szolczanyi, J., Westerman, A., et al. (1987). Laser Doppler measurements of skin vasodilatation elicited by percutaneous electrical stimulation of nociceptors in humans. *Neurosci. Lett.*, *82*, 349–354.

Marinesco, G. (1898). Recherches sur l'atrophie musculaire et la contracture dans l'hémiplégie organique. *Sem. Méd. (Paris)*, *18*, 465–470.

Martin, H. A., Basbaum, A. I., Kwiat, G. C., et al. (1987). Leukotriene and prostaglandin sensitization of cutaneous high-threshold C- and A-delta mechanoreceptors in the hairy skin of rat hind limbs. *Neuroscience*, *22*, 651–659.

McEwen, C. (1940). Herberdens nodes: Heredity in a hypertrophic arthritis of the finger joints (a discussion). *J. Amer. Med. Ass.*, *155*, 2024.

McGuire, J., & Zeek, P. (1932). Pathogenesis of chronic hereditary edema of extremities (Milroys disease). *J. Amer. Med. Ass.*, *98*, 870–873.

Michaud, M. (1870). Note sur la pathogénie du pied bot congénital propos d'un example d'une difformité de ce genre paraissant liée à une lésion congénitale de la moelle épinière. *Arch. Physiol.*, *3*, 586–598.

Milroy, W. F. (1928). Chronic hereditary edema; Milroys disease. *J. Amer. Med. Ass.*, *91*, 1172–1175.

Mitchell, J. K. (1831). On a new practice in acute and chronic rheumatism. *Amer. J. Med. Sci.*, *8*, 55–64.
Möbius, P. J. (1888). Über angeborene doppelseitige Abducens-Facialis-Lähmung. *Münch. Med. Wschr*, *35*, 91–108.
Morley, E. B. (1923). Congenital defect of the pectoralis muscle. *Lancet*, *1*, 1101–1102.
Ormsby, O. S., & Montgomery, H. (1954). *Diseases of the Skin*. Philadelphia: Lea and Febiger.
Osler, W. (1901). Congenital absence of the abdominal muscles with distended hypertrophied urinary bladder. *Bull. Johns Hopk. Hosp.*, *12*, 331–333.
Pitner, S. E., Edwards, J. E., & McCormick, W. C. (1965). Observations on the pathology of Moebius syndrome. *J. Neurol. Neurosurg. Psychiat.*, *28*, 362–374.
Quincke, H. (1893). Über cerebrale Muskelatrophie. *Dtsch. Z. Ner.venheilk.*, *4*, 299–311.
Reed, W. B., Pidgeon, J., & Becker, S. W. (1961). Studies on the triple response of Lewis in patients with spinal cord injury. *J. Invest. Derm.*, *37*, 135–143.
Richards, R. N. (1953). The Möbius syndrome. *J. Bone Jt. Surg.*, *35A*, 437–444.
Schaffer, K. (1897). Zur Lehre der cerebralen Muskelatrophie nebst Beitrag zur Trophik der Neuronen. *Mschr. Psychiat. Neurol.*, *2*, 30–52.
Schiff, J.M. (1858). Lehrbuch der Physiologie des Menschen. Schauenburg.
Schlesinger, H. (1900). Zur Lehre vom angeborenen Pectoralis-Rippendefect und dem Hochstande der Scapula. *Wien. klin. Wschr.*, *13*, 25–31.
Sheldon, W. (1932). Amyoplasia congenita (multiple congenital articular rigidity. arthrogryposis multiplex congenita). *Arch. Dis. Childh.*, 7, 117–136.
Silverman, F. N., & Huang, N. (1950). Congenical absence of the abdominal muscles associated with malformations of the genitourinary and alimentary tracts. *Amer. J. Dis. Childh.*, *80*, 91–124.
Sprofkin, B. E., & Hillman, J. W. (1956). Möbiuss syndrome-congenital oculofacial paralysis. *Neurology (Minneap.)*, *6*, 50–54.
Stecher, R. M., & Karnosh, L. J. (1947). Herberdens nodes. VI. The effect of nerve injury upon the formation of degenerative joint disease of the fingers. *Amer. J. Med. Sci.*, *213*, 181–191.
Steinert, H. (1903). Cerebrale Muskelatrophie. Nest einem Beitrag zur Casuistik der Balkentumoren. *Dtsch. Z. Nervenheilk*, *24*, 1–59.
Stern, W. G. (1923). Arthrogryphosis multiplex congenita. *J. Amer. Med. Ass.*, *81*, 1507–1510.
Sunderland, S., & Ray, L. J. (1952). The effect of denervation on nail growth. *J. Neurol. Neurosurg. Psychiat.*, *15*, 50–53.
Szolcsanyi, J. (1988). Antidromic vasodilation and neurogenic inflammation. *Agents Actions*, *23*, 4–11.
Tait, L. (1871). On the myoidema of phthises. *Dublin Quart. J. med. Sci.*, *52*, 316.
Thompson, M., & Bywaters, E. G. L. (1962). Unilateral rheumatoid arthritis following hemiplegia. *Ann. Rheum. Dis.*, *21*, 370–377.
van Bogaert, L., & Ley, R. A. (1933). Contribution a la sémiologie des troubles pseudocerebelleux et vasomoteurs d'origine paracentrale. *J. Neurol. Psychiat. (Brux.)*, *33*, 171–179.
Winkelman, N. W., & Silverstein, A. (1932). Trophic disturbance of the limbs in retrorolandic lesions. *Res. Nerv. Ment. Dis. Proc.*, *13*, 485–528.
Winkelman, N. W., & Silverstein, A. (1935). Unilateral amyotrophy; its diagnostic importance for cerebral localization. *Amer. J. Syph. Neurol.*, *19*, 58–76.
Winter, S. (1952). Unilateral Heberdens nodes in a case of hemiplegia. *N.Y. St. J. Med.*, *52*, 349–350.

CHAPTER 5

Pain perception and the autonomic nervous system

Numerous painful syndromes have been recognized. In this section, the discussion will be limited to those that are accompanied by disturbances in vasomotor function, sudomotor function, and trophic changes and that are sometimes relieved by surgical approaches targeting the autonomic nervous system.

5.1 Causalgia

Pain syndromes associated with peripheral nerve lesions are particularly common. In areas of incomplete sensory loss due to transection and after surgical repair of a peripheral nerve with partial recovery or in association with neuromas, an overresponse to stimuli may be seen. Such dysesthesias occur after lesions of the nerves in the arm or leg. Careful examination usually shows a heightened threshold to ordinary stimuli in association with pain of peculiar and unpleasant quality, which is confined to the dermatome of the injured nerve. It is thought that the sensory abnormalities are due to incomplete regeneration of pain fibers in the cutaneous sensory network (Weddell, Sinclair, & Feindez, 1948). Numerous examples of this condition have been found many years after injury to the limbs (Herz & Yahr, 1950; White & Sweet, 1955). The majority of patients eventually recover and do not require surgery, but excision of neuromas or neurolysis is not usually effective. When the dysesthetic area becomes sensitive to cold, relief may be obtained by sympathectomy (White & Sweet, 1955). This condition has to be distinguished from causalgia, in which the disorder is not confined to a single dermatome.

Causalgia, also known as complex regional pain syndrome Type II, is characterized by hyperpathia, trophic changes, and vegetative phenomena. It occurs after peripheral nerve injury (White, Heroy, & Goodman, 1948) and usually involves either the arm or leg. There is a partial nerve injury, as a rule, and hyperpathia is most marked in fingers and toes. Trophic changes in the skin are marked but are not different from those seen after

Introduction to Clinical Aspects of the Autonomic Nervous System
DOI: https://doi.org/10.1016/B978-0-323-96011-3.00013-9

peripheral nerve lesions without causalgia. The pain and dysesthesias are not confined to denervated dermatomes but involve a large part of the affected limb. Excessive sweating and vasodilatation occur (De Takats, 1945) and the limb may be either hot or cold, depending mainly on the amount of sweat. Cool limbs are particularly common in the later stages of the disorder (White et al., 1948).

Causalgia is rare and usually seen in association with puncture wounds of nerves and even then it occurs in less than 5% of cases. Causes of causalgia include soft tissue trauma, crush injury, and amputation while historically, the most likely injury resulting in causalgia was due to a high-velocity missile affecting the proximal part of the limb and involving the median or tibial nerves (Richards, 1967).

Because movement causes pain, the limb is kept immobile and joints become stiff. Contact with clothing may be exceedingly unpleasant, and patients may wrap their limbs in cottonwool or moistened towels. Many stimuli applied to other parts of the body, and ordinarily well tolerated, may cause excruciating pain in the affected limb (White & Sweet, 1955). Historically, it was suggested that causalgia was a form of psychoneurosis. This assumption has, however, never been well documented and, on the contrary, psychological studies of patients who have recovered showed no abnormal personality traits (Mayfield & Devine, 1945; Speigel & Milowsky, 1945). Pain is the most characteristic feature of this syndrome and the vasomotor and sudomotor abnormalities are thought to be secondary to it.

The true pathogenesis of causalgia is not understood but several theories have been advanced to explain it. The nerve impulse may be deflected in a crushed part of the nerve. On stimulation of an anterior root, impulses can be recorded in posterior roots of peripherally crushed nerves (Granit, Leksell, & Skoglund, 1944) and this is attributed to abnormal transmission of impulses from motor to sensory fibers at the site of the crush. These impulses are then transmitted back to the spinal cord and are thought to be the basis for causalgia. It has also been suggested that sensory fibers are stimulated at the injury site by impulses traveling to the periphery in sympathetic nerves (Doupe, Cullen, & Change, 1944). Causalgia can be successfully treated in properly selected patients by sympathectomy. Patients respond well if there is relief of pain by sympathetic block and if the pain is aggravated by cold or other environmental factors (White & Sweet, 1955). Patients who have dysesthesias or do not have the complete clinical picture of causalgia do not respond to sympathectomy. These painful conditions are distinguished from causalgia by the absence of burning pain. They occur after trauma, infections, or arthritis without necessarily direct injury to major nerve trunks. Sympathetic dystrophies are not only found in young healthy subjects with normal personalities, but occur in those who may have some personality disorders as well.

Neurophysiologic evidence suggests that normally, an inhibitory influence can abolish stimulus-evoked activity in a sensory system almost as soon as the stimulus is removed. It has been proposed that the brainstem reticular formation acts in a way that allows it to influence this firing pattern by a central biasing mechanism. This is achieved by continuous tonic inhibitory influence on transmission at all levels of the somatic afferent projection system. This bias or tonic inhibition is maintained, in part at least, by continuous sensory input from the skin. Abnormal responses occur in conditions that impair the activity of sensory input or of the central biasing mechanism (Melzack, 1971). Some anesthetic substances abolish the inhibitory influence of the central biasing mechanism so that continuous activity at several synaptic levels of the skin afferent system can occur. A stimulus, therefore, which normally has only a transient effect, can result in persistent activity. Conversely, if there is a loss or decrease in sensory input from the skin the activity of the central biasing mechanism is also decreased. In amputated limbs, the sensory afferent fibers are destroyed and the amount of input to the reticular formation is markedly decreased. This may result in continuous self-perpetuating activity in neuron loops at all levels, which can be repeatedly initiated by minor noxious or other stimuli from the site of injury and may be related to the phenomenon of phantom pain. Emotional stress has the capacity to interfere with the efficacy of the central biasing mechanism also and this may account for the aggravation of spontaneous burning pains associated with causalgia, for example. An explanation for the well-known sustained benefit of analgesic blocks in these situations is that the anesthetic agents break activity in neuron loops. After the block has worn off, peripheral stimulation triggers sustained activity again but there is a time lag for it to spread to a large number of neurons, and this time lag accounts for the pain relief that outlasts the action of local anesthetic. Moreover, in conditions such as causalgia, the temporary removal of discomfort allows the limb to be increasingly used again that increases the pattern of impulses from the periphery particularly from muscles, which then increases the bias in the central biasing mechanism and also delays the resumption of self-sustaining firing in various neuron pools. In patients with causalgia, therefore, the continuous inactivity of the limb aggravates the unfavorable bias of the central biasing mechanism and perpetuates discomfort. It is important, therefore, that at the earliest sign of the development of reflex sympathetic dystrophy (RSD) of any kind, the patient should be treated with local or regional sympathetic blocks or if necessary, both. Excessive delay in treatment, particularly when pain is a prominent feature of the disorder, may lead to permanent disability, no longer responsive to manipulation of the peripheral input by local anesthetic or surgery on the peripheral autonomic nervous system (Bonica, 1973).

5.2 Reflex sympathetic dystrophy—complex regional pain syndrome

The major dysfunctions occurring in RSD also include movement disorder (Selkowitz, 1992). However, the pathogenesis of this clinical manifestation of RSD just as RSD itself remains obscure (Schwartzman, 1993). RSD in children is a more benign disorder most children recover with physical therapy alone (Wilder, Berde, & Wolohan, 1992). In 1994, the International Association for the Study of Pain proposed that the term "Complex Regional Pain Syndrome" be used instead of RSD and proposed a diagnostic criterion. Revised criterion was proposed in 2010 and is commonly referred to as the "Budapest Criteria." Complex regional pain syndrome has two subtypes: type I, formerly known as RSD (without nerve trauma), and type II, formerly known as causalgia (with nerve trauma). There is further subdivision of the syndrome into "warm" versus "cold," and sympathetically maintained versus sympathetically independent. Treatment can be challenging and optimal management is based on an interprofessional approach with physical and occupational therapy, pharmacotherapy, behavioral therapy, and interventions. The review of commonly used therapeutic options including antiinflammatory drugs, anticonvulsants, antidepressants, transdermal lidocaine, opioids, NMDA antagonists, and bisphosphonates is beyond the scope of this chapter.

There is upregulation of sympathetic receptors on nociceptive nerve fibers in complex regional pain syndrome. Sympathetic hyperactivity leads to increased pain and sympathetic sensitivity of nociceptive nerves. Autonomic dysregulation can involve cardiovascular autonomic function and increased heart rate and decreased heart rate variability have been reported in patients with complex regional pain syndrome

(Terkelsen et al., 2012). Changes in peripheral catecholamines sensitivity over time lead to excessive vasoconstriction and hyperhidrosis and explain the cold, clammy extremities seen in the chronic phase of the disease (Goh, Chidambaram, & Ma, 2017).

Nociceptor activation induces early response protein (c-fos) increased expression in spinal cord neurons. The gene encodes FOS-LI, which increases in the superficial laminae of the dorsal horn. There are also changes in preproencephalin messenger RNA and in specific recognition sites that affect neuropeptide gene transcription. Thus, activity in afferent fibers has profound effects on neuronal plasticity after tissue injury and inflammation (Dubner & Ruda, 1992), and this mechanism might not only be involved in RSD of limbs but also in RSD-like conditions such as posttraumatic headache (Appenzeller, 1993). In the latter condition, central sensitization with release of excitatory amino acids including glutamate and neuropeptides such as substance P may play important roles. Elevation of TNF-α, interleukin (IL)-1b, IL-2, and IL-6 have been found in both serum and cerebrospinal fluid. Autoimmunity also contributes to the pathogenesis of complex regional pain syndrome and autoantibodies against beta-2-adrenergic receptor, alpha-1a-adrenergic receptor, and muscarinic-2 receptor have been reported.

Sudeck (1900) described a form of posttraumatic dystrophy in which, after minor trauma, usually in the region of the wrist or ankle joint, atrophy and spotty decalcification of bone occurs. Pain may be severe and abnormal vasomotor and sudomotor activity is seen. Many changes in Sudecks atrophy are due to disuse because of pain. This condition can be successfully treated by sympathectomy (Shumacker & Abramson, 1949; White & Sweet, 1955).

Sudecks atrophy associated with pelvic or lumbar spine lesions has been reported (Cayla & Rondier, 1974). This condition appears to be more common in women; it is often bilateral, nearly all patients have involvement of both the foot and knee region. Though radiologically the hip joint may show osteoporosis, this is usually clinically silent. Pain of causalgic nature is often present and brings the patient to the attention of the physician. Almost all have had surgical procedures including bone grafts, some were found after hip joint surgery and a number of cases occurred after infectious spondylitis, metastatic carcinoma to the pelvis. Stiffness of the joints usually involving the ankle is prominent, though reflex vasomotor abnormalities are less marked than in Sudecks atrophy involving the upper limb. Radiologic investigation of the limb shows typical rarefaction of the bone but this may not appear until several weeks after the beginning of symptomatology. The aggravation of symptoms by stress is not very prominent in this type of Sudecks atrophy but precipitating traumatic factors are almost universally present. The best treatment for this RSD is lumbar sympathetic block. Sometimes better results are obtained by adding a local anesthetic infiltration of the femoral artery perivascular nerve plexuses also.

Other sympathetic dystrophies, which occur in limbs without osteoporosis but are characterized by sweating, pain, and vasomotor changes after minor trauma, have also been recognized (Holden, 1948). In the early stages after injury, there is hyperemia, accompanied by edema and rapid atrophy of muscles. Pain on slight movement is very severe and the limb is kept immobile. In the later stages of the disorder, edema persists, but the limb becomes cyanotic, cold, and clammy. Vasomotor abnormalities have been found in unaffected limbs of these patients. The diagnosis should be made early and treatment by sympathectomy carried out, so that movement of joints becomes possible again and secondary atrophy and demineralization of bones are prevented. Some patients develop swelling of the legs with edema and severe pain on movement after a considerable interval following back surgery. The edema is often responsive to diuretics and there might not be other trophic changes in the affected limb. On occasion, Raynaud phenomenon develops in the hands, but the exact mechanism of the excessive vasoconstriction in the skin of the hands in response to cold has not been determined. This could result however from an abnormal firing of the whole of the postganglionic sympathetic nervous system induced by the initial trauma to the back. Lumbar sympathectomy in such patients is effective in removing the symptoms if it is carried out soon after the clinical onset of the disorder. The effect of surgical sympathectomy on pain and abnormal behavioral responses to ordinary stimuli applied to the skin is not entirely clear. The effect of sympathectomy of the affected limb on pain and abnormal behavioral responses to limb stimuli was assessed in a well-controlled study in a rat model of mononeuropathy produced by loosely applying a ligature to the sciatic nerve (Desmeules, Kayser, & Weil-Fuggaza, 1995). Sympathectomy reduced or prevented the abnormal reaction to cold and to heat, but, the distorted reaction to mechanical pressure was not affected. Surprisingly, the behavioral abnormalities indicating spontaneous pain remained. Sympathectomy, without sciatic nerve ligature, resulted in a reduction of the vocalization threshold to pressure on both hind paws and a short-lasting increased tolerance to cold immersion. Thus, the sympathetic nervous system contributes but is not entirely to blame for the abnormal pain-related behavior after nerve injury. It appears to play a more pivotal role, however, in abnormal pain responses to thermal stimuli.

5.3 Hyperalgesia

Soreness after an injury, hyperalgesia, is universally perceived. The nervous systems processes underlying the perception of this postinjury discomfort are being unraveled and the autonomic structures involved extend centrally and involve peptides in the periphery. After injury, increased sensitivity of pain perception neurons in the spinal cord occurs. After sensitization of these neurons by the injury, nonpainful stimuli evoke exaggerated signaling transmitted to the central nervous system (CNS) and interpreted as pain (Carstens, 1995). Tenderness normally develops at the site of an injury. This gives rise to so-called primary hyperalgesia. But a larger region of uninjured tissue also becomes tender—the area of so-called secondary hyperalgesia. Characteristically hyperalgesia is present when noxious stimuli provoke an exaggerated painful response and nonpainful stimulation such as touch provoke pain (allodynia). Secondary hyperalgesia may occupy a large region. At the site of injury, primary hyperalgesia can be provoked by thermal and mechanical stimuli, and it results in part from the liberation of algesic chemicals from the tissue, which acts on intracutaneous nerve terminals of nociceptive neurons and increases their sensitivity. The region of secondary hyperalgesia is susceptible to mechanical and not thermal stimuli, which evoke pain.

This accepted view has been challenged. Evidence suggests that allodynia and secondary hyperalgesia evoked by mechanical stimuli do not result from enhanced receptor sensitivity in the undamaged area but from an increase in sensitivity of spinal afferent neurons that receive input from the area of secondary hyperalgesia (Willis, 1992). Additional evidence for a central sensitization of neurons and the cellular mechanism responsible for these painful states has also accumulated (Carstens, 1995).

Lewis (1942) proposed that peripheral mechanisms alone account for the phenomenon of secondary hyperalgesia and that impulses along nociceptor afferents are also conducted antidromically along other branches of nociceptor afferents to terminals in the skin, which release algesic substances that sensitize other nociceptors in the region inducing the axon reflex. This sensitization lowers thresholds so that nonpainful stimuli of remote sensitized receptors now result in pain. In this scenario, secondary hyperalgesia is initiated and maintained by sensitization of peripheral nerve terminals. The evidence for this interpretation is based on the fact that infiltrating the skin with local anesthetic above an injury prevents the spread of secondary hyperalgesia antidromically proximal to the site of injury by blocking conduction for some distance in nociceptors and consequently suppressing the release of chemical substances thus avoiding sensitization.

A proposal that secondary hyperalgesia arises centrally due to injury-induced sensitization of pain afferent neurons in the spinal cord is based on the proposition that mild stimuli can cause pain if the receptors in the noninjured remote area have access via interneurons to postinjury sensitized dorsal horn neurons (Hardy, Wolff, & Goodell, 1952). This is supported by the fact that most neurons in the spinal cord that are concerned with pain conduction receive inputs also from mechanoreceptors and the assumption is that these neurons signal pain by higher frequency of impulses if they are sensitized when mechanoreceptors are excited by touch-pressure. Thus, the increased responses of spinal afferents to mechanical stimulation (via interneurons) are now interpreted as painful.

The experimental use of capsaicin injected (or applied) to the skin has allowed a reconciliation of these opposing views (LaMotte, Shain, Simone, & Tsai, 1991; Simone, Sorkin, & Oh, 1991). Capsaicin when applied to the skin evokes a burning sensation followed by a rapidly forming surrounding zone of secondary hyperalgesia (flare). The application of capsaicin, unlike direct injury of the skin. does not, however, cause visible tissue damage and the hyperalgesia disappears after a short time. In this experimental setting, the spread of secondary hyperalgesia proximally was prevented by anesthetizing the skin with local anesthetic proximal to the capsaicin application site supporting a peripheral neurogenic origin of secondary hyperalgesia.

However, anesthetizing the nerves to the forearm before the application of capsaicin prevented the development of secondary hyperalgesia altogether and direct recordings from cutaneous receptors in animals failed to show peripheral sensitization. These observations support a central origin of secondary hyperalgesia. In primates, the application of capsaicin to the skin is followed by increased responsiveness of spinothalamic tract neurons to mechanical stimulation (Simone et al., 1991). Intraneural recordings from single afferent fibers from mechanoreceptors in humans were also carried out. Stimulation of the fibers ordinarily leads to a tactile sensation but after the application of capsaicin stimulation of the same fiber evokes pain (Torebjork, Lundberg, & LaMotte, 1992). The assumption is made that the spinal neurons have become sensitized by the application of capsaicin and stimulation of mechanoreceptors is now perceived as painful.

In addition, damage to superficial tissues by burn, chemical, or nerve injury causes increased sensitivity of spinal neurons involved in pain perception (Carstens, 1995). Also, these spinal neurons are normally subject to subliminal afferent input, which does not cause them to fire but after tissue injury, this subliminal input is sufficient to excite the same neurons. Electrical stimulation of a peripheral nerve proximal to an injury evokes larger responses in spinal neurons than the same stimulation before the injury.

A proposed explanation of the peripheral neurogenic spread of hyperalgesia and of the role of central sensitization reconciles both central and peripheral origins of hyperalgesia (LaMotte et al., 1991). Thus, it is thought that capsaicin activates a set of nerves in the skin that is sensitive to chemicals. Central fibers from these nerves sensitize spinal neurons. This helps injury-released chemicals activate chemosensitive fibers over a large area to activate spinal neurons some with mechanoreceptor afferents remote from the injury site to become sensitized.

This proposal explains mechanical but not thermal hyperalgesia; it depends on a new class of chemoreceptor fibers. The model, however, fits well with the fact that locally anesthetizing the skin is effective in preventing the spread of secondary hyperalgesia and it signs the importance of central mechanisms in the development of secondary hyperalgesia.

The cellular mechanism that underlies pain perception has been studied in animals. Chemical and thermal noxious stimuli affect nociceptors that release a number of neuropeptides including substance-P and CGRP and excitatory amino acids especially glutamate exciting dorsal horn neurons (Carstens, 1995). It is important that the peptides bind to postsynaptic neurokinins receptors and glutamate binds to postsynaptic receptors also. The NMDA receptor is thought to be involved in persistent increases in dorsal horn neuron excitability, which is thought to be the cause of secondary hyperalgesia (Carstens, 1995). The probability that some hyperalgesia states may be treated by NMDA antagonists, which block the cellular events leading to postinjury hyperexcitability is of importance in the treatment of chronic painful states (Dickenson, 1994).

The NMDA receptor requires, apart from binding of glutamate, sufficient postsynaptic depolarization through neurokinin and non-NMDA receptors to remove Mg^{2+} from the ion channel to allow cation fluxes. Ca^{2+} is involved in enhanced cellular excitability. The calcium-dependent processes implicated in neuronal hyperexcitability are: gene induction causing an increased expression of neuromediators such as c-fos and c-jun; c-fos expression is prevented by blocking the NMDA receptor (Zimmermann & Herdegen, 1994). Ca^{2+} activates protein kinase C (PKC) that produces several cellular membrane changes that occur with hyperexcitability. Pretreatment with a ganglioside (GM1) can prevent PKC activation and the development of experimental neuropathic pain (Price, Mao, & Mayer, 1994). Ca^{2+} activates nitric oxide synthase increasing the production of NO. This gaseous neurotransmitter is thought to diffuse back into the presynaptic neurotransmitter terminal and increase cGMP and thus in turn enhance neurotransmitter release. Inhibiting NO production avoids NMDA-induced hyperalgesia and increased excitation of dorsal horn neurons produced by pain-inducing chemicals. Of interest is that cooling of the skin reverses secondary hyperalgesia induced by capsaicin suggesting that established hyperalgesia states may be reversible by interrupting some of these cellular mechanisms (Carstens, 1995) and altitude exposure abolishes the flare induced by capsaicin (Appenzeller & Martignoni, 1996), which implies that some hypoxia associated events interfere with hyperalgesia also. Lastly, the excitotoxic effects of excess glutamate release are well established. These may result in neuronal death (Dubner and Ruda, 1992). Some neurons that succumb to glutamate may be inhibitory thus allowing uninhibited increases in excitability of surviving neurons and this could account for the permanent increase in pain that is sometimes found after nerve injury in for example RSD.

5.4 Raynaud phenomenon

Primary Raynaud phenomenon remains a mystery. It is characterized by episodes of pallor, brought on by cold, in one or more fingers. In the original description, central hyperactivity of the sympathetic nervous system was proposed (Raynaud, 1862). On the other hand, Lewis suggested a local digital vascular fault (Lewis, 1929).

The central and local regulations of capillary blood flow in the finger were studied using the local $Xenon^{-133}$ washout technique. Subjects included women with primary Raynaud's phenomenon, men with vibration-induced white fingers, and controls matched by age and sex. The vasoconstrictor responses to venous stasis of 40 mm of mercury-induced by a local reflex were normal in both Raynaud's phenomenon and vibration-finger patients. Changes in posture from lying to sitting, which induce a centrally mediated sympathetic vasoconstriction, were found to be normal in all groups and were abolished by a proximal nerve block to the appropriate finger. However, vasoconstrictor responses to sitting were increased in Raynaud's phenomenon when compared to responses in controls. Therefore, there are central and local postural vasoconstrictor reflexes also demonstrated in normal fingers (previously shown to be present in the forearm only). In both types of Raynaud's phenomenon, however, there was hyperactivity of the central sympathetic nervous system to orthostatic stress and normal function of the digital arterioles and postganglionic sympathetic nerve fibers (Olsen, Petring, & Rossing, 1987). This is not the result of arteriolar smooth muscle hypertrophy, because no augmented response to venous stasis can be observed, whereas exacerbated responses to venous stasis are found in essential hypertension where the arteriolar smooth muscle becomes hypertrophied (Henriksen, Skagen, & Amtorp, 1981).

The local vasoconstrictor reflex is abolished in patients with autonomic peripheral neuropathy (e.g., in those with diabetic orthostatic hypotension). But this reflex was normal in those with Raynaud's phenomenon, thus excluding dysfunction of the autonomic peripheral innervation in these patients. Moreover, the local reflex is also absent in those with Raynaud's phenomenon associated with generalized scleroderma where the vasoconstriction is obstructive (Kristensen, 1979). Thus, the observed augmented constrictor response to changes in posture (which are centrally mediated) supports the suggestion originally made by Raynaud (1862) that there is a hyperfunction of central sympathetic nervous system structures in subjects with vasospastic phenomena. The response may be due either to constitutional factors that make individuals more susceptible to Raynaud's phenomenon or to the prolonged use of vibrating hand tools, resulting in vibration white-finger syndrome.

In microvascular angina, an impaired forearm vasodilator reserve has also been found (Sax, Cannon, & Hanson, 1987). Patients show abnormalities in vasomotor function not only in the coronary microcirculation, causing angina, but also in asymptomatic forearm blood vessels where the degree of vasodilator impairment (after ischemia) correlated well with that in the coronary circulation.

5.5 Pathophysiology

Whether the pain in sympathetic dystrophy is conducted via the sympathetic nervous system or is a result of cross-excitation from sympathetic fibers to sensory afferents at the site of minor nerve injury is not clear. There have been a number of studies to suggest that pain in these disorders is conducted along sympathetic fibers from a limb (Freeman, Shumacker, & Radigan, 1950; Threadgill, 1947). White and Sweet (1955) and others (Holden, 1948) think, however, that cross-excitation from sympathetic to pain fibers is the main mechanism of sympathetic dystrophy. Some support for this view comes from the production of pain on stimulation at the operation of the peripheral ends of the sympathetic fibers after upper thoracic ganglionectomy. This pain was referred to the arm in patients with causalgia but no pain occurred in those who did not have causalgia (Walker & Nulson, 1948). Traumatic myelopathy may cause reflex sympathetic dystrophy. This is usually associated with incomplete lesions and unexplained pain may appear after a good recovery and may require sympathectomy to prevent serious disability (Wainapel, 1984).

5.6 Associated disorders

Another form of reflex sympathetic dystrophy is the shoulder-hand syndrome. This occurs commonly after myocardial infarction or hemiplegia. The shoulder becomes painful on movement, the hand may be sweaty with vasomotor changes and finally atrophy, particularly of hand muscles, may be seen. The palmar aponeurosis is often thickened in long-standing cases (Bayles, Judson, & Potter, 1950). Thickening of the skin of the palm is also seen in patients with long-standing hemiplegia but the painful shoulder often develops within days after the onset of the weakness and this cannot, as a rule, be prevented by early passive movement of the joint. The cause of the shoulder-hand syndrome is unknown. The pain usually subsides after a variable length of time and sympathectomy is rarely necessary. On rare occasions, sympathectomy leads to neuralgic pain. This usually occurs about two weeks after sympathectomy. The pain is localized to the thighs and is deep and boring in character and often disappears spontaneously within 10 days or two weeks. A few patients, however, have severe pain, which is not relieved by narcotics. A number of patients were markedly improved by the administration of phenytoin or carbamazepine (Raskin, Levinson, Hoffman, Pickett, & Fields, 1974). The most common surgical indication for sympathectomy is the patients with postsympathectomy neuralgia were vascular and have intermittent claudication and leg ulcers. The mechanism of postsympathectomy neuralgia has not been satisfactorily elucidated and the effectiveness of phenytoin and carbamazepine, which alter the excitability of nerve membranes and depress polysynaptic responses does not help in interpreting the pathogenesis of postsympathectomy neuralgia.

RSD is rarely associated with herpes zoster. An 80-year-old woman had a 4-month history of left-arm pain and hyperesthesia, including decreased mobility at the left wrist and left hand joints. The history of discomfort in her upper extremity was preceded by pain in the left side of her neck, shoulder, arm, and hand. It began soon after the appearance, on these same areas, of large vesicular eruptions with classic features of herpes zosters, involving the C7, C8, and T1 dermatomes. She was given a 3-week course of prednisone and some medications for the control of pain. As the herpes rash resolved, the arm became progressively more painful, and fingers gradually became fixed in flexion. While walking, she had to support her arm with her other hand. While resting, she protected her arm with a pillow, for even the weight of the bedsheets caused her discomfort. On examination, her left arm was found to be shriveled, pale, and cyanotic. The skin was glossy and the nails were dystrophic. The digits were barely moveable when the patient tried to move them herself. A left stellate ganglion block was chosen for treatment and was administered every other day for two weeks. Physical therapy followed the injections and led to decreased discomfort and increased mobility in the patient's hand.

Reports of herpes zoster associated with RSD, such as the found in this case, are uncommon. Another similar case, which followed the eruption of herpes zoster in the sacral dermatomes, showed dramatic improvement after propranolol therapy, which was begun 6 weeks after the onset of discomfort.

Unexplained paralysis in patients who show dramatic improvement of weakness after sympathetic blockade must be extremely rare. Four patients had such paralysis, which improved by sympathetic ganglion block or regional intravenous infusion of reserpine, a sympathetic-blocking agent. It is known that in RSD, paresis may be present but is always overshadowed by severe pain, which is also improved, including the paralysis, by sympathetic blockade. In the patients reported, the motor weakness was not associated with pain at the time of examination, although pain may have been the initiating symptom of the patients' disease.

The weakness was monoparetic. Truncal muscles were spared, though in one patient, the facial muscles were involved. It is of interest that in these patients autonomic testing and vasomotor function remained normal, and no evidence was found for a disturbance of peripheral blood perfusion. When the patients were given pilocarpine or edrophonium, no change in muscle strength was observed, but the motor paralysis was markedly improved by sympathetic blockade, suggesting that an increase in sympathetic tone was pathogenetically linked to the weakness. The patients reported were not a homogenous group but had certain common features, which included a slight or moderate monoparesis, exacerbation of weakness by the administration of catecholamine, improvement of weakness by sympathetic blockade, pain that did not accompany the weakness, lowered skin temperature, and a fine postural tremor of the affected limb. There was also an absence of neurological, psychiatric, or vascular disorders that could have explained the clinical symptomatology.

The sympathetic nervous system is known to influence neuromuscular function and muscle contraction. Sympathetic nerves have an antifatigue action, which has been described by Orbeli (1923). Sympathetic stimulation also decreases fatigue of fast-contracting muscles with repetitive stimulation of the motor nerve. Experimentally, however, irritation of sympathetic ganglion cells can produce paralysis of hind legs in rabbits (Wiertz-Hoessels, 1964).

b_2-Adrenoceptors have been found on sarcolemmal membranes of skeletal muscle. Stimulation of these receptors increases contraction of white fast-twitch muscle but reduces the contraction of red slow-twitch muscle fibers. Epinephrine can reduce neuromuscular transmission through its influence on a- and b-adrenoceptors. It is possible that patients with sympathetically modulated paralysis have the disorder because of an abnormality of neuromuscular junction or receptors on skeletal muscle. This is supported by the improvement found with regional intravenous infusion of adrenergic-blocking agents, and the normal results on repetitive stimulation and single-fiber electromyography. It has been suggested that motor paralysis related to sympathetic dysfunction should be termed sympathetic motor paresis. This condition must be rare but is clearly of importance because of its responsiveness to sympathetic blockade or ganglionectomy (Yokota, Furukawa, & Tsukagoshi, 1989).

5.7 Autonomic faciocephalalgia (Hortons syndrome, histaminic cephalalgia, and cluster headaches)

This variety of headache has special clinical features, which set it apart from classical migraine. It is well described by Horton (1956) and also by Kunkle, Pfeiffer, Willhoit, and Hamrick (1952). Characteristically, the headache is unilateral and it is more often found in males than females. The illness usually begins between 20 and 40 years of age. The pain occurs in attacks, is severe, is burning in character, and involves the eye, temple, neck, and sometimes the face. It may extend into the shoulder. The attacks begin and terminate abruptly and do not last much longer than an hour. They occur frequently at night and wake the patient. Characteristically, the pain is associated with tearing, rhinorrhea, nasal obstruction, conjunctival vasodilatation, swelling of temporal vessels, and vasodilatation of the skin of the ipsilateral side. Tenderness of vessels on pressure after an attack is common. The pain is not confined to trigeminal nerve dermatomes but corresponds to the distribution of the external carotid artery and its branches. Trigger zones, like those seen in trigeminal neuralgia, are not found and it differs clinically from migraine because of the sex incidence, absence of prodrome, nausea, and vomiting, and its shorter duration. The origin of the pain is not fully established. It was found that in some patients, pressure on the external carotid artery or its branches alleviated the pain. Elevation of intracranial pressure by the injection of saline, a procedure known to improve pain due to dilatation of intracranial arteries, was without effect in these patients. This suggests that the pain may have originated in extracranial vessels. In other patients, however, rising intracranial pressure was effective in abolishing the pain and the pain was aggravated by head jolt, a procedure known to aggravate pain due to dilatation of intracranial vessels (Wolff, 1963). Although Horton (1941) attributed the pain to a sensitivity of vessels to histamine, there is some doubt about this (Wolff, 1963). Histamine sensitivity cannot explain the episodic and localized nature of the attacks. Reports of therapeutic efficacy of histamine desensitization procedures have not been generally confirmed (Wolff, 1963). Cluster headaches are instead regarded as a migraine variant.

A chronic variant best termed chronic migrainous neuralgia does exist in which the pain continues for long periods of time associated with a persistent ipsilateral Horner syndrome. The pathogenesis of this variety of autonomic faciocephalalgia is not understood, but long-term therapy with lithium is usually effective. In patients with cluster headaches, the cerebral vasculature reacts with excessive vasoconstriction in response to oxygen inhalation. This demonstrated vasomotor abnormality has been therapeutically exploited and the inhalation of oxygen often shortens attacks. Some patients have also noticed beneficial effects from acute exertion and therapeutic exercise is also useful in terminating attacks.

5.8 Raeder's syndrome (paratrigeminal syndrome)

Raeder (1924) described patients with unilateral pain, usually in the distribution of the first division of the trigeminal nerve, associated with an ipsilateral Horners syndrome. A variety of this syndrome has been recognized (Boniuk & Schlezinger, 1961) in which, in addition, paralysis of other cranial nerves, usually the third, occurs. This type of headache has been found in association with inflammation near the internal carotid artery and the Gasserian ganglion at the base of the brain. Although the sympathetic fibers to the face are involved, the resulting Horner syndrome is incomplete because sweating is usually preserved and may be excessive on the involved side. Ptosis and miosis may be transient or remain permanently after many attacks. Facial and corneal sensations are intact. Paratrigeminal neuralgia also seems to be a variant of migraine. Many patients with cluster headaches may show a transient Horner syndrome on the affected side (Wolff, 1963). This type of headache has also been attributed to edema (Kunkle & Anderson, 1960) in the wall of the carotid artery, which affects the sympathetic fibers and interferes, perhaps transiently or permanently, with their function. Edema of this type in intracranial vessels is said to occur in migraine headaches (Wolff, 1963). Treatment is not entirely satisfactory but desensitization with histamine and maintenance on small doses of the drug have been tried (Horton, 1941).

In one case, Raeder's paratrigeminal syndrome occurred in association with an intracranial aneurysm of the supracavernous portion of the internal carotid artery (Law & Nelson, 1968). Because of this, such patients should be carefully evaluated prior to any therapy that might be contraindicated if the disorder is related to an aneurysm.

Secretory fibers to the lacrimal gland, mucosa of the nose, pharynx, palate, and mouth are carried through the greater superficial petrosal nerve, which also contains vasodilator fibers to these structures and the arteries of the pia mater (Chorobski & Penfield, 1932). These fibers are preganglionic and some of them pass to the internal carotid artery and its branches and synapse on small ganglia, which are found on the arterial wall. Periodic discharge of parasympathetic impulses through the greater superficial petrosal nerve could account for many features of cluster headaches and Raeder's syndrome, but not for the unilateral sweating, cutaneous vasodilatation, and Horner syndrome (Gardner, Stowell, & Dutlinger, 1947). Division of the greater superficial petrosal nerve in these conditions has been carried out with some good results (Gardner et al., 1947; White & Sweet, 1955).

Only two women with Raeder's syndrome have been reported. In one, no underlying condition has been recognized. The other woman with Raeder's syndrome had also maxillary sinusitis. After appropriate treatment and drainage of purulent material from the sinus, her symptoms rapidly cleared. She was reexamined later and found symptom-free and neurologically normal except for slight ptosis and miosis on the previously affected side (DeWitt & Thompson, 1975). It is not known why Raeder's syndrome is so rare in women and whether these cases represent a fundamental difference in the pathogenesis of the disorder in females or simply a reporting artifact. Nevertheless, the occurrence of symptoms of Raeder's syndrome in a female should lead to a thorough search of an underlying abnormality that might be amenable to specific treatments.

5.9 Referred pain

Viscera are not sensitive to ordinary stimuli. It can be shown that cutting, crushing, and burning of viscera, which are ordinarily painful superficially, do not cause any sensation in the gut (Lennander, 1901). It was soon discovered, however, that stimuli that are effective in producing pain are those to which the viscera are commonly exposed. Distension (Hertz, 1911), anoxia (Sutton & Leuth, 1930), and metabolites, particularly acids (Moore & Singleton, 1933), are all capable of producing pain. Similarly, distension of the capsule of solid organs like the liver, stretching or crushing of blood vessels, and anoxia of muscles (Wilson & Mussey, 1947) are all painful. Because nerve endings sensitive to pain are not numerous in viscera, the summation of a widespread stimulus appears necessary before pain is felt. The threshold for pain can, however, be considerably lowered by prior irritation of the organ. For example, digital compression of the inflamed appendix causes pain and the normal gastric mucosa becomes sensitive to touch after irritation by mustard (Kinsella, 1948).

Pain from viscera is essentially similar to that arising from deep structures such as muscles, ligaments, joints, and periosteum (Lewis, 1942). The injection of 6% hypertonic saline deeply into skeletal structures produced referred pain in 94% of injections in a study involving 28 normal subjects. The effects varied, however, qualitatively. The most common was referred pain; skin hyperalgesia and muscle spasm were also seen. While the site of reference for a given stimulus was constant for any one individual, it was not always confined to the segment in which the injection was made. It was also observed that preexistent pain in another segment could change the site of referred effects to this segment. When the site of the stimulus was anesthetized prior to the injection of the hypertonic saline, referred effects were often abolished (Hockaday & Whitty, 1967). All the characteristics of referred effects after hypertonic saline injection into muscle or joints have similarities with visceral pain. Thus, it was shown that impulses arising from the body surface may increase some types of mild visceral pain (Cohen, 1944). Peripheral lesions could modify the radiation of anginal attacks. For example, in one patient with myocardial ischemia, pressure on a painful amputation neuroma in a shoulder stump produced anginal pain. In another patient with angina, the pain radiated into the left elbow after a fracture of the joint occurred. Because of these and other similar studies, it has been suggested that cutaneous afferent impulses entering a spinal segment facilitate awareness of visceral discomfort if a simultaneous discharge from a viscus occurs into this same segment (Theobald, 1949; White & Sweet, 1955). If the cutaneous segment is anesthetized, however, visceral pain may still be referred to this segment and anginal pain has on rare occasions been referred to a left phantom limb.

Another type of referred pain has been described (Sterling, 1973). This type of pain is initiated by nontraumatic stimulation of a point on the skin and it is referred to another distant skin area. It appears to be a common phenomenon. Most often the sensation is recognized after unconscious grooming behavior such as scraping fingernails over the scalp, plucking of hairs, and sometimes by minor skin lesions. The stimulus, which elicits the referred sensation, causes a normal sensibility on the site of stimulation but in addition, there is a localized, often sharp feeling at a distant point on the skin. These sensations are not very painful and are often followed by an itchy feeling in the referred area. Though only about half of interviewed normal subjects were able to spontaneously recall this referred cutaneous sensation, the other subjects could repeat this phenomenon subsequent to the inquiry (Sterling, 1973). An adequate stimulus to elicit referred cutaneous sensation must be strong and it depends in part on which of the cutaneous structures are stimulated. For example, pulling on hair is often effective but gentle brushing is not. Careful mapping of referral sites showed that the referred sensation was always ipsilateral; there was never reciprocity between referral and stimulating side; no feeling could be elicited by stimulating the referred side in the original initiating area; and most importantly, the referred sensation was never in the same dermatome as the stimulus.

Similar sensations have been reported after squeezing one or the other testicle, which causes an abnormal crushing feeling superficially in the region of the nipple. Also, in some patients with dilating abdominal aneurysms, testicular pain may appear and precede a fatal rupture of the aneurysm. These phenomena resemble in some way the experimentally produced misdirected reflex responses in frogs who have belly skin transplanted to the back. Stimulation of this skin gives rise to a normal reflex directed to the back and a misdirected reflex toward the belly. Referred sensation is often elicited from sites of irritation where nerve endings are regenerating or degenerating. Whether cutaneous referred sensation arises from abnormal embryonic specification of connections of peripheral nerves or from a temporary abnormality due to abnormal fiber growth is not known. But animal experiments would be necessary to further define the physiology of this rather widespread but not generally recognized phenomenon. Moreover, the findings of referred cutaneous sensation in most subjects suggest that sensory disturbances, which are not confined to dermatomal distributions, should not be glibly dismissed as hysterical.

An alternate theory that endeavors to explain pain can also account for some features of referred pain. It assumes that the substantia gelatinosa functions as a gate control system that modulates the afferent pattern of impulses before they reach the first central transmission (T) cells. The afferent impulse pattern in the dorsal column then acts as a trigger for central control mechanisms that activate structures in the brain. These in turn influence the gate control system in the substantia gelatinosa. Lastly, the T cells induce the firing of systems that allow appropriate responses and perception of pain (Melzack & Wall, 1965). The gate control theory assumes that pain is the result of the interaction between these systems. Referred pain might then be explicable on the following model: the normal activities of the T cells are dominated by their restricted field of input; however, there is an additional diffuse monosynaptic input to these cells, which is recognizable after electrical stimulation of afferents from distant parts (Mendell & Wall, 1965). Normally, it is suggested that this diffuse input is inhibited by the gate mechanism in the substantia gelatinosa. The T cells might, however, be fired by the diffuse monosynaptic input in two ways: either if the input is intense or the gate activity is altered. The gate mechanism may be opened by stimuli from far away or even from the other side of the body since the substantia gelatinosa receives fibers from both sides and through Lissauers tract from a number of adjacent spinal segments as well (Melzack & Wall, 1965). This mechanism might explain the observations that pressure on distant body parts may trigger angina or that phantom pain in a limb may be induced by pressure on the head (Cronholm, 1951).

Several other theories attempt to explain the misrepresentation of painful impulses that arrive at the spinal cord from the viscera and are referred to cutaneous areas. One (Ruch, 1949) suggests that referred pain is due to convergence of visceral and cutaneous afferents on shared neurons of the spinothalamic tract. Because there is a larger number of pain fibers in the posterior roots than axons in the spinothalamic tract, both cutaneous and visceral pain fibers converge on the same neurons in the posterior horns. This theory assumes that these neurons are predominantly activated by stimuli from the body surface, which are also recognized through other sensory modalities such as vision. When visceral impulses activate the secondary sensory neurons, the cortical interpretation of the origin of these impulses is in keeping with that built-up by previous experience and interpreted therefore as arising from the skin. This theory is dependent on the concept of experience, which is difficult to evaluate and is assumed to influence the sensory cortex. This theory does not account for some features of referred pain including the occasional failure of local anesthesia of the area of referral to influence the pain (Hockaday & Whitty, 1967). Another interpretation of referred pain is based on phylogenetic evidence of the distribution of the peripheral nervous system (Sinclair, Weddell, & Feindel, 1948). In fish, there are sensory axons that have one terminal supplying the gut and another the skin, and in the skin of the frog, a single fiber may branch to many widely separated areas (Adrian, Cattell, & Hoagland, 1931). While this single innervation from one axon to both skin and viscera might explain referred pain, such an anatomical arrangement has hitherto not been demonstrated in man.

Extensive experience with visceral pain has clearly shown that the best way to terminate the pain is by interrupting the splanchnic pathways, which mediate impulses from the viscera to the CNS (Wolff & Hardy, 1947). This has also been confirmed at reoperations after splanchnicectomy when a number of normally painful stimuli to the gallbladder or intestines failed to elicit pain (Bentley & Smithwick, 1940; Ray & Neill, 1947). Moreover, after abdominal field block at laparotomies for chronic peptic ulcer, the part of the stomach involved in the ulcer remained sensitive to pressure despite complete anesthesia of the abdominal wall, and this sensitivity was only abolished after additional splanchnic nerve block (Bentley, 1948).

The knowledge of the autonomic innervation of thoracic and abdominal viscera and the pathways, which are related to visceral pain, has been used in surgical attempts to relieve such pain. Historically, sympathectomies were performed for angina pectoris (White & Sweet, 1955). Such operations are palliative and often very effective in removing the pain, at least temporarily, but they do not affect the course of the underlying disorder. Similarly, the surgical relief of pain arising in abdominal viscera is best achieved by splanchnicectomy; however, the most common type of prolonged intractable abdominal pain is associated with malignancy in which invasion of tissues other than the viscera, such as nerves and the abdominal wall itself, is often responsible for the pain. Such sources of pain cannot be influenced by surgery of autonomic nerves or ganglia and require other measures for their relief (White, Smithwick, & Simeone, 1952).

5.10 Erythromelalgia

Erythromelalgia is a rare clinical syndrome characterized by a triad of redness, warmth, and burning pain in the extremities. Feet are most commonly affected but atypical cases involving proximal areas have been reported (e.g., head, neck region, and genitals). It is important to distinguish erythromelalgia in thrombocythemia, a platelet-mediated disorder responsive to aspirin, from primary erythermalgia/erythromelalgia (idiopathic or inherited disorder) and secondary erythermalgia/erythromelalgia associated with a variety of medical conditions.

The inherited form of primary erythromelalgia is an autosomal dominant small-fiber neuropathy caused by a gain-of-function mutation in the SCN9A, SCN10A, and SCN11A gene. These genes encode for the alpha subunit of the voltage-gated NaV 1.7, NaV 1.8, and NaV 1.9 sodium channel expressed in sympathetic ganglia. A gain-of-function mutation in the SCN9A gene causes nociceptive neuron firing at subthreshold stimuli and neuropathic pain (Baker & Nassar, 2020). Primary erythromelalgia often presents in children or adolescents. Classically, symptomatic episodes are triggered by exercise, exposure to heat and warm environments, and wearing tight-fitting shoes or clothes. Each episode may last minutes to days. The symptoms typically get relieved by exposure to cold and elevation of the limb. Treatment includes avoiding triggers and pharmacotherapy. Medications affecting the voltage-gated sodium channels (lidocaine, mexiletine, and carbamazepine) are often the first-line treatment in primary erythromelalgia. Other agents include beta-blockers, calcium channel blockers, tricyclics, and alpha-2-delta ligands. Primary erythromelalgia is difficult to manage and often resistant to treatment. Topical midodrine, 0.2%, may be effective in relieving the pain and well-tolerated (Davis, Morr, Warndahl, & Sandroni, 2015). Other topical treatments include lidocaine patches, compounded topical amitriptyline-ketamine, and topical capsaicin. Alternative therapies for refractory cases are epidural infusions, transcranial magnetic stimulation, injection of botulinum toxin, and sympathectomies. Histopathological findings in primary erythromelalgia are nonspecific and typically do not reveal vascular thrombi. A relative decrease in small nerve fiber density was noted in specimens from 13 of 16 patients in one study (Davis et al., 2006).

Secondary erythromelalgia is mostly associated with myeloproliferative disorders. Other underlying causes include infection (e.g., HIV, syphilis), diabetes mellitus, autoimmune diseases (systemic lupus erythematosus and rheumatoid arthritis), medications (e.g., nifedipine, verapamil, topical isopropanol, pergolide, simvastatin), gout, cancer, multiple sclerosis, thrombotic thrombocytopenic purpura, and various intoxications. Symptoms are often milder compared to primary erythromelalgia and are relieved after the treatment of the underlying disease. In myeloproliferative disorders, platelet activation causes thrombi and occlusion of arterioles and tissue capillaries causing hypoxia. Histopathological studies reveal vascular thrombi. Increased arteriovenous shunting may also be part of the pathogenesis of erythromelalgia. Rapid improvement with aspirin is typical of cases associated with thrombocythemia.

Sweating abnormalities are frequent in erythromelalgia. Thermoregulatory sweat testing results were abnormal in 28 (88%) of 32 patients, and quantitative sudomotor axon reflex test results were abnormal in 22 patients (69%). Abnormalities noted on thermoregulatory sweat testing varied from local hypohidrosis or anhidrosis to global anhidrosis (Davis, Genebriera, Sandroni, & Fealey, 2006).

5.11 Baroreceptor function and pain perception

Carotid and cardiopulmonary baroreceptor reflexes modulate autonomic output and prevent excessive fluctuations in arterial blood pressure to maintain intravascular volume. These baroreceptors also project to critical CNS regions that regulate somatosensory, somatomotor, and CNS arousal. Also, baroreceptor activity modulates the perception of pain and neuroimmune, neuroendocrine, and cognitive responses to physical and psychological stressors. Further research in this critical field will reduce the sequelae of surgical stress and improve chronic pain management and adverse surgical effects (Suarez-Roca et al., 2019). Piezo1 and Piezo2 receptors together are critical mechanoreceptors in baroreflex function. Another ion channel that is relevant is pain perception and regulation of blood pressure is the transient receptor potential vanilloid 1 (TRPV1) channel—the capsaicin receptor—that is expressed predominantly in sensory nerves. TRPV1 is activated by a variety of stimuli, including noxious temperature, changes in pH, as well as osmotic and mechanical stimuli. In the cardiovascular system, sympathetic nerves innervating blood vessels are sensitive to capsaicin, and TRPV1 activation regulates blood pressure (Phan et al., 2020). TRPV1 is also involved in fluid homeostasis and normal osmoregulatory responses. Targeting the TRPV1 channel has led to new therapeutic options for neuropathic pain.

References

Adrian, E. D., Cattell, Mc. K., & Hoagland, H. (1931). Sensory discharges in single cutaneous nerve fibers. *J. Physiol.*, *72*, 377–391.

Appenzeller, O. Post-traumatic headaches. In: Wolffs headache and other head pain. Dalessio, D.J. and S.D. Silberstein (Eds.) (6th ed.). New York and Oxford, Oxford University Press (1993) 365–383.

Appenzeller, O., & Martignoni, E. (1996). The autonomic nervous system and hypoxia: Mountain medicine. *J. Autonom. Nervous Syst.*, *57*, 1–12.

Baker, M. D., & Nassar, M. A. (2020). Painful and painless mutations of SCN9A and SCN11A voltage-gated sodium channels. *Pflugers Archiv: Eur. J. Physiol.*, *472*(7), 865–880.

Bayles, T. B., Judson, W. E., & Potter, T. A. (1950). Reflex sympathetic dystrophy of the upper extremity (hand-shoulder syndrome). *J. Amer. Med. Assoc.*, *144*, 537–542.

Bentley, F. H. (1948). The interpretation of visceral pain. *Ann. Royal College Surg. England*, *3*, 328–335.

Bentley, F. H., & Smithwick, R. H. (1940). Visceral pain produced by baloon distension of the jejunum. *Lancet*, *2*, 389–391.

Bonica, J. J. (1973). Causalgia and other reflex sympathetic dystrophies. *Postgrad. Med.*, *53*, 143–148.

Boniuk, M., & Schlezinger, N. S. (1961). Readers paratrigeminal syndrome. *Trans. Amer. Neurol. Assoc.*, *86*, 165–168.

Carstens, E. (1995). Neural mechanisms of hyperalgesia: peripheral or central sensitization? *News Physiol. Sci. Int. J. Physiol. Produced Jointly Int. Union Physiol. Sci. Amer. Physiol. Soc.*, *10*, 260–265.

Cayla, J., & Rondier, J. (1974). Algodystrophies réflexes des membres inférieurs d'origine vertébro-pelvienne. A propos de 23 cas. *La Semaine des Hopitaux de Paris*, *50*, 275–286.

Chorobski, J., & Penfield, W. (1932). Cerebral vasodilator nerves and their pathway from the medulla oblongata with observations on pial and intracerebral vascular plexus. *Arch. Neurol. Psychiatry (Chicago)*, *28*, 1257–1289.

Cohen, H. (1944). The mechanism of visceral pain. *Trans. Med. Soc. London*, *64*, 35.

Cronholm, B. (1951). Phantom limbs in amputees-study of changes in integration of centripetal impulses with special reference to referred sensations. *Acta Psychiatrica et Neurol. Scandinavica* (Suppl. 72), 1–316.

Davis, M. D., Genebriera, J., Sandroni, P., & Fealey, R. D. (2006). Thermoregulatory sweat testing in patients with erythromelalgia. *Arch. Dermatol.*, *142*(12), 1583–1588.

Davis, M. D., Morr, C. S., Warndahl, R. A., & Sandroni, P. (2015). Topically applied midodrine, 0.2%, an α1-agonist, for the treatment of erythromelalgia. *JAMA Dermatol.*, *151*(9), 1025–1026.

Davis, M. D., Weenig, R. H., Genebriera, J., Wendelschafer-Crabb, G., Kennedy, W. R., & Sandroni, P. (2006). Histopathologic findings in primary erythromelalgia are nonspecific: special studies show a decrease in small nerve fiber density. *J. Amer. Acad. Dermatol.*, *55*(3), 519–522.

De Takats, G. (1945). Causalgic stages in peace and war. *J. Amer. Med. Assoc.*, *128*, 699–704.

De Witt, J. D., & Thompson, S. W. (1975). *Readers syndrome: A case with maxillary sinusitis, Headache* (15, pp. 91–95).

Desmeules, J. A., Kayser, V., Weil-Fuggaza, J., et al. (1995). Influence of the sympathetic nervous system in the development of abnormal pain-related behaviors in a rat model of neuropathic pain. *Neuroscience*, *67*, 941–951.

Dickenson, A. H. (1994). NMDA receptor antagonists as analgesics. In: Progress in pain research and management. H. L. Fields and J. C. Liebeskind (Ed.). Seattle, IASP 1 173–187.

Doupe, J., Cullen, C. H., & Change, G. Q. (1944). Post-traumatic pain and the causalgic syndrome. *J. Neurol. Neurosurg. Psychiatry*, 7, 33–48.

Dubner, R., & Ruda, M. A. (1992). Activity dependent neuronal plasticity following tissue injury and inflammation. *Trends Neurosci.*, *15*, 96–102.

Freeman, L. W., Shumacker, H. B., Jr., & Radigan, L. R. (1950). A functional study of afferent fibers in peripheral sympathetic nerves. *Surgery*, 274–281.

Gardner, W. J., Stowell, A., & Dutlinger, R. (1947). Resection of the greater superficial petrosal nerve in the treatment of unilateral headache. *J. Neurosurg.*, *4*, 105–114.

Goh, E. L., Chidambaram, S., & Ma, D. (2017). Complex regional pain syndrome: A recent update. *Burns Trauma*, *5*, 2.

Granit, R., Leksell, L., & Skoglund, C. R. (1944). Fibre interaction in injured or compressed region of nerve. *Brain*, *67*, 125–140.

Hardy, J. D., Wolff, H. G., & Goodell, H. (1952). *Pain sensations and reactions*. New York: Williams 8 Wilkins.

Henriksen, O., Skagen, K., Amtorp, O., et al. (1981). Augmented vasoconstrictor response to changes in vascular transmural pressure in patients with essential hypertension. *Acta Physiol. Scandinavica*, *112*, 223–229.

Hertz, A. F. (1911). The Goulstonian lectures on the sensibility of the alimentary canal in health and disease. *Lancet, 1*, 1051–1056, 1119–1124; 1187–1193.

Herz, E., & Yahr, M. D. (1950). Painful sensory syndrome during nerve regeneration. *Trans. Amer. Neurol. Assoc., 75*, 213–218.

Hockaday, J. M., & Whitty, C. W. M. (1967). Patterns of referred pain in the normal subject. *Brain, 90*, 481–496.

Holden, W. D. (1948). Sympathetic dystrophy. *Arch. Surg., 57*, 373–384.

Horton, B. T. (1941). The use of histamine in the treatment of specific types of headaches. *J. Amer. Med. Assoc., 116*, 377–383.

Horton, B. T. (1956). Histaminic cephalgia: differential diagnosis and treatment. *Mayo Clinic Proc. Mayo Clinic, 31*, 325–333.

Kinsella, V. J. (1948). *The mechanism of abdominal pain*. Sydney: Australasian Medical Publishing Co. Ltd.

Kristensen, J. K. (1979). Local regulation of digital blood flow in generalized scleroderma. *J. Investig. Dermatol., 72*, 235–240.

Kunkle, E. C., & Anderson, W. B. (1960). Dual mechanism of eye signs of headache in cluster pattern. *Trans. Amer. Neurol. Assoc., 85*, 75–79.

Kunkle, E. C., Pfeiffer, J. B., Jr., Willhoit, W. M., & Hamrick, L. W., Jr. (1952). Recurrent brief headache in cluster pattern. *Trans. Amer. Neurol. Assoc., 77*, 240–243.

LaMotte, R. H., Shain, C. N., Simone, D. A., & Tsai, E. F. P. (1991). Neurogenic hyperalgesia: psychophysical studies of underlying mechanisms. *J. Neurophysiol., 66*, 190–211.

Law, W. R., & Nelson, E. R. (1968). Internal carotid aneurysm as a cause of Readers paratrigeminal syndrome. *Neurology (Mirtneap.), 18*, 43 46.

Lennander, K. G. (1901). Über die Sensibilität der Bauchhöhle und über locale und allgemeine Anästhesie bei Bruch- und Bauchoperationen. *Zentralblatt Fur Chirurgie, 28*, 209–223.

Lewis, T. (1929). Experiments relating to peripheral mechanism involved in spasmodic arrest of the circulation in the fingers. A variety of Raynauds disease. *Heart (British Cardiac Society), 15*, 7–101.

Lewis, T. (1942). *Pain*. London: Macmillan.

Mayfield, F. H., & Devine, J. W. (1945). Causalgia. *Surg. Gynecol. Obstetr., 80*, 631–635.

Melzack, R. (1971). Phantom limb pain: Implications for treatment of pathologic pain. *Anesthesiology, 35*, 409–411.

Melzack, R., & Wall, P. D. (1965). Pain mechanisms: A new theory. *Science (New York, N.Y.), 150*, 971–979.

Mendell, L. M., & Wall, P. D. (1965). Responses of single dorsal cord cells to peripheral cutaneous unmyelinated fibers. *Nature, 206*, 97–99.

Moore, R. M., & Singleton, A. O., Jr (1933). Studies on the pain sensibility of arteries. II. Peripheral paths of afferent neurones from the arteries of the extremities and of the abdominal viscera. *Amer. J. Physiol., 104*, 267–275.

Olsen, N., Petring, O. U., & Rossing, N. (1987). Exaggerated postural vasoconstrictor reflex in Raynauds phenomenon. *Brit. Med. J., 294*, 1186–1188.

Orbeli, L. A. (1923). Die sympathische innervation der Skelettmuskeln. *Izu. Petrog. Nauck Ins. P.F. Lesgafta, 6*, 187–197.

Phan, T. X., Ton, H. T., Gulyás, H., Pórszász, R., Tóth, A., Russo, R., . . . Ahern, G. P. (2020). TRPV1 expressed throughout the arterial circulation regulates vasoconstriction and blood pressure. *J. Physiol., 598*(24), 5639–5659.

Price, D. D., J. Mao, & D. J. Mayer (1994). Central neural progress. In pain research and management. H. L. Fields and J. C. Liebeskind (Eds.). Seattle, IASP 1 61–84.

Raeder, J. G. (1924). Paratrigeminal paralysis of oculopupillary sympathetic. *Brain, 47*, 149–158.

Raskin, N. H., Levinson, S. A., Hoffman, P. M., Pickett, J. B. E., III, & Fields, H. L. (1974). Postsympathectomy neuralgia: amelioration with diphenylhydantoin and carbamazepine. *Ann. Surg.*, *128*(1), 75–78.

Ray, B. S., & Neill, C. L. (1947). Abdominal visceral sensation in man. *Ann. Surg.*, *126*, 709–724.

Raynaud, M. (1862). De l'asphyxie locale et de la gangrène symétrique des extrémités. Paris, Rignoux.

Richards, R. L. (1967). Causalgia, a centennial review. *Arch. Neurol. Psychiatry (Chicago)*, *16*, 339–350.

Ruch, T.C. (1949). Visceral sensation and referred pain. In: Textbook of physiology. J.F. Fulton (Ed.) (16th ed.). Saunders, Philadelphia.

Sax, F. L., Cannon, R. O., III, Hanson, C., et al. (1987). Impaired forearm vasodilator reserve in patients with microvascular angina. *New England J. Med.*, *317*, 1366–1370.

Schwartzman, R. J. (1993). Reflex sympathetic dystrophy. *Current Opin. Neurol. Neurosurg.*, *6*, 531–536.

Selkowitz, D. M. (1992). The sympathetic nervous system in neuromotor function and dysfunction and pain: A brief review and discussion. *Funct. Neurol.*, 7, 89–95.

Shumacker, H. B., Jr., & Abramson, D. I. (1949). Posttraumatic vasomotor disorders with particular reference to late manifestations and treatment. *Surg. Gynecol. Obstetr.*, *88*, 417–434.

Simone, D. A., Sorkin, L. S., Oh, U., et al. (1991). Neurogenic hyperalgesia: Central neural correlates in responses of spinothalamic tract neurons. *J. Neurophysiol.*, *66*, 228–246.

Sinclair, D. C., Weddell, G., & Feindel, W. H. (1948). Referred pain and associated phenomena. *Brain*, *71*, 148–211.

Speigel, I. J., & Milowsky, J. L. (1945). Causalgia: A preliminary report of nine cases successfully treated by surgical and chemical interruption of sympathetic pathways. *J. Amer. Med. Assoc.*, *127*, 9–15.

Sterling, P. (1973). Research note. Referred cutaneous sensation. *Exp. Neurol.*, *41*, 451–456.

Suarez-Roca, H., Klinger, R. Y., Podgoreanu, M. V., Ji, R.-R., Sigurdsson, M. I., Waldron, N., ... Maixner, W. (2019). Contribution of baroreceptor function to pain perception and perioperative outcomes. *Anesthesiology*, *130*(4), 634–650.

Sudeck, P. (1900). Über die akute enzundliche Knochenatrophie. *Archiv für Chirurgie*, *62*, 147–156.

Sutton, D. C., & Leuth, H. C. (1930). Pain. *Arch. Internal Med.*, *45*, 827–867.

Terkelsen, A. J., Mølgaard, H., Hansen, J., Finnerup, N. B., Krøner, K., & Jensen, T. S. (2012). Heart rate variability in complex regional pain syndrome during rest and mental and orthostatic stress. *Anesthesiology*, *116*(1), 133–146.

Theobald, G. W. (1949). The role of the cerebral cortex in the appreciation of pain. *Lancet*, *2*(41–47), 94–96.

Threadgill, F. D. (1947). Afferent conduction via the sympathetic ganglia innervating the extremities. *Surgery*, *21*, 569–574.

Torebjork, H. E., Lundberg, L. E. R., & LaMotte, R. H. (1992). Central changes in processing of mechanoreceptive input in capsaicin-induced secondary hyperalgesia in humans. *J. Physiol. (London)*, *448*, 765–780.

Wainapel, S. F. (1984). Reflex sympathetic dystrophy following traumatic myelopathy. *Pain*, *18*, 345–349.

Walker, A. E., & Nulson, F. (1948). Electrical stimulation of the upper thoracic portion of the sympathetic chain in man. *Arch. Neurol. Psychiatry*, *59*, 559–560.

Weddell, G., Sinclair, D. E., & Feindez, W. H. (1948). An anatomical basis for alterations in quality of pain sensibility. *J. Neurophysiol.*, *11*, 99–109.

White, J. C., Smithwick, R. H., & Simeone, F. A. (1952). *The autonomic nervous system. Anatomy Physiol. Surgical Appl.* (3rd ed.). New York: Macmillan.

White, J. C., & Sweet, W. H. (1955). *Pain, its mechanisms and neurosurgical control.* Springfield, Ill: C.C. Thomas.

White, J. E., Heroy, W. W., & Goodman, E. N. (1948). Causalgia following gunshot injuries of nerves: role of emotional stimuli and surgical cure through interruption of diencephalic efferent discharge by sympathectomy. *Ann. Surg.*, *128*, 161–183.

Wiertz-Hoessels, E. L. (1964). Influence of trophic disturbances on skeletal muscle innervation. *Nature (Lond.)*, *204*, 540–542.

Wilder, R. T., Berde, C. B., Wolohan, M., et al. (1992). Reflex sympathetic dystrophy in children: clinical characteristics and follow-up of 70 patients. *J. Bone Joint Surg. (British volume)*, *74*, 910–919.

Willis, W. D. (1992). *Hyperalgesia and allodynia.* New York: Raven.

Wilson, R. B., & Mussey, R. D. (1947). Nervous pathways involved in pelvic brain. Some anatomic and physiologic aspects. *J. Amer. Med. Assoc.*, *134*, 857–861.

Wolff, H. G. (1963). *Headache and other head pain* (2nd ed.). London: Oxford University Press.

Wolff, H. G., & Hardy, J. D. (1947). On the nature of pain. *Physiol. Rev.*, *27*, 167–199.

Yokota, T., Furukawa, T., & Tsukagoshi, H. (1989). Motor paresis improved by sympathetic block: A motor form of reflex sympathetic dystrophy. *Arch. Neurol.*, *46*, 683–687.

Zimmermann, M., & Herdegen, T. (1994). Control of gene transcription by Jun and Fos proteins in the nervous system: beneficial or harmful molecular mechanisms of neuronal response to noxious stimulation? *Amer. Pain Soc. J.*, *3*, 33–48.

CHAPTER 6

Biofeedback and operant conditioning

If a response of an organism is followed by a positive reinforcer such as money, food, praise, or prize, then the probability that the response will recur is increased. This is the underlying principle of operant conditioning. The same applies if the behavior of the organism is contingent upon avoidance of negative reinforcers. This principle has been shown to work for somatic responses of every kind and almost in every organism and it can also be applied to the autonomic nervous system.

Operant conditioning of heart rate in humans was reported in 1962 (Shearn, 1962). This work has been duplicated in numerous experimental animal and human situations. In early studies, it was necessary to show that the autonomic nervous system obeys the same laws of learning as the somatic nervous system, and no voluntary or somatic responses such as muscle, tension, or respiration, which could interfere with the responses, could be allowed during the experiment. Animals were therefore curarized, which abolished skeletal muscle activity, in order to eliminate all possible somatic influences on the experimental situation. This led to serious problems in the search for positive reinforcement in animals incapable of eating, drinking, seeking shelter or safety; all proven reinforcers (Trowill, 1967). A solution to this difficulty was the use of stimulation of the pleasure center in the hypothalamus as a reinforcer. With this technique, differential learning to speed or slow the heart has been reported. Results were further improved by giving the animal a signal when its heart was doing the correct thing. This feedback seemed to speed and improve the result of conditioning. It was soon shown that visceral learning was much like somatic learning. Conditioning of vasomotor responses was possible so that animals could vasodilate blood vessels in the skin in one ear and simultaneously constrict it in the other (DiCara & Miller, 1968). Similarly, blood pressure increases or decreases could be learned and these changes in pressure were independent of heart rate changes.

In earlier human studies (Engel & Hansen, 1966; Hnatiow & Lang, 1965), positive results were also obtained, but the most important factor

Introduction to Clinical Aspects of the Autonomic Nervous System
DOI: https://doi.org/10.1016/B978-0-323-96011-3.00002-4

in human experiments was the provision of a signal or feedback about the results. The performance in humans, like in animals, is improved if subjects have feedback about correct or incorrect responses. This enhances performance and speeds learning far above what is possible by those who have no knowledge of what it is that is wanted of them. Biofeedback is of proven importance (Brener & Hothersall, 1966) and it has become a standard procedure in operant conditioning of the autonomic nervous system.

Blood pressure changes have been produced using shock avoidance in animals. Making avoidance of an electric shock contingent upon an increase in pressure, the blood pressure could be elevated 20–40 mm of mercury. Reversal of the experimental situation, that is avoiding shock by decreasing the pressure, led to a drop in systemic pressure of 10–20 mm of mercury (Benson, Herd, Morse, & Kelleher, 1969). The reversal of the desired response using the same stimulus assures that the learning process, rather than some other variables, accounts for the observed changes. Biofeedback has been combined with yoga in the treatment of hypertension (Patel & North, 1975). In several studies using either yoga or biofeedback, modification of behavior has been obtained which was designed to maintain a lower level of arousal. This was effective in reducing resting blood pressure over long periods in human subjects. The efficacy of this treatment has been confirmed in other studies. After 6 weeks of treatment by yoga relaxation together with biofeedback, compared to control placebo therapy employing simple general relaxation only, a reduction in blood pressure was obtained in both groups; however, the treated group achieved a significantly greater decrease in blood pressure. In patients achieving a significant reduction in blood pressure, it was also shown that stress caused a smaller rise in systemic pressure than stress in controls. These therapeutic maneuvers are important in the management of hypertension and other disorders because they are without undesirable side effects and are within everyone's capacity. In the control of moderate or severe hypertension, patients were told they would receive money if they came to the laboratory and had their pressure measured at hourly sessions. Pressures were recorded with a cuff system and after some control sessions, conditioning began where a lowered systolic pressure was signaled by a flash of light or brief tone. The patients were told that their signals were good and they should attempt to turn the signals on as frequently as possible. When they achieved 20 signals they were given the money. Average pressure after control sessions was 164.9 mm of mercury and after conditioning sessions 148.4 mm of mercury systolic (Benson, Shapiro, Tursky, & Schwartz, 1971). Control of premature ventricular contractions by keeping heart rate within clearly restricted limits and thereby reducing the incidence of premature ventricular contraction has also been achieved in humans (Weiss & Engel, 1971). Four of five patients who responded initially maintained a decrease in the incidence of premature ventricular contraction for 21 months.

Sympathetic activity in transcendental meditators was measured by the excretion of catecholamines and vanillylmandelic acid. The meditators included 10 long-term meditators and 10 advanced meditators. Both groups subjected themselves to physical exercise of moderate degree and the changes in catecholamine excretion during rest, meditation, or reading before and after exercise were measured. Surprisingly, the effect of transcendental meditation on the autonomic nervous system is one of enhancement of sympathetic activity which is more pronounced in advanced meditators. The transcendental state is, therefore, not merely a trophotropic parasympathetic overactivity, but sympathetic influences can also be detected. Whether a combination of feedback with meditation techniques might change the balance between parasympathetic and sympathetic activity during meditation is not clear (Lang, Dehof, Meurer, & Kaufmann, 1979). Biofeedback has also been used in the treatment of patients with occlusive arterial disease. Such patients were given signals from the electromyogram and skin temperature during several training sessions and their maximum treadmill walking time was measured before and after treatment. In this study, claudication responded to therapy and 9 of 11 patients were able to increase maximum treadmill times by more than 8 minutes. This improvement was associated with significant falls in resting and exercise systolic blood pressures and an increase in postexercise ankle blood pressure and ankle-arm blood pressure ratios. The findings were in keeping with the suggestion that biofeedback training reduces resistance around sites of vascular occlusion and may offer nonsurgical therapy in selected patients with occlusive arterial disease (Greenspan, Lawrence, Esposito, & Voorhees, 1980).

Attempts have been made to clarify the efficacy of various feedback signals. Two different methods have been used. Binary signals give the subject only a clue when his performance exceeds a preset level of the heart rate for example and their success signal is absent or replaced by a different clue when they fail to achieve the criterion level. In this system, the subject knows only if their current heart rate is high or low. They know nothing about their relative degree of success in achieving the desired end or about a possible total failure of control. On the other hand, when information is in an analog form, the subject attends to signals, which vary with, for example, beat-to-beat changes in heart rate, and the criterion level may be displayed at the same time. Analog information, therefore, supplies continuous feedback to the subject and also gives a proportional representation of performance. The relative efficacy of the two methods has been investigated (Lang & Twentyman, 1974). In human subjects, no differences between binary and analog biofeedback procedures were noted for cardiac slowing. However, the analog biofeedback produced a significantly greater instructed acceleration of heart rate than the binary method and it appeared that instructed feedback-assisted speeding and slowing of the heart in humans might be mediated by different physiologic mechanisms. It is surprising that feedback signals presented monaurally have differential effects depending upon whether or not they are presented to the left or right ear. Lateralized specialization of functions has repeatedly been demonstrated. The different effects on cerebral hemispheres during the processing of a variety of stimuli include lateralization for voluntary increases in heart rate by biofeedback signals. Signals given to the right ear are significantly more successful than left ear feedback in increasing voluntarily heart rate. This is surprising since the expected left ear (right hemisphere) superiority in biofeedback training did not materialize. The findings are, however, consistent with a right ear perceptual dominance in the processing of auditory stimuli which would be in keeping with the close relationship of the right ear to the left hemisphere which is dominant for verbal stimuli. Lateralization of cardiovascular function and blood flow has been demonstrated previously and it is therefore not surprising that biofeedback auditory training is also dominantly situated on the right (Greenstadt, Schuman, & Shapiro, 1978).

Other clinical applications of biofeedback techniques have been explored. Electromyographic biofeedback in the treatment of tension headache is eminently successful. Similarly, temperature biofeedback to allow a volitional increase in the temperature of the hands by vasodilatation of hand vessels is successful in the treatment of migraine headaches. The difference in training technique may be based on abnormalities in vasomotor function which have been demonstrated in migraineurs, but not in patients with other headaches. Migraine patients have impaired reflex vasodilatation in response to heat loads. Impairment of this normal vasodilator response can be demonstrated during headache-free intervals (Fig. 6.1) (Elliott, Frewin, & Downey, 1974). Biofeedback has also been tried in Raynaud's disease and studies suggest that the technique might be

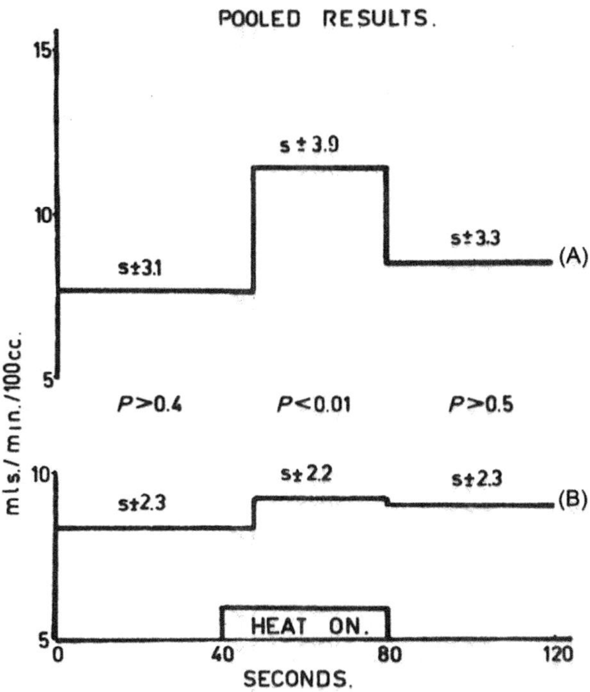

Figure 6.1 Reflex vasodilatation in hands of 10 controls, (A) (upper graph) and 10 migraineurs, (B) (lower) graph). The pooled results show significant vasodilatation in controls only. There was no statistically significant difference in baseline hand blood flows in the two groups. *From Appenzeller et al. (1963).*

useful in reducing changes in skin temperature produced by cold. Biofeedback techniques in behavioral therapy have been well established in insomnia and phobias of various kinds. Changes in the electrical activity of the brain recorded by electroencephalography have also been achieved and some epileptic discharges respond to the biofeedback of the EEG to the patients. Patients with incontinence have become continent again after biofeedback training of contractions of the external and relaxation of the internal anal sphincters.

There are a number of theoretical issues that have not been resolved. One is whether the autonomic nervous system obeys the same laws of learning as the somatic nervous system and is therefore amenable to instrumental conditioning. Biofeedback techniques at present are less concerned with theoretical issues, this simply being a tool to improve results in some pathologic conditions in humans. The declining interest in theoretical problems is perhaps related to the failure to repeat the original observations on curarized animals and some other experiments. Whether or not somatic and autonomic responses are entirely separable might be an academic question. In the central nervous system, a great deal of overlap of somatic and autonomic controls could exist. Some evidence for this comes from animal studies. It was shown that an increase in blood flow in the contralateral immobilized leg can be obtained by electrical stimulation of the appropriate area of the motor cortex and that no change in blood flow in the arm or ipsilateral leg occurred. It seems that the intent or perhaps the effort to make movement may be accompanied by inescapable cardiovascular responses. The central command, therefore, appears to be crucial in both somatic and autonomic responses. An experiment to show this in humans was carried out (Goodwin, McCloskey, & Mitchell, 1972). Subjects had to maintain constant tension in the biceps or triceps muscles of an arm and were given visual feedback of their success. Cardiovascular and respiratory changes were monitored during the task. The experiment was carried out in such a way that the central command for maintenance of tension changed without altering the tension or energy expenditure of the muscles. Comparisons with accompanying cardiorespiratory responses were made. The change in central command was achieved by applying vibration to tendons of agonists or antagonists. This is a stimulus to the primary endings in muscle spindles. If a muscle is held at a given tension by central command and vibration is applied a reflex increase in tension of the muscle occurs by a monosynaptic connection of the muscle spindle. The central command will therefore have to be decreased in order to maintain the tension at the desired level. By applying the vibration to antagonist muscles, the tension in the agonists would be reflexively decreased and the central command would have to be increased in order to maintain a given tension level in the agonist. The results of this experiment were clear-cut. When the central command was increased, cardiorespiratory variables also increased. When the central command was decreased, they also decreased. It is to be noted that these changes were produced without increased work and without a change in tension produced in the muscle. This clearly demonstrates that through a central effort, circulatory and respiratory responses can be dissociated from muscle movements or tension. This supports old observations that both movements and circulatory responses may be initiated from the same central neural structures.

Heart rate variability biofeedback has gained increasing attention as a potential treatment in a variety of chronic diseases. Studies have linked vagally mediated heart rate variability to self-regulatory capacity (Segerstrom & Nes, 2007), emotion regulation, social interactions, and coping skills. The positive effects of heart rate variability biofeedback are related to regular practice. Positive outcomes with heart rate variability biofeedback were reported in patients with chronic pain (Dobbin, Dobbin, Ross, Graham, & Ford, 2013; Hassett et al., 2007); however, another study was negative (Weeks, Whitney, Tindall, & Carter, 2015). Studies have also reported positive outcomes on anxiety, depression, sleep quality, inflammatory state, and cognitive performance (Fournié et al., 2021).

Galvanic skin responses (GSR) are indices of peripheral sympathetic nervous activity, reflecting sweat gland function. GSR is a sensitive measure of emotional arousal and attention. Studies have demonstrated the benefit of GSR biofeedback therapy for the management of epilepsy, possibly via modulation of thalamocortical sensory information as well as changes in activity in the ventromedial prefrontal cortex/orbitofrontal cortex (Nagai, Critchley, Featherstone, Trimble, & Dolan, 2004). An increase in sympathetic activity reduces cortical excitation (Nagai, Goldstein, Critchley, & Fenwick, 2004), and GSR biofeedback training leads to a reduction of cortical potential amplitude both in healthy participants and patients with epilepsy (Nagai, Critchley, Rothwell, Duncan, & Trimble, 2009).

6.1 Interoception

Interoception refers to the process by which the nervous system senses and integrates information about the inner state of the body. Interoceptive information is often first processed in subcortical structures. Interoceptive afferent inputs are conveyed by lamina I in the dorsal horn or the nucleus of the solitary tract and are relayed either directly or via the parabrachial nucleus to the hypothalamus, amygdala, and eventually the insular cortex (via the thalamus). Some afferent signals can induce immediate reflex adjustments of the end-organs via projections from specific nucleus of the tractus solitarius subnuclei to premotor sympathetic neurons of the rostral ventrolateral medulla including the C1 area, vagal nuclei, or respiratory pattern generators. Another major input to the central autonomic network originates from forebrain areas involved in the processing of interoceptive signals, including the amygdala and the bed nucleus of the stria terminalis, anterior insula, and anterior cingulate cortex. The anterior insula receives inputs via the mid-insula from the dorsal posterior insula, which is the crucial cortical node in the interoceptive system. The anterior insula integrates interoceptive signals with emotional and cognitive processing and is involved in the attention of the physiological state of the body. The ventral anterior insula (frontal insular cortex) and the pregenual anterior cingulate cortex contain large layer V projection neurons, called von Economo neurons, which project to the periaqueductal gray, parabrachial nucleus, and other areas of the central autonomic network. Recent models of interoception and emotion incorporate the Bayesian concepts of predictive coding. In one such model, projection neurons in the anterior insula and pregenual anterior cingulate cortex targeting autonomic control areas generate predictions about physiological states that are sent via corticocortical projections to the posterior insular cortex. In the posterior insula, these visceromotor simulations are compared to afferent interoceptive inputs, and this comparison generates a prediction error signal that is sent back to the anterior insula and anterior cingulate cortex to refine interoceptive predictions. These prediction errors trigger both emotional feelings and homeostatic responses (Barrett & Simmons, 2015; Seth & Friston, 2016). Interoceptive information may not require higher levels of cortical processing but in other cases, interoceptive signals may engage higher-order processing at perceptual, cognitive, and/or affective levels, thus rising to the level of conscious awareness (Denton et al., 1999). Regulatory signals of interoception also involve descending signals from the brain to the periphery. In nonneural pathways, regulatory signals are delivered to the periphery via the vascular or lymphatic systems, whereas in neural pathways, the final effectors are sympathetic or parasympathetic ganglion neurons, which receive input from the brain through the spinal or vagal/cranial efferent nerves and directly synapse with internal organ nonneural cells. The impact of the regulatory signals may be best measured by the responses in the target organs (Chen et al., 2021).

Biofeedback training, like relaxation techniques, provides no shortcut to a cure; like dieting or exercise, it must be learned and practiced and the overemphasis of the use of this technique in many disorders and in all patients should be discouraged. Though theoretically, every human being is inherently capable of mastering the technique whereby feedback instruments accelerate the learning process, in practice, few are capable of self-regulating both psyche and soma. Moreover, this technique is clearly of no use when permanent alterations due to disease have occurred. In psychosomatic illness, organic damage may eventually be irreversible.

Nevertheless, biofeedback training leads to a more responsible attitude for the subjects' own health and it, like meditation, yoga, or progressive relaxation exercises, provides patients with a technique of perceiving their bodily functions and shifts the responsibility for health maintenance at least in part onto the patient. It also encourages the perception of the relationship between health and life span and decreases the expectation of instant pharmacologic cures for many psychosomatic symptoms.

One of the debates about interoceptive awareness is whether it is good to have more of it or less, and the brain in some cases may be "hyperfocused" on bodily sensations. Studies should further investigate neural circuits involved in interoception, bio/neurofeedback, and cognitive training at the neuroanatomical and functional levels as well as the potential therapeutic role of biofeedback and/or cognitive coping training in the management of autonomic disorders including disorders of orthostatic intolerance (e.g., postural tachycardia syndrome), orthostatic hypotension, and syncope.

References

Barrett, L. F., & Simmons, W. K. (2015). Interoceptive predictions in the brain. *Nat. Rev. Neurosci.*, *16*(7), 419–429.

Benson, H., Herd, J. A., Morse, W. H., & Kelleher, R. T. (1969). Behavioral induction of arterial hypertension and its reversal. *Amer. J. Physiol.*, *217*(1), 30–34.

Benson, H., Shapiro, D., Tursky, B., & Schwartz, G. E. (1971). Decreased systolic blood pressure through operant conditioning techniques in patients with essential hypertension. *Science (New York, N.Y.)*, *173*(3998), 740–742.

Brener, J., & Hothersall, D. (1966). Heart rate control under conditions of augmented sensory feedback. *Psychophysiology*, *3*(1), 23–28.

Chen, W. G., Schloesser, D., Arensdorf, A. M., Simmons, J. M., Cui, C., Valentino, R., ... Langevin, H. M. (2021). The emerging science of interoception: sensing, integrating, interpreting, and regulating signals within the self. *Trends Neurosci.*, *44*(1), 3–16.

Denton, D., Shade, R., Zamarippa, F., Egan, G., Blair-West, J., McKinley, M., ... Fox, P. (1999). Neuroimaging of genesis and satiation of thirst and an interoceptor-driven theory of origins of primary consciousness. *Proc. Natl. Acad. Sci. USA*, *96*(9), 5304–5309.

DiCara, L. V., & Miller, N. E. (1968). Instrumental learning of vasomotor responses by rats: learning to respond differentially in the two ears. *Science (New York, N.Y.)*, *159* (3822), 1485–1486.

Dobbin, A., Dobbin, J., Ross, S. C., Graham, C., & Ford, M. J. (2013). Randomised controlled trial of brief intervention with biofeedback and hypnotherapy in patients with refractory irritable bowel syndrome. *J. Royal Col. Phys. Edinburgh*, *43*(1), 15–23.

Elliott, K., Frewin, D. B., & Downey, J. A. (1974). Reflex vasomotor responses in the hands of patients suffering from migraine. *Headache*, *13*(4), 188–196.

Engel, B. T., & Hansen, S. P. (1966). Operant conditioning of heart rate slowing. *Psychophysiology*, *3*(2), 176–187.

Fournié, C., Chouchou, F., Dalleau, G., Caderby, T., Cabrera, Q., & Verkindt, C. (2021). Heart rate variability biofeedback in chronic disease management: A systematic review. *Complem. Therap. Med.*, *60*, 102750.

Goodwin, G. M., McCloskey, D. I., & Mitchell, J. H. (1972). Cardiovascular and respiratory responses to changes in central command during isometric exercise at constant muscle tension. *J. Physiol.*, *226*(1), 173–190.

Greenspan, K., Lawrence, P. F., Esposito, D. B., & Voorhees, A. B. (1980). The role of biofeedback and relaxation therapy in arterial occlusive disease. *J. Surg. Res.*, *29*(5), 387–394.

Greenstadt, L., Schuman, M., & Shapiro, D. (1978). Differential effects of left vs right monaural biofeedback for heart rate increase. *Psychophysiology*, *15*(3), 233–238.

Hassett, A. L., Radvanski, D. C., Vaschillo, E. G., Vaschillo, B., Sigal, L. H., Karavidas, M. K., ... Lehrer, P. M. (2007). A pilot study of the efficacy of heart rate variability (HRV) biofeedback in patients with fibromyalgia. *Appl. Psychophysiol. Biofeedback*, *32* (1), 1–10.

Hnatiow, M., & Lang, P. J. (1965). Learned stabilization of cardiac rate. *Psychophysiology*, *1* (4), 330–336.

Lang, P. J., & Twentyman, C. T. (1974). Learning to control heart rate: binary vs analogue feedback. *Psychophysiology*, *11*(6), 616–629.

Lang, R., Dehof, K., Meurer, K. A., & Kaufmann, W. (1979). Sympathetic activity and transcendental meditation. *J. Neural Transm. (Vienna, Austria: 1996)*, *44*(1–2), 117–135.

Nagai, Y., Critchley, H. D., Featherstone, E., Trimble, M. R., & Dolan, R. J. (2004). Activity in ventromedial prefrontal cortex covaries with sympathetic skin conductance level: A physiological account of a "default mode" of brain function. *Neuroimage*, *22* (1), 243–251.

Nagai, Y., Critchley, H. D., Rothwell, J. C., Duncan, J. S., & Trimble, M. R. (2009). Changes in cortical potential associated with modulation of peripheral sympathetic activity in patients with epilepsy. *Psychosomat. Med.*, *71*(1), 84–92.

Nagai, Y., Goldstein, L. H., Critchley, H. D., & Fenwick, P. B. (2004). Influence of sympathetic autonomic arousal on cortical arousal: implications for a therapeutic behavioural intervention in epilepsy. *Epilepsy Res.*, *58*(2–3), 185–193.

Patel, C., & North, W. R. (1975). Randomised controlled trial of yoga and bio-feedback in management of hypertension. *Lancet*, *2*(7925), 93–95.

Segerstrom, S. C., & Nes, L. S. (2007). Heart rate variability reflects self-regulatory strength, effort, and fatigue. *Psychol. Sci.*, *18*(3), 275–281.

Seth, A. K., & Friston, K. J. (2016). Active interoceptive inference and the emotional brain. *Philos. Trans. Royal Soc. Lond. Ser B Biol. Sci.*, *371*, 1708.

Shearn, D. W. (1962). Operant conditioning of heart rate. *Science (New York, N.Y.)*, *137* (3529), 530–531.

Trowill, J. A. (1967). Instrumental conditioning of the heart rate in the curarized rat. *J. Comparat. Physiol. Psychol.*, *63*(1), 7–11.

Weeks, D. L., Whitney, A. A., Tindall, A. G., & Carter, G. T. (2015). Pilot randomized trial comparing intersession scheduling of biofeedback results to individuals with chronic pain: Influence on psychologic function and pain intensity. *Amer. J. Phys. Med. Rehabil. Assoc. Acad. Physiatrists*, *94*(10 Suppl 1), 869–878.

Weiss, T., & Engel, B. T. (1971). Operant conditioning of heart rate in patients with premature ventricular contractions. *Psychosomat. Med.*, *33*(4), 301–321.

CHAPTER 7

Aging and exercise

7.1 Aging and the autonomic nervous system

7.1.1 Introduction

Quality of life is the primary goal of the management of the elderly. The cure of disease is not always possible, and alleviation of symptoms can often be achieved by simple measures. Aging is generally thought to be associated with a decline of autonomic nervous system function. Research on this and other nervous system functions, for example, odor perception, has shown that decline with advancing age is by no means uniform. Differences between age groups are well-established, but most studies have ignored substantive differences within age groups. This has led to the establishment of the gerontology of the usual, while totally neglecting the role of psychosocial and physiologic factors, which impinge upon the aging process and may slow or accelerate it. To understand the aging of the autonomic nervous system or its resistance to age, it is necessary to explore the central fact of heterogeneous responsiveness of the organism.

Heterogeneity in the nervous system is evident in the structure and function of the aging brain. Neuronal cell loss, which is general in some regions, is highly selective in other regions; for example, neurofibrillary tangles, even in the absence of dementia, are primarily found in some cells only. Neuropsychological assessment shows a general decline in intellectual function, yet some intellectual abilities remain unimpaired and some even improve throughout the life span. Surveys of whole-brain sections stained for myelin show so-called myelogenetic cycles that reflect a continuously evolving myelination of various tracts in the central nervous system of man up to the eighth decade.

In the autonomic nervous system, there are indications that some function declines, yet others become exaggerated. For example, adrenergic discharge in elderly subjects is often markedly enhanced, and the adrenergic innervation of blood vessels is increased in vasa nervorum (Appenzeller, Dhital, Cowan, & Burnstock, 1984). The autonomic function is also modulated by factors that are extrinsic to the aging process,

Introduction to Clinical Aspects of the Autonomic Nervous System
DOI: https://doi.org/10.1016/B978-0-323-96011-3.00004-8

for example, physical fitness and the secretion of stress hormones. To focus solely on average deficits leads to a picture of inescapable decline with advancing age. But it is necessary to explore what sustains autonomic nervous system function in trained individuals of advanced age and what debilitates medicated or ordinary aging subjects. This may lead to insights on the management of normal aging and improved quality of life.

Given the evidence in favor of heterogeneous effects of aging, it is puzzling that claims are still being made that age-related deficits of autonomic nervous system function show uniformity across the diversity of clinical deficits. If this uniformity were indeed found, then the observed effects of aging of the autonomic nervous system would be contrary to those in other parts of the nervous system.

A closer examination of the changes reported suggests that many are the result simply of a sedentary way of life rather than the aging process. Autonomic deterioration is promoted by disease. Some 80% of people over the age of 65 years have one or more diseases, and almost all subjects of that age have some pathology that has not progressed to manifest clinical symptoms and is as yet not classifiable as disease. Moreover, those who do have clinically recognizable diseases are often encumbered by medications that further interfere with autonomic function.

It is well recognized that diversity is more marked with increasing age. Thus, statistical normality is a physiologic aberration. While it is indisputable that advanced age is a risk factor for chronic diseases and decay in autonomic function, it is clear, however, that people age physiologically at different rates, and that this makes it difficult to define the population for a particular study of normal aging.

Since most older people have a diversity of chronic diseases, the study of healthy older subjects is the study of an exceptional group, and even more exceptional is the study of the trained elderly. To use such groups for the assessment of autonomic normality is not justifiable. Nevertheless, to ignore the heterogeneity of the elderly is to support the gerontology of the usual and ignore the gerontology of the ideal.

7.1.2 Nutritional effects on autonomic nervous system function

A decrease in protein intake with age was found. In men, this is 1.1 g/day/year, but less in women (0.6 g/day/year). Daily energy intake is also lost to a greater extent in men (12 kcal/year) than women (4 kcal/year) (Garry, Rhyne, & Haliova, 1989). These decreases in nutrition with advancing age may affect neurotransmitter function and, through their influence on energy metabolism, impair autonomic function. It is not, however, clear from such reports that longitudinal study in healthy elderly men and women does allow for the heterogeneous influences on autonomic function in individual aging subjects.

A protein meal consumed at breakfast induces more fatigue and sleepiness than an isocaloric carbohydrate meal. In the evening, however, these effects are reversed when a carbohydrate meal induces more fatigue. Young and old subjects have significant differences in calorie and carbohydrate intake and mood and activity, even when selecting nutrients from identical available foods and while living in an identical controlled environment (Lieberman, Wurthman, & Teichert, 1989). The interrelation between protein and carbohydrate foods and behavior is well-documented and is mainly due to changes that these specific nutrients induce in serotonin synthesis and secretion. Since serotoninergic neurons are believed to be important in the regulation of behavior, sleep, pain sensitivity, depression, and autonomic function, changes in nutritional habits with age are clearly of importance in autonomic responsiveness. The L-tryptophan/large neutral amino acids ratio represents the relative availability of plasma L-tryptophan for crossing the blood-brain barrier and is thought to be the best indicator of brain serotonin synthesis. For example, the consumption of a carbohydrate-rich meal results in a rise in the plasma tryptophan/large neutral amino acid ratio. This increase results in a subsequent increase in brain tryptophan uptake and serotonin synthesis (tryptophan is the dietary precursor of serotonin). On the other hand, protein intake increases large neutral amino acid levels more than tryptophan levels, and this subsequently decreases tryptophan in the brain. Consequently, serotonin synthesis in the brain is unchanged or decreased. Unfortunately, brain serotonin levels after carbohydrate or protein intake have not been measured in humans. However, good evidence from animal studies suggests that similar changes in brain tryptophan uptake and serotonin synthesis can be expected in the human brain as well. Thus, this important neurotransmitter in the autonomic nervous system might equally be affected by meal-type and frequency and across ages. The central issue of heterogeneous responses, for example, in fit or unfit elderly individuals in brain synthesis of neurotransmitters important in autonomic function, has not been explored.

7.1.3 Fluid intake and thirst

Aging is associated with a decline in homeostatic capabilities to regulate fluid metabolism. Statements such as "thirst is less common in old age than in the young, being frequent in children, diminishing in adults, and becoming rare in old age," are found in the literature without actual supportive data. Both eating and drinking depend on normal functioning of a number of peripheral and central receptors that monitor deficiencies and translate appropriate stimuli into consummatory behavioral activities. Studies on thirst have shown that the ability in the elderly is clearly decreased (Rolls, 1989). Elderly individuals become readily dehydrated in the most benign dehydrating circumstances. For example, mild water restriction prior to laboratory procedures, fever, infection, even mild diarrhea, or the use of diuretics. In a hospital setting, these effects can be cumulative and are not adequately corrected by the sluggish response to stress of elderly subjects until profound or even fatal disorders may occur (Leaf, 1984). Since intravascular volume is important in vasomotor responses, this propensity to dehydration may have profound effects on autonomic function in elderly individuals.

Renal function is also decreased with advancing age. For example, glomerular filtration and tubular secretion rate correlate with the decrease in creatinine clearance that occurs in the elderly (Montamat, Cusack, & Vestal, 1989). It should be noted that changes in water and electrolyte homeostasis, nutritional, and environmental factors profoundly affect the pharmacodynamics and drug interactions in the elderly. Thus, many drugs affecting autonomic function have exaggerated or deleterious effects in the elderly because of these complicating factors and may seriously impede the quality of life in old subjects. Drug therapy, particularly with substances affecting autonomic function, metabolism, and homeostasis in the elderly, is vastly complicated by many factors, which are unique to this age group as a whole, but even further so by the often-heterogeneous responses due to behavioral and environmental influences in old age. Aging is associated with a progressive decline in plasma renin, angiotensin II, and aldosterone levels, and increases in atrial natriuretic peptide. These changes promote salt wasting by the kidney. Baroreceptor-mediated increase in plasma vasopressin associated with orthostatic stress is blunted in many healthy elderly individuals. These physiologic changes predispose elderly people to volume contraction and hypotension.

7.1.4 Vasomotor control

Transient or prolonged systemic hypotension may give rise to a significant fall in blood flow, particularly to a diseased part of the vasculature supplying the brain. Focal cerebral hypoperfusion resulting from a combination of occlusive vascular disease and orthostatic hypotension is a treatable cause of transient ischemic attacks and may be important in the prevention of permanent focal neurologic deficits in some patients (Dobkin, 1989). Advancing age is associated with a decrease in baroreceptor function, and this may further be aggravated by the use of a variety of hypnotic or psychotherapeutic drugs, in addition to hypotensive medications. This age-related decay in baroreflex activity occurs independently of the increasing hypertension commonly found in older individuals and is not related to atherosclerosis, which might impair baroreflex activity due to a generalized increase in the rigidity of the vascular tree. Since the maintenance of blood pressure does not only depend on baroreflex activity but also on catecholamine and plasma renin, the responses of these substances in the elderly have also shown differences from responses in young subjects. Urinary excretion of dopamine and the dopamine/norepinephrine and epinephrine ratios rise with advancing years. Aldosterone excretion is also higher in older subjects. Thus, orthostatic hypotension, a common clinical finding in elderly subjects, may be multifactorial and closely related to autonomic activity in these age groups.

Norepinephrine levels increase with increasing age. Measurement of norepinephrine responses to stress and the effect on blood pressure in various age groups have shown that young subjects have lower mean arterial blood pressures while resting than old subjects and lower circulating levels of norepinephrine. After standing and isometric exercise, there is an increase in circulating levels of norepinephrine, which is less in young than in old subjects. Similarly, the cold pressor test, a common test to assess the integrity of the sympathetic vasomotor nerves, gives a much smaller pressor response in young than in old subjects. This, however, may reflect the better training status of young individuals and the detraining effects of old age. These findings support the idea that stress placed upon the sympathetic nervous system causes less secretion of norepinephrine and smaller blood pressure responses in young than in old individuals (Palmer et al., 1978). Despite apparent increases in sympathetic tone with aging, cardiac and vascular responsiveness is diminished and infusions of beta-adrenergic agonists result in smaller increases in heart rate, left ventricular ejection fraction, and cardiac output in older individuals compared with younger men.

Abnormal vasomotor responses have been found in a large number of elderly patients and aging is associated with increased vascular stiffness and peripheral resistance, and reduced alpha-adrenergic-mediated vasoconstriction. Analysis of blood pressure responses to the Valsalva maneuver may reveal blunted late phase II and absent or weak phase IV overshoot. Hogikyan et al. demonstrated that the cardiovascular noradrenergic impairment associated with aging may be reversible by suppression of sympathetic nervous system activity with guanadrel, which suggests heightened sympathetic nervous system activity and possibly postsynaptic receptor desensitization (Hogikyan & Supiano, 1994). Some of the physiologic changes associated with aging may be reversible. In patients with an acute stroke, the overshoot is often absent, but normal vasomotor responses for age may return in the chronic stage. This is attributed to a disturbance of central structures either in the brainstem or hypothalamus (Johnson, Smith, Spalding, & Wollner, 1965). Such disturbed vasomotor responses are also seen when these structures are involved by tumors or acted upon by a variety of drugs (Barraclough & Sharpey-Schafer, 1963).

The impairment of baroreceptor function over the age of 40 in the recumbent position is further aggravated in upright elderly subjects, and this probably is the result of pooling of the blood below the heart in an incompletely constricted vasculature (Appenzeller, 1974). A morphologic basis for the observed deterioration in baroreflex activity with age has been suggested. The white rami communicantes in the paravertebral sympathetic chain in humans show a reduction in internodal length with advancing years. This observation is interpreted to show that Wallerian degeneration and segmental demyelination occur in white rami communicantes with increasing frequency in the elderly and that regeneration does not keep pace with successive degenerative events. The demonstrated irregularities in internodal length of white rami communicantes and the shortening of internodes with advancing years could result in temporal dispersion of impulses with an impaired ability of the nerve to transmit synchronous volleys, and this might be important in baroreflex activity. In addition, a progressive loss of larger myelinated fibers and an increase in the number of smaller fibers presumably reflecting collateral sprouting is also found more frequently in older subjects, and this could have a similar effect upon conduction in myelinated fibers in the sympathetic chain (Appenzeller & Ogin, 1973).

7.1.5 Autonomic control of cerebral blood flow

The metabolic control of cerebral blood flow is most important in autoregulation and depends primarily on arterial pCO_2 and pH. Vasoconstriction results from a fall in pCO_2 and dilatation from a rise in pCO_2. Evidence has also been found that sympathetic pathways to the cerebral blood vessels must be intact for normal autoregulation, and some patients with autonomic dysfunction have impairment of cerebral autoregulation because of abnormalities in the autonomic innervation of their cerebral blood vessels. Cerebral autoregulation is effective within a mean arterial pressure range of 80–150 mmHg and is well preserved in healthy aging but the lower pressure threshold may be higher for individuals with hypertension. Impairment of cerebral autoregulation might account for the occurrence of blocked autonomic reflexes previously demonstrated in elderly subjects. Measurement of cerebral blood flow with concomitant assessment of arterial pressure in a number of elderly subjects with postural hypotension has shown that symptomatic individuals had either bilateral or unilateral impairment of cerebral autoregulation. On the other hand, those who did have impaired baroreflexes and postural hypotension but remained asymptomatic had normal cerebral autoregulation. A partial explanation can now be offered for the common clinical experience when small drops in arterial blood pressure may become symptomatic in the elderly.

Whether or not impaired autoregulation can be related to mental function and perfusion pressure in the elderly has not been decided. Chronic cerebral hypoperfusion is associated with the development of periventricular white matter hyperintensities with associated deleterious effects on cognition and gait (Novak & Hajjar, 2010). The importance of symptomatic manifestations of decreased cerebral blood flow, such as transient dizziness in those with cardiac arrhythmias and the different influences of drugs affecting blood pressure regulation commonly used in the aged, might be explicable on impaired autoregulation (Wollner et al., 1979). In a study of middle cerebral artery velocities at high altitude (Wood, Appenzeller, & Appenzeller, 1989), a clear decrease was found as a function of age. Yet, these same subjects had a significantly decreased altitude-induced erythrocytosis, which should have increased middle cerebral artery velocities. None of the elderly participants in this high-altitude research expedition was symptomatic. This suggests that the responses of the cerebral circulation to hypoxia and hypocapnia and the autonomic innervation are heterogeneous across age groups and depend, in this instance, on physical fitness.

The vasoconstrictive response of cerebral vessels to hypocapnia was measured in healthy volunteers of various ages at sea level and a clear reduction in vasoconstriction in the hypocapnic state occurred with advancing ages. In this instance, the impaired responsiveness of the aged cerebrovasculature was attributed to minor atherosclerosis with loss of elasticity of cerebral vessels rather than to an impairment of autonomic function. A number of factors may thus contribute to the normal autoregulatory failure of cerebral circulations with advancing age (Yamaguchi et al., 1979).

7.1.6 Thermoregulation

A recognized hazard during the cold months in temperature climates is accidental hypothermia, which affects the elderly and infirm. Otherwise healthy older people may have low body temperatures in a cold environment and function reasonably well (Taylor, 1964). Serious hypothermia with impaired consciousness in the aged seems to be associated with particular social conditions, concomitant disease, and sometimes medication, but whether these are directly related to symptomatic hypothermia is not clear. Importantly, behavioral thermoregulation may even be crucial for survival. For example, rats respond to an elevated core temperature by grooming to promote evaporation, locomotion to find a cooler place, and reduction of muscle activity heat production with postural extension (Roberts, 1988). Thermoregulatory behaviors are modulated by circadian and sleep-state processes.

Exposure of older people to cold causes in them a smaller increase of oxygen consumption; the fall in rectal temperature is greater, and in skin temperature less than in young similarly exposed people (Horvath, Radcliffe, Hutt, & Spurr, 1955). How much of this is directly attributable to malfunction of central thermoregulatory structures, perhaps induced by hypothalamic blood-flow changes with age, abnormalities in efferent neural pathways, or blood vessels directly concerned with thermoregulatory mechanisms, remains to be determined. The change in temperature does not need to be extreme to see differences between the aged and the adult. DeGroot and Kenney showed in humans that core body temperature dropped with mild ambient cooling in the aged, while it remained high in young subjects. In this study, the shivering response occurred at a much lower core body temperature in elderly subjects and the elderly subjects showed a progressive decline to almost 0.2°C below their baseline value by the time shivering set in (Degroot & Kenney, 2007).

Epidemiologic studies have shown that the incidence of ischemic heart disease is higher in winter than in summer months. A survey to assess the risk of ischemic heart disease during winter months with sudden cold snaps and snowfalls was carried out in persons aged 65 or older. Surprisingly, cold snaps had no effect on the daily rate of sudden death, but increased rates of sudden death were found after a snowfall. Thus, the conclusion was reached that the increased frequency of death, presumably from ischemic heart disease in winter, particularly among the elderly, is due to factors other than short-term cold stress and is probably not related to impairment in temperature regulation (Anderson & Rochard, 1979). This, of course, is not surprising, since the normal increase in oxygen consumption with exposure to cold is less in the elderly than in young individuals. Thus, a tendency to decrease the risk of myocardial ischemia from sudden exposure to cold is evident because of a relative decrease in oxygen demand from tissues.

When the minimum daily temperature falls from 17°C to −5°C, the death rate from coronary and cerebral thrombosis rises linearly as the air temperature falls. For example, daily deaths per million increase from 4.9 to 6.9 from myocardial infarction and from 3.2 to 4.8 from stroke. The cold-related mortality from these vascular causes is very large compared with that from accidental hypothermia, which accounts for only 0.016 daily deaths per million. An explanation for cold-related thrombosis has been offered. An increase in platelet counts and platelet volume can occur 6 hours after mild surface cooling, which causes little fall in core temperature of experimental subjects (Keatinge et al., 1984). An increase in packed cell volume by 7% and increased platelet count produced a 15% increase in the fraction of plasma volume occupied by platelets. While few of these changes were observed in the first hour of the 6-hour experiments, whole-blood viscosity increased by 21%, and plasma viscosity usually increased, together with a rise in arterial pressure, from an average of 126/69 to 138/87 mm Hg. Accompanying these changes in blood constituents and viscosity were also an increase in plasma cholesterol concentration and in high- and low-density lipoprotein fractions, whereas total lipoprotein and lipoprotein fractions were unchanged. While these experiments were done in young individuals, and increases in platelets, in red cells, and viscosity of the blood were achieved in the presence of normal thermoregulatory adjustments to mild surface cooling, it is not clear whether the results may provide an explanation for rapid increases in coronary and cerebral thrombosis in cold weather in the elderly.

Nevertheless, the raised arterial pressure and perhaps the increased cholesterol concentration may contribute to the increase in thrombotic events in vascular beds already affected by atherosclerosis in those at risk. Important and useful advice that can be given to the elderly on behavioral thermoregulatory activity, including appropriate protective clothing during outdoor sojourns and limitation of outdoor activities in very cold spells, including readjustment of heating patterns in homes to restrict cold areas, might prevent some death from coronary and cerebral thrombotic events in winter.

Hypothermia in old people is far more prevalent than generally recognized, and it is considered to be a frequent cause of death in winter. Though generally, exposure to cold in older people is thought to be the prime cause of hypothermia, other contributing factors include social influences, such as inadequate income, physical inability to tend to heating appliances, and isolation. Moreover, it has been shown that some older people do not feel cold as intensely as young subjects and are therefore not motivated to avoid excessively cold ambient temperatures. A study of elderly women confirms that they experienced comfort and warmth in cool environments and that this resulted from their lower warmth and cold thresholds. This change in temperature thresholds may contribute to the high incidence of accidental hypothermia in senescence (Watts, 1971). It must, therefore, be stressed that hypothermia in the elderly occurs in quite moderate ambient conditions, where the patient may be found with a low central body temperature in a well-insulated bed. In such cases, heat production is inadequate to maintain central body temperature (Exton-Smith, 1973).

Temperature regulation in survivors of accidental hypothermia of the elderly has been studied (MacMillan et al., 1967). In such subjects, the resting temperature is low, and when they are reexposed to cold, their temperature falls progressively, and at an abnormally fast rate. These abnormalities are attributed to impairment in the normally occurring increases in heat production and decreases in heat loss on exposure to cold. The disturbance in temperature regulation and, therefore, in autonomic function, was present even 3 years after recovery from accidental hypothermia. Since all examined cases showed impaired thermoregulatory activity, it was thought that survivors of accidental hypothermia are particularly liable to recurrences on even moderate cold exposure. Appropriate care to reduce this risk and early diagnosis before the temperature drops to dangerously low levels might prevent some cold weather deaths.

There is a decline in heat production with advancing age. The rate of decline varies among species, but a fixed minimum of heat production is usually reached near senescence and appears to be independent of species, reaching a value of about 20 calories/kg/day at the age of approximately 100 years in man. It has been suggested that survival is related to this minimum heat production, which is necessary for life of homeotherms (Calloway, 1974). It is not known, however, whether the minimum heat production declines uniformly with advancing years and is independent of other factors than those associated with aging. The heterogeneity of heat production is almost certainly influential in changing minimum calorie values per kilogram, body weight per day, in different individuals.

7.1.7 Ethnic thermoregulatory differences

Indigenous people exposed to extreme seasonal cold in the Arctic or sub-Arctic regions maintain higher hand and foot temperatures than Europeans under the same conditions. Experimentally, this can be demonstrated by finger or hand immersion in cold water. The higher skin temperature of cold-region natives ensures that on immersion of their hands heat losses are greater. Arctic natives have greater pain tolerance when their extremities are exposed to cold than do people of European or African ancestry. They also show more frequent and earlier cold vasodilatation. Similar findings have been reported from studies of indigenous people of the Altiplano of Peru. Data have also been accumulated to show that cold vasodilatation can be used as a test to detect genetic admixture in populations of the Arctic, where the frequency of Caucasian genes may differ from less isolated population groups to more isolated groups. Thus, it has been suggested that differences in thermal responses to cold-water exposure are related to different amounts of genetic admixture in each village, rather than to the effect of distributions of thermal responses or other undetected variables (Hurlich & Steegmann, 1979). Of interest is also that high-altitude natives of the Tibetan plateau maintained considerably warmer hands and fingertips as shown by thermography than Caucasians at the same altitude and ambient temperature (Appenzeller & Wood, 1988). Cross-sectional differences in various age groups have not been hitherto demonstrated in high-altitude natives or lowlanders. On the other hand, heat production by contracting muscles seems to have a close relationship to adaptation to cold. Thus, skeletal and heart muscle contractions in cold acclimatized animals are accompanied by greater heat production than in animals that are not cold-acclimatized. This muscular adaptation to cold might be an important source of heat, and it could be of fundamental importance in survival in cold climates (Ivanov, 1980). Clearly, even elderly individuals who, through physical training, have acquired a muscle mass larger than average might be better able to survive cold weather than their sedentary counterparts of the same age.

Responses to cold exposure vary in different races and individuals of different ages. In indigenous Andean people, exposure to cold shows that they maintain substantially warmer extremity temperatures than subjects of Caucasian descent. This was confirmed in 2-hour nude exposures to cold in natives from Chinchero and Nunoa (Peru). Studies on surface temperature of the hand and on blood pressure and heart rate responses to cold in highland Quechua Indians were carried out. During cold immersion of the hand for 20 minutes, women maintained warmest skin temperatures and most marked cold-induced vasodilatation and reactive hyperemia. They also had the least increase in systolic cold-pressor responses and heart rates during cold-water immersion. White men participating in the same study were least tolerant to cold. Thus, Andean Indians are genetically more tolerant to cold exposure than whites (Little, Thomas, & Larrick, 1973). This is also true of Tibetan highland indigenous people (Appenzeller & Wood, 1988).

Baldness usually accompanies advancing age. It is, however, useful in heat regulation in those exposed to extreme ambient temperatures. It is well-known that natives of very cold regions of the globe are rarely if ever bald, whereas those indigenous to hot climates usually are afflicted by baldness. The effect of environmental temperature on sex-hormone secretions, which have a controlling influence on baldness, and genetic variations in responsiveness to these hormones by the hair follicles in cold or warm climates has, however, not been defined.

7.1.8 Sweat-gland activity in the aged

The total number of eccrine glands is fixed throughout life; however, older adults exhibit a lower sweat output per activated gland in response to a given pharmacological stimulus or passive heating compared with younger adults. The decline in sweat rate with aging may be due to decreased aerobic fitness and/or sensitivity of sweat glands to cholinergic stimulation rather than age (Best, Caillaud, & Thompson, 2012; Kenney & Munce, 2003).

Circadian rhythms in dry-heat loss due to radiation, convection, and conductance are thought to be responsible for the known circadian variation in body temperature in humans. Studies in hot environments in which temperature and heat were constantly maintained were done on a number of individuals to determine the circadian rhythms of sweating. Measurements from the anterior side showed that sweating rates declined remarkably during the period of 2.00–5.00 hours, whereas at other times of the 24-hour cycle, sweating remained generally high in this hot environment. Moreover, the reduction in sweat rates seemed independent of sleep or sleeplessness (Tokura et al., 1979). When thresholds for sweating and vasodilatation were examined over 24 hours, it became clear that these vary considerably and are entrained to variation in internal body temperature in humans. Thus, the circadian rhythm of body temperature is connected through a reference point, which is shared by both vasomotor and sudomotor control (Wenger et al., 1976). Because advancing age changes both circadian thermoregulatory and vasomotor function, it is likely that sudomotor activity in the aged is markedly different over a 24-hour period, when compared with the sudomotor activity of young individuals. The interpretation of sweat tests, therefore, must consider not only the known circadian rhythmicity in sweat-gland activation, but also the changes brought by age in the distribution and activity of these glands.

7.1.9 Neurogenic control of the circulation

Since the discovery of 5-hydroxytryptamine (serotonin, 5-HT), advances have been made in the understanding of the role of this substance in cardiovascular regulation. Changes in the dynamics in brain serotonin biosynthesis lead to alteration in cardiovascular function. Activation of cerebral serotonin receptors produces pressor effects in some normotensive subjects, but depressor effects in other normotensive species. Reductions in the level of serotonin can prevent the onset of hypertension in experimental hypertensive animals and lower the blood pressure in other animals with established experimental hypertension. It has been suggested that the ability of brain serotonin to influence arterial blood pressure is mediated through serotoninergic neuronal systems and their effect on efferent sympathetic activity. In this context, the manipulation of brain serotonin metabolism by dietary factors (see above) and the decreased sympathetic tone that is known to accompany physical training may be important in the heterogeneity of responsiveness of vasomotor activity in various age groups.

The assumption is often made that a clear-cut relationship between various neural structures concerned in the regulation of the circulation and with modulation of baroreceptor reflexes exists. Nevertheless, studies on the organization of central autonomic pathways suggest that their activity may be poorly understood. In the brain, a decline in dopamine and norepinephrine is related to a loss of dopaminergic and noradrenergic neurons in the substantia nigra and locus coeruleus. There is also a decrease in choline acetyltransferase and acetylcholinesterase with aging. The clinical implications of these changes are not fully understood. In humans, the control of variables and measurements are made difficult by the limits of invasive procedures, but some progress has been made by pharmacological manipulations of baroreflexes. When the heart-rate intervals are plotted against systolic pressure during baroreceptor activity, a measure of baroreceptor activation can be obtained (baroreflex-cardiovagal sensitivity). Thus, the function of baroreflexes and their effect on the cardiovascular system can be assessed during normal behavior and during sleep. Such measurements have shown that in normal subjects there is an increased baroreceptor sensitivity during sleep, but this sensitivity is maximal during REM sleep (Smyth, Sleight, & Pickering, 1969a, 1969b). Using the same technique, it was possible to show that the sensitivity of baroreflexes is decreased in hypertensive patients and that a decreased sensitivity also accompanies advancing years in otherwise normal subjects (Gribbin, Pickering, Sleight, & Peto, 1971). What is of great interest also in the context of heterogeneity of responses is the finding that baroreflex sensitivity decreases during exercise and more so with increasing workloads. For example, at a heart rate of 150 beats/min, no effect on baroreflexes was detectable in spite of an increase of 25–30 mm of mercury in arterial pressure (Bristow, Honour, Pickering, Sleight, & Smyth, 1971).

7.1.10 Continuous sympathetic nerve discharges

Autonomic neural function can be evaluated in humans by direct nerve impulse recordings made through tungsten microelectrodes (see Wallin, 1983) (see Chapter 2). A relationship between plasma levels of norepinephrine and sympathetic bursts in the nerves to blood vessels of forearm muscles has been found in subjects at rest. Since there is an increasing level of norepinephrine with age, the number of sympathetic bursts also increases, but the plasma level of norepinephrine is probably determined by impulse activity in the sympathetic nerve supplying a particular vascular bed. Nevertheless, that activity has no great influence on the setting of blood pressure on a long-term basis. It is therefore not surprising that levels of norepinephrine have not been satisfactorily correlated with blood pressure levels. These electrophysiologic findings suggest, however, that marked variations within age groups in both numbers of sympathetic discharges in nerves supplying blood vessels to muscles are likely to be found. The variations depend on many factors, including physical training of the elderly. In a large study of 40 subjects aged 18–75, the aging-associated increase in sympathetic bursts to muscle blood vessels at rest was confirmed. However, in the elderly, there is less increase in muscle sympathetic bursts while standing (Iwase, Mano, & Watanabe, 1991). This could be related to the more frequent occurrence of symptomatic orthostatic hypotension in the elderly. Unfortunately, the above study did not address the level of fitness of the elderly nor the effect of training on the number of sympathetic bursts.

7.1.11 Pupillary changes with age

The normal pupil is 2.5–5.5 mm in diameter. The pupil is smaller in infancy and widens progressively until adolescence when it attains its largest diameter and becomes narrower again in old age. The magnitude of the pupillary responses parallels the size of the pupil for the most part, and most active pupillary reactions are obtained in young people (Larsson & Osterlind, 1944).

Pupil cycle time is a simple way of measuring autonomic reflex activity (Martyn & Ewing, 1986). The normal pupil cycle time was found to be 946 ± 120 ms when the average time for 100 cycles of dilatation and constriction was determined. The upper limit of normal is 1150 ms. There is apparently no correlation with age in normal subjects (19–65 years; mean, 34 years). However, when pupil cycle time is tested at high altitude, a clear acceleration was found in all subjects; moreover, this accelerated response was significantly enhanced in older subjects and clearly related to advancing age. This test illustrates well the heterogeneity of responsiveness and also shows an improved functional capacity of the iris sphincter at high altitude with advancing age. Ordinary test responses under usual test conditions may have significant deviations from normality under unusual circumstances in ordinary subjects (Appenzeller & Martignoni, 1994). Light reflex amplitude is affected by age and disease. Age reduces the size of the reflex pupillary constriction to light, but only as a consequence of the progressively smaller pupil with advancing years.

Enhanced mydriasis, in response to phenylephrine, a directly active alpha-adrenoceptor agonist, and hydroxyamphetamine, is found with advancing age (Smith & Smith, 1983). This change in responsiveness in older individuals parallels the decrease in the size of the pupil in darkness in old subjects, which is sometimes called "senile miosis"; however, heterogeneous responsiveness across age groups is clearly found. The pupillary diameter in darkness also decreases with advancing age and in the eighth decade is only about half that found in the second decade of life (Smith, 1988).

7.1.12 Neurogenic bladder dysfunction in the aged

A number of nonneurogenic causes of incontinence in the elderly, such as stress incontinence in women and incontinence associated with bladder infection and vesical carcinoma, have been recognized. Nevertheless, the commonest cause of incontinence in the aged is of neurogenic origin (Brocklehurst, 1967). This may be related to lesions of the central nervous system, psychotherapeutic drugs, which inhibit the reticular formation or cause dysfunction of the sacral reflex arcs.

In the management of incontinence of the aged, the capacity of the bladder, the quantity of residual urine, and the presence or absence of repeated spontaneous bladder contractions have been considered. In older people without incontinence, the bladder capacity diminishes, there is some residual urine, and occasionally there are large uninhibited contractions of the bladder resulting in incontinence. In elderly subjects, on the other hand, who are continuously incontinent, all these factors become exaggerated, but there is a great deal of overlap between normal elderly individuals who may have urinary urgency and nocturia with occasional incontinence and patients who are permanently incontinent. Cystometrograms in elderly incontinent patients usually show repeated contractions at small increments of fluids, with evacuation at low volumes.

7.1.13 Aging and disorders of sexual function

Studies on the sexual behavior of normal elderly men are scarce. In a study by Rowland et al., penile sensitivity, response to penile ischemia, and somatosensory evoked potentials showed age-related changes; however, self-reported erectile capacity, ratings of overall sex life, and levels of testosterone did not change over different age groups (Rowland, Greenleaf, Dorfman, & Davidson, 1993). Decreasing erectile capacity in aging men may be related to decreasing sensory/neural and autonomic functioning but factors other than the frequency of and potency for sexual response are important to the overall rating of sex life. The performance in this area of those who have had a myocardial infarction is also clouded in uncertainty. Advice about sexual activity after such events varies from total abstention to "do as you like" counsel. In a survey of sexual activity of postmyocardial infarction patients attending an exercise rehabilitation center for 3 years, half of the large group claimed unchanged sexual performance and activity. In the remaining half, sexual activity was reduced, including more passive sexual roles. In this group, there was a recurrence of angina or ventricular premature beats during intercourse in a relatively small number of patients. The personalities of the two postmyocardial infarction groups had no significant differences, though a questionnaire completed by their wives suggested that those with decreased sexual activity had increased difficulty in adjusting to life and work and that neuroticism, including depression, was prevalent. The attendance at the rehabilitation program was better for those who engaged in normal sexual activity after myocardial infarction. The better adjustment of these patients suggests that self-esteem and effective participation in exercise rehabilitation programs may in part be based upon normal sexual activity; moreover, their better adjustment may also be associated with an improved prognosis. After myocardial infarction for most men, there is no contraindication to normal sexuality (Kavanagh & Shephard, 1977).

Sexual changes in women during aging are closely associated with hormonal changes. In women, postmenopausal changes influence sexual function with menopausal changes including decreased vaginal lubrication, vasomotor symptoms, neurologic and psychosexual changes, which include mood changes, anorgasmia, decreased libido, and impaired sexual performance. Female sexual dysfunction also increases with age. One cause of such may be the increased incidence of pelvic floor disorders, which are increased with age and following childbirth (Walsh & Berman, 2004).

7.1.14 Variations in sinus rhythm

Children and young adults show changes in heart rate, and if these are larger than 0.12 seconds between the longest and the shortest cycles, the condition is called sinus arrhythmia. This is normal in young subjects, but it diminishes with advancing years. Sinus arrhythmia reflects vagal tone during each respiratory cycle and is mediated by the Bainbridge reflex.

The damping of sinus arrhythmia seen with advancing years is also associated with a decrease in heart rate variability with deep breathing and Valsalva maneuver. This disappearance of sinus arrhythmia may be linked to a gradual increase with age in the circulating norepinephrine concentrations (Ewing et al., 1978) and changed adrenoceptor function (Box & Cox, 1964), which reflects reduced responsiveness to adrenergic stimulation (Rowe & Troen, 1980).

The measurement of heart rate responses to a number of stimuli have become standard methods for assessing autonomic cardiovascular function and generally involve the measurement of reflex changes in heart rate in response to Valsalva maneuver, deep breathing, and changes in posture with head-up tilt. Other tests include compression and decompression of the carotid sinus and injection of a variety of pharmacologic pressor agents. It is clear that the age-related changes in autonomic function can modify the results of such tests, and results should be compared to normative values for age and sex.

7.1.15 Orthostatic stress, age, and autonomic function

Standing after recumbency in healthy subjects causes characteristic changes in heart rate and blood pressure. The heart rate increases abruptly toward a primary peak about 3 seconds after standing and a further increase to a maximum at about 12 seconds occurs. Then, a relative minimum heart rate is reached at 20 seconds after standing, and eventually the heart rate rises again. The first heart rate peak is due to a sudden inhibition of cardiac vagal tone and is thought to be initiated by the exercise reflex, which is triggered by voluntary (static) muscle contractions. The more gradual secondary heart-rate increase, which starts around 5 seconds after standing and gradually rises toward the secondary peak, is a result of further reflex inhibition of cardiac vagal tone and is attributed to a decreased activation of arterial baroreceptors by a transient fall in arterial pressure due to gravity. The subsequent fall in heart rate is the result of the recovery of arterial pressure and again is due to a vagal inhibition of the sinus node mediated by arterial baroreflexes. This relative bradycardia is a vagal reflex, but its manifestation depends on a preceding sympathetically mediated vasoconstriction.

The initial blood pressure response on standing shows an immediate pressure increase, which is followed by a fall in blood pressure with a nadir within 10 seconds and recovery of BP within 30 seconds.

The immediate rise in blood pressure is most likely due to a mechanical increase in total peripheral vascular resistance by the contracting muscles that resist gravity on standing. The temporary fall in systolic and diastolic pressure, following the initial rise in blood pressure, is attributed to the compression of capacitance vessels (veins) by the contracting muscles in the legs and abdomen, causing a shift of blood toward the heart and thus increasing right atrial and ventricular filling pressures associated with rapid vasodilatation in the active muscles with a pronounced fall in systemic vascular resistance. This, presumably, activates cardiopulmonary receptors and results in a reflex release of vasoconstrictor tone. A further mechanism contributing to the fall in blood pressure is activation of arterial baroreceptor mechanism by the immediate and transient arterial pressure increase on standing, which contributes to the reflex release in vasoconstrictor tone, accounting for the striking fall in blood pressure after the initial rise. The subsequent recovery and overshoot of arterial pressure, which occurs after about 7 seconds, is the consequence of a decrease in activation of arterial baroreceptors and cardiopulmonary receptors.

In healthy subjects, there is a rapid and characteristic counterregulation of the circulation to the orthostatic stress. The circulation usually adjusts within a minute of standing. Initial orthostatic hypotension, which is defined as a transient BP decrease ≥40 mmHg systolic BP and/or ≥20 mmHg diastolic BP within 15 seconds, is commonly seen with active standing in healthy individuals (between 25% and 70%). After the first minute, only minor changes in heart rate and blood pressure occur normally. The normal steady-state blood pressure, in response to standing, is an increase in diastolic pressure by only about 10%, with little or no change in systolic pressure. Also, the steady-state heart-rate increase is only about 10 beats/min (see Wieling, 1988).

The relation of the secondary heart-rate increase peak on standing to age shows that advancing years decreases significantly the secondary heart-rate response peak (Dambrink & Wieling, 1987). On the other hand, when the test was applied to a number of endurance-trained subjects, no change with advancing age could be found, and the additional stress of high-altitude exposure caused an equal increase in the secondary heart-rate peak elevation in young and old participants in a high-altitude research expedition. This illustrates the concept of heterogenetic autonomic responses depending on physical fitness. Advancing age, even with the addition of hypoxic stress, does not diminish cardiovascular responses in selected aging subjects (Appenzeller & Martignoni, 1994).

7.1.16 Aging and adrenoceptor-mediated responses

Evidence that aging decreases β_1-adrenoceptor responsiveness is available, but information on β_2 and α_2-adrenoceptors is controversial. The number of ß receptors in cardiac myocytes is unchanged with advancing age, but the affinity of ß receptors for agonists is reduced. Age-related changes in β_1-adrenergic function may be related to desensitization of the adenylate cyclase system in response to a chronic increase in sympathetic activity in the elderly. Beta blockade with propranolol at rest and during exercise is less effective in elderly subjects when compared with ß blockade in young individuals. There is also a consistent right shift of the dose–response curve for isoprenaline-induced increases in heart rate with aging. Therefore, the assumption is made that β_1-adrenoceptors are of less importance in influencing heart rate in elderly subjects. No useful information is available on β_2-adrenoceptors, responsiveness, and aging, and controversy continues on the effect of aging on α_1-adrenoceptors. The effect of phenylephrine, for example, decreases with age, but small doses of epinephrine do not change the blood pressure of young subjects but increase the pressure of older individuals. On the other hand, α-adrenoceptor antagonists like phentolamine and prazosin have equal effects in old and young subjects.

Both human and animal tissue experiments have shown a decreased efficiency of adrenoceptor responses with advancing age. In sympathetic ganglia of aged humans, degenerative changes are seen (Appenzeller & Richardson, 1966), and such ganglia are less sensitive to blocking agents in older subjects. Isolated human arteries and veins show no change in contractile capacity from elderly subjects in response to norepinephrine when compared with the responses of young controls. However, responses from older animals clearly show a species difference and decreased potency in responsiveness when compared with responses of younger animals.

The observed effect of aging upon adrenoceptor function is probably the result of multifactorial and complex changes and interactions. The relative importance of many factors and differences between regional sympathetic innervation and animal species studied remain to be elucidated. The evidence for an increased sympathetic activity with age is mainly based on increased plasma catecholamine concentrations in old people and an increase in impulse traffic in recording from cutaneous and muscle sympathetic nerves in older subjects. There is no good evidence as yet for heterogeneous responses in different elderly subjects. There is some evidence, however, that the clearing of catecholamines from the circulation is changed, and neural secretion is also increased in elderly subjects (Christensen, 1982). These age-related changes have been attributed to compensatory phenomena for failing physiologic mechanisms, for example, an increased sympathetic drive to a less efficient old myocardium. The accepted increase in circulating catecholamines with advanced age may also result from a decrease in excretion or removal from the circulatory by a fall in uptake into sympathetic nerves or other sympathetic effector organs. Thus, uptake of catecholamines may be less efficient with age, or there may be less sympathetic nerve endings due to subclinical autonomic degeneration in elderly people, though this does not seem to be the case with perivascular noradrenergic nerve fibers in animals (see above). The metabolism does not seem to be influential in changing adrenergic function with advancing years, though some evidence suggesting that serum monoamine oxidase activity increases with age is available. On the other hand, another important enzyme in catecholamine inactivation, catecholamine-O-methyl transferase, shows no significant change with age.

7.1.17 Postprandial hypotension

This is a serious complication in patients with autonomic failure. However, the elderly without evident autonomic disease are also subject to postprandial hypotension (Lipschitz et al., 1983). In such individuals, a predominant fall in systolic blood pressure occurs after food. Elderly subjects often take drugs that impair autonomic function and thus become more susceptible to significant hypotension and possible cerebral ischemia. The postprandial hypotension is particularly associated with carbohydrate meals (Appenzeller & Goss, 1970), and profound postural hypotension may follow such meals.

7.1.18 The impact of sex on sympathetic neurocirculatory regulation

Age-related alterations in neurocirculatory regulation differ between men and women. Young women have a lower blood pressure than men, but women demonstrate larger blood pressure increases with age. By age 60 years, average blood pressure is greater in women. Young men demonstrate a positive relationship between resting muscle sympathetic nerve activity and vascular resistance, whereas this relationship is absent in young women due to β_2 adrenergic-mediated vasodilation (ß blockade restores the association). While both sexes demonstrate an increase in resting muscle sympathetic nerve activity with age, the change is greater in women and postmenopausal women demonstrating a positive relationship between muscle sympathetic nerve activity and vascular resistance with a decline in β_2 adrenergic vasodilatory mechanisms (Klassen, Joyner, & Baker, 2021).

7.1.19 Cellular senescence and the autonomic nervous system

Cellular senescence is a permanent arrest of cell proliferation in replication-competent but apoptosis-resistant cells (Kirkland & Tchkonia, 2017). Stresses associated with senescence telomere shortening, DNA damage, oncogenic insults, metabolic stresses, epigenetic changes, and mitochondrial dysfunction. Yu et al. demonstrated that the density of sympathetic fibers was significantly elevated in aged mouse tissues and human colon adenoma tissues compared to the sympathetic fiber densities in young mouse tissues and human nonlesion colon tissues. They also demonstrated that senescent cells recruited and promoted the outgrowth of sympathetic fibers in vitro and in vivo in aged tissues via the secretion of netrin-1. The elevated sympathetic activity contributed to impaired cognitive function in naturally aged mice, which can be reversed by blocking nonselective β receptors (Yu et al., 2020).

7.2 Exercise and the autonomic nervous system

This section will summarize the effects of acute exercise and exercise training on the autonomic nervous system.

7.2.1 Acute exercise

The central autonomic network is involved in the cardiovascular response during exercise with activation of hypothalamic, cardiovascular, and cardiorespiratory centers. Group III and IV muscle afferents provide feedback to central structures via the "exercise pressor reflex" to match systemic oxygen delivery with metabolic demand (Fu & Levine, 2013). There is vasoconstriction in inactive muscles and vasodilation in active muscles via the production of vasodilatory metabolites (i.e., adenosine, prostaglandins, nitric oxide). In patients with autonomic failure with cardiovascular noradrenergic impairment, there is blunted α-adrenergic vasoconstriction in nonexercising skeletal muscles and hypotension can occur during exercise and may be the initial manifestation of the disease. Conversely, patients with afferent baroreflex failure have an excessive increase in blood pressure during exercise due to the lack of feedback regulation of heart rate and blood pressure via the arterial baroreflex. During acute exercise, there is also an increase in cardiac output, oxygen uptake, and arterial pressure. The increase in heart rate and blood pressure at the beginning and during exercise is due to the resetting of the arterial baroreflex. Stimulation of mechanosensitive and metabosensitive (group III and IV) skeletal muscle afferents increases substance P input at the nuclei of the solitary tract, which reduces baroreflex buffering of sympathetic outflow and contributes to an upward resetting of the baroreflex operating point during exercise (i.e., exercise pressor reflex) (Rowell & O'Leary, 1990).

After acute exercise, heart rate deceleration is mostly driven by vagal reactivation at first and sympathetic withdrawal later in recovery (Fu & Levine, 2013; Goldberger et al., 2006). Atropine abolishes heart rate recovery while pyridostigmine enhances heart rate recovery postexercise. Postexercise hypotension is characterized by a reduction in blood pressure below the values observed either immediately prior to exercise or on a control day. The existing literature on postexercise hypotension is full of a wide range of varied protocols, methods, and analyses. This phenomenon can occur in healthy individuals and diseased states such as hypertension or autonomic failure. Postexercise hypotension has been found following various types of exercise. While a short bout of exercise can produce postexercise hypotension, more prolonged and/or higher intensity exercise may result in a greater decrease in blood pressure and longer postexercise hypotension duration (Pescatello et al., 2004). The pathophysiology of postexercise hypotension is not fully understood but it is thought that the combination of centrally mediated decreases in sympathetic nerve activity with reduced signal transduction from sympathetic nerve activation into vasoconstriction, as well as local vasodilator mechanisms, contributes to the fall in arterial blood pressure seen after exercise (Halliwill, Buck, Lacewell, & Romero, 2013). The circulating adrenaline released from the adrenal medulla has a moderate vasodilatory effect. Studies measuring circulating adrenaline found elevated or unchanged levels during the hypotensive period. The fact that postexercise hypotension persists during adrenaline infusion and β-receptor blockade suggests that any role of adrenaline is minimal. Similar findings have been reported with other vasodilatory substances such as adenosine, potassium, and atrial natriuretic peptide with levels found to be either increased or unchanged during postexercise hypotension (MacDonald, 2002).

Patients with a variety of autonomic disorders have blunted or abnormal cardiovascular responses to exercise and low VO_2max. Decreased ability to release norepinephrine from sympathetic nerves was reported in patients with Parkinson's disease with cardiac sympathetic denervation (Imrich et al., 2009). Another study suggested that patients with PD with cardiac sympathetic denervation may have reduced cardiac contractility (Nakamura et al., 2010). A comprehensive evaluation with cardiac sympathetic neuroimaging, autonomic physiology, and catecholamine neurochemistry is necessary to investigate the effect of cardiac noradrenergic deficiency on exercise tolerance in Parkinson's disease. In pure autonomic failure, hypotension during exercise and blunted heart rate response have been reported but supine exercise (i.e., recumbent cycle ergometer, rowing, swimming) may prevent exercise-induced hypotension. In one study, water drinking did not change the abnormal cardiovascular responses to supine cycling exercise but improved orthostatic tolerance postexercise in patients with pure autonomic failure (Humm, Mason, & Mathias, 2008). In patients with multiple system atrophy, the fall in blood pressure during exercise was more pronounced in the cerebellar subtype compared to the parkinsonian subtype (Smith & Mathias, 1996). Heart rate increased similarly in both MSA subtypes and systemic vascular resistance fell similarly, but cardiac index rose more in the parkinsonian subtype compared to the cerebellar subtype. Resting plasma norepinephrine levels were subnormal in both forms of multiple system atrophy and did not increase with exercise (Smith & Mathias, 1996).

During outdoor running activities such as running on mountain trails, induced fatigue influences all biomechanical parameters, especially exercise intensity and this depends on the particular environment. Nevertheless, studies on continuous effects are hitherto lacking. Thus, wearable inertial sensors might be appropriate to use especially in field studies (Apte et al., 2021). Fatigue can occur with hypoxia, with reduced oxygen availability, and in response to physiological and pathological conditions. At altitude, ambient hypoxia results from a drop in the oxygen pressure in the atmosphere with elevation. Hypoxia also occurs when metabolic needs are not met at the cellular level. Both ambient hypoxia or exercise-induced muscle hypoxia elicits adaptive responses in blood vessels in the skeletal muscles, which eventually cause blood vessel proliferation or regression in skeletal muscles. Finally, this leads to the release of myokines. These molecules can affect distant tissues and change the capillary and muscle interface. Fatigue is also associated with exercise-induced hypoxia.

Hypoxia is defined here as a reduction of oxygen availability that can occur in the skeletal muscle tissue of an individual exposed to ambient hypoxia and during physical exercise if the oxygen supply to contracting myofibers cannot match their increased metabolic needs. The superimposition of these two stressors—ambient hypoxia exposure and exercise-induced local hypoxia—can exacerbate the hypoxic stress experienced by the skeletal muscles. The capillary-to-myofiber interface serves as the site for exchanging oxygen, nutrients, metabolic heat, and waste between the blood and myofibers. It affects the functional capacity of the skeletal muscle. The capillary microvasculature is a highly adaptive tissue with remarkable plasticity that can grow or regress in response to various physiological, pathological, and environmental stressors. This is termed angio-adaptation. Skeletal muscle angio-adaptation involves complex and dynamic molecular and cellular responses in mountaineers, athletes, and the clinic (Lemieux & Birot, 2021).

7.2.2 Exercise training

Exercise training is recommended for improvement of physical and mental health and is a key factor in both primary and secondary prevention of cardiovascular and neurological diseases. Exercise training has been shown to increase vagal modulation and baroreflex sensitivity and decrease sympathetic tone in different populations such as chronic heart failure, post-myocardial infarction, hypertension, and obstructive sleep apnea among others. The benefits are observed soon after the initiation of exercise training and are maintained as long as the activity is continued. Exercise training is associated with a decrease in plasma norepinephrine and proBNP levels in patients with chronic heart failure. Change in resting plasma norepinephrine level was the most powerful predictor of cardiac mortality at follow-up (Rengo et al., 2014). Chronic increases in nitric oxide with exercise training lead to functional and histological alterations of vascular endothelium. The best type of exercise training (aerobic, resistance, combination of different exercise regimens) and the optimal intensity of exercise training to achieve improvement in cardiovascular autonomic regulation in different diseases are still unanswered issues.

Improvement in baroreflex function with exercise training could be the result of increased blood vessel compliance and distensibility. Exercise training also induces structural neuroplasticity within the central autonomic network, specifically the nucleus of the solitary tract. Integrative feed-forward and feedback adjustments of circulatory control during exercise are mediated by the interaction between the nucleus of the solitary tract and the paraventricular nucleus of the hypothalamus. Exercise-related peripheral information conveyed in part by the baroreceptors located in the carotid sinuses and the aortic arch to the nucleus of the solitary tract is transmitted directly and/or indirectly via catecholaminergic projections to the paraventricular nuclei of the hypothalamus. This results in the simultaneous activation of oxytocin and vasopressin neurons projecting back to the nucleus of the solitary tract causing an increase in blood pressure (Michelini & Stern, 2009). Central modulation in the inhibitory and excitatory autonomic pathways secondary to exercise training also involves increase nitric oxide level in the paraventricular nucleus of the hypothalamus and reduced central angiotensin II level (Zheng, Li, Cornish, Zucker, & Patel, 2005). Exercise training promotes cardiac remodeling, increases red blood cell volume, blood, and plasma volume, improves renal-adrenal function, and enhances vascular structure and function as well as glucose delivery.

7.2.3 Thermoregulatory response to exercise

Exercise is associated with considerable heat loads. In prolonged endurance events, the amount of sweat secretion and heat generation is largely dependent on body weight and on ambient conditions. Increases in heat loss via cutaneous vasodilation and sweating induced by the activation of the autonomic nervous system facilitate the increases in dry heat exchange (primarily convection and radiation) and evaporative heat loss. Heatstroke does rarely occur in athletes if ambient conditions, such as high humidity and temperature, are combined with prolonged and high levels of exertion (Appenzeller, 1988). In spontaneous heatstroke, when it occurs, as it frequently does, for example, during the annual religious pilgrimage to Mecca (the Hajj), the disorder is frequently related to unaccustomed physical exertion. It was observed that individuals with preexisting disease or advanced age are more prone to heatstroke. It was also observed that many patients with heat stroke are comatose and have constricted pupils. A study to assess the role of stress hormones and endocrine responses in heatstroke showed inordinately large increases in opioids in the circulation during heat stroke, returning to normal levels after recovery. Stress hormones, such as prolactin, cortisol, and growth hormone also increased, but proportionately much less than ß-endorphin-ß-lipotropin in the circulation (Appenzeller, Khogali, & Carr, 1986). Upon the cessation of exercise, there is suppression in the heat loss responses; however, this leads to a sustained elevation of muscle and core temperatures above resting levels for as long as 90 minutes postexercise. The intensity of exercise and ambient temperature conditions influence the magnitude of the elevation in tissue temperatures and the time required for recovery. The alteration in postexercise thermoregulation has been associated with altered baroreflex activity, which contributes to vasodilation and reductions in mean arterial blood pressure.

Though much work has focused upon the pituitary-adrenal response to hyperthermia, the magnitude of this response may reflect the severity of the acute thermal challenge (Osgood, Szyfezbein, & Murphy, 1984). Plasma levels of ß-endorphin do not always reflect levels in the human brain or in CSF. The small pupils in pilgrim victims of heatstroke, however, suggest that similar elevation of endogenous opioids to the levels found in their circulation may also have been present in their brains. This hypothesis is supported by the finding that the normal rise in body temperature with exercise in men is abolished by naloxone, a specific opioid antagonist (DeMeirleir, Arentz, & Hollman, 1985).

During stress, such as fever or exercise (Carr, Bullen, & Skrinar, 1981), it has been observed that central and peripheral opioid systems are mobilized and produce physiologic effects. It is suggested that the effectiveness of naloxone in reversing certain forms of shock and coma (as seen in heat stroke) apparently results from its normalization effect on the sympathetic-parasympathetic balance (Holaday & Faden, 1983). This balance may also be deranged during heatstroke, as suggested by the characteristically dry skin and miosis of the pupils and by precipitation of experimental heatstroke by atropine (Hubbard, Matthew, & Francesconi, 1982). The finding of high opioid levels in the circulation in heatstroke victims at the Mecca pilgrimage suggests a possible therapeutic potential for opioid antagonists in the management of patients with this disorder.

7.2.4 Exercise, autonomic nervous system, opioids, and fatigue

Advances in opioid research have kindled interest in the physiologic function of opioids in animals and humans. It has been proposed that endogenous opioids are important in pain modulation and that they influence physiologic function, including sleep and fatigue, respiration, thermoregulation, and sexual and endocrine activity. New studies in humans and animals continue to expand the hypothetical influence of these substances (Appenzeller, 1981).

Assays of ß-endorphin and other endogenous opioids in urine, plasma, and cerebrospinal fluid have helped define the effects of stress, pain, and other behavioral manipulations on their levels. However, a direct relationship between a given opioid level and modification of behavior or physiologic function is hard to document. While the number of identified opioids and knowledge of their relationship to experimental manipulation has increased, causal interactions between different opioids or hormone levels in various body compartments and behavioral or physiologic changes remain highly speculative.

Endurance training has been recommended for a number of diseases, and countless healthy persons take up strenuous sports in an attempt to improve their physical appearance, stave off the ravages of age, and also to improve occasional or serious depressive illness, or ward off excessive fatigue (Appenzeller, 1981). Because of the reputed beneficial effects of opioids, measurement of their level before and after a long endurance race in highly trained athletes was carried out. Men and women who participated in the ultrarun, an annual event held in a wilderness area, were found to have a significant increase in circulating ß-endorphin-ß-lipotropin at the end of the race, compared with prerace ß-endorphin-ß-lipotropin levels. Furthermore, no increases in the plasma content of these substances occurred at 5K or at 30K of this 46-K race. Continued training and participation in this event led, in the same runners, to a significant decrease in run-induced elevation of immunoreactive ß-endorphin- ß-lipotropin plasma levels. The stress of the race, however, judged by the increases in serum myoglobin in the same runners, did not decrease. Adaptive phenomena to sustain high-energy expenditures necessary for successful long-distance running presumably change the transient acute exercise-induced plasma elevation in ß-endorphin- ß-lipotropin or facilitate their removal from the circulation. Many studies have since confirmed these findings. However, a uniform picture of the exercise-induced changes in opioids has not emerged because of different exercise protocols and the impossibility to control for psychological factors in wilderness running and in competition. In highly trained habitual exercisers, the exercise-induced release of ß-endorphin- ß-lipotropin decreases with time, and other evidence suggests that highly trained endurance athletes have lower resting levels of these substances (Appenzeller, Appenzeller, Standefer, Skipper, & Atkinson, 1984).

Anecdotal accounts of an increased pain threshold and resistance to fatigue after physical activity as well as testimonials of a runner's high have appeared. The mechanisms for these effects of exercise remain, however, speculative. Behavioral changes are difficult to relate to the effects of exercise on plasma opioids. Clearly, extrapolation to opioid levels in the CSF or to receptor activity in the brain would be more relevant. Nevertheless, interrelationships may exist between neuropeptides in various compartments. Thus, it has been suggested (Carr, Bergland, & Hamilton, 1982) that peptides from the circulation may cross the normally tight blood-brain barrier by reflux, so that elevation in plasma opioids may be reflected in the CSF. The possibility exists also that increases in blood levels of one or more opioids may cause secretion into the blood of other peptides, which in turn may reflux across the blood-brain barrier by other means.

Both CSF and plasma levels of ß-endorphin rise after endotoxin administration, which is associated with a rise in body temperature (Carr et al., 1982). The assumption, therefore, that body temperature elevation may have disrupted the blood—brain barrier and allowed reflux of stress-induced opioid secretion into the CSF in the runners at the end of the race, is reasonable. Daily endurance training causes associated diurnal increases in body temperature. Whether such exercise-induced heat loads have an effect on the blood-brain barrier permeability to plasma opioids, similar to fever, is not certain. The possibility exists, however, that the decreasingly acute elevation in exercise-induced plasma ß-endorphin-ß-lipotropin levels in the circulation after a race might be due to increasing reflux of opioids into the CSF and thus inhibit further stimulation to release by a negative feedback mechanism.

The behavioral effects of opioids on pain threshold (Goldfarb, Sforzo, & Flynn, 1984), perceived exertion (Haier, Quaid, & Mills, 1981), temperature regulation, and other central nervous system functions have been assessed by the use of the specific opioid antagonist naloxone. The effect of naloxone is, however, not confined to opioid receptors alone. Naloxone did not change mood elevation after exercise, but that cannot be taken as evidence to prove that endogenous opioids in the brain or CSF are not responsible for the behavioral changes attributed to exercise. Nevertheless, the transiently elevated circulating opioids after physical exertion remains speculatively linked to the documented increased pain threshold, to mood changes, and, in chronic exercisers, to the beneficial behavioral changes attributed to training.

7.2.5 Sympathetic activity

Circulating catecholamines increase markedly (450% above baseline levels) but return to baseline within 1 hour after the race. Simultaneously measured dopamine-ß-hydroxylase showed no change throughout the study. The dissociation of catecholamines and dopamine-ß-hydroxylase during such events remains unexplained but has also been found in other stressful situations, such as during parturition (Appenzeller & Schade, 1979).

The cold pressor test administered to runners before the race showed an impairment of sympathetically mediated vasoconstriction. After the race, a fall in blood pressure, rather than the expected increase after immersion of the hand in ice water, was observed. The baseline studies showed only minimal responses in diastolic blood pressure (normal 15 mm of mercury or more) (Table 7.1) (Appenzeller, Appenzeller, et al., 1985). Cardiac output, which normally increases during the cold pressor test, remained unchanged before and after the race.

The diving response, consisting of bradycardia and an increase in blood pressure, showed the expected bradycardia (larger at the completion of the run), but again, no sympathetically mediated vasoconstriction (Appenzeller, Spar, & Umsted, 1986). These tests, taken together, suggest different sympathetic responses to the stress of a race in well-trained endurance subjects. The reason for the dampening of sympathetic vasoconstrictor responses in endurance-trained subjects remains to be elucidated.

Table 7.1 Blood pressure and cardiac output before and after cold exposure in endurance-trained subjects.

Cold pressor test Blood pressure $n = 43$		**Cold pressor test Cardiac output (L/min) $n = 10$**	
At rest	**After race**	**At rest**	**After race**
Systolic		Baseline	
$+7.6 \pm 2.2^a$ $P < .01$	-11.8 ± 2.3^a $P < .001$	7.9 ± 2.0	11.2 ± 4.0^b
Diastolic		Cold exposure	
$+6.7 \pm 1.4^a$ $P < .001$	-3.3 ± 1.2^a $P < .05$	7.8 ± 1.9	10.7 ± 3.6^b

$P < .001$.
[a]Paired *t*-test before and after race.
[b]Paired *t*-test N.S.

7.2.6 Parasympathetic function

It is well recognized that a preponderance in parasympathetic tone characterizes the endurance-trained state (resting bradycardia). Tests of parasympathetic function in endurance-trained subjects also reveal, however, significant deviation from normal. Thus, heart-rate variability was within the range found in severely disabled patients with diabetic neuropathy, and the percentage of athletes with abnormal results reaches 90% (Appenzeller, Hudson, & Johnson, 1986).

7.2.7 Vasopressin and atrial natriuretic factor

These hormones are closely related to fluid homeostasis and to the regulation of blood pressure. Normally, vasopressin and atrial natriuretic factor have an inverse relationship, but after a race, significant increases in both hormones occur in endurance-trained subjects (Solomon, Corwin, & Appenzeller, 1987). Normally, during endurance events of the magnitude to which these athletes submitted themselves, considerable dehydration occurs. This dehydration should lead to a suppression of atrial natriuretic factor release. The mismatch in signals (contracted intravascular volume due to dehydration and lack of suppression of atrial natriuretic factor release) must be explained.

Hypothetically, rhythmic contraction of muscles during the run causes neurogenic afferent stimuli to release the peptide, irrespective of other vascular or volume-related stimuli.

7.3 Exercise and aging

7.3.1 Introduction

In 1772, the English physician William Heberden described a patient with heart disease who improved his condition on his own. The patient whose daily activities were severely restricted by chest pain assigned himself the task of sawing wood for half an hour every day. Within weeks, the patient's chest pain lessened significantly, and he was able to resume many of his normal activities (Tarlov, Ware, & Greenfield, 1989). To this day, the benefits of exercise, particularly of prolonged exertion, in curing or preventing disease or prolonging life, remain subjects of interest and debate. Concerns about the rising costs of health care have only intensified the interest.

To be able to move is an important aspect of the quality of life of ordinary people and also for the attainment of Olympic gold of elite athletes. The ability to move is for the most part dependent on skeletal muscles and their intact innervation and blood supply. Both aging and disease reduce skeletal muscle mass, slow conduction in motor nerves, and reduce blood flow. This is followed by a decrease in work capacity and a deterioration in the quality of life of the elderly. But transiently impaired muscle function may also decrease the chances of an Olympic gold medal for a world-class athlete.

Life-long fitness may offer a way to promote well-being and decrease the need for costly curative medicine. A retrospective cohort study enrolled patients at a tertiary care academic medical center from January 1, 1991, to December 31, 2014, with a median follow-up of 8.4 years. The goal of the study was to assess the association between all-cause mortality and cardiorespiratory fitness in patients undergoing exercise treadmill testing. Patients were stratified by age- and sex-matched cardiorespiratory fitness into performance groups: low (<25th percentile), below average (25th–49th percentile), above average (50th–74th percentile), high (75th–97.6th percentile), and elite (≥97.7th percentile). Risk-adjusted all-cause mortality was inversely proportional to cardiorespiratory fitness and was lowest in elite performers (Mandsager et al., 2018). Results have raised hopes that preventive medicine and sports can lower morbidity; they may help preserve good mental and physical health and the ability to move until shortly before death without necessarily extending life span. Beyond mortality benefit, exercise helps preserve bone density in elderly women, improves cardiovascular function even in diseased hearts, and preserves muscle mass and peripheral nerve conduction velocities of older people (Bicknell, Appenzeller, & Barrett, 1984; Jokl, 1984).

The ultimate abridged morbidity is, of course, frequently newsworthy when an athlete dies within a few minutes during or immediately after intense physical activity. However, studies of fit populations might improve our understanding on how to avoid the prevalent and costly extension of morbidity, often by well-intentioned medical measures. Another hint might come from the study of the age-related decline in body function in well-trained habitual exercisers. Since the nervous system is especially vulnerable to this age-related decay, it is a sensitive indicator of possible beneficial effects of sports on aging.

Exercise has an effect on the human genome. The length of telomeres, the 5′-TTAGGGn-3′ tandem repeats at the ends of mammalian chromosomes, is one indicator of biological age. In sedentary individuals, telomere length decreases with age, but longer telomeres are found in moderately active individuals. Most studies have shown that regular physical activity increases telomerase activity and conserves telomere length compared to other types of training. However, longitudinal investigations may be needed to confirm these preliminary results (Sellami, Bragazzi, Prince, Denham, & Elrayess, 2021).

7.3.2 Exercise and neurodegeneration

Alzheimer's disease is more common in subjects with no or only one healthy lifestyle factor. The risk of Alzheimer's was 37% lower in those with two to three and 60% lower in those with four to five beneficial lifestyle factors. Healthy lifestyle factors were swimming, running, "gym workouts" that last at least 1 hour/day for 6 days/week (Dhana, Evans, Rajan, Bennett, & Morris, 2020).

The protective effects of physical activity against Parkinson's disease and mortality have been investigated. A total of 45,923 individuals were identified in one study that reported a dose-response association between physical activity and all-cause mortality in Parkinson's disease. An inverse association between physical activity and all-cause mortality and an inverse dose-response association between the total physical activity amount and mortality was found in both men and women. Those who maintained physical activity before and after their Parkinson's diagnosis showed the greatest mortality reduction, while physical activity after the Parkinson's disease diagnosis was also associated with reduced mortality. Activity modification to increase and maintain physical activity would be beneficial for Parkinson's disease management; however, randomized clinical trials for these associations have not as yet been performed (Yoon, Suh, Yang, Han, & Kim, 2021).

7.3.3 Field studies on mountain neurology to mimic aging

Aging is generally thought to be associated with a decline in nervous system function, but such a decline in function is by no means uniform. Memory deficits are common, peripheral nerve motor conduction velocities slow with age (Bicknell et al., 1984), and autonomic function such as muscle sympathetic nerve activity increases at rest (Iwase et al., 1991). Differences between age groups are well established, but most studies have hitherto ignored substantive differences within age groups. This has led to the establishment of the gerontology of the usual, which neglects the psychosocial and physiologic factors that can slow or accelerate the aging process (Appenzeller, Wood, & Appenzeller, 1991). Luckily, nature provides a kind of time machine that can accelerate the aging of the nervous system, consistently and reversibly: exposure to high altitude. Hypoxia at altitude closely mimics advancing age and the associated decay in body function (Appenzeller, Greene, Qualls, & Martignoni, 1994; McFarland, 1963). Examples of this reversible aging of young subjects at altitude are found in the decline in writing ability and high-frequency hearing loss indistinguishable from presbycusis (McFarland, 1963). Moreover, the study of the nervous system of normally aged people in their natural habitat is complicated by the ubiquitous presence of age-associated disease, for example, atherosclerosis increasing nervous system deficits independently of age. Therefore the study of trained sojourners transiently aged at altitude may reveal the relationship of fitness to the aging-associated decay in body function and eventually to abridgment of morbidity.

Research on the effects of altitude has been conducted in many laboratories, where altitude can be simulated by decreasing pressure in hypobaric chambers or by lowering the percentage of oxygen in inspired air (Wood, 1988). Many important contributions to the understanding of altitude effects have been made from such studies, but laboratory studies, in general, do not mimic exactly the physiologic and clinical responses of humans in the field at altitude. One reason is that it is difficult, if not impossible, to study the effects of exercise in the cramped quarters of an altitude chamber. In addition, prolonged altitude exposure for weeks on end of larger numbers of subjects is not easily accomplished in the confinement of an altitude chamber. The nervous system effects of visual and other sensory stimulation provided by mountains may also have important effects that cannot be mimicked in an altitude chamber. Lessons from the field, therefore, provide more realistic and added dimensions to those learned in the laboratory and give the opportunity to compare the responses to altitude and exercise of nonadapted and adapted sojourners, athletes, and natives. Therefore, mountain neurology is particularly relevant in assessing the role of sports in aging and in the abridgment of morbidity.

One feature of aging is a decrease in cerebral blood flow (Wood et al., 1989). It is not clear, however, whether this is due to aging, age-associated cerebrovascular disease, or brain atrophy often accompanying advanced age. An increase in cerebral blood flow occurs in young and old sojourners on acute altitude exposure (Appenzeller, Greene, and Appenzeller) and results from the hypoxia (Severinghaus, Chiodi, & Eger, 1966) but it could also be related to a shear stress-induced vasodilatation (Melkumyants, Balashov, & Khayutin, 1995) because of altitude-associated tachycardia. The cerebrovascular reactivity to CO_2 tells a similar story. It ordinarily decreases with advancing age. In several groups of trained runners, however, the cerebral vasoreactivity to CO_2 showed no relationship to age at either low or high altitude supporting the contention that endurance training retards the usual age-related decay in function (Anand, Greene, & Appenzeller, 1995).

The neural control of renal arteries also seems to age transiently on altitude exposure. The effect is measured by applying sinusoidal neck suction to manipulate baroreceptors at the carotid bifurcation. In the renal circulation, flow velocities are significantly reduced at the peak of neck suction and while altitude has no effect on this flow reduction, advancing age is associated with significantly smaller vasoreactivity (reduction in flow velocities) at altitude only (unpublished observations). This again points to hypoxia as a necessary addition to reveal neurogenic renal vasoconstrictor aging in fit individuals.

Orthostatic tolerance changes at altitude suggesting transient aging of autonomic responses to the upright posture. In endurance-trained subjects, age has no effect on these responses at low altitude, and young and older athletes show the same orthostatic responses when exposed to altitude (Appenzeller et al., 1991).

Competitors in a 100-mile (161k) trail race at moderate altitude were studied before and after the race. The American competitors were compared with 5 Tarahumara Indians (renowned for their ceremonial ultralong-distance trail runs) who participated in the same race. In this study of heart rate variability, there was a significant decline in total power with advancing age before the race at an altitude of 3100 m. After the race at the same altitude, there was a significant further reduction in total power in all runners but the younger competitors (Tarahumaras) aged significantly more than the American competitors (Appenzeller, Qualls, & Martignoni, 1995). Thus, ultralong-distance runners at altitude show the same age-related changes in heart rate variability as untrained individuals at sea level and long competition at altitude with significant added hypoxia leads to a further aging of heart rate variability. No ethnic differences were apparent in these endurance-trained athletes but the Tarahumaras aged more because of their relative youth. This suggests that transient aging at altitude, like natural aging at sea level, is limited by the allotted life span. If one is say 50 years old, one can age at altitude approximately an additional 30 years, but at 20, an additional 60 years of age-related decay is possible.

Reflex tachycardia occurs on swallowing. Studies of this central autonomic (brain stem) reflex were carried out on well-trained sojourners and natives to assess the effects of age and ethnicity on this reflex at low and high altitudes. In untrained subjects, an age-related decay in reflex response occurs at low altitude. In well-trained subjects, this central autonomic nervous reflex is preserved into advanced age. But after altitude exposure to 3700 m for 13 days and upon arrival at 5100 m, this central autonomic reflex is abolished even in well-trained sojourners. While high-altitude natives have normal swallow-induced tachycardia at 5100 m, altitude-adapted sojourners show no return of this reflex after 13 days at high altitude when many other altitude-induced adaptations have occurred. Brain stem hypoxia, it seems, causes transient aging of this response, revealing an age-related decay in function that is not apparent in trained subjects at sea level.

Smell perception declines with advancing age. This is in keeping with the general decline in sensory function found in cross-sectional studies on aging. However, in a sample of endurance-trained subjects, no such age-related decline was found at low altitude. Only when the same subjects were tested at high altitude could a marked decline in smell perception with advancing age be seen, similar to that reported at sea level in untrained elderly subjects. Thus, training may help preserve olfaction, against the usual decay in function with advancing age (Wood, Wood, & Appenzeller, 1992).

Acute mountain sickness is a disorder characterized by headache, palpitation, lassitude, nausea, sometimes vomiting. Measurement of cranial blood flow in endurance-trained subjects shows an altitude-related increase in internal carotid and preauricular artery flows over a prolonged period (Appenzeller, Greene, & Appenzeller, 1985). Anecdotal evidence suggests that endurance training assures immunity to acute mountain sickness, but whether this immunity is related to the increase in flows remains speculative. Nevertheless, it is known that altitude headache responds to the ingestion of glucose, which in turn causes, in normal subjects, activation of the sympathetic nervous system (Fig. 7.1). But sympathetic activity is attenuated in endurance-trained subjects (see above). The possibility exists, therefore,

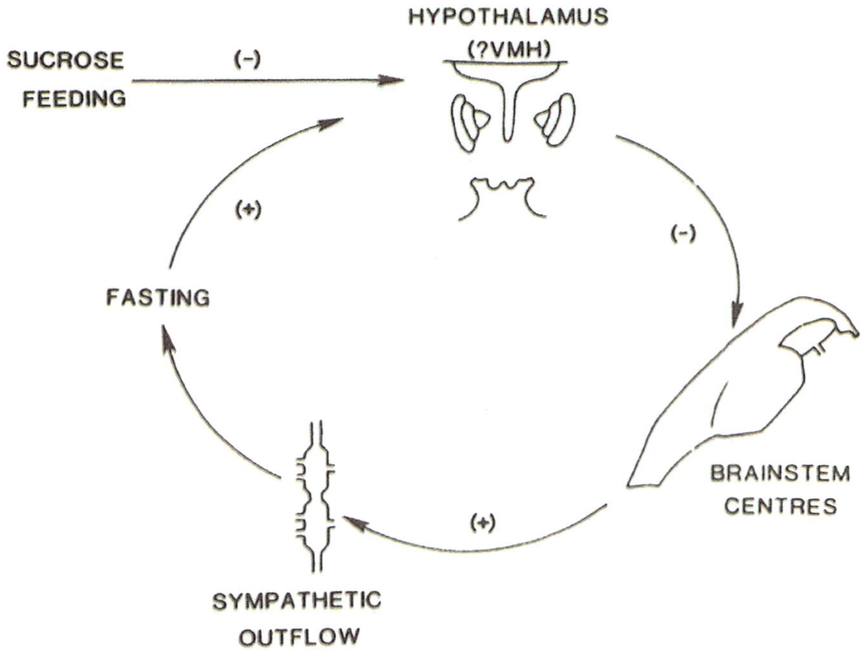

Figure 7.1 Schematic representation to show hypothetical dietary effects on sympathetic activity. Sucrose inhibits (−) hypothalamic activity; this, in turn, removes the inhibitory effects of the hypothalamus on brain-stem autonomic centers, which then activate (+) sympathetic outflow. Fasting has the opposite effect by activating the hypothalamus, causing a decrease in sympathetic outflow. *From Appenzeller (1983).*

that the relative immunity to altitude illness, in spite of an increasing cerebral blood flow in endurance-trained subjects, might result from lack of extracranial vasoconstriction (a steal phenomenon), relatively decreasing intracranial blood flow and thus alleviating brain swelling.

These observations at sea level and in the time machine provided by altitude exposure suggest that exercise may play a role in staving off aging of the nervous system. Together with other studies of the effects of fitness, they imply that lifestyle changes can result in significant abridged morbidity of the nervous system shortly before the natural life span ends. Sports for the masses, not just for elite athletes competing in the Olympic Games, seem to be a desirable public health goal and may impact health care costs for the elderly.

There are plenty of avenues for further study. Lifelong exercise may help reduce the incidence of parkinsonism (Sasco, Paffenbarger, & Gendre, 1992) and perhaps other age-associated degenerative diseases of the brain. Lifelong training seems to be beneficial; it constitutes a complex set of activities, however, that need to be investigated and quantified to optimize its reputed benefits in achieving the goals of robust health until shortly before death (abridged morbidity).

7.3.4 A call for "freestyle neurology"

Pheidippides (530–490 BCE) completed his run from Marathon to Athens and appraised the Athenians of their victory over Persia in the Battle of Marathon. After sharing his good news, he collapsed and died (Appenzeller, 2021). The term "freestyle neurology" is used here for research in which one practices what one preaches, engaging in sport and adventure that offers a chance to advance knowledge about exercise and the autonomic nervous system, unimpeded by the confines of federal funding or the commercial restrictions that come with funding from the pharmaceutical industry.

References

Anand, I., Greene, E. R., Appenzeller, P., et al. (1995). Middle cerebral artery (MCA) flow velocities (FV) and cerebrovascular reactivity to CO_2 at altitude: Effects of age and ethnicity. *Clin. Auton. Res.*, *5*, 330.

Anderson, T. W., & Rochard, C. (1979). Cold snaps, snowfall and sudden death from ischemic heart disease. *Can. Med. Ass. J.*, *121*, 1580–1583.

Appenzeller, O. (1974). Baroreflex activity and transient ischemic attacks. In *Proceedings X International Congress of Neurology, International Congress Series* (Vol. 319, pp. 521–523).

Appenzeller, O. (1981). What makes us run? *N. Engl. J. Med.*, *305*, 578–580.

Appenzeller, O. (1983). Influences of physical training, heat acclimation and diet on temperature regulation in man. In M. Khogali and, & J. R. S. Hales (Eds.), *Heat Stroke and Temperature Regulation* (pp. 283–292). Sydney, New York, London: Academic Press.

Appenzeller, O. (1988). Neurology of endurance training. In O. Appenzeller (Ed.), *Sports Medicine. Fitness, training, injuries* (3rd, pp. 35–71). Baltimore: Urban and Schwarzenberg.

Appenzeller, O. (2021). Reflections on the past three decades of high-altitude exercise research and the autonomic nervous system. *Clin. Auton. Res.*, *31*(1), 31–33.

Appenzeller, O., Appenzeller, J., Skipper, B., et al. (1985). Attenuation of cold pressor test; evidence for decreased sympathetic tone in endurance-trained subjects. *Ann. Neurol.*, *18*, 150.

Appenzeller, O., Appenzeller, J., Standefer, J., Skipper, B., & Atkinson, R. (1984). Opioids and endurance training; longitudinal study. *Ann. Sports Med.*, *2*, 22–25.

Appenzeller, O., Dhital, K. K., Cowan, T., & Burnstock, G. (1984). The nerves to blood vessels supplying blood to nerves: The innervation of vasa nervorum. *Brain Res.*, *304*, 383–386.

Appenzeller, O., & Goss, J. E. (1970). Glucose and baroreceptor function in cerebrovascular disease and in other disorders with baroreceptor reflex block. *Arch. Neurol.*, *23*, 137–146.

Appenzeller, O., Greene, E. R., & Appenzeller, M. (1985). Doppler common carotid and preauricular blood flows and resistances at altitude: A study of endurance-trained subjects. *Ann. Sports Med.*, *2*, 120–124.

Appenzeller, O., Greene, R., Qualls, C., & Martignoni, E. (1994). Valsalvas maneuver and intracranial circulation: Effects of altitude and age. *Neurology*, *44*, A220.

Appenzeller, O., Hudson, T., & Johnson, C. (1986). Parasympathetic function; endurance training mimics diabetic vagal neuropathy? *Muscle Nerve*, *9*, 253.

Appenzeller, O., Khogali, M., Carr, D. B., et al. (1986). Makka Hajj: Heatstroke and endocrine responses. *Ann. Sports Med.*, *3*, 30–32.

Appenzeller, O., & Martignoni, E. (1994). Life-styles and cardio-cerebrovascular risks. In G. Nappi, I. Richichi, G. Micieli, et al. (Eds.), *Neurocardiology* (pp. 13–18). London: Smith-Gordon.

Appenzeller, O., & Ogin, G. (1973). Myelinated fibres in the human paravertebral sympathetic chain: Quantitative studies on white rami communicantes. *J. Neurol. Neurosurg. Psychiat.*, *36*, 777–785.

Appenzeller, O., Qualls, C., & Martignoni, E. (1995). Frequency domain analysis of heart rate variability (FDHRV): Effects of altitude and prolonged exertion. In J. R. Sutton, C. S. Houston, & G. Coates (Eds.), *Proceedings of the 9th International Hypoxia Symposium at Lake Louise, Canada* (p. 313).

Appenzeller, O., & Richardson, E. P., Jr (1966). The sympathetic chain in patients with diabetic and alcoholic polyneuropathy. *Neurology (Minneap.)*, *16*, 1205–1209.

Appenzeller, O., & Schade, D. (1979). Neurology of endurance training. III. Sympathetic activity during a marathon race. *Neurology*, *29*, 540.

Appenzeller, O., Spar, J., & Umsted, D. (1986). Diving response; effects of endurance training. *Ann. Sports Med.*, *3*, 19–21.

Appenzeller, O., & Wood, S. C. (1988). Pentoxifylline: Thermoregulatory and vasomotor effects at high altitudes. *Ann. Sports Med.*, *4*, 279–285.

Appenzeller, P., Wood, S. C., & Appenzeller, O. (1991). Testing autonomic function: Clinical effects of age and stress. In M. Yoshikawa, M. Uono, H. Tanabe, & S. Ishikawa (Eds.), *New trends in autonomic nervous system research* (951, pp. 196–199). Excerpta Medica Int. Congress Series.

Apte, S., Prigent, G., Stöggl, T., Martínez, A., Snyder, C., Gremeaux-Bader, V., & Aminian, K. (2021). Biomechanical response of the lower extremity to running-induced acute fatigue: A systematic review. *Front. Physiol.*, *12*, 646042.

Barraclough, M. A., & Sharpey-Schafer, E. P. (1963). Hypotention from absent circulatory reflexes. Effects of alcohol, barbiturates, psychotherapeutic drugs and other mechanisms. *Lancet*, *1*, 1121–1126.

Best, S., Caillaud, C., & Thompson, M. (2012). The effect of ageing and fitness on thermoregulatory response to high-intensity exercise. *Scand. J. Med. Sci. Sports*, *22*(4), e29–e37.

Bicknell, J. M., Appenzeller, O., Barrett, D., et al. (1984). Neurology of endurance training: Nerve conduction velocity in marathon runners; effects of conditioning and exertion. *Ann. Sports Med.*, *1*, 12–16.

Box, G. E. P., & Cox, D. R. (1964). An analysis of transformations. *J.R. Statist. Soc.*, *26*, 211–252.

Bristow, J. D., Honour, A. J., Pickering, G. W., Sleight, P., & Smyth, H. S. (1971). Effect of bicycling on the baroreflex regulation of pulse interval. *Circ. Res.*, *33*, 582–592.

Brocklehurst, J. C. (1967). The management of incontinence. *Postgrad. Med. J.*, *43*, 527–533.

Calloway, N. O. (1974). Heat production and senescence. *J. Am. Geriatrics Soc.*, *22*, 149–150.

Carr, D. B., Bergland, R. M., & Hamilton, A. (1982). Endotoxin-stimulated opioids peptide secretion: Two secretory pools and feedback control in vivo. *Science (New York, N.Y.)*, *217*, 845–848.

Carr, D. B., Bullen, B. A., & Skrinar, G. S. (1981). Physical conditioning facilitates the exercise-induced secretion of beta-endorphin and beta-lipotropin in women. *N. Engl. J. Med.*, *305*, 560–563.

Christensen, N. J. (1982). Sympathetic nervous activity and age. *Eur. J. Clin. Res.*, *12*, 91–92.

Dambrink, H. H. A., & Wieling, W. (1987). Circulatory response to postural change in healthy male subjects in relation to age. *Clin. Sci.*, *72*, 335–341.

Degroot, D. W., & Kenney, W. L. (2007). Impaired defense of core temperature in aged humans during mild cold stress. *Am. J. Physiol. Regul. Integr. Comp. Physiol.*, *292*(1), R103–R108.

DeMeirleir, K., Arentz, T., & Hollman, W. (1985). The role of the endogenous opiates in thermal regulation of the body during exercise. *Br. Med. J.*, *290*, 739–740.

Dhana, K., Evans, D. A., Rajan, K. B., Bennett, D. A., & Morris, M. C. (2020). Healthy lifestyle and the risk of Alzheimer dementia: Findings from 2 longitudinal studies. *Neurology*, *95*(4), e374–e383.

Dobkin, B. H. (1989). Orthostatic hypotension as a risk factor for symptomatic occlusive cerebrovascular disease. *Neurology*, *39*, 30–34.

Ewing, D. J., Campbell, I. W., et al. (1978). Immediate heart rate response to standing: Simple test for autonomic neuropathy in diabetes. *Br. Med. J.*, *1*, 145–147.

Exton-Smith, A. N. (1973). Accidental hypothermia. *Br. Med. J.*, *4*, 727.

Fu, Q., & Levine, B. D. (2013). Exercise and the autonomic nervous system. *Handb. Clin. Neurol.*, *117*, 147–160.

Garry, P. J., Rhyne, R. L., Haliova, L., et al. (1989). Changes in dietary patterns over a 6-year period in an elderly population. *Ann. N Y. Acad. Sci.*, *561*, 104–112.

Goldberger, J. J., Le, F. K., Lahiri, M., Kannankeril, P. J., Ng, J., & Kadish, A. H. (2006). Assessment of parasympathetic reactivation after exercise. *Am. J. Physiol. - Heart Circ. Physiol.*, *290*(6), H2446–H2452.

Goldfarb, A. H., Sforzo, G. A., & Flynn, M. G. (1984). Effect of exercise intensity upon serum beta-endorphin. In *Proceedings of ACSM Mid-Atlantic Meeting*, February 2, 1984.

Gribbin, B., Pickering, T. G., Sleight, P., & Peto, R. (1971). Effect of age and high blood pressure on baroreflex sensitivity in man. *Circ. Res.*, *29*, 424–431.

Haier, R. J., Quaid, K., & Mills, J. C. (1981). Naloxone alters pain perception after jogging. *Psychiat. Res.*, *5*, 231–232.

Halliwill, J. R., Buck, T. M., Lacewell, A. N., & Romero, S. A. (2013). Postexercise hypotension and sustained postexercise vasodilatation: What happens after we exercise? *Exp. Physiol.*, *98*(1), 7–18.

Hogikyan, R. V., & Supiano, M. A. (1994). Arterial alpha-adrenergic responsiveness is decreased and SNS activity is increased in older humans. *Am. J. Physiol.*, *266*, E717–E724.

Holaday, J. W., & Faden, A. I. (1983). Spinal shock and injury: Experimental therapeutic approaches. *Adv. Shock Res.*, *10*, 95–98.

Horvath, S. M., Radcliffe, C. E., Hutt, B. K., & Spurr, G. B. (1955). Metabolic responses of old people to a cold environment. *J. Appl. Physiol. (Bethesda, Md.: 1985)*, *8*, 145–148.

Hubbard, R. W., Matthew, C. B., & Francesconi, R. (1982). Heat stressed rat: Effects of atropine, desalivation or restraint. *J. Appl. Physiol. (Bethesda, Md.: 1985)*, *53*, 1171–1174.

Humm, A. M., Mason, L. M., & Mathias, C. J. (2008). Effects of water drinking on cardiovascular responses to supine exercise and on orthostatic hypotension after exercise in pure autonomic failure. *J. Neurol. Neurosurg. Psychiatry*, *79*(10), 1160–1164.

Hurlich, M. G., & Steegmann, A. T., Jr (1979). Hand immersion in cold water at 5°C in sub-arctic Algonkian indian males from two villages: A European admixture effect? *Hum. Biol.*, *51*, 255–278.

Imrich, R., Eldadah, B. A., Bentho, O., Pechnik, S., Sharabi, Y., Holmes, C., & Goldstein, D. S. (2009). Functional effects of cardiac sympathetic denervation in neurogenic orthostatic hypotension. *Parkinsonism Relat. Disord.*, *15*(2), 122–127.

Ivanov, K. P. (1980). The muscle heat production after adaptation to cold. *Med. Biol.*, *58*, 76–81.

Iwase, S., Mano, T., Watanabe, T., et al. (1991). Age-related changes of sympathetic outflow to muscles in humans. *J. Gerontol.*, *46*, M1–M5.

Johnson, R. H., Smith, A. C., Spalding, J. M. K., & Wollner, L. (1965). Effect of posture on blood pressure in elderly patients. *Lancet*, *1*, 731–733.

Jokl, E. (1984). Olympic medicine and sports cardiology. *Ann. Sports Med. Res.*, *1*, 127–169.

Kavanagh, T., & Shephard, R. J. (1977). Sexual activity after myocardial infarction. *Can. Med. Assoc. J.*, *116*, 1250–1253.

Keatinge, W. R., Coleshaw, S. R. K., et al. (1984). Increases in platelet and red cell counts, blood viscosity, and arterial pressure during mild surface cooling: Factors in mortality from coronary and cerebral thrombosis in winter. *Br. Med. J.*, *289*, 1405–1408.

Kenney, W. L., & Munce, T. A. (2003). Invited review: Aging and human temperature regulation. *J. Appl. Physiol. (1985)*, *95*(6), 2598–2603.

Kirkland, J. L., & Tchkonia, T. (2017). Cellular senescence: A translational perspective. *EBioMedicine*, *21*, 21–28.

Klassen, S. A., Joyner, M. J., & Baker, S. E. (2021). The impact of ageing and sex on sympathetic neurocirculatory regulation. *Semin. Cell Dev. Biol.*, *116*, 72–81.

Larsson, S., & Osterlind, G. (1944). Studies on the cause of senile miosis and rigidity of the pupil. *Acta Ophthal.*, *211*, 25.

Leaf, A. (1984). Dehydration in the elderly. *N. Engl. J. Med.*, *311*, 791–792.

Lemieux, P., & Birot, O. (2021). Altitude, exercise, and skeletal muscle angio-adaptive responses to hypoxia: A complex story. *Front. Physiol.*, *12*, 735557.

Lieberman, H. R., Wurthman, J. J., & Teichert, M. H. (1989). Aging, nutrient choice, activity and behavioral responses to nutrients. *Ann. N. Y. Acad. Sci.*, *561*, 196–208.

Lipschitz, L. A., Nyquist, R. H., et al. (1983). Postprandial reduction in blood pressure in the elderly. *N. Engl. J. Med.*, *309*, 81–83.

Little, M. A., Thomas, R. B., & Larrick, J. W. (1973). Skin temperature and cold pressor responses of Andean Indians during hand immersion in water at 4°C. *Hum. Biol.*, *45*, 643–662.

MacDonald, J. R. (2002). Potential causes, mechanisms, and implications of post exercise hypotension. *J. Hum. Hypertens.*, *16*(4), 225–236.

MacMillan, A. L., Corbett, J. L., Johnson, R. H., Crampton Smith, A., Spalding, J. M. K., & Wollner, L. (1967). Temperature regulation in survivors of accidental hypothermia of the elderly. *Lancet*, *2*, 165–169.

Mandsager, K., Harb, S., Cremer, P., Phelan, D., Nissen, S. E., & Jaber, W. (2018). Association of cardiorespiratory fitness with long-term mortality among adults undergoing exercise treadmill testing. *JAMA Network Open*, *1*(6), e183605.

Martyn, C. N., & Ewing, D. J. (1986). Pupil cycle time: A simple way of measuring an autonomic reflex. *J. Neurol. Neurosurg. Psychiat.*, *49*, 771–774.

McFarland, R. A. (1963). Experimental evidence of the relationship between ageing and oxygen want: In search of a theory of ageing. *Ergonomics*, *6*, 339–366.

Melkumyants, A. M., Balashov, S. A., & Khayutin, W. M. (1995). Control of arterial lumen by shear stress on endothelium. *News Physiol. Sci.*, *10*, 204–210.

Michelini, L. C., & Stern, J. E. (2009). Exercise-induced neuronal plasticity in central autonomic networks: Role in cardiovascular control. *Exp. Physiol.*, *94*(9), 947–960.

Montamat, S. C., Cusack, B. J., & Vestal, R. E. (1989). Management of drug therapy in the elderly. *N. Engl. J. Med.*, *321*, 303–310.

Nakamura, T., Hirayama, M., Yamashita, F., Uchida, K., Hama, T., Watanabe, H., & Sobue, G. (2010). Lowered cardiac sympathetic nerve performance in response to exercise in Parkinson's disease. *Mov. Disord.*, *25*(9), 1183–1189.

Novak, V., & Hajjar, I. (2010). The relationship between blood pressure and cognitive function. *Nat. Rev. Cardiol.*, 7(12), 686–698.

Osgood, P. F., Szyfezbein, S. K., & Murphy, J. L. (1984). Beta-endorphin release and stress analgesia following burn injury in the rat. *Proc. Am. Burn. Assoc.*, *16*, 52.

Palmer, G. J., Ziegler, M. G., et al. (1978). Response of norephinephrine and blood pressure to stress increases with age. *J. Gerontol.*, *33*, 482–487.

Pescatello, L. S., Guidry, M. A., Blanchard, B. E., Kerr, A., Taylor, A. L., Johnson, A. N., & Thompson, P. D. (2004). Exercise intensity alters postexercise hypotension. *J. Hypert.*, *22*(10), 1881–1888.

Rengo, G., Pagano, G., Parisi, V., Femminella, G. D., de Lucia, C., Liccardo, D., & Leosco, D. (2014). Changes of plasma norepinephrine and serum N-terminal pro-brain natriuretic peptide after exercise training predict survival in patients with heart failure. *Int. J. Cardiol.*, *171*(3), 384–389.

Roberts, W. W. (1988). Differential thermosensor control of thermoregulatory grooming, locomotion, and relaxed postural extension. *Ann. N. Y. Acad. Sci.*, *525*, 363–374.

Rolls, B. J. (1989). Regulation of food and fluid intake in the elderly. *Ann. N. Y. Acad. Sci.*, *561*, 217–225.

Rowe, J. W., & Troen, B. R. (1980). Sympathetic nervous system and ageing in man. *Endocrinol. Rev.*, *1*, 167–179.

Rowell, L. B., & O'Leary, D. S. (1990). Reflex control of the circulation during exercise: Chemoreflexes and mechanoreflexes. *J. Appl. Physiol. (1985)*, *69*(2), 407–418.

Rowland, D. L., Greenleaf, W. J., Dorfman, L. J., & Davidson, J. M. (1993). Aging and sexual function in men. *Arch. Sex. Behav.*, *22*(6), 545–557.

Sasco, A. J., Paffenbarger, R. S., Gendre, I., et al. (1992). The role of physical exercise in the occurrence of Parkinsons disease. *Arch. Neurol.*, *49*, 360–365.

Sellami, M., Bragazzi, N., Prince, M. S., Denham, J., & Elrayess, M. (2021). Regular, intense exercise training as a healthy aging lifestyle strategy: Preventing DNA damage, telomere shortening and adverse dna methylation changes over a lifetime. *Front. Genet.*, *12*, 652497.

Severinghaus, J. W., Chiodi, H., & Eger, E. I. (1966). Cerebral blood flow in man at high altitude. *Circ. Res.*, *19*, 274–282.

Smith, G. D., & Mathias, C. J. (1996). Differences in cardiovascular responses to supine exercise and to standing after exercise in two clinical subgroups of Shy-Drager syndrome (multiple system atrophy). *J. Neurol. Neurosurg. Psychiatry*, *61*(3), 297–303.

Smith, S. A. (1988). Pupillary function in autonomic failure. In R. Bannister (Ed.), *Autonomic failure* (2nd ed., pp. 393–412). Oxford: Oxford University Press.

Smith, S. A., & Smith, S. E. (1983). Evidence for a neuropathic actiology in the small pupil of diabetes mellitus. *Br. J. Ophthalmol.*, *67*, 89–93.

Smyth, H. S., Sleight, P., & Pickering, G. W. (1969a). Reflex regulation in arterial pressure during sleep in man. *Circ. Res.*, *24*, 109–121.

Smyth, H. S., Sleight, P., & Pickering, G. W. (1969b). Reflex regulation of arterial pressure during baroreflex sensitivity. *Circ. Res.*, *24*, 109–121.

Solomon, S., Corwin, E., & Appenzeller, O. (1987). Atrial natriuretic factor and vasopressin in endurance trained subjects. *Ann. Sports Med.*, *3*, 163–166.

Tarlov, A. R., Ware, J. E., Greenfield, D., et al. (1989). The medical outcomes study: An application of methods for monitoring the results of medical care. *JAMA: J. Am. Med. Assoc.*, *262*, 925–930.

Taylor, G. F. (Ed.), (1964). Letters to the editor. Old people in the cold. *Br. Med. J.*, *1*, 428.

Tokura, H., Ohta, T., et al. (1979). Circadian change of sweating rate measured locally by the resistance hygrometry method in man. *Experientia*, *35*, 615–616.

Wallin, G. (1983). Intraneural recording and autonomic function in man. In R. Bannister (Ed.), *Autonomic failure. A textbook of clinical disorders of the autonomic nervous system* (pp. 36–51). Oxford, New York: Oxford University Press.

Walsh, K. E., & Berman, J. R. (2004). Sexual dysfunction in the older woman: an overview of the current understanding and management. *Drugs Aging*, *21*(10), 655–675.

Watts, A. J. (1971). Hypothermia in the aged: A study of the role of cold-sensitivity. *Environ. Res.*, *5*, 119–216.

Wenger, C. B., Roberts, M. F., et al. (1976). Nocturnal lowering of thresholds for sweating and vasodilation. *J. Appl. Physiol. (Bethesda, Md.: 1985)*, *41*, 15–19.

Wieling, W. (1988). Standing, orthostatic stress, and autonomic function. In R. Bannister (Ed.), *Autonomic failure* (2nd ed., pp. 308–320). Oxford: Oxford University Press.

Wollner, L., McCarthy, S. T., et al. (1979). Failure of cerebral autoregulation as a cause of brain dysfunction in the elderly. *Br. Med. J.*, *I*, 1117–1118.

Wood, L., Wood, S., Appenzeller, P., et al. (1992). Olfaction age and altitude. In J. R. Sutton, J. Coates, & C. S. Houston (Eds.), *Proceedings of the 7th International Hypoxia Symposium at Lake Louise, Canada* (p. 313).

Wood, S. C. (1988). Preface. *Ann. Sports Med.*, *4*, 165.

Wood, S. C., Appenzeller, O., & Appenzeller, P. (1989). Transcranial measurement of cerebral blood flow in sojourners and natives at high altitudes. *Ann. Sports Med.*, *4*, 289–293.

Yamaguchi, F., Meyer, J. S., et al. (1979). Normal human aging and cerebral vasoconstrictive responses to hypocapnia. *J. Neurol. Sci.*, *44*, 87–94.

Yoon, S. Y., Suh, J. H., Yang, S. N., Han, K., & Kim, Y. W. (2021). Association of physical activity, including amount and maintenance, with all-cause mortality in Parkinson disease. *JAMA Neurol.*

Yu, A. Q., Wang, J., Zhou, X. J., Chen, K. Y., Cao, Y., Wang, Z. X., & Mao, Z. B. (2020). Senescent cell-secreted netrin-1 modulates aging-related disorders by recruiting sympathetic fibers. *Front. Aging Neurosci.*, *12*, 507140.

Zheng, H., Li, Y. F., Cornish, K. G., Zucker, I. H., & Patel, K. P. (2005). Exercise training improves endogenous nitric oxide mechanisms within the paraventricular nucleus in rats with heart failure. *Am. J. Physiol. - Heart Circ. Physiol.*, *288*(5), H2332–H2341.

CHAPTER 8

Autonomic neuropathies

Damage to autonomic nerves can be due to diverse causes, such as an inherited disorder such as a Hereditary Sensory Autonomic Neuropathy, metabolic derangements as in diabetes, or accompanying a primarily peripheral nerve injury. The clinical manifestations of autonomic neuropathies can also be diverse and involve various systems such as cardiovascular, sudomotor, gastrointestinal, and urogenital systems.

8.1 Blood supply to peripheral autonomic nerves

Blood vessels might be of pathogenetic significance in certain neuropathies where autonomic symptoms are often the overwhelming clinical difficulties, for example, in diabetic neuropathy. The circulation in nerves is not passive but is regulated by adrenergic and nonadrenergic neuroeffector mechanisms. The localization of various peptides and serotonin within perivascular nerves supplying vasa nervorum is of relevance not only to normal peripheral nerve function, but also in the maintenance of normal blood–nerve barrier function. In experimental and human neuropathies, advances have been made in the definition of the role of these perivascular nerves and in the changes found in these nerves in various neuropathies.

8.1.1 Historical aspects

The functional significance of vasa nervorum was noted in the 17th century (Spiegel, 1627). In the 18th century (Haller, 1756; Ruysch, 1701), these blood vessels were rediscovered, but it was not until later that these blood vessels were studied experimentally (Isenflamm & Doerffler, 1768). Details about the origin of vasa nervorum, their distribution within the nerve, and their physiologic role were published throughout the 19th century (Bichat, 1830; Ranvier, 1878; Zuckerkandl, 1885). Vascular pathology in these vessels and the effect on peripheral, autonomic, and cranial nerve function were not as yet recognized.

During World War I, there were a large number of nerve injuries. The possibility that vasa nervorum and nervi nervorum may be important in nerve regeneration, which was of great interest at the time, was not

Introduction to Clinical Aspects of the Autonomic Nervous System
DOI: https://doi.org/10.1016/B978-0-323-96011-3.00011-5

considered until well after the conclusion of the war when an examination of the blood supply of large peripheral nerves of the upper limbs in humans was undertaken (Ramage, 1927). With the advent of World War II, interest in the vasa nervorum was kindled again by numerous injuries to peripheral nerves. Physiologists noticed for the first time, a possible functional significance of these blood vessels. The pathologic significance of vasa nervorum in peripheral nerve disorders was then also appreciated. Many subsequent studies have characterized intricate details of the vascular topography, and physiologic studies have recognized the importance of vasa nervorum for electric conduction in peripheral nerves. But the neurogenic regulation of these vessels, pivotal in the control of vascular tone and in the responses to changing functional demands, has been neglected.

8.1.2 Vascular anatomy

The early study of the anatomic distribution of vasa nervorum was mainly carried out in fetal material. Injection techniques, using plastic dyes, histochemical methods, and microangiography, have shown fine details of the arrangements of these vessels (Bell & Weddell, 1984; Nobel & Black, 1974). These blood vessels have also been studied in vivo and contrast-enhanced ultrasound can also detect the vasa nervorum (Qureshi, Saleem, Ahrar, & Raja, 2017). Extensive reviews of the anatomy of the circulation of cranial, peripheral, and autonomic nerves have been published (Sunderland, 1978). Descriptions of the sciatic vasa nervorum in mammals and humans have also been published (Bell & Weddell, 1984).

All blood vessels, which enter and terminate within a nerve, are termed vasa nervorum. In addition, extraneural (nutrient) and intraneural blood vessels are recognized. The intraneural vessels can be subdivided into epineural, perineural, and endoneural vasa nervorum. These blood vessels all arise from the nutrient arteries. The nutrient arteries may arise as branches of the main arteries of limbs and thus provide a direct conduit of blood to the nerve. Less common are nutrient arteries that are branches of subsidiary vessels that supply extraneural tissues, primarily muscles and skin. Variations in the origin of nutrient arteries are found in various species, and there are usually asymmetries on the two body sides. Nevertheless, the origin of these arteries is consistent for any particular nerve (Sunderland, 1945). In humans, the length of extraneural arteries varies from 5 to 15 mm, but branches of the radial artery supplying the median nerve in the forearm may reach up to 25 mm (Sunderland, 1978). This variability in the lengths of nutrient arteries accounts for long segments of main nerve trunks being without external blood supply. The circulation to these segments, however, is easily maintained by descending intraneural vessels. The size of the nutrient artery or group of nutrient blood vessels determines the size and number of other arteries along the nerve (Sunderland, 1978). Once nutrient arteries reach a nerve trunk, they distribute within the nerve and create four distinct intraneural plexuses of vasa nervorum, which are arranged concentrically. These intraneural plexuses are parallel to the long axis of the nerve. The first plexus is on the surface of the nerve, followed by epineural, perineural, and endoneural plexuses. These blood vessels are generally of small caliber, 92% being smaller than 10 mm in diameter (Marcarian & Smith, 1968). Because there are no postnatal developmental changes in the vasculature of peripheral nerves (Sunderland, 1978), the decrease in the density of endoneural capillaries with age is the result of increasing myelination and a consequent separation of the capillaries (Lang, 1962). Epineural blood vessels are branches of nutrient arteries. They divide after reaching the nerve into ascending and descending branches and communicate by an epineural plexus of longitudinally oriented blood vessels. Within the nerves, glomerular-like loops and arteriovenous shunts have been described (Lundborg, 1970). The epineural vasa nervorum are of arteriolar size, but in the median and sciatic nerves, larger vessels are also found.

Ultrastructural examination of perineural blood vessels has confirmed the anatomical features described by using light microscopy (Key & Retzius, 1876). The perineurium itself consists of flattened cells arranged in layers, with basement membranes on both sides. The basement membranes are closely apposed towards the pointed ends of the cells. These perineural cells are encircled by a fibrous outer layer and merge gradually into the connective tissue of the epineurium. Within the perineurium, an interfascicular plexus of arterioles, capillaries, and venules can be found. Short perineural sleeves accompany the entering blood vessels into the endoneurium (Waggener, Bunn, & Beggs, 1965). Studies of the outer perineural nerve sheath in the maintenance of the blood–nerve barrier have suggested that this barrier is anatomically situated in the perineural sleeves, which accompany the penetrating vessels (Olsson, 1972). However, the importance of the innervation of such sleeves and the possible role in maintenance of the blood–nerve barrier of this innervation has not been assessed. The perineural vasa nervorum are for the most part of arteriolar and precapillary size. These vessels take an oblique course in penetrating the perineurium. This arrangement results in a valve-like mechanism that closes when there is swelling of the fascicle (Lundborg, 1970).

In the endoneurium, the vasa nervorum branch frequently and symmetrically, consisting only of capillaries. Widespread anastomoses from longitudinal branches occur proximally and distally. Connecting transverse vascular channels, which run parallel with nerve fibers, are also found. These endoneural capillaries drain into venules and veins, which are located for the most part in the perineurium only.

It is suggested that venous drainage accompanies the intraneural arterial pattern. The veins exit from the nerve at the site of arterial entries. The blood from nerve trunks is thought to drain directly into the veins of muscles or through neurovascular bundles. There are great differences in venous anatomy between species. Differences also exist in the innervation of veins (regardless of species) and possibly also in the role of veins in the maintenance of a normal blood–nerve barrier.

8.1.3 Vasa nervorum: structural aspects

Vasa nervorum have the same structure as other blood vessels, but ultrastructural features are less well-defined. There are endothelial-type junctions at points of cellular contacts (Olsson & Reese, 1969). There is a relatively large extracellular space in the surrounding connective tissue, and outside this zone, connective tissue fibers parallel the endoneural blood vessels. Outside the endoneurium are connective tissue cells, fibrocytes, macrophages, and mast cells, which are embedded in collagen. The number of mast cells varies in different peripheral, somatic, and autonomic nerves. The number and distribution of cells also vary in different species. Mast cells are thought to be modulators of blood-vessel caliber, and they are frequently found along vasa nervorum; however, their role in health and disease states is debated. Mast cells both synthesize and are storage sites of vasoactive substances, which can be released in pathologic reactions of peripheral, autonomic, and cranial nerves to injurious agents. These substances can also contribute to edema of nerves by changing the permeability of the blood–nerve barrier.

The destination of nutrients is the nerve fiber. In order to reach these fibers from the blood, nutrients must pass through the vessel wall, enter the extracellular space, and penetrate the basement membranes of Schwann cells. This arrangement differs from that in the central nervous system, where a small extracellular space and the close apposition of glial cells to vessel walls make the distance of penetration of nutrients much shorter. An important barrier to the passage of many molecules, particularly proteins, is the tight junction between endothelial cells of the endoneural capillaries. No such junctions are found in the epi- and perineural vessels, so that the leakage of many substances into the surrounding tissue proceeds unimpeded.

8.1.4 The blood–nerve barrier

The permeability of vasa nervorum regulates the entry of metabolites, toxins, and other substances to the nerve fibers and is an important but not exclusive function of the blood–nerve barrier in health and disease. The distribution of proteins, toxins, dyes, and other noxious substances was studied (Waksman, 1961). For example, diphtheria toxin and toxoid pass unhindered into guinea pigs' nerves, but not into the nerves of rabbits. That is because the blood–nerve barriers of these two species differ; hence, their vulnerability to disease also differs.

Using the mast-cell degranulator compound 48/80, it was shown that vasa nervorum are important in the blood–nerve barrier function of rabbits and guinea pigs. This compound causes degranulation of all mast cells in these animal nerves. However, mast cells found within the endoneurium of normal rat peripheral nerves are not degranulated. This preservation of mast cell granules in the endoneurium is thought to be due to an inability of compound 48/80 to pass the blood–nerve barrier in the endoneurium, because when injected directly into the endoneurium, this compound degranulates endoneural mast cells. In experimental isonicotinic acid hydrazide peripheral neuropathy, endoneural degranulation of mast cells is found, indicating a disruption of the blood–nerve barrier in this experimental disease (Olsson, 1968). Because the permeability of vasa nervorum contributes to the formation of edema, by inference, the innervation of these blood vessels may also play an important role in the formation of nerve edema. The permeability of vasa nervorum to serum proteins has been studied using injections of proteins made traceable with dyes or fluorescent or radioactive markers. The use of peroxidase tracers has made it possible to examine the permeability of vasa nervorum at an ultrastructural level as well (Mellick & Cavanagh, 1968). Because both albumin and gamma globulin travel unimpeded to the connective tissue around blood vessels, it can be inferred that both penetrate the epineural vasa nervorum. These proteins are found in the extracellular space and in the innermost parts of the perineurium. Passage of small amounts of other serum proteins across the vasa nervorum has also been found (Mancini, 1963). In other vascular beds that allow the unhindered passage of these proteins, the blood vessels have a slow flow of blood, and, by inference, this is probably also true of vasa nervorum. Whether the perivascular nerve fibers of vasa nervorum influence the passage of proteins from these blood vessels into the nerve is not known. The reported species differences in permeability of vasa nervorum may depend on the innervation of these vessels or on morphologic characteristics of their endothelial junction, which vary from species to species. Lymphatics are also important in the reabsorption of proteins, but whether nervi nervorum influence lymph flow has not been studied. The presence of lymphatics in the epineurium of some peripheral nerves has been reported (Sunderland, 1965).

8.1.5 Innervation of vasa nervorum and the nervi nervorum

In addition to traditional neurotransmitters mediating vasoconstriction through noradrenergic fibers and vasodilatation through parasympathetic acetylcholine-containing fibers, peptidergic and serotoninergic fibers have been found in perivascular plexuses. The innervation of vasa nervorum by adrenergic and cholinergic nerve has been demonstrated using conventional histologic techniques in combination with fluorescence histochemistry (Amenta, Mione, & Napoleone, 1983). A technique for the examination of epi- and perineurium of peripheral, autonomic and cranial nerves by wholemount stretch preparations made it possible to apply selective and sensitive immunohistochemical methods. This technique allowed the assessment of the extent of the innervation of vasa nervorum and of the neurotransmitter content of nervi nervorum, including peptides and serotonin (Appenzeller, Dhital, Cowen, & Burnstock, 1984). These studies have confirmed earlier anatomic descriptions (Hromada, 1963) that have shown four distinct regions of peri- and nonperivascular innervation associated with nerve trunks. These regions are the perivascular innervation of extraneural nutrient arteries and veins; the perivascular innervation of intraneural blood vessels, that is, the vasa nervorum; and the nonperivascular nerve fibers (nervi nervorum), which originate from the nerve trunk and from perivascular nerve plexuses and which are distributed to the epi-, peri-, and endoneural connective tissues. Lastly, there are intrafascicular nerve fibers containing specific neurotransmitters (Figs. 8.1–8.3).

Figure 8.1 Catecholamine fluorescence shows four regions of peri- and nonperivascular innervation typically associated with a nerve trunk (in this case the sciatic nerve of the rat). (A) extraneural (nutrient) artery; (B) epineural vasa nervorum; (C) nonperivascular perineural nerve fibers (nervi nervorum); (D) intrafascicular nerve fibers within the main nerve trunk; calibration bars = 25 mm. *From Dhital, K., Lincoln, J., Appenzeller, O., & Burnstock, G. (1986). Adrenergic innervation of vasa and nervi nervorum of optic, sciatic, vagus and sympathetic nerve trunks in normal and streptozotocin-diabetic rats.* Brain Res., 367*(1–2), 39–44.*

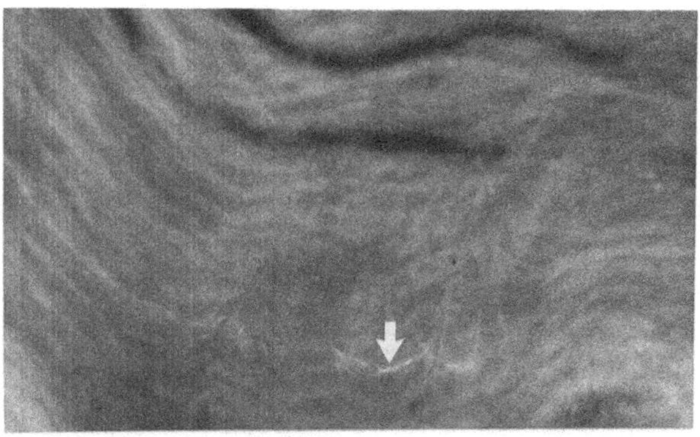

Figure 8.2 Distribution of norepinephrine (NE)-containing nerves in whole-mount preparations of normal human sural nerve. (A) Fluorescent NE-containing nerves forming a perivascular plexus on a nutrient artery. (B) NE-containing nerves forming a perivascular plexus in a perineurial artery (arrowhead) and in the nervi nervorum (arrows). (C) Endoneural vein showing sparse NE-containing nerve fibers. (D) Auto fluorescent fascicular preparation showing one NE-containing intrafascicular nerve fiber. (E) Perifascicular NE-containing nerve fiber in a fascicular preparation, which also shows high background autofluorescence. (F) NE fluorescence in a nonperivascular nerve fiber of the nervi nervorum in the epineurium. Bars = 10 mm. *From Lincoln, J., Milner, P., Appenzeller, O., et al. (1993). Innervation of normal human sural and optic nerves by noradrenalin and peptide containing nervi vasorum and nervorum: Effect od diabetes and alcoholism.* Brain Res., 632*, 48–56.*

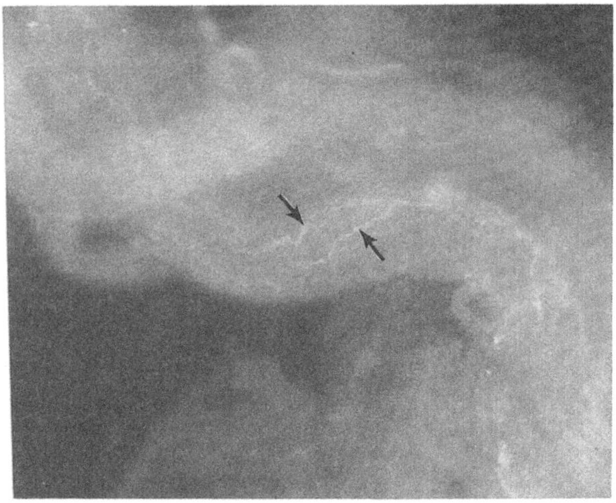

Figure 8.3 Diagrammatic representation of the arterial (A) and venous (B) innervation of the vasa nervorum. There is great variability in the densities of perivascular nerves but the rank order depicted here is representative of most nerves.

8.1.6 Noradrenergic innervation of intraneural blood vessels

The density of adrenergic nervi nervorum and nerves innervating the vasa nervorum (nervi vasorum) varies from nerve to nerve and in experimental conditions. Thus, histochemical descriptions in streptozotocin-diabetic rats show the density of these nerves to be markedly increased in the sciatic and vagus nerves and in the paravertebral sympathetic chain of diabetic animals (Fig. 8.4). In contrast, the optic nerve meninges of diabetic animals showed a substantial loss of peri- and nonperivascular adrenergic nerve fibers (Fig. 8.4). Noradrenaline assays confirmed the increase in the vagus nerve and the decrease in the optic nerve, but the diabetic sciatic nerves had significantly lower noradrenaline levels than controls. Thus, in streptozotocin diabetes, a simultaneous decrease in adrenergic fibers around blood vessels in the optic nerve sheath, with an increase in adrenergic fibers in other nerve sheaths in the same animals, is found (Fig. 8.5).

The majority of the total sciatic trunk noradrenergic content is in the intrafascicular nerve bundles, whereas in the vagus, most of the noradrenaline is found in epi- and perineural tissue. In the sciatic nerve, the intrafascicular noradrenergic nerve bundles are heavier than those in the nerve sheath and account for a greater percentage of total nerve weight. The opposite situation is present in the vagus; thus, relatively small changes in intrafascicular nerve fiber content in the sciatic nerve will be more marked

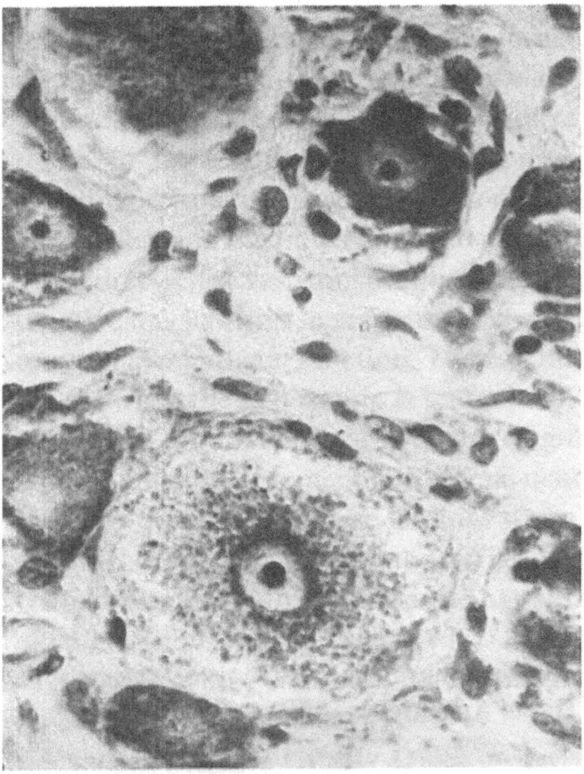

Figure 8.4 Peri- and nonperivascular catecholamine-positive nerve fibers in the vasa nervorum and nervi nervorum of the sciatic, vagus and optic nerves of control and streptozotocin-diabetic rats. (A) Control sciatic vasa nervorum, (B) diabetic sciatic vasa nervorum showing increased perivascular adrenergic innervation; (C) control vagal vasa nervorum; (D) diabetic vagal vasa nervorum. Note increase in adrenergic nerves in vasa nervorum and nervi nervorum (large arrows). Mast cells are indicated by small arrows. (E) Control optic vasa nervorum; (F) diabetic optic vasa nervorum showing loss of adrenergic innervation. Open arrows indicate the outer margins of vessel walls. Calibration bars 25 mm. *From Dhital, K., Lincoln, J., Appenzeller, O., & Burnstock, G. (1986). Adrenergic innervation of vasa and nervi nervorum of optic, sciatic, vagus and sympathetic nerve trunks in normal and streptozotocin-diabetic rats.* Brain Res., *367(1–2), 39–44.*

than similar small changes in smaller diameter nerves such as the vagus. Thus, streptozotocin-induced diabetes may cause loss of intrafascicular noradrenergic nerve fibers in the sciatic nerve and mask the histochemical changes observed in the epi- and perineurium (Dhital, Lincoln, Appenzeller, & Burnstock, 1986). Similar differential susceptibility to streptozotocin in rats has been demonstrated in other tissues. Whether

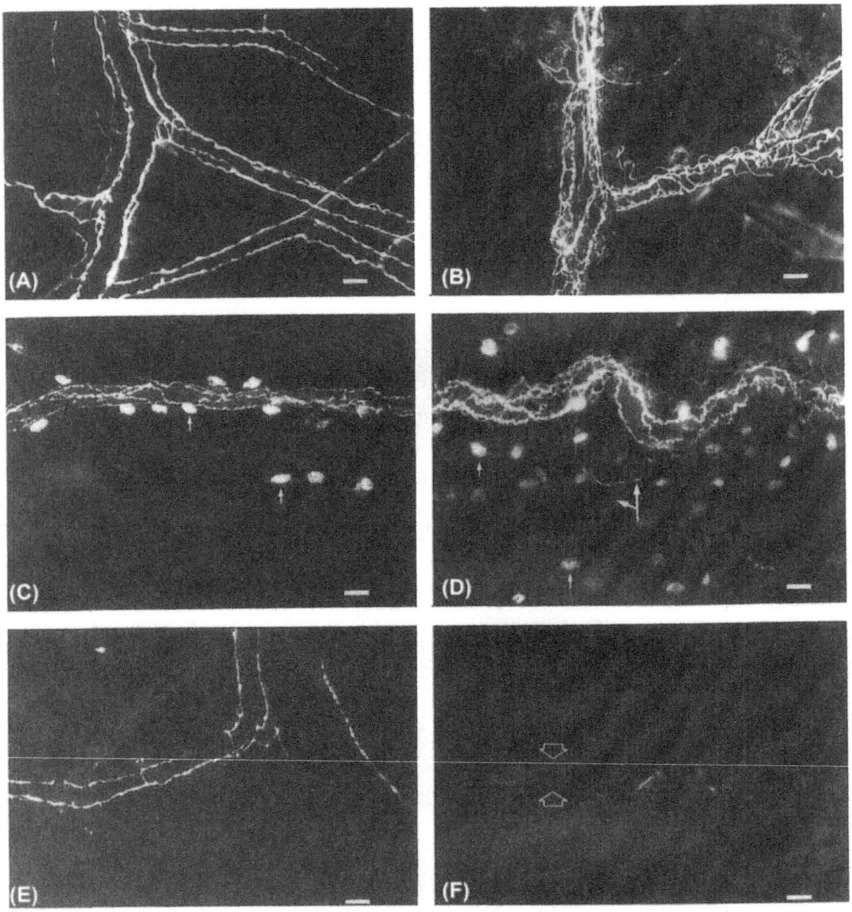

Figure 8.5 Norepinephrine (NA) content in whole segments of sciatic, vagus and optic nerves of control and diabetic animals expressed in mg/g tissue (left side of figure) and ng/cm length (right side of figure) shown as mean + SEM. Number of animals in each experiment indicated below each bar. Open bars (controls), shaded bars (diabetic animals). *nd*, Not detected. $^*P < .05$. *From Dhital, K., Lincoln, J., Appenzeller, O., & Burnstock, G. (1986). Adrenergic innervation of vasa and nervi nervorum of optic, sciatic, vagus and sympathetic nerve trunks in normal and streptozotocin-diabetic rats.* Brain Res., 367*(1–2), 39–44.*

such differential reactions are the result of tissue function, varied susceptibility to the noxious agents, or depend on the origin of noradrenergic fibers, has not been determined. In these experimental studies, an increase in the number of mast cells in the vicinity of vasa nervorum was also found in the diabetic condition. The close association between these cells and the walls of small blood vessels was established, and the suggestion that mast cells play a role in the regulation of microcirculation was made. In rats, the mast cells contain, in addition to histamine, 5-hydroxytryptamine (Parratt & West, 1957). The increased number of these cells found in streptozotocin diabetes may reflect a pathologic phenomenon associated with the microcirculatory dysfunction in these animals. Though streptozotocin animal models for human diabetes do not mimic all the features of the human disease, these findings support a role for vasa nervorum in the genesis of some clinical symptoms of diabetic peripheral and cranial nerve involvement in humans.

8.1.7 Vasoactive intestinal polypeptide

Vasoactive intestinal polypeptide (VIP), a neuropeptide that functions as a neurotransmitter and neuromodulator, is an important vasoactive peptide found in most vascular beds (Edvinsson, 1985). Nerve fibers showing VIP-like immunoreactivity (VIP-LI) have been demonstrated in walls of vasa nervorum in animals and in sural nerves of humans (Appenzeller et al., 1984) (Figs. 8.6 and 8.7). Blood vessels from rabbits and rats contain a sparse network of fibers, which tend to accompany the larger blood vessels found in the sciatic, vagus, and sympathetic nerve trunks. The density of this innervation increases with the age of animals.

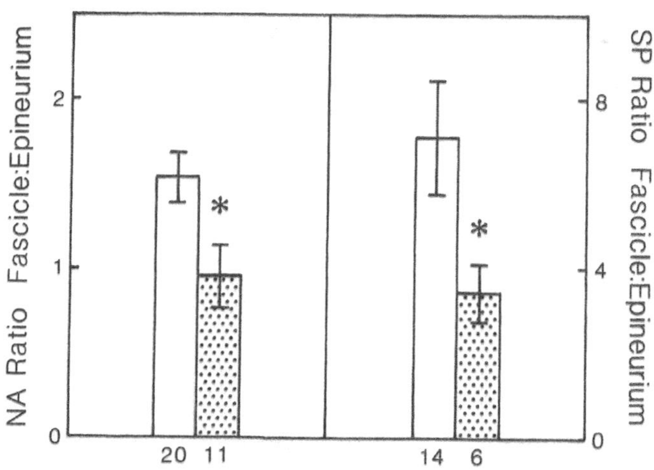

Figure 8.6 VIP-like immunoreactive nerve fiber. Rat sciatic nervi nervorum. Calibration bar 25 mm. *From Dhital, K.K., & Appenzeller, O. (1988). Innervation of vasa nervorum, Chapter 11. In G. Burnstock & S.G. Griffith (Eds.),* Noradrenergic innervation of blood vessels *(191–211). Boca Raton, FL: CRC Press.*

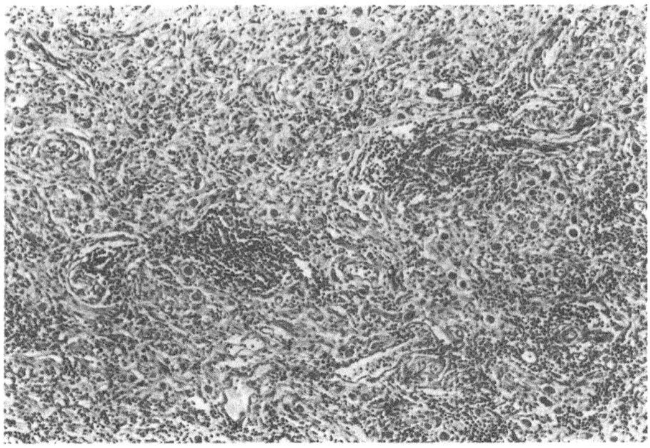

Figure 8.7 VIP-like immunoreactive nerve fibers. Human sural nerve perivascular nerve plexus vasa nervorum.

VIP may play an important role in the reaction to nerve injury. Sciatic nerve transection causes a marked increase in VIP content along the entire proximal stump of the rat nerve. By contrast, there is a significant fall in the level of Substance-P (SP). Changes in the levels of these peptides can be observed within 3 days and for at least 30 days after nerve section. These changes are exaggerated if the experiments are carried out in neonatal rats. Confirmation of these findings in nerves can be obtained from the examination of the level of VIP in hind-paw skin, which is supplied by the distal stump of the sectioned nerve and which shows a significant fall or undetectable level of the peptide (Anand, Antia, & Blanck, 1983). The origin of the VIP accumulation in the proximal stump of sciatic nerves is not fully known, but experiments imply that the increase in VIP-LI might derive from sympathetic ganglion cells where the peptide has been colocalized in acetylcholinesterase (AChE)-staining postganglionic sympathetic neurons (Lundberg et al., 1979). It has also been shown that sciatic nerve section causes an ipsilateral increase in VIP levels in the dorsal lumbar cord (McGregor et al., 1984). Moreover, the injection of *Mycobacterium leprae* into the hindlimbs of mice, which damages sciatic nerves, also causes an increase in VIP in dorsal lumbar cords of these animals. Dorsal rhizotomy, however, decreases VIP levels in the dorsal horn of rats and cats (Yaksh, Abay, & Go, 1982). Observations on streptozotocin-diabetic rats show an increase in VIP in the diabetic nerve, but Substance-P remains normal. During crush-induced Wallerian degeneration, there is a severe reduction of both VIP and SP in the distal segment.

In such animals, there is also an accumulation of water in the endoneural compartment, which has been attributed to an impaired extraction by the perineurium, which may be enhanced by the increase in VIP content of the diabetic nerve. However, these changes were thought to be unlikely to contribute to nerve fiber damage (Anand et al., 1988).

The administration of capsaicin to neonatal rats induces loss of unmyelinated afferent fibers and of VIP-containing fibers in the dorsal horn and depletes afferent SP fibers in sciatic nerves (Gibson, McGregor, & Bloom, 1982). In experiments where neonatal rats were given systemic capsaicin or treated by local application of capsaicin to a sciatic nerve prior to sectioning the nerve, the increase in VIP content of the proximal sciatic stump remained unchanged. On the other hand, treatment of rats in the neonatal period with 6-hydroxydopamine or antinerve-growth-factor antiserum induced, with the ensuing sympathetic denervation brought about by the administration of these substances, a significant decrease of VIP in the distal stump of the sectioned nerve. These treatments and consequent sympathetic denervation also induced an elevation in VIP content proximal to the nerve injury. This sympathectomy-induced decrease in VIP, while not completely eliminating VIP levels, supports the idea that a majority of postneurectomy increases in VIP levels are related to sympathetic neurons, probably situated in the paravertebral sympathetic chain. Immunohistochemical studies of wholemount epi- and perineural stretch preparations of sciatic nerves support the idea that the VIP accumulation after neurectomy occurs in nervi nervorum, which presumably originate in the sympathetic nervous system (Fig. 8.8) (Dhital & Appenzeller, 1988). VIP is a vasodilator and potentiates the formation of edema (Williams, 1982). These findings may be of importance in the changes in the blood—nerve barrier in neuropathies of various types and in the increase in epineural permeability and nerve water after nerve injury (Olsson & Kristensson, 1979).

VIP may regulate local blood flow and vessel permeability, and it could be important in the organizational changes and regenerative processes in both peripheral and central nervous systems. It is also likely that these responses to injury are nonspecific. Moreover, its importance in clinical situations, particularly in causalgia, is not known. Injury to nerves in humans produces sympathetic reflex dystrophies, including causalgia, and the vasomotor system is markedly affected in the involved limb, but the vasa nervorum in this condition need further study.

Paravertebral sympathetic chain neurons probably give rise to perivascular VIP-LI nerve fibers, but the possibility that these fibers might also innervate

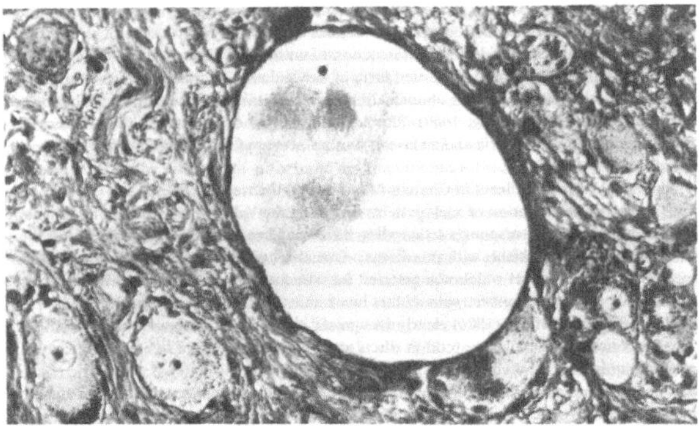

Figure 8.8 Rat sciatic nerves. (A) Perivascular noradrenergic innervation of vasa nervorum. (B) After neonatal sympathectomy with 6-hydroxydopamine. (C) Perivascular VIP-like immunoreactive nerve fiber. (D) Increase in VIP-like immunoreactivity after neurectomy. Calibration bars 25 mm.

cranial nerve vasa nervorum or the nerve trunks themselves must be considered. There is increasing evidence for a central origin of VIP-LI fibers innervating the cerebral vasculature (Kobayashi, Kyoshima, Olschowka, & Jacobowitz, 1983). Similarly, it is possible that the VIP-LI fibers innervating cranial nerve vasa nervorum may also have a central origin. Some evidence suggests that adrenergic perivascular plexuses around vasa nervorum derive mainly from proximal blood vessels, where the noradrenergic innervation is very dense (Cowen, Haven, & Wen Quin, 1982). Some support for an alternative view also exists. That view suggests that at least the innervation of arteriae nervorum might derive from the nerves they supply with blood.

In immature animals, vasomotor responses to physiologic conditions and probably also to noxious stimuli are different from those found in mature animals. The finding of fewer VIP-containing perivascular fibers in peripheral and autonomic nerves in young animals implies that some early vasomotor activity might result from an immature innervation of vasa nervorum (Lanpus, Collins, Bloodworth, & Wood, 1962). VIP appears to have an important role in normal function and in the reaction to injury in many vascular beds. Studies on VIP-containing perivascular nerve fibers around vasa nervorum in somatic and autonomic nerves imply a similar role for this peptide in this vascular bed. The functional role of the VIP-LI nervi nervorum has not been investigated, but it is possible that local pain or sensitivity of partially traumatized or diseased nerves might be related to these peptide-containing fibers, perhaps analogous to the pain sensitivity of dural coverings of the brain.

8.1.8 Substance-P

Substance-P (SP), an undecapeptide, is important in the transmission of nociception in the central nervous system (Pernow & Substance, 1983). Advances in immunohistochemical techniques have allowed the visualization of SP-containing nerve fibers around autonomic and somatic vasa nervorum (Fig. 8.9). They have also made possible the visualization of a delicate nonvascular plexus of SP-containing fibers in the epi- and perineurium. This plexus presumably forms a part of a network of nervi nervorum. The distribution and densities of these SP-containing nerve fibers, like those found around cerebral arteries and veins, varies with the species examined. The distribution of SP-containing immunoreactive perivascular nerve fibers around vasa nervorum at various levels of the nerves, however, has not been studied. As in other vascular beds, these nerves may be more prominent around larger vasa nervorum; for example, SP-containing fibers are very dense around the circle of Willis and become less dense

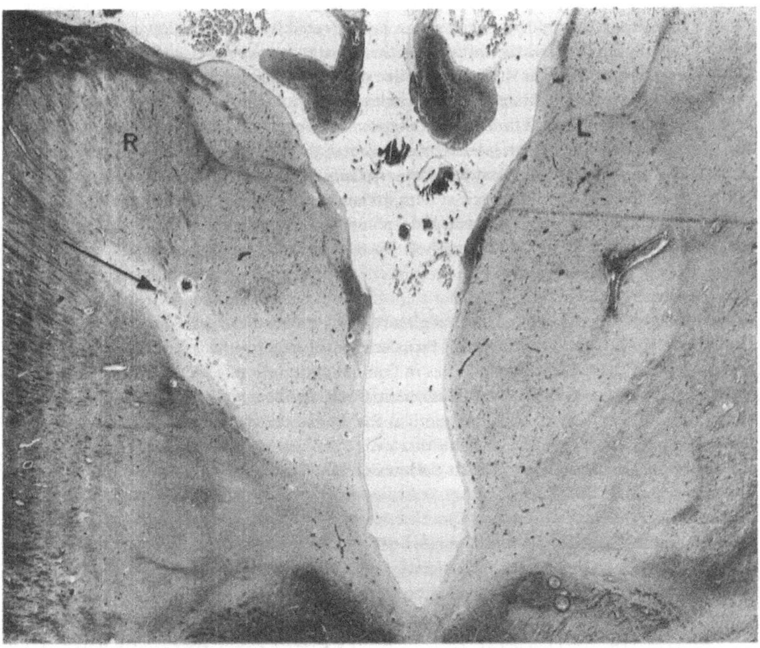

Figure 8.9 SP-like immunoreactive nerve fibers, guinea-pig vagus. Perivascular nerve fibers of vasa nervorum (large arrow), nervi nervorum (small arrow). Calibration bar 25 mm. *From Dhital, K.K., & Appenzeller, O. (1988). Innervation of vasa nervorum, Chapter 11. In G. Burnstock & S.G. Griffith (Eds.),* Noradrenergic innervation of blood vessels *(191–211). Boca Raton, FL: CRC Press.*

more distally. These fibers may mediate vascular pain perception, as in certain blood vessels in the brain, which are innervated by SP-containing nerve fibers that originate in the trigeminal system (Moskowitz, 1984).

The origin of SP-containing fibers around vasa nervorum and in nervi nervorum has not been examined. Whether capsaicin administration causes a substantial loss of SP fibers in nerves around vasa nervorum and in nervi nervorum is not known, but the suggestion that SP-containing nerve fibers have a sensory role in peripheral nerves seems reasonable. In the rabbit, SP-containing nervi nervorum and fibers supplying vasa nervorum increase in density with advancing age. SP has both vasodilator and vasoconstrictor actions on vascular smooth muscle (Hellstrand & Jarhult, 1980). It is also thought to be important in reflex vasodilatation (Burnstock, 1977) and more recent studies have shown that SP-induced vasodilatation is endothelium-dependent (Zawadski, Furchgott, & Cherry, 1981). There is no correlation between the density of SP-containing nerve fibers and the vascular responsiveness to SP, which supports a sensory rather than primary neurovascular regulatory role for this peptide. The best view at present is that SP does not play a role in direct short-term autoregulation of vascular tone, but that it regulates blood flow in response to noxious stimuli or during pathologic conditions, such as ischemia, claudication, or after aneurysmal rupture. A role for SP in unilateral headache in humans has also been proposed (Moskowitz, 1984).

8.1.9 Calcitonin gene-related peptide

Calcitonin gene-related peptide (CGRP) is the main product of calcitonin gene expression in nonthyroidal tissue. It is a 37-amino acid residue peptide with potent vasodilator properties.

Immunohistochemical studies have shown CGRP to coexist with SP in various peri- and nonperivascular tissues. CGRP-like immunoreactive nerve fibers have been demonstrated in a few peripheral nerves of the rat and guinea pig around vasa nervorum and in nervi nervorum (Fig. 8.10). CGRP exists in two forms, α-CGRP and β-CGRP, that share more than 90% homology. In the peripheral nervous system, α-CGRP is found mainly in sensory neurons that are closely associated with blood vessels, at the junction of the adventitia and the media passing into the muscle layer, where CGRP release is localized appropriately to act as a potential vasodilatory agent. CGRP may play a role in the regulation of blood pressure and CGRP

Figure 8.10 CGRP-like immunoreactive nerve fibers, guinea-pig vagus. Perivascular nerve fibers of vasa nervorum (large arrows). Nervi nervorum (small arrows). Calibration bar 25 mm. *From Dhital, K.K., & Appenzeller, O. (1988). Innervation of vasa nervorum, Chapter 11. In G. Burnstock & S.G. Griffith (Eds.),* Noradrenergic innervation of blood vessels *(191–211). Boca Raton, FL: CRC Press.*

knockout mice have elevated daytime blood pressure, higher urinary and plasma norepinephrine concentrations, increased baroreflex-mediated vagal activation after α adrenergic receptor stimulation, and significant aorta stiffening (Mai, Wu, Diedrich, Garland, & Robertson, 2014).

8.1.10 Neuropeptide-Y

Neuropeptide-Y (NPY) is a 36-amino acid residue peptide with potent vasomotor functions and is contained in perivascular nerve fibers throughout the circulatory system. Using immunohistochemical techniques, NPY has been found in nerve fibers innervating vasa nervorum and within nervi nervorum. The distribution of NPY is similar to that of norepinephrine (NA)-containing fibers (Figs. 8.11 and 8.12), a feature consistent with studies on other vascular beds. There is much evidence for the coexistence of NPY with NA in perivascular nerves (Edvinsson, Edman, Ottoson, & Uddman, 1985). NPY is a potent vasoconstrictor, and it has been shown to reduce cerebral blood flow (Tuor, Kelly, & Edvinsson, 1985). The demonstration of NPY in perivascular nerves supplying vasa nervorum and within nervi nervorum strongly implicates this peptide in the maintenance of vascular tone in nerve trunks in normal and pathologic states.

A differential susceptibility of the adrenergic innervation of vasa and of nervi nervorum in streptozotocin-diabetic rats has been reported

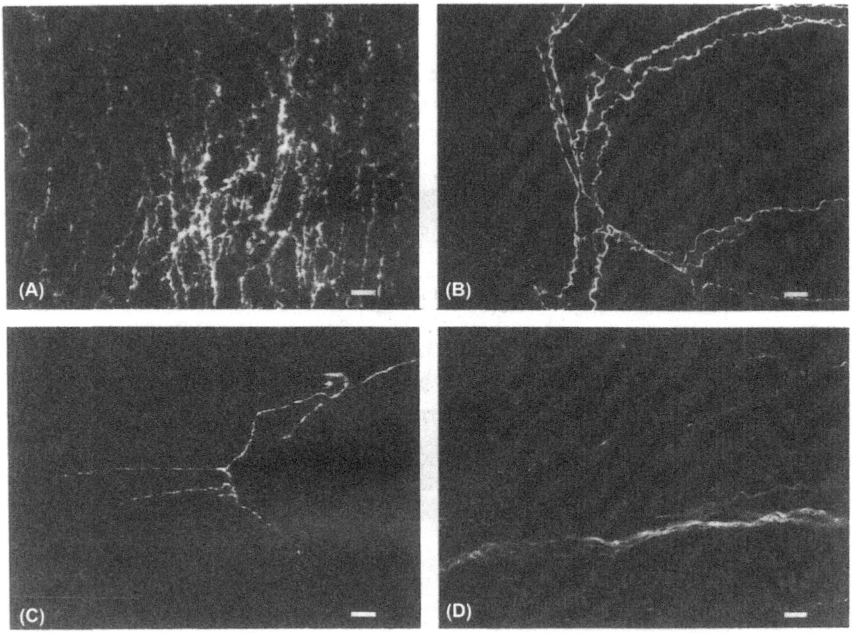

Figure 8.11 NPY-like immunoreactive perivascular nerve plexus nutrient artery rat sciatic nerve. Calibration bar 25 mm. *From Dhital, K.K., & Appenzeller, O. (1988). Innervation of vasa nervorum, Chapter 11. In G. Burnstock & S.G. Griffith (Eds.),* Noradrenergic innervation of blood vessels *(191–211). Boca Raton, FL: CRC Press.*

(Dhital, Lincoln, Appenzeller, et al., 1986). NPY and VIP-like nerve profiles of vasa and nervi nervorum at 8 weeks after the induction of diabetes by streptozotocin were examined in the intraneural circulation and in nervi nervorum in optic sympathetic nerve trunks, vagus and sciatic nerves, and in the circle of Willis. A loss of NPY-like nerve profiles was found in the sympathetic, vagus, and sciatic nerves. The optic nerve and cerebral blood vessels remained unaffected by the streptozotocin-diabetic state. Simultaneously, marked proliferation of VIP-like nerve fibers was found in all tissues examined in diabetic animals (Appenzeller, Milner, & Lincoln, 1989). These observations indicate a differential vulnerability of peptide-containing nerve fibers after 8 weeks of streptozotocin diabetes, mimicking that of the adrenergic innervation. However, NPY-like nerve fibers showed an inverse vulnerability to that of the adrenergic innervation in streptozotocin-diabetic rats. The proliferation of VIP-like nerve fibers is, however, similar to that found in other autonomic nerves in these animals.

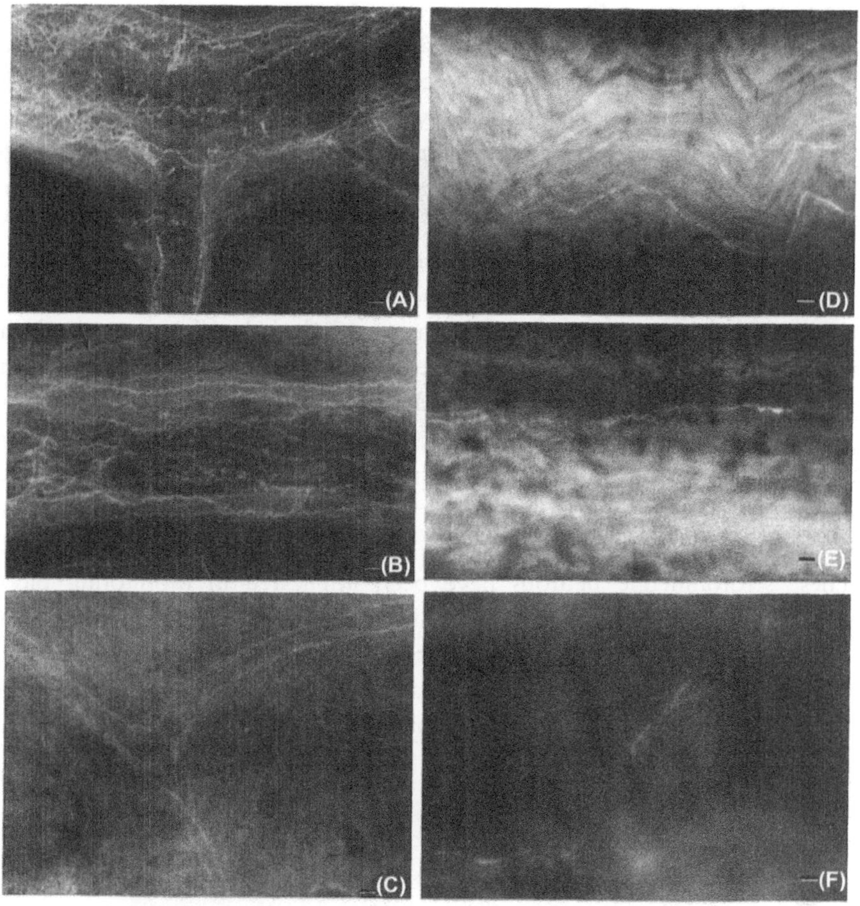

Figure 8.12 NPY-like immunoreactive nervi nervorum (arrow) and perivascular nerve plexus (arrowheads) human sural nerve.

8.1.11 Serotonin

Serotoninergic fibers have been demonstrated in perivascular plexuses of blood vessels supplying blood to the brain (Griffith, Lincoln, & Burnstock, 1982; Edvinsson, Degueurce, & Duverge, 1983). Serotonin (5-HT) is known to be a vasoconstrictor and vasodilator, depending on the vascular bed and its concentration (Vanhoutte, Cohen, & van Neuten, 1984). 5-HT-LI fibers innervate the major resistance vessels, and it is suggested that these fibers play a role in the normal regulation of the cerebral circulation and may also be pathogenetically important in disease. In vasa nervorum, single serotoninergic fibers are present around some blood vessels, and they are more frequent in the sciatic and vagus nerves, particularly in older animals (Appenzeller, Dhital, Cowen et al., 1984) (Fig. 8.13). The role of serotoninergic fibers in normal vasomotor function and in peripheral nerve function and dysfunction remains, however, unknown.

A projection of serotoninergic fibers from the central nervous system into the perivascular network of the vertebral-basilar circulation in spinal cord blood vessels and around the anterior spinal artery has been documented by histofluorescence techniques. In lumbar segments of the spinal cord, serotonin-containing nerve fibers have also been found, which connect to descending serotoninergic pathways of the intermediolateral cell column and the ipsilateral dorsal roots. These fibers are thought to originate from the raphe nuclei in the brain, but definitive evidence for this hypothesis has not been presented. Serotoninergic fibers that leave the central nervous system through dorsal spinal roots and innervate as yet unidentified peripheral nerve structures have also been found. Serotoninergic cell bodies and their dendrites have been shown to be in close contact with brain-stem parenchymal vessels. These serotoninergic neurons have been implicated in monitoring blood temperature and mechanical deformation of vessel walls (DiCarlo, 1984). The possibility that the perivascular serotoninergic fibers around vasa nervorum take origin from dorsal root ganglia has not been excluded. There is substantial evidence for a central origin of 5-HT-LI nerve fibers that innervate

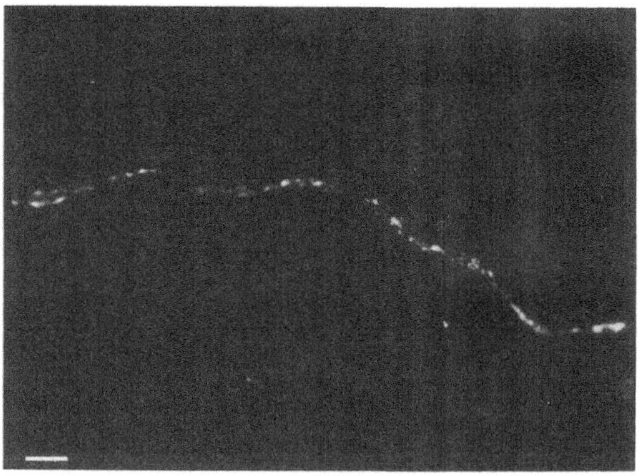

Figure 8.13 Serotonin-like immunoreactive nerve fibers, guinea pig sciatic nerve. Perivascular nerve plexus vasa nervorum (large arrow). Nonperivascular nervi nervorum (small arrow). Calibration bar 25 mm. *From Dhital, K.K., & Appenzeller, O. (1988). Innervation of vasa nervorum, Chapter 11. In G. Burnstock & S.G. Griffith (Eds.),* Noradrenergic innervation of blood vessels *(191–211). Boca Raton, FL: CRC Press.*

cerebral blood vessels (Scatton, Duverger, LHeureux, & MacKenzie, 1985), but other immunohistochemical studies suggest an additional peripheral origin for these fibers. These studies have demonstrated 5-HT-LI in the superior cervical and nodose ganglia and disappearance of perivascular serotoninergic fibers after the removal of the superior cervical ganglia (Matsuyama, Matsumoto, & Shiosaka, 1984). By analogy, the 5-HT-LI perivascular nerve fibers innervating vasa nervorum may also originate from a peripheral source other than the main nerve trunks.

An important role for the central serotoninergic neuron system supplying fibers to some cerebral vessels has been shown in the vascular reactivity to carbon dioxide (Hanko, Sundler, Hardebo, & Owman, 1985). Serotonin has also been shown to induce ion-dependent increases in blood—brain barrier permeability (Olesen, 1985). Therefore, serotoninergic innervation of vasa and nervi nervorum may be important in maintaining the integrity of the blood—nerve barrier and in regulating the peripheral and cranial nerve environment.

Serotonin is proposed as a possible initiator of cold-induced vasoconstriction in Raynaud phenomenon. Ketanserin, a serotonin antagonist, has been used to alleviate the clinical symptoms induced by digital artery constriction in patients suffering from this condition (Seibold & Jageneau, 1984). The role of serotoninergic perivascular nerves and of vasa nervorum in the genesis of ischemic neuropathies, or in alterations in the blood—nerve barrier remains, however, speculative.

The perivascular nerve plexuses in peripheral vasa nervorum closely mimic the neurotransmitter and peptide content of the perivascular nerve plexuses of the central nervous system. Proposals have been made that the intimate morphologic arrangements between peptide immunoreactive dendritic profiles and capillary basement membranes might provide a morphologic basis for a chemosensory function in the brain. This may be involved in the regulation of blood flow and of the blood—brain barrier. The close ultrastructural relationship found in normal and abnormal nerves between Schwann cells and small unmyelinated fibers, which do not form part of Remark bundles, suggests a similar role for these nerves in the control of the blood—nerve barrier.

8.1.12 Vasa nervorum and peripheral nerve ischemia

Clinical and experimental observations show a relationship between vasa nervorum and peripheral nerve dysfunction in pathologic states. The effects of ischemia are multifactorial and probably combine with the production of edema in ischemic nerves to produce dysfunction. Neurophysiologic studies show that complete deprivation of blood supply by compression leads to transmission block. Whether this is the result of ischemia, as originally proposed, or more recently attributed to mechanical damage to the myelin sheath by compression is still disputed (Gilliatt & Harrison, 1984). In human experiments, the inflation of a blood-pressure cuff above systolic blood pressure around a limb results in a centripetal paralysis of both motor and sensory function, beginning in the fingertips about 15 minutes after the production of ischemia. But, direct nerve compression, as well as ischemia, complicate the interpretation of results of such experiments. How ischemia itself alters peripheral nerve function requires further research. Ischemia is associated with anoxia, changes in carbon dioxide, electrolyte content, and metabolic acidosis. In vitro, mammalian nerves in a totally anoxic environment continue to conduct impulses when stimulated for up to 1 hour. When the nerve is then washed in an oxygen-free medium, it becomes responsive again for shorter periods of time (Fox & Kenmore, 1967). One interpretation of this experiment is that anoxia alone is not the cause of the conduction block, but rather that the anoxia causes the release of inhibitors, which affect impulse conduction and which can be removed by an anoxic washing solution. It is not, however, known what component of the many structures that make up a peripheral nerve release the inhibitors that produce this effect.

Controversy about the morphologic changes induced by experimental ischemia in peripheral nerves reflects the different species examined by various investigators and the diverse methods used to study morphologic consequences. Swelling of axis cylinders with vacuolization of the myelin sheath after chronic ischemia has been reported (Denny Brown & Brenner, 1944). More recently, these findings have been confirmed, but it has been suggested that a direct pressure effect on myelinated fibers of the nerve trunks may be important in the early conduction block.

The blood vessels to nerves must clearly exert an important role in regenerating nerve fibers after ischemia and in the recovery of function. But how the nerves to these blood vessels might affect ischemic or crushed peripheral nerves remains unknown. Moreover, it is difficult to recognize either morphologic or functional differences between the rates of regeneration of axons in nerves that were crushed and in those that were both crushed and devascularized. Severe structural lesions are difficult to produce in experimental animals by ischemia, perhaps because even after ligating the main nutrient blood vessels to a nerve, small nerve fibers might derive significant nutrients by diffusion from surrounding well-vascularized tissues (Blunt, 1957).

8.1.13 Vasa nervorum in human autonomic neuropathies

Vasa nervorum participate in arteriosclerosis. In this condition, peripheral nerve dysfunction occurs clinically, and pathologic changes are also found in sural nerves (Chopra & Hurwitz, 1967). The lumen of vasa nervorum in the sural nerve is markedly narrowed when obliterative arteriosclerotic disease is found in large vessels of the lower limbs (Eames & Lange, 1967). On ultrastructural examination, endoneural capillaries show thickening of the basement membrane and irregular endothelial cells, including small projections of cytoplasmic processes into the lumen, as well as lipid inclusions in the cytoplasm. Proliferation of endothelial cells has also been found, and thickened perivascular collagen surrounds capillaries. Ultrastructural observations of epineural blood vessels have shown similar findings, and, in addition, abundant intercellular collagen is found in the media and smooth muscle cells and on the luminal side of the internal elastic lamina. Clinical accompaniments of these serious occlusive vascular disorders have been documented and include sensory deficits, paresthesia, impaired vibration sense, and diminution of ankle jerks (Miglietta, 1967). A decrease in conduction velocities was also found. Pathologically segmental demyelination is thought to be the initial change, but Wallerian degeneration is also found.

In autonomic neuropathies associated with thromboangiitis obliterans, small arteries of the extremities are found to contain diffuse inflammatory proliferative lesions and segmental occlusions in the affected vessels. All three vascular coats are involved. Sometimes the vasa nervorum may be affected in different peripheral nerves, particularly in the lower extremity. Occlusion of vasa nervorum caused by thickened walls and by thrombosis is associated with severe pain and dysfunction of peripheral nerves (Krucke, 1955). This pain is not entirely explicable by the impaired blood supply to tissues of the limb but must also result from the involvement of peripheral nerves themselves. Perhaps the vasa nervorum or nervi nervorum play a role in nociception from nerves. The paroxysmal pain is severe and may not be accompanied by intermittent claudication, and the nerves themselves become tender to slight pressure in this condition. Significant arterial embolization to the main vessels of a limb can be associated with persistent sensory and motor deficits. This has been found after embolectomy when the loss of the limb has been prevented by timely surgical intervention (Haimovich, 1950). Vasa nervorum are also involved in periarteritis nodosa and in rheumatoid arthritis. Clinically, there is usually a mononeuritis multiplex in periarteritis nodosa. Involvement of large nerve trunks rather than asymmetric peripheral neuritis is the usual clinical finding, though the latter may also occasionally occur in this condition. There is tenderness of the nerve trunks apart from pain and paresthesia. The tenderness usually occurs distal to the involvement of the nerves, which suggests that the nervi nervorum might play an important role in the genesis of pain in inflammatory mononeuritis multiplex.

Morphologic changes in vasa nervorum occur in diabetes mellitus. The peripheral neuropathy in this disease has even been attributed to these vascular changes (Fagerberg, 1959). Sural nerve biopsies from patients with diabetes mellitus have not all shown vascular changes in the vasa nervorum, and ischemia may not be important in all patients with diabetic neuropathy (Thomas & Lascelles, 1965). Numerous studies have examined peripheral nerves at autopsy or on sural nerve biopsy in patients with diabetes mellitus, and it was assumed that a unitary hypothesis might explain the clinical and pathologic findings in peripheral nerves in this common human disease. The probability that both ischemia and metabolic changes contribute to varying degrees in different patients and in different peripheral nerves to the clinical deficits remains the best explanation at present. Sympathetic and parasympathetic denervation is a feature of clinical diabetic neuropathy. A quantitative ultrastructural investigation of the components of blood vessel denervation of vasa nervorum in diabetes mellitus was performed. While it was not possible to determine the neurotransmitter content of the nerves of vasa nervorum or nervi nervorum from this study, epineural arterioles in the sural nerve showed, nevertheless, a morphometric difference in the pattern of their innervation in six diabetic patients compared with six nondiabetic patients of comparable age. The perivascular axons were found at a greater distance from the nearest smooth muscle cells in the diabetic group, and the mean distance of such axons from the effector sites in the vascular smooth muscle cells was nearly twice that in the control group (Grover Johnson, Baumann, & Imparato, 1981). This ultrastructural study clearly indicated that in human diabetic neuropathy, the autonomic terminal innervation of vasa nervorum in the sural nerve is affected, but whether there is a differential susceptibility of different nerves or of vasa nervorum remains to be determined.

Histochemical, immunohistochemical, and neurochemical techniques were used to examine the innervation of epineurial nerve sheaths and fascicular nerve bundles of human sural and optic nerves from controls and patients with diabetic and alcoholic peripheral neuropathy. The neurochemical analysis for norepinephrine (NE), vasoactive intestinal polypeptide (VIP), neuropeptide-Y (NPY), and substance P (SP) in fascicular and epineurial preparations from the sural and optic nerves confirmed light microscopic observations. Age and sex had no effect on transmitter levels. In patients with peripheral neuropathy, there was a significant reduction in SP fascicular/epineurial ratio in both the optic nerve, which was histologically normal, and in the sural nerve, which showed evidence of pathology. The NE fascicular/epineurial ratio was significantly reduced in sural nerves from patients with peripheral neuropathy with a more marked reduction in diabetic peripheral neuropathy (Fig. 8.14) (Lincoln, Milner, & Appenzeller, 1993).

The vasa nervorum can be the site of amyloid deposition, and amyloid peripheral neuropathy has been attributed to ischemia or associated pressure by amyloid deposits in the parenchyma of nerves, but the effect of amyloid on perivascular plexuses and their peptide content was not examined (Appenzeller, 1983).

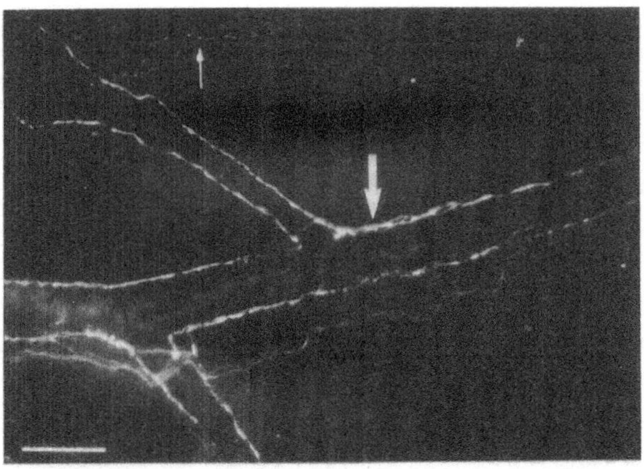

Figure 8.14 Comparison of the fascicular/epineurial ratios for NE and SP of human sural nerves obtained from controls (clear bars) and from specimens with histologic evidence of neuropathy (spotted bars). Data from neuropathic diabetic and alcoholic nerves are combined Results are given as mean ± SEM. The number of patients from which sufficient material was available is given under each bar. $^{*}P < .05$ Student's two-tailed *t*-test. *From Lincoln, J., Milner, P., Appenzeller, O., et al. (1993). Innervation of normal human sural and optic nerves by noradrenalin and peptide containing nervi vasorum and nervorum: Effect od diabetes and alcoholism.* Brain Res., 632, *48–56.*

8.2 Pathogenesis of autonomic failure in peripheral nerve disease

Cholinergic postganglionic sympathetic unmyelinated fibers form part of peripheral nerves and innervate sweat glands. These fibers degenerate, and demyelination of preganglionic sympathetic efferent fibers is also found. These pathologic findings account for the loss of sweating, but hyperhidrosis may be seen, as a part of causalgia, after partial nerve injuries; it may also be seen in malignant and toxic neuropathies. Compensatory sweating may appear when the degree of sweating loss on the extremities and trunk (particularly in diabetic autonomic neuropathy) is widespread.

Postural hypotension is caused by damage or loss of small-diameter myelinated and unmyelinated fibers in afferent and efferent nerves of the baroreflex arcs. Such damage can be seen in diabetes mellitus, amyloidosis, and also in tabes dorsalis, among others. Small fibers are also affected in acute inflammatory neuropathies with segmental demyelination that involve myelinated autonomic fibers in the vagus and sympathetic pathways. When the splanchnic vascular bed is extensively denervated, orthostatic hypotension is marked, because the splanchnic blood vessels lose tone. These blood vessels are essential in maintaining blood pressure by neurogenic vasoconstriction.

The impairment of heart-rate control (heart-rate variability) is predominately due to parasympathetic denervation. It is seen in patients with autonomic neuropathy, including diabetes mellitus and respiratory neuropathy (Appenzeller, Parks, & MacGee, 1968).

Disturbance of bladder function, impotence, and pupillary abnormalities are the clinical manifestations of peripheral nerve disease when autonomic fibers to these structures are involved. On the other hand, autonomic dysfunction, particularly orthostatic hypotension, is not common in dying-back neuropathies, which affect mostly large-diameter long peripheral nerve fibers. There is, therefore, a relative sparing of shorter autonomic fibers, with preservation of autonomic function.

8.3 Innervation of vasa nervorum and the nervi nervorum in human sural nerves

The vasa nervorum and nervi nervorum were studied at the ankle in 21 normal human sural nerves obtained 8–24 hours after sudden death (ages 14–96). Plexuses of dense catecholamine containing perivascular nerves of extraneural nutrient arteries were found; veins were not innervated. Epineural arteries had a similar innervation. Perineural arteries showed scant parallel noradrenergic fibers, and only occasional veins were sparsely innervated. The endoneural blood vessels showed a scanty innervation, including the veins, with noradrenergic perivascular nerve plexuses. Nervi nervorum-containing noradrenaline were present in the epineurium, less commonly found in the perineurium, and nonexistent in the endoneurium. NPY was perivascular, with occasional nervi nervorum containing this putative neurotransmitter. VIP was found in perivascular plexuses of nutrient arteries and in occasional nervi nervorum. CGRP, CCK and 5-HT were found in myelinated nervi nervorum (Figs. 8.15 and 8.16) (Appenzeller, Lincoln, & Zumwalt, 1989).

Similar studies in sural nerves of fetuses showed no noradrenergic perivascular nerve plexuses. In 8 patients with advanced neuropathies, including multiple symmetric lipomatosis and alcoholic and diabetic neuropathy, nonspecific denervation of vasa nervorum was found; conversely, adrenergic and VIP-ergic nervi nervorum were nonspecifically preserved in such

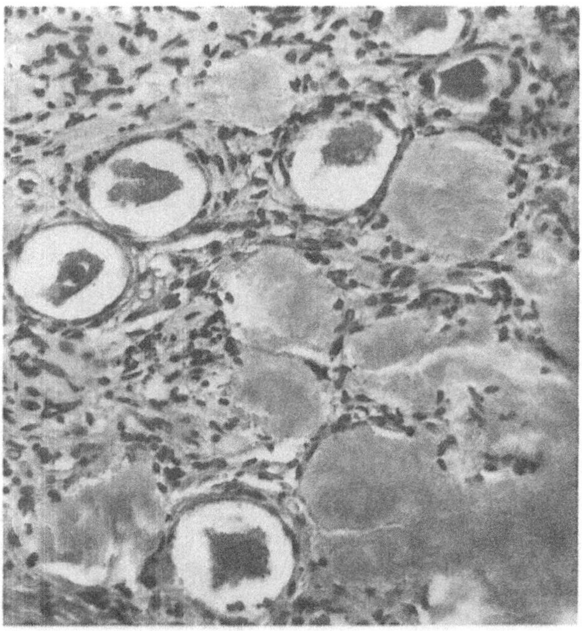

Figure 8.15 Catecholamine-fluorescent varicose perivascular plexus of vasa nervorum and nervi nervorum (arrow) in normal sural nerve (sudden death same age as the patient in Fig. 8.19).

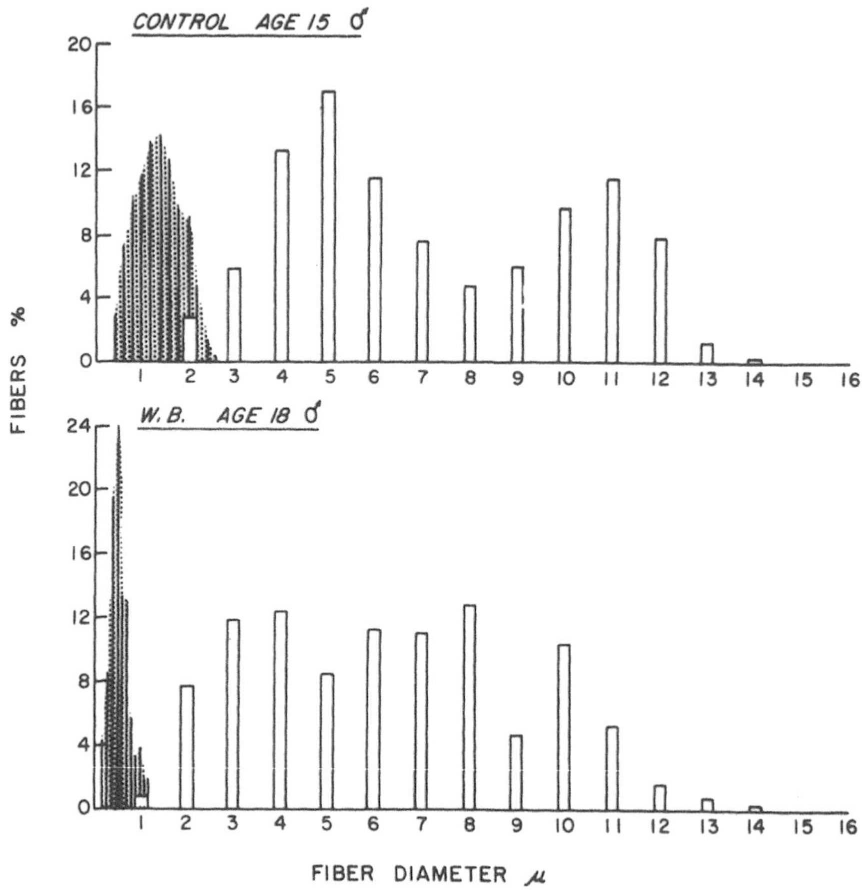

Figure 8.16 Neuropeptide-Y-like fluorescent perivascular varicose fibers of vasa nervorum (arrow) in normal sural nerve (sudden death).

nerves. In acute inflammatory neuropathy, the perivascular innervation and nervi nervorum appeared normal in distribution and in transmitter content. In minimally symptomatic diabetic neuropathy and in Navajo neuropathy, a reduction in perivascular innervation of vasa nervorum was found. Quantitative analysis of catecholamine content of these nerves confirmed the immunohistochemical findings (Appenzeller, Lincoln, & Zumwalt, 1989) (Figs. 8.17 and 8.18).

The autonomic innervation of the vasa nervorum and of the nervi nervorum needs further study in larger numbers of patients with neuropathies to delineate the possible role of neurotransmitters of the autonomic perivascular nerves in the genesis of such neuropathies.

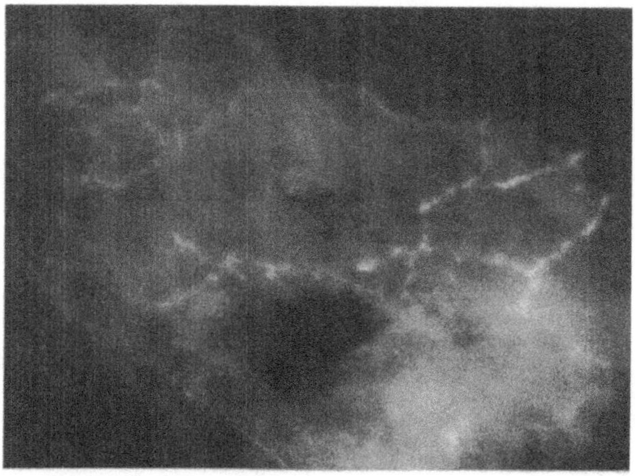

Figure 8.17 Norepinephrine (NA) content of normal (from cases of sudden death) and neuropathic sural nerves. The NA content of nerves and epineurium is significantly correlated, making the ratio of these values essentially independent of the delay of removing the nerves after death. All neuropathic nerves fall below the regression line of controls, implying relative preservation of norepinephrine in epineurium.

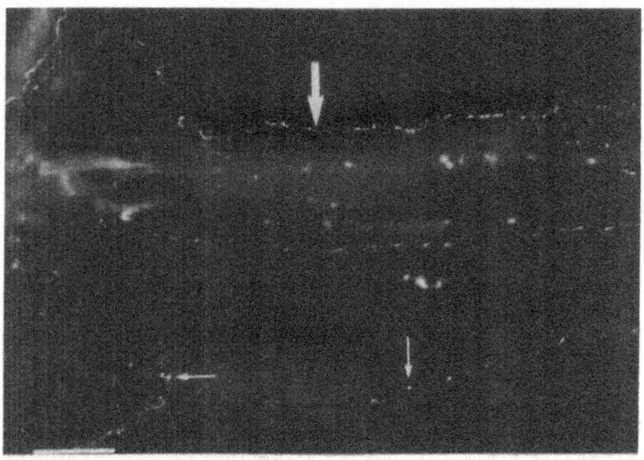

Figure 8.18 A significant difference ($P < .02$) in the ratio of norepinephrine (NA) content of nerves and epineurium, between controls and neuropathic human sural nerves.

Similar studies on mummified sural nerves (approximately 2500 years old) from upper Egypt and from Peru showed a well preserved peptidergic innervation of the vasa nervorum and of the nervi nervorum. In addition, nitric oxide synthase (neuronal isoform) was also present; subsequently also found in present-day biopsy material (Fig. 8.19). PGP9.5, an axonal marker, was also present in mummified sural nerves.

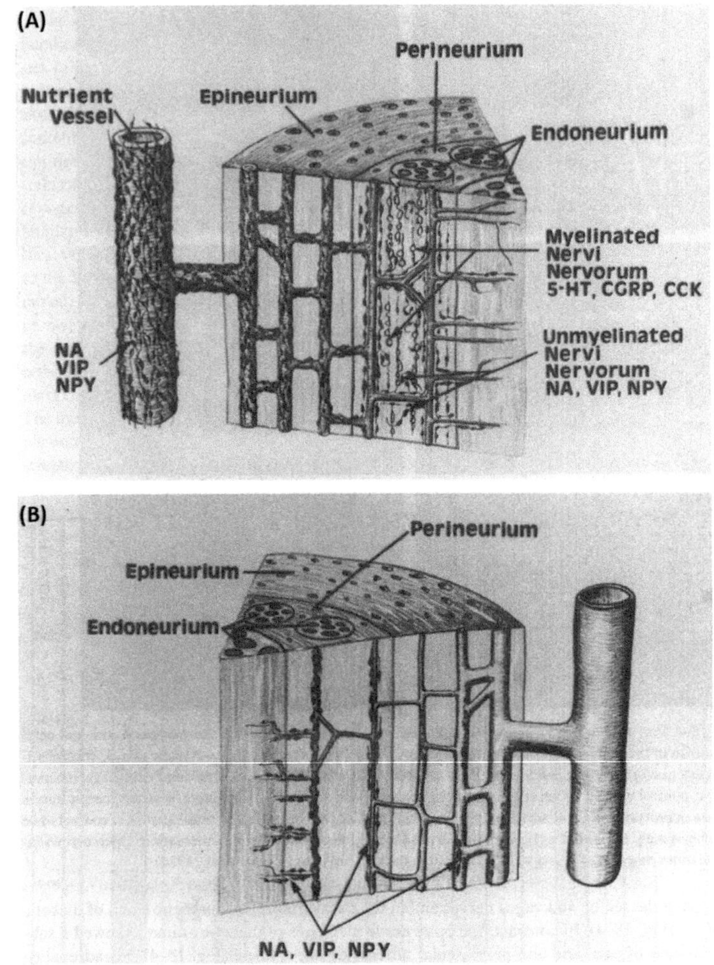

Figure 8.19 Immunohistochemistry—nitric oxide synthase (NO) (neuronal isoform). Upper panels: -present day human sural nerve. Lower panels: mummified human sural nerve (Oasis of Kharga, Upper Egypt c. 1500 BCE; courtesy, Smithsonian Institution, Washington, DC). Fluorescent fibers containing NO can be seen in longitudinal and cross-sections of both nerves. (Calibration bars = 10 mm). *Courtesy Prof. P. K. Thomas, London.*

8.4 Human autonomic neuropathies

8.4.1 Presenting clinical features

A detailed autonomic history provides valuable information in patients presenting with autonomic neuropathy. A key part of taking the history is to delineate the temporal course of the patient's symptoms. Some neuropathies present with an acute onset (e.g., immune-mediated neuropathies), whereas others are insidious with slow progression (e.g., hereditary neuropathies). A complete autonomic review of system should be performed to investigate different autonomic functions including thermoregulation, cardiovascular regulation, gastrointestinal motility, urinary storage and micturition, vasomotor activity, glandular secretion, sexual symptoms, and pupillomotor symptoms. None of these clinical manifestations are specific for any particular autonomic neuropathy, but the sum of these disorders may be seen to varying extent in different patients and, more characteristically, in some types of autonomic peripheral nerve diseases. History taking should also focus on modifying factors such as changes in symptoms with posture (supine, sitting, standing), and the effect of the time of the day and/or meals on symptoms.

A detailed family history and medical history of associated illnesses may provide clues regarding the underlying etiology (i.e., long-standing poorly controlled diabetes may suggest diabetic autonomic neuropathy, a history of bilateral carpal tunnel syndrome, heart failure with preserved ejection fraction, and alternating constipation with diarrhea suggests possible amyloid neuropathy). Clinical manifestations of somatic nerve involvement include increased pain sensitivity or loss of function with hyposensitivity and even a total loss of sensation. The combination of subacute severe ataxia with burning pain, hyperesthesia, and allodynia suggests involvement of large-, small-, and medium-size neurons commonly seen with sensory neuronopathies (e.g., Sjögren's syndrome). Isolated damage of small sensory fibers can cause spontaneous and stimulus-evoked distal skin pain and sensory loss (classic "painful feet"); however, other symptoms such as deep aching and neuropathic itch are common. The medical history should include a complete list of prescribed and over-the-counter medications and supplements because of their potential effects on the autonomic nervous system. History taking is the most important part of the autonomic evaluation. Specific syndromes can be recognized as they cluster into patterns of presentation. Furthermore, it guides the clinical evaluation and testing in patients with suspected autonomic neuropathies (Lamotte & Sandroni, 2021).

8.4.2 Clinical evaluation

A detailed complete neurological evaluation is an indispensable part of the evaluation in patients with suspected autonomic neuropathies. The sensory examination is particularly important and it looks for specific patterns such as the classic symmetric stocking-and-glove distribution involving small and large fibers, the pure distal loss of small fiber sensory modalities, or the patchy and proximal distributions of sensory abnormalities suggesting a ganglionopathy.

Measurement of blood pressure and heart rate should include standing and comparison to baseline supine values (at least 5 minutes of supine rest). An important clinical distinction is that of neurogenic orthostatic hypotension—a cardinal manifestation of cardiovascular sympathetic adrenergic failure. A decreased heart rate response to OH has been shown to be a surrogate marker of neurogenic OH (Norcliffe-Kaufmann et al., 2018). In a prospective study of 423 patients, neurogenic OH was reliably distinguished from other causes of OH when the ratio of orthostatic heart rate (HR) changes over systolic blood pressure (SBP) changes (ΔHR/ΔSBP) at 3 minutes of tilt was < 0.5 beats/min/mmHg (Norcliffe-Kaufmann et al., 2018). This ratio has been validated at the bedside when patients were actively standing (Fanciulli et al., 2020).

Beyond blood pressure and heart rate measurements, the bedside autonomic evaluation includes assessment of pupillary light reflex and pupillary response to accommodation, assessment of skin coloration of the face, hands, and feet as well as assessment of sudomotor function (e.g., dryness of the skin, asymmetry in sweating, trophic changes with loss of hair/nails or atrophic skin).

8.4.3 Paraclinical evaluation

Objective testing, guided by the history and examination, is extremely helpful in establishing the diagnosis of peripheral autonomic neuropathy. Standardized autonomic function testing with an autonomic reflex screen characterizes the pattern of autonomic involvement and permits prompt and appropriate treatment. The autonomic reflex screen tests postganglionic sympathetic sudomotor, cardiovagal, and cardiovascular adrenergic functions (see chapter on autonomic function testing). Other autonomic tests are not performed routinely. These tests may provide valuable information regarding the involvement of afferent sensory nerves (e.g., cold pressor test and handgrip test exhibit an exaggerated pressor response in patients with impairment of the afferent limb of the baroreflex) or the enteric nervous system (e.g., gastrointestinal scintigraphy studies). Skin biopsy provides a quantitative assessment of cutaneous autonomic innervation (e.g., standard punch biopsies 3 mm in diameter from the distal leg, distal thigh, and proximal thigh sites). Skin biopsy has become an important diagnostic tool in familial amyloid polyneuropathy with visualization of amyloid under polarized light with Congo red staining but also for the diagnosis of synucleinopathies in vivo. At this time, analysis of sweat gland density is limited to research laboratories.

In patients presenting with concomitant symptoms of large fiber polyneuropathy, a nerve conduction study with electromyogram will help document the extent of sensory and/or motor deficits, and categorize the type of injury (demyelinating vs axonal).

First-line laboratory testing for autonomic neuropathies includes fasting glucose or oral glucose tolerance test, thyroid function tests, kidney function tests, vitamin B12, serum, and urine protein electrophoresis with immunofixation. If there is a suspicion of autoimmunity, testing includes antinuclear antibodies (ANA), extractable nuclear antigen antibodies (if ANA are positive), erythrocyte sedimentation rate, C-reactive protein, lip biopsy, and autoimmune dysautonomia panel including α-3 nicotinic acetylcholine ganglionic antibodies. Other tests to exclude secondary causes of autonomic dysfunction include screening for adrenal insufficiency, anemia, cardiomyopathy or arrhythmia, and pheochromocytoma. Second-line laboratory testing should be guided by clinical suspicion (e.g., paraneoplastic antibodies, IgA tissue transglutaminase antibodies).

8.4.4 Classification

Here, we present a classification of autonomic neuropathies based on the underlying etiology. We summarize the etiologies, presenting clinical features, suggested testing, and treatments of selected autonomic neuropathies in Table 8.1.

8.4.4.1 Immune-mediated and inflammatory autonomic neuropathy

Most cases of subacute autonomic failure (onset to peak deficit < 3 months) are attributable to neurological autoimmunity when toxic and metabolic causes are excluded. Immune-mediated autonomic neuropathies include autonomic neuropathies associated with sensory and motor neuropathy (e.g., acute inflammatory neuropathies such as Guillain–Barre syndrome), dysautonomia associated with malignancy (e.g., paraneoplastic autonomic neuropathies), and seropositive or seronegative autoimmune autonomic ganglionopathy (AAG).

Autonomic over- or underactivity is common in Guillain–Barre syndrome and occurs in about two-thirds of patients. Sympathetic overactivity

Table 8.1 Classification of autonomic neuropathies: etiology, clinical presentation, testing, and treatment.

Etiology	Clinical presentation	Testing[a]	Treatment
Immune-mediated			
AAG	Acute; nOH; GI hypomotility; hypotonic bladder; anhidrosis; mydriasis; genitourinary symptoms	Labs: + α-3 nAch abs (50%); pupillometry, GI motility and urodynamic studies can be helpful	Immunotherapy 12w treatment trial of IVIG 0.4 g/kg/w or 1 g IVMP/w; consider long-term immunosuppressant if response to trial (azathioprine, mycophenolate, or rituximab)
Acute cholinergic neuropathy	Acute; sicca syndrome; hypotonic bladder; GI hypomotility; erectile failure; mydriasis	Labs: + α-3 nAch abs; pupillometry, GI motility and urodynamic studies can be helpful	Immunotherapy—limited data. Similar to AAG.
Acute sympathetic neuropathy	Acute; previous viral infection; nOH; ejaculatory dysfunction; sphincter dysfunction; Horner; anhidrosis, nOH	Labs: + α-3 nAch abs	Immunotherapy—Limited data. Similar to AAG.
Acute immune GI dysmotility	Acute/subacute (< 6 months); previous viral infection; GI hypomotility; intestinal pseudo-obstruction; achalasia; gastroparesis/slow intestinal transit; pyloric stenosis; rarely anal spasm	Labs: α-3 nAch abs negative; ± neuronal voltage-gated calcium and potassium channel antibodies; GI motility studies are abnormal	Immunotherapy—Limited data. Similar to AAG.

Guillain–Barre syndrome	Acute (<4 weeks); viral infection; somatic (motor and/or sensory) involvement is frequent; pure dysautonomia possible; afferent autonomic involvement with tachycardia, labile hypertension, bradycardia; dysrhythmia, GI dysmotility and urine retention	CSF: lymphocytic pleocytosis; EMG/NCS: demyelinating neuropathy vs. axonal neuropathy	Immunotherapy IVIG 0.4 g/kg/5 d or PLEX
PParaneoplastic			
Panautonomic	Subacute; generalized autonomic failure; limbic encephalitis (ANNA-1, LGI1/CASPR2, PCA-2); brainstem encephalitis (PCA-2); gait disorders (AP3B2); cognitive disorders, chorea, ataxia (CRMP-5); encephalopathy, myeloneuropathy (NIF); stiff-person syndrome / rigidity (Gly-R); psychosis, seizure, dyskinesia (NMDA)	Labs: Ach-R, ANNA-1, AP3B2, NIF, PCA-2, voltage-gated potassium channel (LGI1/CASPR2), CRMP-5, Gly-R, NMDA antibodies; CT scan CAP and PET-CT to look for malignancy	Treatment of underlying tumor; immunotherapy (limited data—IVIG, IVMP, PLEX)
Enteric	Subacute; GI hypomotility; gastroparesis; associated features with meningoencephalomyelitis (GFAP), encephalopathies, gait disorders, etc.	Labs: Ach-R, ANNA-1, AP3B2, GFAP, NIF, CRMP-5, DPPX (DPP6), LGI1/CASPR2; CT scan CAP; PET-CT to look for malignancy	Treatment of underlying tumor; immunotherapy (limited data—IVIG, IVMP, PLEX)

(*Continued*)

Table 8.1 (Continued)

Etiology	Clinical presentation	Testing[a]	Treatment
LEMS	Acute/subacute; muscle weakness with oculobulbar weakness; dry mouth; constipation; erectile dysfunction; anhidrosis; nOH	Labs: + P/Q voltage-gated calcium channel antibodies; EMG/NCS with repeat stimulation = increment; CT scan CAP to look for malignancy	Treatment of underlying tumor; 3,4-DAP; immunotherapy
Associated with immune disorders			
Sjögren's syndrome	Subacute/chronic; sicca syndrome; autonomic manifestations vary from OI to nOH, tachycardia; sensory neuronopathy with ataxia	Lip biopsy; Schirmer test; anti-Ro (SSA)/anti-La (SSB) antibodies; ANA; ENA; RF; saliva flow rate; EMG/NCS (sensory neuronopathy, length-independent pattern); rheumatology evaluation	Trial of immunotherapy (i.e., IVIG) + treatment of Sjögren syndrome
Systemic Lupus Erythematous and MCTD	Subacute/chronic; rash; alopecia; arthritis; neuropsychiatric manifestations; autonomic manifestations vary from OI to autonomic failure	ANA; ENA; RF; APLS; Complement (C3, C4, CH50); CBC (hemolytic anemia), lymphopenia, thrombocytopenia; rheumatology evaluation	Immunotherapy for SLE or MCTD
Celiac disease	Subacute; GI symptoms with diarrhea; bloating; constipation; gluten intolerance; skin changes; migraine; cerebellar ataxia	Celiac antibodies screening with intestinal biopsy; CBC; LFTs; ANA; ENA; RF	Gluten free diet

Rheumatoid arthritis	Subacute / chronic; polyarthritis; variable autonomic involvement	Anticyclic citrullinated peptide antibodies; ANA; ENA; RF; rheumatology evaluation	Immunotherapy for RA
Metabolic/ nutritional			
Diabetes	Chronic; history of diabetes; distal painful small and/or large fiber neuropathy; retinopathy; nephropathy; foot ulcer; cardiac disease; variable autonomic involvement (tachycardia; OI; AF)	Every pattern can be seen on autonomic testing; Fasting glucose, 2-hour glucose tolerance, HbA1c; EMG/NCS; cardiology, nephrology and ophthalmology evaluation	Treatment of diabetes
TIND	Acute; following rapid glycemic control (insulin, oral hypoglycemic agent, diet); concomitant diabetic retinopathy; ↓A1c > 1%/ month; orthostatic intolerance, nOH, bloating, sweat changes, and sexual dysfunction	A1c with rapid decline; EMG/ NCS; ophthalmology evaluation	Treatment is supportive with the goal of stable glucose control
Chronic kidney disease	Chronic; history of chronic kidney disease ± dialysis; erectile dysfunction; GI hypomotility; hypo-/anhidrosis; hypotension during dialysis; OI; OH; silent myocardial infarction	Kidney function tests; nephrology evaluation	Treatment is supportive with treatment of underlying kidney disease

(*Continued*)

Table 8.1 (Continued)

Etiology	Clinical presentation	Testing[a]	Treatment
Ethyl alcohol	Chronic; history of chronic alcohol abuse; erectile dysfunction is most frequent; GI with dyspepsia, constipation, diarrhea; cardiovascular involvement with OI and OH is possible	Liver function test; CBC (macrocytosis); EMG/NCS to look for length-dependent axonal polyneuropathy	Treatment is supportive with abstinence from alcohol and vitamin supplementation (thiamin)
Infectious			
HIV	Acute/subacute/chronic; history of HIV; opportunistic infections; associated with distal symmetric polyneuropathy; dry eyes or dry mouth; constipation; erectile dysfunction; OI; rare OH	HIV testing; EMG/NCS	Antiretroviral therapy
Leprosy	Subacute/chronic; history of leprosy; discolored patches of skin; painless ulcer; decrease sensitivity to pain and temperature; focal anhidrosis; blunted tachycardia; rare nOH; erectile dysfunction, cardiac autonomic neuropathy	Biopsy lesion / nerve; EMG/NCS with mononeuropathy	Antibiotics

Chagas disease	Subacute/chronic; Palpitation, syncope, and risk for sudden death; mild dysmotility to severe megaesophagus and megacolon; OH; cardiac conduction system and myocardial damage	Trypanosoma cruzi antigens/lysate-based enzyme-linked immunosorbent assay; echocardiogram; EKG; Holter-EKG	Benznidazole, Nifurtimox
Botulism	Acute; muscle weakness; cholinergic failure; dysphagia; mydriasis; urinary retention; dry mouth; respiratory distress	Neurotoxin in the serum, stool, or contaminated food or by culturing *Clostridium botulinum* from the stool. EMG/NCS: low CMAPs on NCS. High-frequency RNS shows increment	Antitoxin and supportive care
Toxic			
Heavy metals	Chronic; history of exposure to heavy metals (thallium, arsenic, mercury); usually with sensorimotor symptoms	Heavy metal screening; EMG/NCS	No specific treatment; symptomatic management
Chemotherapy	Subacute; history of exposure to chemotherapy with neurotoxic properties (e.g., cisplatinum, vincristine, paclitaxel, taxol, doxorubicin, cytosine arabinoside, PD-1 inhibitors); usually with sensorimotor symptoms	EMG/NCS	Discontinue offending agent

(*Continued*)

Table 8.1 (Continued)

Etiology	Clinical presentation	Testing[a]	Treatment
Systemic disorder			
Light chain Amyloidosis	Subacute/chronic; 20% with neuropathy; wide spectrum of symptoms: cardiovascular, GI, genitourinary; weight loss and fatigue (50%); periorbital/facial purpura; hepatomegaly; macroglossia; small fiber neuropathy usually associated with large fiber neuropathy; 75% of patients with AF and painful length-dependent sensory-motor neuropathy; 25% with pure autonomic failure or with somatic neuropathy without autonomic symptoms.	Immunofixation of serum and urine with free light chain assay; fat aspirate; bone marrow/skin/minor salivary gland/nerve biopsy for Congo red staining; mass spectroscopy; echocardiogram with strain assessment; troponin T and N-terminal pro BNP; EMG/NCS; EKG	Stem cell transplantation; chemotherapy; daratumumab (CD-38 directed monoclonal antibody)
Porphyria	Acute/paroxysmal; severe and diffuse abdominal pain; constipation; nausea; confusion; psychiatric disturbances; autonomic instability (tachycardia, hypertension)	Urinary and stool porphyrins, %-aminolevulinic synthetase level	Alkaline heme, small interfering RNA (givosiran)

Hereditary			
TTR amyloidosis (familial)	Chronic; length-dependent small fiber sensorimotor polyneuropathy with loss of pain and temperature; heart failure; arrhythmia; nephrotic disease; bilateral carpal tunnel syndrome; autonomic involvement is variable with possible AF, erectile dysfunction, constipation, genitourinary abnormalities	TTR gene analysis; numerous variants, most common = Val30Met mutation in Portugal, Brazil, Sweden, and Japan; Val122Ile mutations = most common in African American; EMG/NCS; EKG	Suppression of TTR synthesis (interfering RNA = Patisiran, antisense oligonucleotide = Inotersen); TTR stabilization (Tafamidis); liver transplantation
Fabry disease	X-linked; acroparesthesia; fatigue; hearing loss; corneal opacity; renal dysfunction; skin (angiokeratomas); modest autonomic dysfunction with hypohidrosis; impotence; gastrointestinal dysfunction	α-galactosidase enzyme activity in the blood	Enzyme replacement therapy
HSAN	Chronic; variable phenotypes—juvenile to adult presentation; profound distal sensory loss, acral mutilation; absence of tears and recurrent pneumonias (HSAN-III); ulcers; variable autonomic and motor disturbances; involvement of afferent limb of baroreflex with hypertensive crisis (HSAN-III)	Genetic testing: ELP1/IKBKAP, SPTLC1, HSN2, NTRK1; cold pressor test and measure of vasopressin characteristic of afferent baroreflex failure in some patients	No specific treatment; carbidopa helpful to prevent excessive blood pressure variability

(*Continued*)

Table 8.1 (Continued)

Etiology	Clinical presentation	Testing[a]	Treatment
CANVAS	Cerebellar ataxia, neuropathy, vestibular areflexia; cold feet, erectile dysfunction, postural lightheadedness, constipation dry mouth or eyes? drop attack	Genetic testing: AAGGG repeat expansion in the second intron of replication factor C subunit 1 (RFC1) gene	No specific treatment
Unknown etiology			
Ross syndrome	Chronic; Ross triad = segmental anhidrosis ± compensatory hyperhidrosis, hyporeflexia, tonic pupils	TST with segmental anhidrosis ± hyperhidrosis. QSART and TST with postganglionic pattern of sudomotor involvement. ARS with mild cardiovagal and adrenergic impairment (25% of patients); ophthalmology evaluation with dilute pilocarpine test	No specific treatment. Prevention of heat stroke (cooling vest, misting fans, hydration)
Holmes Addie	Chronic; hyporeflexia, tonic pupils	Ophthalmology evaluation; dilute pilocarpine test	No specific treatment. Possible evolution to Ross syndrome.

α-3 nAch abs, Alpha-3 nicotinic acetylcholine ganglionic antibodies; *AAG*, autoimmune autonomic ganglionopathy; *Ach-R*, acetylcholine receptor α-1 subunit; *AF*, autonomic failure; *ANA*, antinuclear antibodies; *ANNA-1*, antineuronal nuclear antibody 1; *AP3B2*, adaptor protein 3B2; *APLS*, antiphospholipid syndrome; *ARS*, autonomic reflex screen; *CANVAS*, cerebellar ataxia, neuropathy, and vestibular areflexia syndrome; *CASPR2*, contactin-associated protein-2; *CBC*, complete blood count; *CMAP*, compound muscle action potential; *CRMP-5*, collapsin response-mediator protein-5; *CT CAP*, computed tomography scanner of the chest-abdomen-pelvis; *DPPX*, dipeptidyl peptidase-like 6; *EKG*, electrocardiogram; *EMG/NCS*, electromyogram/nerve conduction study; *ENA*, extractable nuclear antigen antibodies; *GFAP*, glial fibrillary acidic protein; *GI*, gastrointestinal; *Gly-R*, glycine receptor; *HSAN*, hereditary sensory and autonomic neuropathy; *IVIG*, intravenous immunoglobulin; *IVMP*, intravenous methylprednisolone; *LEMS*, Lambert–Eaton myasthenic syndrome; *LFTs*, liver function tests; *MCTD*, mixed connective tissue disorder; *NIF*, neuronal intermediate filament; *NMDA*, n-methyl-D-aspartate; *nOH*, neurogenic orthostatic hypotension; *OI*, orthostatic intolerance; *PCA-2*, Purkinje cell cytoplasmic antibodies 2; *PET*, positron emission tomography; *PLEX*, plasmapheresis; *QSART*, quantitative sudomotor axon reflex test; *RF*, rheumatoid factor; *RNA*, ribonucleic acid; *SLE*, systemic lupus erythematous; *TIND*, treatment induced neuropathy of diabetes; *TST*, thermoregulatory sweat test; *TTR*, transthyretin.

[a]Autonomic reflex testing and thermoregulatory sweat test (if available) are helpful in establishing the diagnosis of peripheral autonomic neuropathy. Autonomic reflex screen should be performed expeditiously in all cases of suspected autonomic neuropathy to characterize the pattern of autonomic involvement and permit prompt and appropriate treatment.

is common during the acute phase with hypertension, hyperhidrosis, and tachycardia (autonomic storm), while parasympathetic failure is more evident during recovery. Environmental or physiologic stimuli can trigger unpredictable deterioration, including hypoxia and fever. The intravenous administration of drugs, intubation, and physical therapy to the chest may cause deterioration in cardiovascular autonomic function. The careful monitoring of autonomic cardiovascular dysfunction in patients with the Guillain–Barre syndrome may significantly reduce the mortality from this recoverable neurologic disease (Dalos, Borel, & Hanley, 1988). Labile hypertension with tachycardia, bradycardia, and hyperhidrosis reflect impairment of the afferent limb of the baroreflex in Guillain–Barre syndrome. Bowel and bladder involvement tend to be mild. Elevated muscle sympathetic nerve activity has been documented during the acute phase of Guillain–Barre syndrome. Autonomic dysfunction may be prominent in acute inflammatory neuropathy but is rarely so in chronic inflammatory neuropathy. Heart-rate variability changes, orthostatic hypotension, and changes in sweating and sphincter disorders result from demyelination of fibers in the vagus and glossopharyngeal nerves and from pathology in the innervation of sphincters. Preganglionic sympathetic efferent fibers are also involved. Clinically, syncope, sometimes with irreversible damage to the brain or sudden death due to cardiac arrhythmia, has been reported to be associated with autonomic disorders in these conditions (Appenzeller & Marshall, 1963). Takotsubo cardiomyopathy may also occur as a complication; however, this is a rare complication usually associated with the severe axonal type of Guillain–Barre syndrome (Puentes, Teijelo, Stiep, Mannava, & Margolesky, 2021).

Chronic inflammatory demyelinating neuropathies are rarely associated with symptomatic autonomic dysfunction, but this can be demonstrated on special tests. This difference is related to the acuteness of the process, particularly the sudden conduction block, which occurs in the acute form of the disease. Figueroa and colleagues investigated the autonomic function in 47 patients meeting chronic inflammatory demyelinating polyneuropathy (CIDP) criteria. The authors found that autonomic symptoms were uncommon, mainly gastrointestinal (19% of patients) and genitourinary (17% of patients). Autonomic impairment on autonomic reflex screen was frequent (22/47; 47%) but very mild (CASS, 0.8 ± 0.9; CASS ≤ 3, all cases). Deficits were predominantly sudomotor (16/47) and cardiovagal (10/47) with relative adrenergic sparing (4/47). Sudomotor impairment was predominantly distal and postganglionic. Disease duration, severity of somatic sensorimotor symptoms, and small-fiber deficits did not predict the presence of autonomic deficits (Figueroa et al., 2012).

An acquired pandysautonomic syndrome was delineated in 1969 (Young, 1969). During the illness, postural hypotension, fixed heart rate, absence of sweating, dryness of eyes, nasal and oral mucous membranes, mid-position nonreactive pupils, and absent bowel motility, including a hypotonic bladder, are found. The glucose tolerance curve is sometimes diabetic in type and the CSF protein may be elevated. This panautonomic dysfunction occurs in a setting of normal mental function, extraocular motility, muscle strength, coordination, and normal deep tendon and superficial reflexes. Normal sensation is preserved and a number of other laboratory studies in such patients are normal also. The disease is usually self-limited, most patients showing clinical recovery, but some have slight residual autonomic dysfunction, which is usually not disabling. A sural nerve biopsy carried out during the symptomatic phase of the disease was reported to have shown no abnormalities (Young, 1969). A biopsy carried out some years after the onset of symptoms and at a time when autonomic dysfunction was still present but not disabling showed the fiber size spectrum of myelinated fibers to be slightly shifted to the left (Fig. 8.20). Ultrastructural examination revealed numerous small normal myelinated fibers of various calibers; however, the large myelinated fibers were not present in normal numbers and small myelinated fibers showed some redundant basement membranes. Unmyelinated fibers were found in larger than normal amounts and the majority were enclosed in bands of Schwann cells, the cytoplasm of which was thrown into stacks or plate-like folds (Fig. 8.21). Some axons had numerous vesicles and dense bodies within the cytoplasm, but most nonmyelinated axons were ultrastructurally normal (Fig. 8.22). A marked shift in the fiber size spectrum of unmyelinated axons larger than 1 m in diameter was seen (Fig. 8.20). Sympathetic and parasympathetic functions were acutely and predominantly affected in all patients who prior to the illness were normal. No patient had evidence of intoxication, though in some it was actively thought and excluded. One of 10 cases in the literature has apparently recovered completely; the others remain with slight evidence of residual deficits. Though previous descriptions of acute peripheral autonomic failure have appeared in the literature, these were usually in association with carcinoma (Ivy, 1961).

The pathogenesis of such complete sympathetic and parasympathetic postganglionic denervation is not entirely clear, but the clinical features

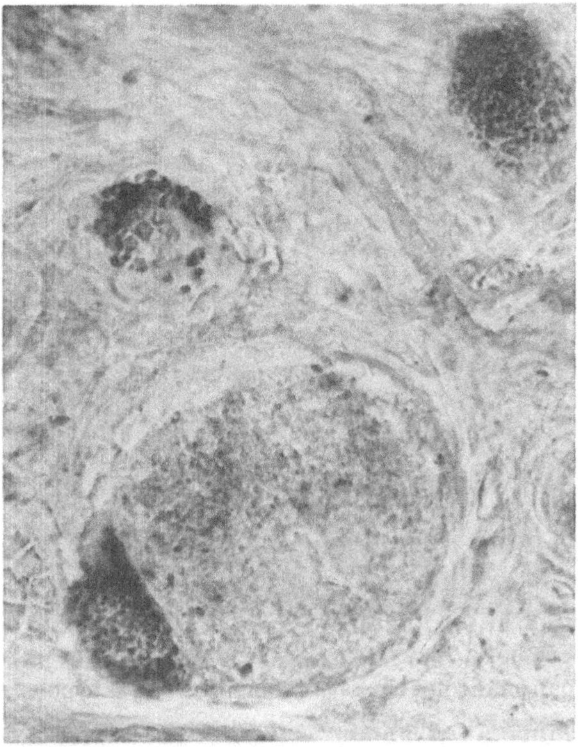

Figure 8.20 Fiber size spectrum of myelinated fibers (open bars) and unmyelinated fibers (solid bars) of control and patient showing a slight shift to the left of myelinated fibers and a marked shift to the left in fiber size distribution of unmyelinated fibers with almost complete absence of axons larger than 1 mm. *From Appenzeller, O., & Kornfeld, M. (1973). Acute pandysautonomia clinical and morphologic study.* Arch. Neurol., *29, 334–339.*

together with the documented recovery of function and regeneration of unmyelinated fibers suggest an immunologic basis of the disorder. The high CSF protein level without cellular reaction in several cases implies a relationship to the Guillain–Barre syndrome and further supports this interpretation. The association of pandysautonomia with malignant disease is also consistent with this view.

Acute pandysautonomia appears similar also to an experimental disease produced in rabbits by the injection of human sympathetic tissue mixed with Freund's adjuvant. In this animal disease, an abnormality of reflex vasomotor function related to temperature regulation was found. The disorder was transient and was associated with circulating antibodies to a

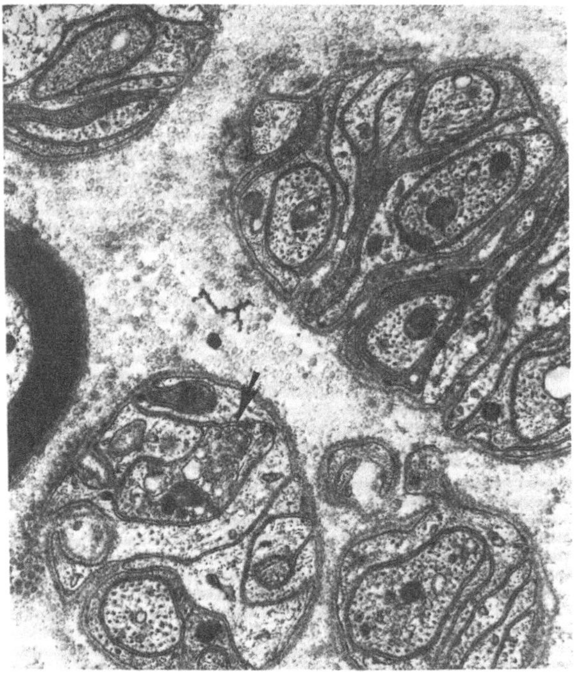

Figure 8.21 Cytoplasm of Schwann cells enclosing unmyelinated axons forms multiple flat folds. Axons measure less than 0.8 mm in diameter. Redundant basement membranes in the center; a fibroblast in right upper corner (® 27, 500). *From Appenzeller, O., & Kornfeld, M. (1973). Acute pandysautonomia clinical and morphologic study.* Arch. Neurol., *29, 334–339.*

sympathetic tissue antigen. It was concluded that the etiology of experimental autonomic neuropathy is not known. The clinical similarity of the two disorders suggests, however, that the pathogenesis of acute pandysautonomia might be closely related to that of experimental autonomic neuropathy. Though the production of experimental autonomic neuropathy preceded the clinical recognition of acute pandysautonomia by some years, it might prove a useful model for the study of this interesting disease of unmyelinated nerves (Appenzeller & Kornfeld, 1973).

The typical syndrome associated with autoimmune autonomic ganglionopathy (AAG) occurs in previously healthy individuals with onset of autonomic failure over the course of a few days or weeks. Symptoms may follow a viral prodrome, immunization, or surgery. Women are more often affected compared to men. Patients typically present with symptoms related to sympathetic failure (e.g., orthostatic hypotension and anhidrosis), parasympathetic

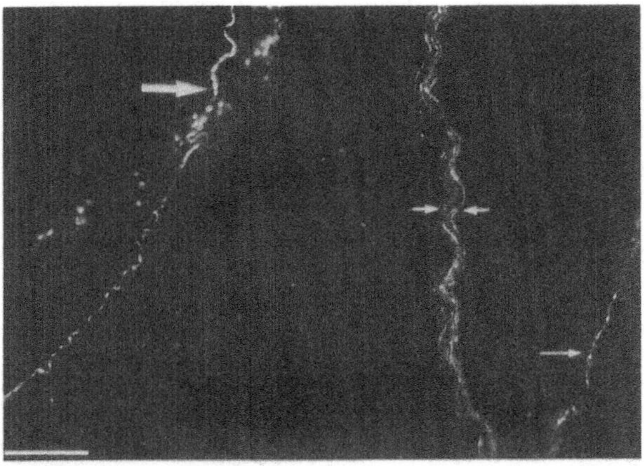

Figure 8.22 Budded bands of Schwann cells of unmyelinated type and one axon, probably regenerating (arrow) with abundant vesicular and tubular profiles and two dense bodies (® 31, 200). *From Appenzeller, O., & Kornfeld, M. (1973). Acute pandysautonomia clinical and morphologic study.* Arch. Neurol., *29, 334–339.*

failure (e.g., impaired heart rate variability, dry mouth, and impaired pupil constriction), and gastrointestinal dysmotility. A chronic progressive course that may resemble degenerative forms of autonomic failure has been described. About 50% of patients with AAG have antibodies against the neuronal nicotinic acetylcholine receptor found in autonomic ganglia. Higher levels of serum ganglionic acetylcholine receptor antibodies correlate with more severe autonomic deficits, and some patients improve with therapeutic plasma exchange or IV immunoglobulin. Quantitative autonomic biomarkers should be used to define initial deficits, guide therapeutic decisions, and document treatment response (Koay et al., 2021). Karagiorgou and colleagues demonstrated that a cell-based assay for detection of serum ganglionic acetylcholine receptor antibodies was more specific to AAG than a radioimmunoassay precipitation assay (Karagiorgou et al., 2022).

Paraneoplastic autonomic neuropathies may have a presentation resembling AAG. Paraneoplastic autonomic neuropathy may manifest as a subacute generalized autonomic failure with orthostatic hypotension, anhidrosis, dry mouth, erectile failure, and gastrointestinal dysmotility. Prominent symptoms of gastrointestinal dysmotility may be seen in the limited paraneoplastic enteric neuropathy. Paraneoplastic neuropathies may also occur as part of a more widespread neurologic disorder affecting somatic peripheral nerves or the central nervous system. Although paraneoplastic dysautonomia is most often associated with the well-characterized onconeural antibodies anti-Hu, several other antibodies have been reported (Table 8.1). Anti-Hu is highly associated with underlying malignancy, particularly small cell lung carcinoma (SCLC). Autonomic manifestations often occur before the detection of the underlying malignancy, but can also appear at the time of cancer diagnosis, or even after the tumor is treated. Lambert—Eaton myasthenic syndrome (LEMS) presents clinically with symmetrical proximal weakness, reduced or absent reflexes, ptosis, and autonomic dysfunction in about 75% of patients. Autonomic tests show cholinergic and adrenergic abnormalities. Almost all patients show autoantibody against PQ-type voltage-gated calcium channels.

Morvan's syndrome may manifest with peripheral, central, and autonomic nervous system impairment. Clinical features of peripheral involvement include muscle stiffness, myotonia, cramps, myokymia, and weakness. Autonomic involvement is characterized by hyperhidrosis, fever, hypersialorrhea and increased lacrimation, tachycardia, cardiac arrhythmias, severe constipation, and urinary incontinence. The detection of voltage-gated potassium channel antibodies (VGKC-complex antibodies) is diagnostic, and these antibodies have been demonstrated in the serum and on neuronal dendrites in the hippocampus, thalamic neurons, and peripheral nerves. Although these antibodies are directed against CASPR2, LGI1, or both, CASPR2 antibody's predominance is found in association with thymoma.

8.4.4.2 Autonomic neuropathies associated with systemic autoimmunity

Sjögren's syndrome is characterized by sicca symptoms (i.e., dry eyes and dry mouth) and lymphocytic infiltration of exocrine glands. Sjögren's syndrome has been associated with small-fiber neuropathy and sensory ganglionopathy, and patients often report symptoms of autonomic dysfunction. Findings on autonomic reflex screen are variable including nonlength-dependent or length-dependent postganglionic sympathetic sudomotor impairment, decreased cardiovagal function, and impaired sympathetic noradrenergic function. Autonomic features may be the first manifestations of Sjögren's syndrome.

Other rheumatological disorders such as rheumatoid arthritis, scleroderma, systemic lupus erythematous, and psoriatic arthritis have been associated with autonomic neuropathy.

Dysautonomia can be seen in patients with gluten sensitivity and celiac disease (gluten-sensitive enteropathy). Postural nausea may be the primary symptom for referral. A gluten-free diet is recommended; however, it is still debated if this treatment is associated with improvement in autonomic symptoms. The role of antigliadin antibodies in the pathophysiology of autonomic neuropathy associated with celiac disease is unknown. In a nationwide population-based sample of patients with biopsy-verified celiac disease, there was an increased risk of autonomic neuropathy of 4.2 (1.4–12.3; $P = .009$) (Thawani, Brannagan, Lebwohl, Green, & Ludvigsson, 2015).

8.4.4.3 Metabolic/nutritional

Diabetes mellitus is the most common cause of autonomic neuropathy is the United States. Virtually all organs innervated by the autonomic nervous system may be involved in patients with diabetes. The severity and extent of impairment is usually related to the duration of diabetes and the extent of hyperglycemia. Cardiovascular autonomic neuropathy is associated with increased mortality in patients with diabetes (Rathmann, Ziegler, Jahnke, Haastert, & Gries, 1993). Clinical manifestations may resemble the ones associated with hypoglycemia with postural dizziness, nausea, vertigo, weakness, presyncope, or syncope. Exercise intolerance and painless myocardial ischemia are manifestations of cardiovascular autonomic neuropathy in diabetes. On testing, reduction in heart rate responses to deep breathing (respiratory sinus arrhythmia) is often the first abnormality; however, diabetic autonomic neuropathy also often causes impairment in postganglionic sympathetic sudomotor and adrenergic functions. Orthostatic hypotension from cardiac autonomic neuropathy is typically a late-occurring event. Therefore, early orthostatic hypotension within the first 5 years of diabetes diagnosis should prompt evaluation for other etiologies. In postural hypotension associated with diabetes, resting catecholamines and total platelet alpha$_2$-adrenoceptor numbers were not remarkably different than in those without this complaint; however, the percentage of high-affinity platelet alpha$_2$-adrenoceptor was lower in those with diabetic autonomic neuropathy (Senard, Barb, & Estan, 1990). Thus, platelet and vascular adrenergic receptors are not uniformly affected in diabetic orthostatic hypotension.

The prognosis of diabetic patients with symptoms of autonomic failure is much worse. Several long-term studies have been published all agreeing on this point (Sampson, Wilson, & Kargianis, 1990; Navarro, Kennedy, & Loewenson, 1990). However, not all abnormal autonomic functions have the same prognostic implication. Thus, isolated abnormalities without symptoms have no effect on prognosis. Moreover, postural hypotension, diarrhea, and gustatory sweating often persist for long periods without increases in intensity (Sampson, Wilson, Kargianis, et al., 1990). In addition, optimization of the treatment of diabetes improves autonomic and somatic nerve function (Kennedy, Navarro, & Goetz, 1990).

Hypoglycemia unawareness has been attributed to autonomic disturbances in patients with diabetic neuropathy. However, this ANS dysfunction appears to be limited to those with short-duration diabetes; in those with long-lasting diseases, hypoglycemia unawareness does not correlate with ANS dysfunction (Hepburn, Patrick, & Eadington, 1990). In another study, no correlation between unawareness of hypoglycemia and autonomic neuropathy could be found (Ryder, Owens, & Hayes, 1990).

In diabetes, comparisons of cardiovascular reflex tests with other tests often give nonhomogenous results. Diabetic foot ulcers seem to occur only in patients with abnormalities in sweating but venous occlusion plethysmography may show normal responses in the hands of such patients. In addition, some patients have normal autonomic tests of cardiovascular reflexes and impaired vasoconstrictor responses (Ryder, Kennedy, & Newrick, 1990; Gilmore, Allen, & Hayes, 1990). Thus, when assessing autonomic function, the general or local effects of reflex stimulation must be kept in mind and local stimuli might be more informative than general tests of reflex function (Ewing, 1991).

Several other tests have been employed to assess autonomic function in diabetes. The cough test is simple and reproducible but not widely used (Cardone, Paiusco, & Marchetti, 1990). A long QT interval has been used to gage the possibility of sudden death in diabetics. While it did not prove useful in this respect, it had value in predicting the presence of autonomic involvement (Ewing & Neilson, 1990).

Abnormalities in heart rate variability on power spectral analysis have been demonstrated in patients with diabetes and these have also been found on 24-hour Holter recordings of the EKG (Malpas & Maling, 1990). These efforts have not shown their superiority to ordinary tests of reflex cardiovascular autonomic function.

Parasympathetic pupillary dysfunction may precede sympathetic denervation as assessed by light reflex latencies to dilatation and constriction (Lanting, Bos, & Aartsen, 1990).

Erectile failure in diabetes has been correlated with penile cholinergic nerve damage in diabetic patients with erectile dysfunction (Blanco, De Tejada, & Goldstein, 1990). Retrograde ejaculation occurs rarely and may be related to sympathetic dysfunction. In female patients, atrophic vaginal wall and lack of lubrication characterize vaginal autonomic neuropathy resulting in painful intercourse.

Pathological examination of vagal nerves from patients with diabetic gastroparesis has shown myelin damage but surprisingly, regenerating fibers were also present suggesting that the notion of autovagotomy as a cause of gastroparesis is no longer tenable (Britland, Young, & Sharma, 1990). In addition, reduced gall bladder contractions after different stimuli (Fiorucci, Bosso, & Scionti, 1990) and decreased pancreatic polypeptide secretion have been found in autonomic neuropathy (Carra, Brambilla, & Meschi, 1990). Abnormalities in the gastrointestinal tract in patients with diabetes are well recognized. In addition to impaired gastric emptying (gastroparesis), patients may report diabetic diarrhea. Gastroparesis can be managed with the use of erythromycin, which has motilin-like activity because it binds to motilin receptors and acts as a motilin agonist (Janssens, Peeters, & Vantrappen, 1990). Pyridostigmine has a prokinetic effect on the gut and may be helpful in patients with gastroparesis and/or chronic constipation.

In diabetic polyneuropathy, studies have shown that some patients have no overshoot at the end of the Valsalva maneuver and, therefore, presumably did not constrict their arterioles and veins in response to the falling cardiac output. In some of these patients, normal vasodilatation in the hand in response to heating the body was found. Vasoconstriction after a deep breath was also noted. These intact peripheral vasomotor responses to the hand were interpreted to show that the baroreceptor pathways may also be interrupted on the afferent side in diabetes (Sharpey-Schafer & Taylor, 1960). Numerous studies have since confirmed these original findings (Ewing, Campbell, Burt, & Clarke, 1973). In patients with suggestive symptoms of diabetic autonomic neuropathy, an abnormal response to the Valsalva maneuver was found in 62% and the blood pressure response to sustained handgrip was abnormal in 41%. A significant correlation between the two tests was also found. Electrodiagnostic studies of motor nerve conduction velocities showed a greater slowing in the lower limbs in these patients than in subjects with normal vasoconstrictor responses. Peripheral vascular and sweat gland reflexes in patients with diabetic neuropathy were impaired. Vasoconstriction was inadequate in responses to cough, a single deep breath, and sudden noise. These deficits were accompanied by abnormal reflex sweat gland activation. Some evidence for changes in central nervous conductivity in patients with diabetic neuropathy was also found (Goadby & Downman, 1973).

Orthostatic hypotension, with or without hypertension in the recumbent position, may be a disabling complication of diabetes, and investigation of the renin-angiotensin system to explain these paradoxical blood pressure responses in diabetic patients has been carried out (Christlieb, Choodappasetty, & Braaten, 1974). Though in some patients with orthostatic hypotension not associated with diabetes mellitus an increase in plasma renin activity to the stimulus of upright posture was found, this was not the case with diabetic orthostatic hypotension. In diabetic patients, there was a decreased renin release with upright posture. A review of the literature of the renin-angiotensin response in patients with idiopathic orthostatic hypotension showed that 10 of 15 patients studied had an increase in plasma renin activity when upright and 4 of 4 patients showed an increase of plasma renin activity in response to sodium depletion. On the other hand, a patient with amyloidosis and orthostatic hypotension failed to increase plasma renin activity like patients with diabetes. It has been suggested that the associated nephropathy in diabetes mellitus may be the cause of the failure to increase plasma renin activity in response to upright posture and sodium depletion. Other possible defects in blood pressure homeostasis in diabetics such as catecholamine deficiency and perhaps blood volume abnormalities may combine with the renin-angiotensin system to make orthostatic hypotension disabling in patients with diabetic neuropathy. In diabetic autonomic neuropathy associated with postural hypotension, a resting tachycardia and painless or silent myocardial infarction may occur. The differential diagnosis from hypoglycemia and from symptomatic orthostatic blood pressure falls due to drugs including hypotensive agents, diuretics, tricyclics, phenothiazines, vasodilators, and glyceryl trinitrate need to be kept in mind. Moreover, insulin itself may aggravate postural hypotension in the absence of hypoglycemia presumably due to decreased venous return or an alteration in capillary permeability with a consequent decrease in plasma volume. The appearance of congestive heart failure or nephrotic syndrome due to renal involvement by diabetes can profoundly affect symptomatic manifestations of postural hypotension. The mechanisms of symptomatic orthostatic hypotension in diabetic neuropathy may involve one or more factors that account for the variability of symptoms and also affects treatment depending on dominant influences. Commonly, there is damage to sympathetic vasoconstrictor fibers, particularly in the splanchnic bed, muscle, and skin. There may, in addition, be a decrease in plasma renin responses to standing probably resulting in part at least from sympathetic denervation of the juxtaglomerular mechanisms. A decrease in recumbent and standing norepinephrine levels has been found and contributes to the hypotension. In some patients, there is a hyperadrenergic postural hypotension where norepinephrine levels are elevated on standing. This is probably the result of hypovolemia and is usually accompanied by an unexpected tachycardia on standing.

Resting tachycardia can be seen in patients with diabetic autonomic neuropathy. This tachycardia has been attributed to damage of parasympathetic fibers and it is thought to represent early cardiac involvement by diabetic peripheral neuropathy. When in the later stages of the development of diabetic autonomic neuropathy, sympathetic impairment of cardiac innervation also occurs and the heart rate usually drops to lower levels at rest, which are still significantly higher than in controls.

There is an increased incidence of myocardial infarction and sudden death in patients with diabetes. This has been attributed to autonomic neuropathy causing damage to autonomic afferents from the myocardium. The patients are thought to be deprived of ischemic pain, leading to infarction which otherwise might have been prevented by the onset of angina. This is clearly not an entirely acceptable explanation because patients with advanced autonomic neuropathy may still have typical cardiac pain during infarction and suffer from angina. In some patients with diabetes who succumb to sudden death, abnormal autonomic reflexes may account for their demise (Ewing, Campbell, et al., 1980). That arrhythmia and intrinsic vascular disease, perhaps with electrolyte abnormalities, are more important factors in the genesis of myocardial infarction, and sudden death in diabetics is supported by studies of the function of the transplanted human heart. The surgical technique of orthotopic cardiac transplantation retains the posterior wall of the recipient atria. The donor's heart is sutured to the atria and great vessels and the valves, ventricles, the arterial supply and much of the atrial muscle including the sinoatrial node remain untouched. The transplanted heart is totally denervated and in long-term studies has been shown not to become innervated again. This heart, because of an absence of vagal tone, has, like the diabetic heart, a high resting rate without sinus arrhythmia. On exercise, there is only a slow increase in heart rate, but peak rates are not achieved until after about 5 minutes, and slowing of the heart rate on cessation of activity is comparably delayed. The changes in heart rate are probably caused by circulating catecholamine levels. Many studies have confirmed the suspicion that in humans the denervated transplanted heart depends functionally on the Frank–Starling mechanism (force of contraction is closely related to the stretching of heart muscle). The denervated transplanted heart, therefore, functions well enough, but the lack of sympathetic drive limits an increase in stroke volume on exertion, and the absence of vagal tone is evidenced by resting tachycardia. Most levels of activities undertaken by such patients, however, are well performed, and the demands on the heart are met by adjusting stroke volume and by the effects of circulating catecholamines.

Patients with diabetic autonomic neuropathy and other neuropathies have an enhanced vascular sensitivity to α-adrenergic agonists (e.g., norepinephrine). The cause of this enhanced sensitivity is not fully understood but has been attributed to denervation supersensitivity. Beta-receptor agonist infusion also shows enhanced responses in patients with autonomic neuropathy, but the effects of β-receptor stimulation on hormonal-metabolic responses in diabetic autonomic neuropathy need further studies.

The responses to infusion of epinephrine (β-adrenergic receptor stimulation) were studied in diabetic patients with autonomic neuropathy and compared with responses in diabetics without neuropathy and controls. In patients with autonomic neuropathy, significant decreases in arterial pressure were found, which remained unchanged in the other study groups. The cardiac output increased to the same extent in the three groups, so that the decrease in blood pressure in those with autonomic neuropathy was the result of a significantly larger decrease in total peripheral vascular resistance. Also, there was a significant increase in heart rate during the infusion in those with neuropathy, but not in diabetics without neuropathy. The hormonal responses were characterized by a greater increase in glucose, a faster glucose-appearance rate, and higher levels of lactate, glycerol, and free fatty acids in the patients with autonomic neuropathy than in the other two groups. In diabetic autonomic neuropathy, therefore, β-receptor function in response to epinephrine infusion is enhanced.

In some studies, the arterial norepinephrine concentration has been shown to rise during epinephrine infusion, which suggests that sympathetic nerves are activated by epinephrine infusion. The decrease in arterial pressure in patients with diabetic autonomic neuropathy could result from an absence of sympathetic nerve activation, and excessive β-receptor-mediated decrease in vascular resistance may also account for the fall in blood pressure in the study group. Diabetic patients without neuropathy have increased glycemic responses to epinephrine infusion, and this is attributed to an increase in insulin secretion, as the blood glucose levels rise. In patients with diabetic autonomic neuropathy, the enhanced glycemic response to epinephrine may also be due, in part, to an increase in hepatic β-adrenergic sensitivity, because hepatic glycogenolysis and epinephrine effect are most likely the result of β-adrenergic activation. In the muscle of patients with diabetic autonomic neuropathy, increases in glycogenolysis and lactate formation, through a β-receptor-mediated mechanism, may contribute to the glycemic responsiveness in such individuals. The enhanced lactate response is thought to be a reflection of an increase in muscular glycogenolysis.

The concentration of free fatty acids and glycerol reflects lipolysis, and the response to epinephrine infusion was greater in patients with neuropathy. This implies an increase in adipose tissue sensitivity to epinephrine in these patients and suggests a denervation supersensitivity, not only of the vasculature but also of adipose tissue (Hilsted, Richier, & Madsbad, 1987).

The degree of autonomic cardiovascular involvement in diabetes has been measured by a number of methods, all of which give some evidence of autonomic neuropathy (Ewing et al., 1980) (refer to Chapter 23). Blood pressure responses to muscular exercise can also be measured. Sustained isometric exercise causes an increase in heart rate and an increase in cardiac output including a blood pressure rise, but no change in peripheral vascular resistance. One way of assessing this response uses a handgrip dynamometer, which is calibrated to demand 30% of the individual's maximum voluntary contraction. The blood pressure is measured during the handgrip. Patients with autonomic neuropathy have an abnormally low diastolic blood pressure response. A rise in diastolic pressure of 16 mmHg or more is normal, 11–15 mmHg borderline, and less than 10 mmHg evidence of autonomic neuropathy.

It should be clearly understood that symptoms of autonomic neuropathy may not be present when clear evidence of abnormal cardiovascular reflexes is obtainable on autonomic testing. Abnormal autonomic responses in diabetics occur during ketoacidosis without symptoms of autonomic neuropathy. In addition, tests of parasympathetic function have shown widespread involvement and are more commonly abnormal than those of sympathetic function. Progression with time has been established for both parasympathetic and sympathetic impairment in many patients. When autonomic dysfunction is detectable either by special tests or clinically, the mortality is much higher. In an early study, 21 patients who died while being observed over a period of 5 years did have autonomic neuropathy (a calculated mortality rate of 56%). Half died from renal failure and the other half from causes probably related to autonomic involvement (Ewing et al., 1980).

An association of severe diabetic autonomic neuropathy with iritis has been reported. Iritis is an immunologic disorder suggesting that the neuropathy might be associated with immune dysfunction (Guy, Richards, & Edmonds, 1984). This could also explain the immunocyte infiltration in sympathetic ganglia of some patients with autonomic neuropathy (Fig. 8.23).

One feature of diabetic autonomic neuropathy is the dysfunction of vagal nerves. In patients with gastrointestinal symptoms and diabetes, the changes may be similar to those found after a surgical vagotomy. In diabetic autonomic neuropathy, gastric acid secretion may not be stimulated via the vagus or by hypoglycemia but it can still be produced in response to humoral agents such as pentagastrin. Vagal cholinergic mechanisms are involved in the stimulation of gastric secretion by pancreatic polypeptide hormone, which is activated spontaneously and during food intake. Reliably, vagal nerve activity and pancreatic polypeptide stimulation can be induced by insulin hypoglycemia and this is completely abolished by vagotomy. Examination of insulin hypoglycemia and its effect on pancreatic polypeptide secretion in juvenile diabetics have shown severe impairment in those with autonomic neuropathy when compared to

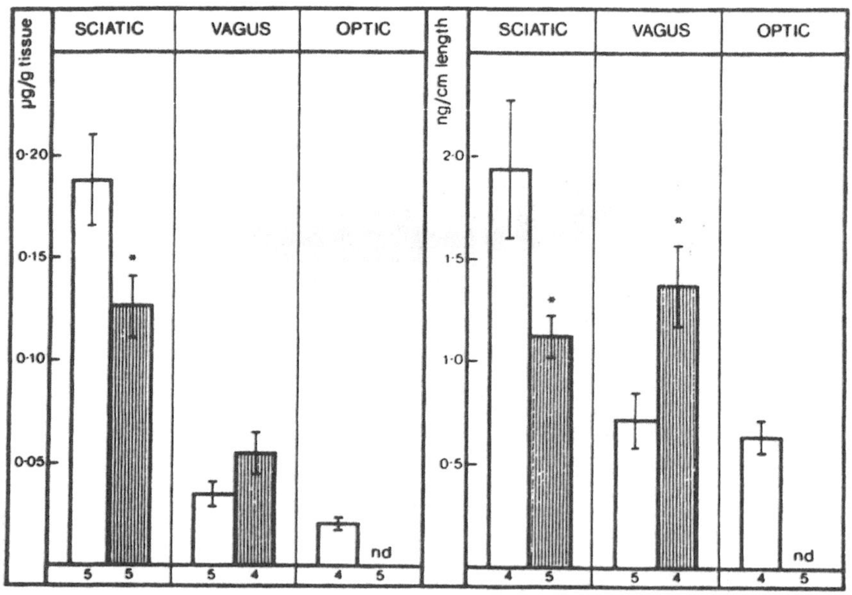

Figure 8.23 Paravertebral sympathetic chain from a patient with longstanding diabetic neuropathy and sudden death. There is widespread, mainly perivascular, lymphocytic infiltration suggesting the possibility of an immunologic reaction. *Courtesy of J. Appenzeller, M.D.*

diabetics without clinical evidence of neuropathy. A short history of a few years of diabetes usually accompanies a normal pancreatic polypeptide response to insulin-induced hypoglycemia but a significantly decreased response occurs with increasing duration of diabetes. The degree of impairment of pancreatic polypeptide stimulation by hypoglycemia could be correlated with impairment of vibration sense, suggesting that it may serve as an early sign of diabetic autonomic neuropathy. Since the insulin-induced hypoglycemia and consequent release of pancreatic polypeptide depend upon an intact vagal nerve supply to the pancreas, it is likely that abnormal responses result from vagal neuropathy. It cannot be completely excluded, however, that the abnormality is caused by another as yet unrecognized complication of diabetes since it was clearly related to the duration of the disease (Krarup et al., 1979).

If comparisons of autonomic ganglion cells are made from patients with diabetes and alcoholism with and without neuropathy, remarkable changes in the sympathetic neurons from those with neuropathy can be seen. A large number of abnormal giant sympathetic neurons, distended by a refractile material, are found. This material is eosinophilic and periodic acid-Schiff positive but does not stain with fat stains. The normally occurring pigment granules eventually disappear from affected cells, and these giant neurons are seen in all stages of degeneration, eventually leaving behind an empty gliocyte capsule (Figs. 8.24–8.26). A direct relation between these

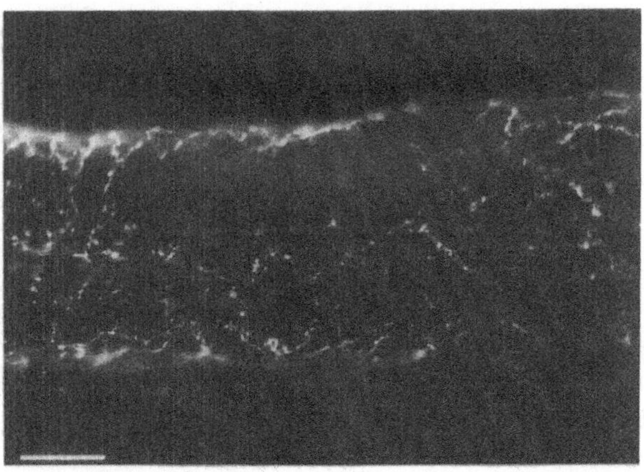

Figure 8.24 Giant sympathetic neuron from a patient with diabetic polyneuropathy (hematoxylin and eosin). *From Appenzeller, O., & Richardson, Jr. E. P. (1966). The sympathetic chain in patients with diabetic and alcoholic polyneuropathy.* Neurology (Minneapolis), 16, *1205–1209.*

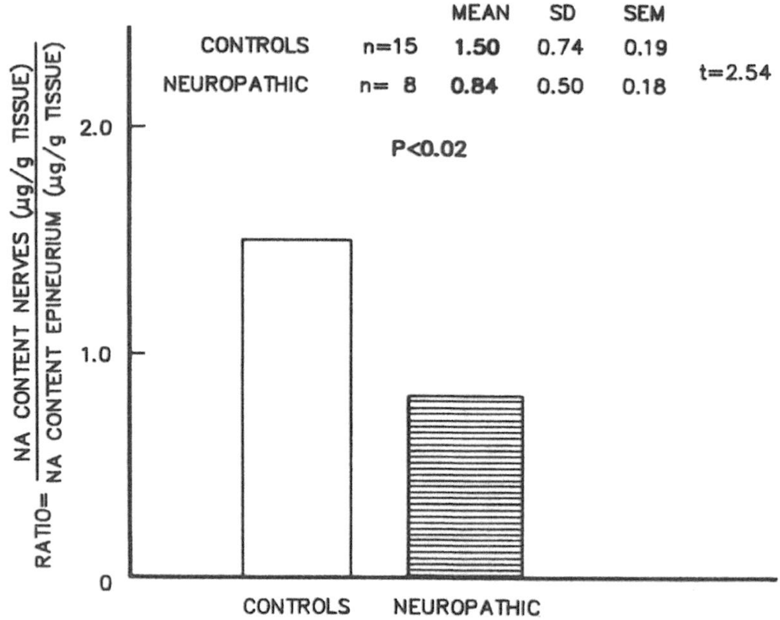

Figure 8.25 Giant sympathetic neuron from a patient with diabetic polyneuropathy showing abnormal accumulation of a refractile material (Sudan black B). *From Appenzeller, O., & Richardson, Jr. E. P. (1966). The sympathetic chain in patients with diabetic and alcoholic polyneuropathy.* Neurology (Minneapolis), 16, *1205–1209.*

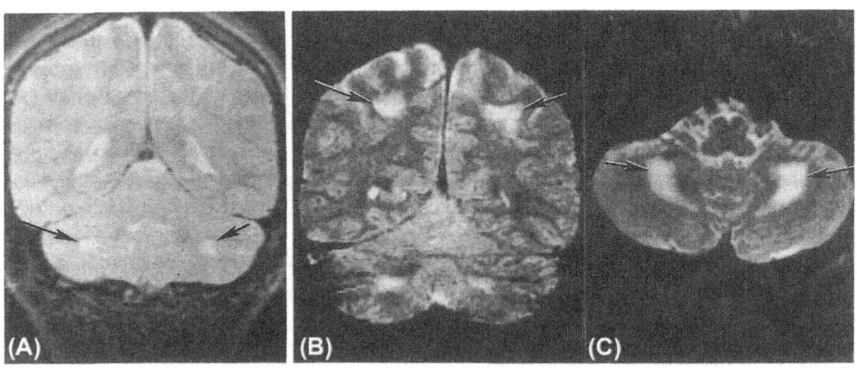

Figure 8.26 Empty gliocytic capsule. The giant neuron has disappeared but a strand of PAS-positive material and the nucleolus can be seen. From a patient with diabetic polyneuropathy (periodic acid-Schiff). *From Appenzeller, O., & Richardson, Jr. E. P. (1966). The sympathetic chain in patients with diabetic and alcoholic polyneuropathy.* Neurology (Minneapolis), 16, *1205–1209.*

morphological changes in patients with diabetic and alcoholic neuropathy and the observed autonomic disturbances have not yet been demonstrated. In the feet of diabetics with peripheral neuropathy, however, reflex vasomotor responses are completely absent, and the pattern of vascular responsiveness is characteristic of denervation hypersensitivity (Moorhouse, Carter, & Doupe, 1966). In such patients, the vessels of the feet exhibit an autonomous vascular tone that is independent of body temperature. This tone is only augmented by local cold and may be responsible for the discomfort and pain that accompanies exposure in such patients.

8.4.5 Treatment-induced neuropathy of diabetes

Treatment-induced neuropathy of diabetes (TIND) presents with acute/subacute neuropathic pain in a length-dependent fashion in patients with diabetes mellitus (Gibbons & Freeman, 2010). Treatment-induced neuropathy of diabetes is an iatrogenic complication secondary to aggressive glycemic control, and, specifically, the risk of TIND increases with HbA1C changes >1% per month. Clinically, there is allodynia, hyperalgesia, and a variety of autonomic symptoms including orthostatic intolerance, orthostatic hypotension with or without syncope, sweating abnormalities, and erectile dysfunction. TIND occurs in both type 1 diabetes (mean age of 25 years, 79% female) and type 2 diabetes (mean age >50 years, equal distribution between males and females). TIND has been described in patients treated with insulin, oral hypoglycemic medications, or diet control alone. TIND It can occur in parallel with worsening diabetic retinopathy. The pathophysiology is not fully understood and may be a combination of failure of axonal transport, endoneural ischemia, and cytokine-induced inflammation. The prevalence of TIND in the general population is unknown, but it is estimated that up to 10% of patients with diabetic neuropathy have symptoms due to TIND. Treatment is supportive with the goal of stable glucose control.

8.4.6 Autonomic neuropathy in systemic disorders

Autonomic dysfunction also has repeatedly been found in chronic renal insufficiency. Evaluation of normal subjects and comparison with those with renal insufficiency before and after dialysis and of patients who have been dialyzed for some time, including those with diabetic end-stage renal disease, showed autonomic abnormalities in uremia independent of other complications. Dialysis transiently improves autonomic deficits, particularly in nondiabetic end-stage renal disease. By contrast, diabetic patients with chronic renal failure tend to deteriorate in spite of adequate dialysis and show, in addition to parasympathetic dysfunction, which predominates in the nondiabetic uremic patients, deteriorating sympathetic disorders (Heidbreder & Heidland, 1985).

Baroreceptor function in uremic patients has been reportedly shown to be abnormal. They also differ from those in patients with primary hypertension. In some uremic patients, a neurogenic component to their hypertension may be present similar to that found in experimental baroreceptor afferent denervation (Levy, Lilley, et al., 1978).

In chronic hemodialysis patients, severe hypotension is a frequent complication, and this has been attributed to volume depletion. However, such patients have clinical and electrodiagnostic evidence of peripheral nerve disease and autonomic dysfunction may play a part in these hypotensive episodes. The Valsalva maneuver and the response to hypotension induced by the inhalation of amyl-nitrite were studied in a group of chronic renal failure patients on hemodialysis. Autonomic dysfunction was found particularly in patients with clinical and electrodiagnostic evidence of peripheral nerve disease. In these patients, the overshoot after Valsalva maneuver was absent, and no increase in heart rate in response to amyl-nitrite-induced hypotension was found (Appenzeller & Lueker, 1972; Kersh et al., 1974).

The activation of the sympathetic nervous system in cirrhosis of the liver has been well recognized (Robertson et al., 1993) but it is not known whether this is a feature of cirrhosis or the often-accompanying alcoholism. In those with alcoholic cirrhosis, the sympathetic nerves to the kidney, heart, and mesenteric circulation responded differentially to clonidine. The nerves to the hepatomesenteric circulation were particularly vulnerable to the sympatholytic effects of the drug (Esler, Dudley, & Jennings, 1992).

Postural hypotension has also been demonstrated in Wernicke disease. Contrary to expectation, in view of the sites of lesions in the hypothalamus, midbrain tegmentum, and fourth ventricle roof nuclei, the physiological data suggest that the disturbance responsible for the postural hypotension lies in the efferent sympathetic nervous system, probably in sympathetic neurons, intermediolateral column, or peripheral nerves (Birchfield, 1964). The presence of abnormal sympathetic neurons on histological examination in alcoholic peripheral neuropathy would appear to implicate the paravertebral sympathetic chain as the most likely site of the disturbance.

Other common causes of metabolic/nutritional autonomic neuropathy are summarized in Table 8.1.

8.4.6.1 Infectious

Infectious causes of autonomic failure are rarely considered but are an important clinical problem (Ewing, 1991). Table 8.2 summarizes the main infectious causes of autonomic dysfunction.

Autonomic dysfunction is relatively common with human immunodeficiency virus (HIV); however, the overwhelming symptoms of the illness often overshadow autonomic failure (Freeman, Roberts, & Friedman, 1990). Symptomatic autonomic neuropathy has been associated with increased morbidity. Heart rate variability has been shown to be reduced in HIV-infected individuals even in the early stages of infection (Mittal, Wig, Mishra, & Deepak, 2004).

American trypanosomiasis, also called Chagas disease, is a parasitic infection caused by Trypanosome Cruzi. In severe infection, Chagas disease causes cardiac failure. However, heart failure caused by Chagas disease differs from ordinary heart failure in that sympathetic nerve activity decreases in the former and increases in the latter (Iosa, Dequattro, & Lee, 1990). Diffuse damage of the autonomic nervous system, mainly the parasympathetic branch, is one of the hallmarks of Chagas disease and affects the heart and digestive tracts. Chagas cardiomyopathy may present with arrhythmias, symptoms of heart failure with fatigue and dyspnea, conduction blocks with or without syncope, peripheral embolism, and sudden death. Gastrointestinal manifestations of enteric nervous system dysfunction in Chagas disease include dysphagia due to megaesophagus, dyspepsia, and chronic constipation or diarrhea secondary to megacolon.

Leprosy is a chronic infectious disease caused by *Mycobacterium leprae.* Small nerve fiber involvement is common in the initial stages of the infection,

Table 8.2 Main infectious causes of autonomic neuropathy.

Virus
HIV
Herpes viruses
Bacteria
Borrelia budorgferi
Mycobacterium lepra
Corynebacterium diphteriae
Clostridium tetani
Clostridium botulinum
Parasite
Trypanosoma cruzi (Chagas disease)

whereas large-diameter myelinated fibers may be involved later. Leprosy neuropathy can present as isolated mononeuritis, mononeuritis multiplex, and polyneuritis. Autonomic dysfunction, including anhidrosis, localized alopecia, and reduced heart rate variability, has been reported in leprosy.

Lyme disease is caused by *Borrelia burgdorferi*. Although cases of neuroborreliosis primarily affecting the autonomic nervous system are rare, cases presenting with urinary retention and intestinal pseudo-obstruction have been reported. Following Lyme infection, complex regional pain syndromes and reflex sympathetic dystrophy with regional sympathetic hyperactivity have been reported in some patients and may be related to the persistence of an inflammatory response. Posttreatment Lyme disease syndrome or chronic Lyme disease is controversial, and there are overlapping symptoms with postural tachycardia syndrome and chronic fatigue syndrome.

Botulism is an intoxication caused by *Clostridium botulinum* toxin. Botulism is characterized by facial and bulbar motor weakness, ptosis, ophthalmoplegia, paralysis, and autonomic manifestations. On examination, mydriasis poorly reactive to light and accommodation is often seen. Other autonomic manifestations include orthostatic hypotension, urinary retention, erectile dysfunction, constipation, hypo-/anhidrosis, and dry mouth. Long-lasting autonomic dysfunction has been reported in adults (Potulska-Chromik et al., 2013). A variant of this condition due to intoxication with botulinum B has preponderant effects upon cholinergic autonomic function (Jenzer, Mumenthaler, Ludin, & Robert, 1975). This condition is characterized by a benign but protracted course of paresis of accommodation, pupillary abnormalities, impaired salivary and lacrimal secretion, constipation, and disturbances of micturition, all due to peripheral cholinergic dysfunction. Orthostatic hypotension and apathetic behavior are also present, but this cannot be definitely attributed to botulinum toxin. A striking feature of this disorder is the duration of symptoms, which can vary from 38 to 80 days. The neurologic examination is otherwise normal. Normal electromyographic findings document normal neuromuscular transmission. No specific treatment is necessary, and administration of antitoxin is not recommended. This condition must be distinguished from acute pandysautonomia, which on clinical and morphologic grounds, is a different entity.

8.4.6.2 Toxic autonomic neuropathies

Autonomic dysfunction in alcoholic peripheral neuropathy occurs, though postural hypotension is less disabling. Chronic alcoholic abusers have evidence of vagal denervation, shown by abnormal heart-rate responses and some sympathetic dysfunction. Pathologically, giant sympathetic neurons and relative preservation of the density and fiber diameters of myelinated nerves in the splanchnic, vagus, and sympathetic afferents have been found. In severe cases, degeneration in the vagus and sympathetic nerves occurs. The vagus is involved earlier and there is relative preservation of the sympathetic vasomotor control, in keeping with the predominant dying back features of this neuropathy, where short- and smaller-diameter myelinated fibers in the sympathetic nervous system are spared until later in the illness. Disturbance of esophageal motility in chronic alcoholics with peripheral neuropathy is a result of damage to the vagus nerve.

Autonomic dysfunction in peripheral nerve disorders associated with vincristine, perhexiline maleate, heavy metals such as thallium, arsenic, organic solvents, mercury, acrylamide, intermittent acute and variegate porphyrias, and vitamin B6 intoxication, where peripheral vasomotor failure leads to the burning hands-burning feet syndrome, have been reported. Immune checkpoint inhibitors have been associated with autonomic ganglioneuropathy with symptoms of cardiovascular autonomic failure but also autonomic enteric neuropathy.

8.4.6.3 Hereditary autonomic neuropathies

The major feature of hereditary sensory and autonomic neuropathies (HSANs) is the loss of large myelinated and unmyelinated fibers. Multiple types and subtypes of HSANs have been described.

8.4.6.3.1 HSAN I

Hereditary sensory and autonomic neuropathy type I is the most common HSANs. HSAN1 is genetically heterogeneous and most cases have an autosomal dominant inheritance. Mutations in the SPTLC1, SPTLC2, ATL1, ATL3, and DNMT1 genes have been reported in families with HSAN1. Symptom onset ranges from early childhood to late adulthood. Patients have prominent sensory loss followed by weakness and foot ulcers, whereas autonomic involvement is variable. Analysis of nerve conduction studies showed a predominately axonal and sensory peripheral neuropathy, whereas a motor neuropathy was also present in most patients, and a minority had conduction studies suggestive of demyelination.

8.4.6.3.2 HSAN 2

Hereditary sensory and autonomic neuropathy type 2 is an autosomal recessive disorder and various genetic mutations have been reported. Loss of pain, temperature, pressure, and touch sensation with large and small-fiber sensory involvement is characteristic. Mutilation of the fingers and toes occurs as the disease progresses.

8.4.6.3.3 HSAN 3

Autonomic dysfunction dominates the clinical picture in hereditary sensory and autonomic neuropathy type 3 (familial dysautonomia or Riley-Day syndrome). It is transmitted as an autosomal recessive trait and is mainly seen in children of Ashkenazi Jewish descent. HSAN3 is due to the truncated form of I kappa B kinase complex-associated protein coded by a gene that is located on chromosome 9q3. Autonomic crises are characterized by nausea and vomiting, irritability, tachycardia, hypertension, facial flushing, bronchorrhea, and diminished oral coordination resulting in swallowing and speech dysfunction. Triggers of these sympathetic storms include physical or emotional stress. Other autonomic manifestations include orthostatic hypotension, hypersialorrhea, gastrointestinal motility dysfunction, bladder dysfunction, decreased or absent tearing, pupil dilation, hypo-/anhidrosis, and episodic hyperhidrosis. Vomiting and autonomic crises may require intravenous hydration and carbidopa may be helpful to control symptoms. Other medications used to treat autonomic crises include diazepam, clonidine, or pregabalin (Axelrod, 2004). In an extensive review of this subject, Riley (1957) noticed that the condition had been reported as early as 1927 (Vollmer, 1927) although clinical details were insufficient to make the diagnosis certain.

The absence of tears when crying is an important identifying feature of this disease. A normal infant, however, does not produce tears before 6–8 weeks of life so their absence cannot be used as an early diagnostic feature. Although tears are absent in most cases with familial dysautonomia, an occasional patient may produce a few drops while crying, but this is obviously less than normal. On biopsy of lacrimal glands, no abnormalities are found (Dunnington, 1954) and tearing can be induced by acetyl-b-methylcholine and neostigmine methylsulphate (prostigmine) (Kroop, 1956). In such patients, however, mechanical stimuli, which normally induce tearing, such as irritation of the middle turbinates or strong odors, are without effect and the children appear undisturbed by these maneuvers. The introduction of filter paper into the conjunctival sac (Schirmer's test) also fails to produce tearing (Dunnington, 1954). In most patients with familial dysautonomia, the moisture of the eyes is normal and corneal ulceration is only occasionally seen. The defect seems to lie mainly in the production of tears after emotional stimuli or mechanical irritation.

Disturbances of vasomotor control are particularly striking in familial dysautonomia. Such patients have unexpected swings of blood pressure, and emotion induces both systolic and diastolic hypertension, whereas changes of posture from recumbent to upright are associated with profound falls in blood pressure. This lability of arterial pressure is particularly marked in older children and in some it is associated with lack of hypertension after the cold pressor test. A further manifestation of disturbed vasomotion is skin blotching. This is evident by red, often sharply demarcated, and sometimes symmetrical macules over the upper thorax and face, but occasionally involving the whole body. These macules are often induced by eating or excitement. The blotches fade rapidly after the meal. There is also marked vasoconstriction of extremity vessels with emotion, which may give rise to persistent cyanosis (Day & Klingman, 1939; Engel & Aring, 1945). During crises, patients are irritable, have screaming attacks, cannot concentrate, and show hypersensitivity of their hair, genitalia, and feet. The hair on the head may stand on end. Eating may cause sweating on the head and back but not on the palms or soles of the feet.

Abnormal ventilatory and circulatory responses to hypoxia have also been found. In six patients with familial dysautonomia, hypoxia produced a decrease rather than a normal increase in the slope of the CO_2 response curve. With hypoxia, bradycardia and hypotension occurred rather than the tachycardia and hypertension normally expected and no increase in ventilation was seen. In these patients, administration of pure oxygen showed that peripheral chemoreceptor activity was induced by the hypoxia but obscured by a low central respiratory drive. This was attributed to the hypotension induced by the hypoxia, which exaggerated the cerebral hypoxia and thus depressed the central respiratory drive. It was suggested that these findings might explain the high death rate during anesthesia and the prolonged breath-holding and syncopal attacks often seen in patients with familial dysautonomia (Edelman, Richards, & Fishman, 1967).

Excessive perspiration in such patients is often seen and this becomes particularly marked when they are excited. The sweat electrolytes are, however, normal (Di SantAgnese, Darling, Perera, & Shea, 1953).

The control of body temperature is abnormal and such infants have hyperthermia with infections or may run fevers without any obvious cause. It is not, however, clear whether this is due to abnormalities of central thermoregulatory structures or to the abnormalities in vasomotor control. Moreover, etiocholanolone levels during bouts of unprovoked fevers in such patients have not so far been investigated.

There are feeding difficulties dating from birth and, in several infants, tube feeding was necessary for the first few months. Many older children seem to chew excessively before swallowing their food and the majority of patients drool. The latter is not due to excessive salivation, since this was found normal in most patients tested and only increased in a few (Lourie, 1943). Altered esophageal motility has been demonstrated in some patients by fluoroscopy (Linde & Westover, 1962), and there is an absence of fungiform papillae on the tongue (Smith & Dancis, 1963). The intestinal function is usually normal although some patients have repeated attacks of abdominal cramps, nausea, and vomiting (Engel & Aring, 1945). Most reported patients had no deep tendon reflexes. In some, sluggish jerks, particularly knee jerks have, however, been found. These children are clumsy and slow in acquiring motor skills and are delayed in their motor milestones. The patients learn to talk later than usual and fine articulation is disturbed. There are abnormalities in voice production and the speech is often nasal, monotonous, and tends to fade. Hearing is normal but language performance is poor (Riley, 1957). More than half of the patients with familial dysautonomia have scoliosis (Riley, Freedman, & Langford, 1954) and in some pes cavus is found. Many patients walk better on their toes and most cannot walk on their heels, but it is not clear whether this is due to weakness. Seizures are not a feature of this disorder.

The most striking disturbance in sensation is the relative indifference to pain. This is often noticed by the parents of affected children because they do not complain of painful injuries as often as expected and fractures may first be noted by the swelling they produce. This indifference to pain is not associated with anesthesia and has been interpreted as a lack of awareness of discomfort (Riley, 1957). There is, however, corneal anesthesia in most cases. A number of hyperesthetic and dysesthetic areas are found, particularly to pressure and pinprick. Proprioceptive impairment in the lower limbs is held responsible for the wide-based gait and positive Romberg sign and Charcot joints affecting the knees have been found (Brunt, 1967).

Psychological disturbances in such patients are very common, but it is difficult to differentiate those caused by the abnormal autonomic discharge in response to stress from disturbances in behavior.

There are other features of this disease that occur with sufficient frequency to deserve mention, although they may not be directly related to disturbed nervous function. Bronchopneumonia is a common occurrence, and many are subject to episodes of excessive bronchiolar secretions, evidenced by the transient appearance of rales and rhonchi over the lung fields (Moloshok & Moseley, 1956). It is not certain whether these pulmonary episodes are related to poor swallowing. Most patients are of small stature and are underweight; some are dwarfed. There is a delay in maturation of the skeleton, as judged by X-rays, and at least one boy had no pubertal changes at the age of 17 (Engel & Aring, 1945) although other cases were normal in this respect. Most patients with familial dysautonomia have a distinct general appearance, with a bird-like face and protruding upper teeth. Because of their unsteadiness, they tend to hold their hands upon walking.

Most cases have occurred in Jewish children. There are, however, some on record who probably had dysautonomia and were not of Jewish extraction (Riley et al., 1954; Linde, 1955). Occasionally, this disease appears in siblings (Riley, 1952) and consanguinity has been noted in some families. It is of interest that a man who had nine healthy children with his first wife had two subsequent children affected by the disease with a second wife (Martin, 1953) and this is of some significance in genetic counseling.

In the past, the diagnosis of familial dysautonomia was based largely on the unusual combination of symptoms and signs of vegetative and nervous system dysfunctions. However, biochemical tests have been used to suggest the diagnosis. It can be shown that in this disease, the urinary excretion of homovanillic acid is increased and that vanillylmandelic acid is decreased. This abnormality has been detected within the first 3 weeks of life by chromatography of the urine (Geltzer, Gluck, Talner, & Polesky, 1964). This is suggestive of a deficiency in the release or production of endogenous epinephrine and norepinephrine (Smith & Dancis, 1963). This might be the result of impaired function of the enzyme dopamine-ß-hydroxylase (DBH), which could cause deficiencies in norepinephrine stores. Other studies, however, have shown that catecholamine stores in the adrenal gland, at any rate, are in fact elevated in familial dysautonomia and that the defect probably involves an impaired release of catecholamines (Smith & Dancis, 1967). In animals, stress increases plasma levels of DBH and this occurs even after adrenalectomy (Weinshilboum, Kvetnansky, & Axelrod, 1971), but chemical sympathectomy decreases DBH levels (Weinshilboum & Axelrod, 1971). Because sympathetic stimulation releases DBH and norepinephrine from sympathetic nerves, the serum level of this enzyme is taken as an indicator of peripheral adrenergic activity. DBH decrease in the plasma in dysautonomia is consistent with other findings of abnormal catecholamine metabolism but it cannot distinguish between a deficient norepinephrine synthesis and a defect in catecholamine release. Moreover, the DBH deficiency is not found in all patients with dysautonomia (Freedman, Ebstein, Goldstein, Axelrod, & Dancis, 1975). During frequent emotional crises, norepinephrine and dopamine plasma levels can be markedly increased, but epinephrine is less elevated. The vomiting is most closely related to high dopamine levels, whereas hypertension can best be correlated with norepinephrine levels. Valium is effective during crises and it relieves vomiting. This action may be related to it interfering with the effects of dopamine and norepinephrine because its use is usually associated with a decrease in plasma dopamine levels (Pearson, 1979). Genetic testing provides a definitive diagnosis. The disease is due to a founder mutation in the ELP1 gene (9q31), encoding elongator-1 protein, resulting in a truncated unstable protein. ELP1 is implicated in the migration, survival, and myelination of neurons during development and, in humans, the mutation affects the development of primary sensory and autonomic neurons.

The cholinergic dysfunction in familial dysautonomia, which includes parasympathetic pupillary supersensitivity and hypolacrima, can be corrected by either the topical application of acetyl-13-methylcholine or the infusion of this substance (Smith, Hirsch, & Dancis, 1965). Corneal hypoesthesia is also suggestive of acetylcholine dysfunction. The normal human and bovine cornea contain large amounts of acetylcholine and its synthesizing enzyme cholinesterase transferase. Though the function of acetylcholine in this site has not been clearly established, it might be related to transmitter activity of corneal sensory nerves. In familial dysautonomia, therefore, the ocular deficits are due to sensory, motor, and secretory system impairment of the cholinergic system. In one case of familial dysautonomia, the eye was examined postmortem and the cornea and iris contained no detectable choline acetyltransferase (Mittag, Mindel, & Green, 1974). Normal levels of this enzyme were present in the ciliary body and retina. In another dysautonomic patient, a lacrimal gland biopsy showed that it contained no detectable level of choline acetyltransferase, whereas human and other animal tissues had significant levels of this enzyme in lacrimal glands. These results suggest that a deficiency or absence of choline acetyltransferase in cholinergic nerve terminals might be the basis of at least the ocular abnormalities in familial dysautonomia. A unifying hypothesis attributing the peripheral sympathetic dysfunction to a cholinergic defect at the ganglionic level might account for many of the other clinical features of the disorder but cannot explain the sensory deficits constantly found in this disease.

Infusion of norepinephrine in such patients causes an exaggerated elevation of the blood pressure and, instead of the usual bradycardia, which accompanies this elevation in normal subjects, a tachycardia is observed (Smith & Dancis, 1964). The normal bradycardia with norepinephrine-induced hypertension can be blocked by atropine, and this appears, therefore, to be mediated by parasympathetic mechanisms (Goldenberg, Pines, Baldwin, Greene, & Roh, 1948). These results can be interpreted to mean either a deficient parasympathetic response to hypertension or an increased chronotropic effect of norepinephrine on the heart. In support of the former is the known supersensitivity of the pupil to conjunctival instillation of small amounts of acetyl-b-methylcholine (Mecholyl), which is consistent with parasympathetic denervation (Smith & Dancis, 1963). This supersensitivity is present within 24 hours of birth in patients with familial dysautonomia (Geltzer et al., 1964).

The effect of postural changes and the application of cold to a limb on the forearm circulation in patients with familial dysautonomia has been studied (Mason, Kopin, & Braunwald, 1966). It was found that these patients have no reflex venous or arteriolar constriction in response to postural changes and cold. This abnormal vasomotor activity could well account for the postural hypotension, which is such a distinctive feature of the disease.

A widely used test for the release of endogenous epinephrine is the insulin tolerance test. In normal fasting subjects, the administration of 0.1 U/kg of body weight will cause a 50% reduction of the blood sugar within half an hour and by two hours, the blood sugar level is back to normal. In a patient with familial dysautonomia, this test produced an 80% fall in blood sugar level within 30 minutes and it was still 40% below the baseline 2 hours later. The administration of small amounts of epinephrine at that time restored the blood sugar to normal (Geltzer et al., 1964). These results are also compatible with the interpretation that the endogenous production of epinephrine is physiologically inadequate, or it may be inappropriate to stressful situations.

Another abnormality in such patients is the absence of the flare component of the triple response of Lewis elicited after intracutaneous injection of histamine. This may be related to their partial insensitivity to pain, since the flare has been shown to be absent in a patient with congenital universal insensitivity to pain in whom there were no small, thinly myelinated fibers in the dorsal root ganglia (Swanson, Buchan, & Alvord, 1963).

The pathological basis of this rare congenital syndrome remains to be elucidated. Several cases have been examined at necropsy. Abnormalities of the central nervous system, which could be related to the disease, have been found. In one case, a striking thalamic cyst (Fig. 8.27) was seen, and, in another, degenerative changes in the reticular formation of the pons and medulla were found (Cohen & Solomon, 1955). In other cases, however, no changes in the central nervous system were seen. Incomplete examination of autonomic ganglia has shown alterations that cannot, however, account for all the symptoms and signs of the disease.

Other pathologic studies in some cases of familial dysautonomia showed scanty submucosal axons in the tongue and a total absence of taste buds. But in an older girl, though submucosal axons and neurons were found their number was decreased. Though the tongue is often smooth on inspection, small papillae can be seen on histologic examination. A decreased number of neurons in dorsal root ganglia with occasional areas of degeneration was found and the volume of the ganglia was only half of normal. Similarly, sympathetic ganglia were small and had a reduced number of sympathetic neurons. The ciliary ganglion was also smaller than normal and contained only few neurons, some of which showed degenerative changes (Pearson, Axelrod, & Dancis, 1974). The sural nerve was carefully examined in one case (Aguayo, Nair, & Bray, 1971). A marked decrease in the number of unmyelinated fibers and a total absence of myelinated fibers greater than 12 m in diameter was found. Examination of single teased fibers in this nerve showed abnormally short internodes on fibers 6–12 m in diameter. Subsequent examination of other sural nerves in such patients confirmed the decrease in unmyelinated axons but the most marked reduction in myelinated fibers was found in those with diameters smaller than 5 m. Catecholamine fluorescence showed no norepinephrine-containing axons in the nerve or in perivascular nerve plexuses (Pearson et al., 1974). Examination of paraspinous muscles obtained during corrective orthopedic surgery for scoliosis showed histochemical fiber type grouping but only slight muscle fiber atrophy without evidence of acute denervation (Pearson et al., 1974).

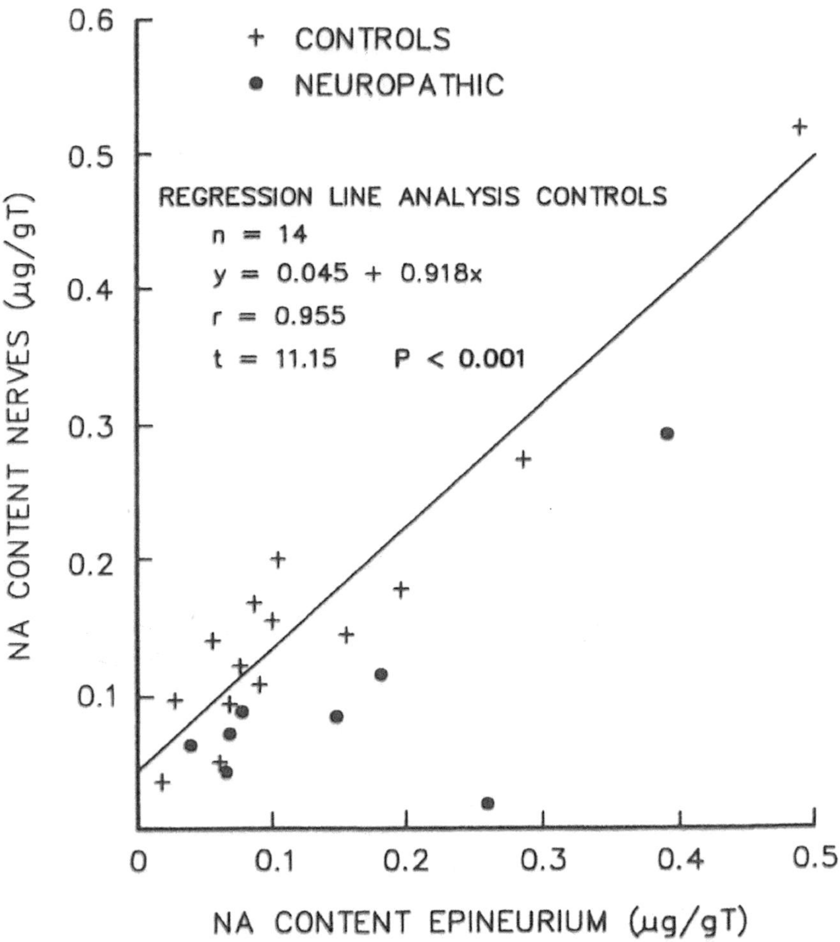

Figure 8.27 Section through dorsomedial nuclei of the thalamus showing a cyst (arrow). (Loyes stain). *From Engel, G. L., & Aring, C. D. (1945). Hypothalamic attacks with thalamic lesion. I. Physiologic and psychologic considerations.* Arch. Neurol. Psychiatry (Chicago), 54, *37–43.*

Clinical-pathological correlation is now possible to explain some, but not all, features of familial dysautonomia. The impaired taste can be attributed to reduction or defects in sensory lingual taste buds. The changes in the sural nerve, if representative of the whole peripheral nervous system, can account for postural hypotension, altered excretion of catecholamine metabolites, adrenergic denervation supersensitivity, skin botching, and impaired thermoregulation. The loss of unmyelinated C fibers in peripheral nerves is consistent with impairment of pain and temperature sensation and the absent corneal and axonal reflexes. Because large myelinated fibers are rare, the tendon stretch reflexes, which depend on muscle spindle afferents subserved by larger myelinated fibers, could be impaired or abolished.

Hypothetically, nerve growth factor has been implicated in the pathogenesis of familial dysautonomia (Aguayo et al., 1971; Pearson et al., 1974). The suggestion has been made that a developmental arrest of the migration of neurons from the neural crest could explain the absence of unmyelinated fibers and of large myelinated fibers including muscle spindle afferents. Whether or not nerve growth factor is responsible for the developmental arrest is not known. There is, however, a great similarity of the morphologic findings in familial dysautonomia and those experimentally induced by anti-nerve growth factor antiserum. Further study of patients with familial dysautonomia should yield important information about the trophic relationships of nerves and tissues and perhaps lead to palliative if not curative therapy for this disabling disease. Nerve growth factor (NGF) has now been measured in cultured skin fibroblasts from controls and from five patients with familial dysautonomia (Schwartz & Breakefield, 1980). Only approximately 10% of active NGF/mg of immunoreactive protein was found in cells from dysautonomic patients. If the fibroblasts of both controls and dysautonomic patients were treated with isoproterenol a beta-adrenergic agonist, a 17- to 170-fold increase in cyclic adenosine monophosphate (cAMP) content of both cells was found 10 minutes later. After exposure for 3—4 hours to this agent, the immunoreactive B-nerve growth factor increased 50%—300% in control fibroblasts, but no change was seen in dysautonomic cells. These findings are consistent with the suggestion that the molecular basis for the genetic defect in familial dysautonomia could be in the processing of precursors or in the structure of the active B subunit of NGF. Though most studies have been hitherto focused on the autonomic nervous system and its transmitters, the occurrence of similar though not identical symptoms in patients with familial hyperbradykininemia (Streeten, Kerr, Kerry, Prior, & Dalakos, 1972) suggests that other substances might now be fruitfully examined in attempts to unravel the true pathogenesis of this disease.

In a number of other studies, it has been shown that at least in animal brain-derived cells, B-NGF varies with the intracellular cAMP content. When such cells are treated with ß-agonists such as isoproterenol, an increase in cAMP occurs and the levels of B-NGF concurrently increase also. The proposal has been made that cAMP-dependent protein kinase activates the enzymatic processes, which lead to pro-B-NGF conversion into B-NGF. Thus, the finding of an absence of an increase in NGF in dysautonomic fibroblasts compared to control human fibroblasts after hours of incubation with a ß-adrenergic agonist suggests that the defect results from a change in the enzyme, which is responsible for the processing of the precursor molecules and which is dependent upon cAMP protein kinase activation. Alternatively, there may be a defect in the structure of the precursor molecule for B-NGF. The hypothesis may explain the presence of an immunologically recognizable protein (NGF) in familial dysautonomia, which, however, is biologically less active and is converted less efficiently to the active form after increase in cAMP. There are, therefore, clear differences in immunologic and biologic properties of B-NGF in dysautonomic skin fibroblasts.

Clinical manifestations of familial dysautonomia can result from other inherited diseases. If the diagnosis is to be made in acutely ill infants, this possibility needs to be kept in mind. Metabolic derangements can produce phenocopies of familial dysautonomia and one example is hyperammonemia due to propionyl coenzyme A carboxylase deficiency (Harris, Yang, Wolf, & Snodgrass, 1980). A male infant with vomiting and coma without ketoacidosis was found to have abnormal responses to methacholine and histamine and abnormal urinary catecholamine excretion. After an episode of hyperammonemia, a liver biopsy showed a partial deficiency of carbamoyl phosphate synthetase activity. When the patient was given a protein-restricted diet, supplemented with a mixture of ketoacid analogs of essential amino acids, ketosis and acidosis occurred. Subsequently, the demonstration of a deficiency of propionyl coenzyme A carboxylase deficiency led to the correct diagnosis. The similarity of presentation at birth of this case and familial dysautonomia suggests that appropriate metabolic studies be carried out in selected patients to exclude possible disorders, which might be treatable by dietary manipulation.

There is no curative treatment for familial dysautonomia. Symptomatic treatment can be given with sedatives and phenothiazine drugs for gastrointestinal and behavioral disorders. Corneal ulceration may require protection of the cornea or grafting and attacks of pulmonary infection appropriate antibiotic therapy. Hypotensive agents in hypertensive crises should not be used because of the abnormal vasomotor function in these patients.

8.4.6.3.4 HSAN 4

Hereditary sensory and autonomic neuropathy type 4 (HSAN4) is also called congenital insensitivity to pain with anhidrosis. HSAN4 has been linked to mutations in the NTRK1 gene. Autonomic symptoms include thermoregulatory impairment with anhidrosis, leading to episodic hyperthermia that can be associated with seizures.

Autonomic manifestations with hypohidrosis or impairment in cardiovascular function may also be encountered in other hereditary sensory and autonomic neuropathies (e.g., HSAN 5, HSNA6, and HSAN8).

8.4.6.3.5 Fabry disease

Fabry disease is an X-linked lysosomal storage disorder characterized by cell storage of glycolipid ceramide trihexoside within the nervous system, including neurons of the autonomic nervous system and a dying-back process is more prominent in small myelinated and unmyelinated fibers. These pathologic features are consistent with autonomic dysfunction that has been demonstrated in this disease (Cable et al., 1980). Severe pain that begins in the distal extremities is the hallmark of early neurological involvement in Fabry disease, and hypohidrosis is also a common presenting symptom. Hypohidrosis in Fabry disease may be related to Gb3 deposition causing sweat gland dysfunction as well as damage to autonomic small fibers. Dysfunction of the central autonomic network may also contribute to hypohidrosis in Fabry disease. Abnormal pupillary responses to pilocarpine, reduced saliva production and tear formation, and abnormal cardiovascular responses including decreased reflex rises in plasma noradrenaline were also found in nine men with Fabry disease. Sympathetic and parasympathetic function was impaired, and this was thought to represent small nerve-fiber dysfunction in peripheral nerves. This disease is a generalized metabolic disorder, with autonomic nervous system involvement; however, clinically, autonomic dysfunction is rarely incapacitating. In this disease, the classic clinical manifestation of corneal whirled deposits and the reddish-brown angiokeratomas are not troublesome, but in some pain may be severe, often leading to multiple drug use and even abuse, depression, and sometimes suicide. In a series of seven such patients (Filling Katz, Merrick, & Fink, 1989) good relief of the pain was obtained with carbamazepine, but the preexisting autonomic dysfunction was exacerbated in two, causing ileus, urinary retention, and gastrointestinal disturbances on relatively low doses of the drug. When treating the pain associated with severe autonomic disturbances in this condition, carbamazepine must be used cautiously because it may seriously worsen already crippling autonomic deficits.

8.4.6.3.6 Cerebellar ataxia, neuropathy, vestibular areflexia (CANVAS)

Autonomic dysfunction is also a major feature of cerebellar ataxia, neuropathy, vestibular areflexia—"CANVAS" syndrome (Wu et al., 2014). CANVAS is due to AAGGG repeat expansion in the second intron of replication factor C subunit 1 (RFC1) gene. In a case series of 23 patients, 83% had evidence of autonomic dysfunction; all patients had at least one autonomic symptom and 91% had more than two symptoms. Cold feet (78%), erectile dysfunction (78% of men), light-headedness (65%), constipation (65%), and dry mouth or eyes (52%) were the most common symptoms reported. Thirty percent of patients fulfilled criteria for orthostatic hypotension. Autonomic dysfunction in CANVAS may be related to a primary ganglioneuropathy involving the autonomic, vestibular, facial, trigeminal, and sensory ganglia and their projections. Some patients with CANVAS also report syncope as well as drop attack with abrupt loss of tone. The drop attacks may be similar to the ones seen in Tumarkin's otolithic crises in patients with late or end-stage endolymphatic hydrops (usually idiopathic, i.e., Meniere's disease). Impairment of the vestibulo-sympathetic reflex may also contribute to syncope in these patients.

8.4.6.3.7 Charcot-Marie-Tooth disease

Abnormal vascular reflexes in Charcot-Marie Tooth disease have been found (Appenzeller, 1963). Further examination of such patients including the Valsalva maneuver, pulse rate, and systolic blood pressure responses to standing have confirmed abnormal vascular reflex function in 6 of 17 adult patients with this disease. One also had giant nerve bundles in the myenteric plexus of the bowel, which was resected for coexistent regional ileitis. This pathologic finding supports autonomic nervous system involvement in this disease (Brooks, 1980). Clinically, Charcot and Marie (1886) clearly recognized the vasomotor abnormalities in all their five patients noting the bluish or reddish discoloration and mottling of the feet and legs commonly found in such individuals.

8.4.6.4 Amyloid neuropathy

The amyloidoses are a group of disorders characterized by the deposition of insoluble and degradation-resistant fibrillar protein in a beta-pleated sheet structure within various tissues and organs. Amyloidoses associated with neuropathy include the AL type (immunoglobulin light or heavy chain) or familial type (due to mutations in transthyretin, apolipoprotein

A-I, or gelsolin). Commonly involved organs are the heart, kidneys, nervous system, liver and gastrointestinal tract but lung, muscle, and soft tissue can also be affected.

Autonomic features frequently accompany amyloid neuropathy. Autonomic manifestations include orthostatic hypotension, constipation, early satiety with gastroparesis, urinary retention, and erectile dysfunction. The involvement of somatic sensory fibers causes loss of temperature sensation and pain in a length-dependent pattern progressing to cause numbness, neuropathic pain, weakness, and imbalance. Heart involvement often presents with heart failure with preserved ejection fraction. Testing of the autonomic nervous system can help in differentiating autonomic involvement from hypotension from heart failure. The diagnosis of systemic amyloidosis requires a clinical presentation suggestive of an amyloidosis syndrome with tissue confirmation of Congo red-positive extracellular deposits showing apple-green birefringence under polarized light microscopy. Amyloid typing on tissue samples (e.g., by mass spectrometry proteomic-based analysis) allows for the identification of the different forms of amyloid neuropathies. Autonomic symptoms are most prominent in the light chain and inherited autosomal dominant transthyretin types of systemic amyloidosis.

AL amyloidosis is caused by a clonal plasma cell disorder and is characterized by multiorgan involvement and a faster loss of organ function compared to other systemic amyloidosis types. AL amyloidosis is most often seen in patients older than 50 years. The treatment goal in AL amyloidosis is the elimination of the production of the amyloidogenic light chains by the plasma cell clone and relies on the use of chemotherapeutic agents (e.g., bortezomib-based regimens).

Hereditary transthyretin variant amyloidosis is the most common form of hereditary amyloidosis. There have been numerous mutations reported and some mutations predominately cause polyneuropathy, while other mutations are cardiomyopathic or cause mixed phenotypes. V30M is the most common TTR mutation worldwide. The landscape of treatment for hereditary transthyretin amyloidosis (hATTR) has evolved considerably over the past few years. Liver transplantation has long been considered a standard treatment option for hATTR because replacing the primary source of mutant transthyretin (TTR) reduces its production. Liver transplantation is associated with improved long-term survival, as well as improvement in sensory and motor impairment. Nevertheless, progression of peripheral neuropathy and cardiomyopathy has occurred following liver transplantation and genotype has been shown to influence posttransplantation survival (Banerjee et al., 2017). Furthermore, liver transplantation is associated with the adverse effects and complications of long-term immunosuppression. Oral administration of tafamidis meglumine, a kinetic TTR stabilizer, prevents misfolding and deposition of mutated TTR (Bulawa et al., 2012). Tafamidis has been approved in European countries for patients with hATTR with stage 1 symptomatic polyneuropathy (low disease burden) and the treatment was approved in the United States for hATTR cardiomyopathy. Two studies have demonstrated less neurologic deterioration with tafamidis in early-stage hATTR polyneuropathy using 20 mg once daily compared with placebo; however, the results did not reach statistical significance (Coelho et al., 2012; Coelho et al., 2013). Furthermore, outcomes were mixed in patients with mid- and late-stage Val30Met hATTR or with non-Val30Met hATTR with possible worsening in autonomic function in patients with baseline dysautonomia and worsening in cardiac function in patients with baseline cardiac disease (Cortese et al., 2016). Diflunisal is a nonsteroidal antiinflammatory drug with TTR-stabilizing properties. Diflunisal has demonstrated potential benefit in hATTR compared to placebo; however, adverse events such as renal failure or thrombocytopenia may limit its use. Two small oligonucleotide-based gene therapies, one small interfering RNA triweekly intravenously administered (patisiran) and one antisense once weekly subcutaneously administered (inotersen), are approved in the United States for treatment of hATTR. Importantly, both agents have demonstrated efficacy in patients with early- and late-stage disease and improve, halt, or slow neuropathy progression.

In a review of 138 cases with amyloidosis, orthostatic hypotension was found in 11; 2 of these patients had peripheral neuropathy. Abnormal sweating was not a prominent symptom, but it was impaired in all six patients in whom it was tested. Necropsy examination was performed in two cases and in one, amyloid deposits were found in both peripheral and autonomic nerves. In another patient with widespread amyloidosis of the central and peripheral nervous systems, orthostatic hypotension was not noted during life (Fig. 8.28). The cause of orthostatic hypotension in primary amyloidosis is likely to be a disturbance of function of peripheral structures concerned with vasomotor control. The symptomatic manifestation of this, however, may depend on the degree of incapacity of the patient and the extent of nervous system involvement. A clinicopathological study of one patient with primary amyloidosis and one with secondary

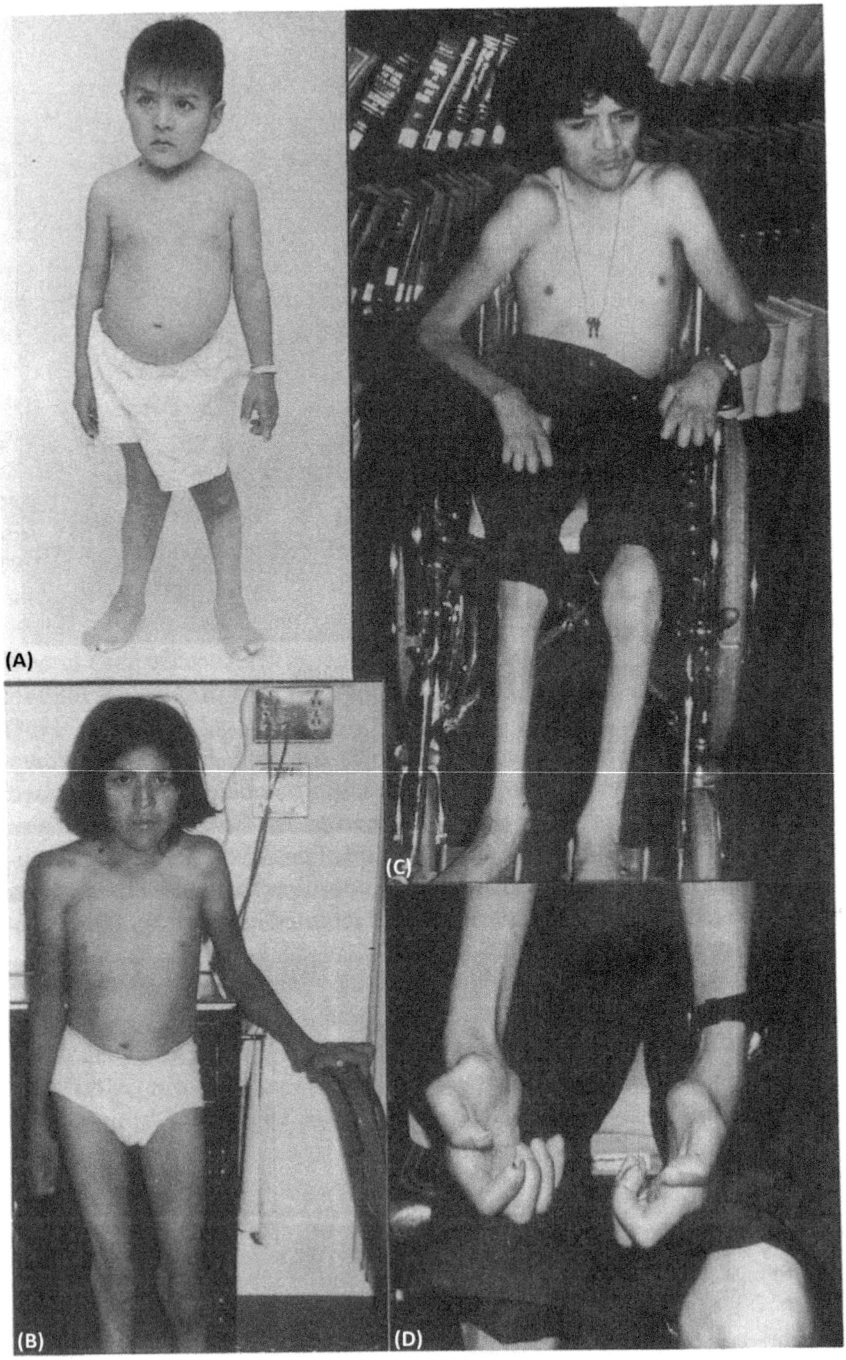

Figure 8.28 Posterior root ganglion infiltrated with amyloid which has replaced some ganglion cells. The patient did not have symptomatic postural hypotension. *From Appenzeller and True (1967).*

amyloidosis showed in the primary variety restricted pathologic changes in the autonomic nervous system, which were not associated with symptoms, whereas in secondary amyloidosis disabling autonomic disturbances with widespread destruction of autonomic nerves were found (Nordborg, Kristensson, Olsson, & Sourander, 1973). In the dominantly inherited form of amyloidosis (Andrade, 1952) postural hypotension and widespread sweating abnormalities may be disabling features in some but not all patients (Dyck & Lambert, 1969). Similar disturbances of vasomotor function are also found in alcoholics, porphyria, and a variety of other neuropathies (Barraclough & Sharpey-Schafer, 1963).

8.4.6.5 Other causes of autonomic neuropathy

8.4.6.5.1 Ross syndrome

The cardinal features of Ross syndrome with tonic pupils, hypo-/areflexia, and segmental anhidrosis were first reported in 1958 (Ross, 1958). A case series of 26 patients with Ross syndrome found that the most common initial reported manifestation was an abnormal segmental sweating response in 16 patients (described as hyperhidrosis in 12 patients and anhidrosis in 4 patients) while a tonic pupil was the initial manifestation in 10 patients. Other commonly reported symptoms included fatigue, chronic cough, and increased urinary frequency. An associated autoimmune disorder was only identified in one patient, suggesting that Ross syndrome is not associated with autoimmunity in most cases. Positive autoantibodies were found in a minority of patients, often with unclear clinical significance. Distributions of anhidrosis encountered were segmental ($n = 15$), widespread ($n = 7$), and global ($n = 4$). Anhidrosis progressed slowly over time and sudomotor dysfunction was predominantly (post)ganglionic. Cardiovagal and adrenergic functions were preserved in most patients (Lamotte et al., 2021). Immunohistochemical study of anhidrotic and hyperhidrotic skin in 12 subjects with Ross syndrome revealed a complex and progressive involvement of cutaneous sensory and autonomic innervation underlying the impairment of heat production and heat dissipation through both loss of sweating and loss of cutaneous blood flow regulation (Nolano et al., 2006). In this study, the persistence of a sudomotor vasoactive intestinal peptide-immunoreactive innervation, although deranged and poor differentiated Holmes–Adie from Ross patients.

Orthostatic hypotension can be a feature of some patients with the Holmes–Adie syndrome. At any time during their life, they may develop symptomatic orthostatic hypotension. On examination, the patients are neurologically normal except for tonic pupils and impairment of deep tendon jerks. Motor nerve conduction velocities are normal but autonomic studies confirm the presence of a baroreflex impairment, which may involve impairment of afferent pathways (Johnson, McLellan, & Love, 1971). The one patient studied had urinary and plasma renin-sodium relationships, which were entirely normal. Other autonomic function tests including the analysis of the blood pressure responses to isometric handgrip stress, rapid withdrawal of blood, and prolonged orthostasis confirmed the afferent baroreflex block previously described.

8.4.6.6 Others

There are rare patients with multiple mucosal neuromas who have dysautonomia and an abnormal triple response of Lewis after an intradermal injection of histamine (Horstink et al., 1974). Patients have painless grayish-yellow tumors on the conjunctivae, tongue, oral cavity, pharynx, and larynx. They are usually arranged symmetrically about the midline, typically along the anterior dorsal area of the tongue. Such patients have a marfanoid habitus, thick everted lips, and multiple endocrine tumors, including medullary thyroid carcinoma and pheochromocytoma. Pes cavus, funnel chest, myelinated fibers in the cornea, skin pigmentation, hypertrophic nerves, and abnormalities of Auerbach's and Meissner's plexuses and myopathy have been described. The clinical features start in early childhood, but endocrine tumors make their appearance only in adolescence. The condition shows an autosomal dominant hereditary pattern. Several neural crest-derived structures are involved in this disease, including dental abnormalities and abnormalities in the autonomic nervous system. A number of abnormal responses have been documented, including the following: clinically impaired lacrimation, parasympathetic denervation supersensitivity of the pupils with intact sympathetic responses, orthostatic hypotension with an intact overshoot in response to Valsalva maneuver, normal responses to vagal stimulation (eyeball pressure and carotid sinus massage), impaired pilomotor activity, absent reflex vasodilatation with preservation of reflex vasoconstriction in hand blood vessels, abnormal glucose levels in response to intravenous insulin administration, and dermographia. Sweat and salivary gland function were normal. A sural nerve biopsy showed degeneration and regeneration of unmyelinated fibers similar to the degeneration and regeneration found in some patients with acute pandysautonomia.

Acrodynia (pink disease) is a syndrome with characteristic clinical features, but its etiology is not fully understood. The disease has become rare, but in the past was particularly common in England and Australia. In Europe, it was described by Feer (1923), who called the disorder infantile vegetative neurosis. Acrodynia is practically confined to infancy and appears to be associated in some cases with repeated ingestion or contact with mercury. The number of infants exposed to sufficient amounts of mercury and for long periods is, however, far greater than the number of patients with the disease. It remains, therefore, uncertain whether the disorder is an unusual reaction to mercury or is simply mercury poisoning. The clinical manifestations include listlessness, with lack of interest in play, progressing to restlessness and irritability. Associated upper respiratory tract infections are often seen, and recurring widespread rashes are found. The tips of the fingers and toes become pink, and a dusky color spreads later to the palms and soles. The extremities become cold and clammy and the cheek and nose may appear bright red. These changes are the most distinctive features of the disease and are responsible for its name. When the syndrome becomes apparent, there is profuse and persistent sweating. Desquamation of the skin of the palms and soles may occur at intervals. Swelling of the fingers, which is due to hyperplasia and hyperkeratosis of the skin, is seen. Fever appears only with intercurrent infections. The children experience extreme pain in their hands and feet, and older children will complain of burning sensations in their extremities. Photophobia is very common, and characteristically children bury their heads in a pillow to shield their eyes. There is hypotonia of limbs, allowing the patients to assume unusual postures for hours. Loss of normal teeth is a characteristic feature of the disease, sometimes accompanied by necrosis of the mandible. The gums become swollen and inflamed and associated excessive salivation is seen. Fingernails appear discolored and may drop off, and occasionally gangrene of fingers and toes has been observed. The hair falls out and is often pulled out by the child. Large quantities of water are drunk, anorexia is frequent, and diarrhea is the rule. Prolapse of the rectum is a common complication. There are accompanying tachycardia and hypertension. Tendon reflexes usually disappear in the course of the illness, and the child refuses to walk, although there is no paralysis. Characteristic pathological features have not been found at necropsy, but the disorder seems to affect terminal sprouts of peripheral nerves, judging from the symptomatology. Laboratory investigations are not helpful, because mercury levels are not a reliable indication of the presence of this disease. The disease can be avoided if children are shielded from contact with compounds containing mercury. The link between pink disease and mercury is complicated by the absence of this disorder in recent epidemics of mercury poisoning in Japan and in other parts of the world. An exception is an epidemic in Argentina that resulted in the exposure of 7000–10,000 babies to phenylmercury fungicide, which was used to treat the cloth of diapers (Gotelli, Astolfi, Cox, Cernichiari, & Clarkson, 1985).

Peripheral neuropathy in patients with chronic hypoxia due to respiratory tract disease at sea level (Appenzeller et al., 1968) has hitherto not been found in subjects living under chronic hypoxic conditions at altitude. A recent epidemiologic survey of Andean altitude natives, however, has shown a high incidence of burning-hands-burning-feet syndrome most likely due to peripheral nerve involvement in these subjects. Maladaptation of Andean altitude natives to chronic hypoxia has manifested itself in many ways but peripheral nerve dysfunction in 39% compared to 12% at sea level (León-Velarde & Arregui, 1994) is a finding that, because of its symptomatology, suggests predominant autonomic neuropathic features that need to be further examined.

Impaired reflex vasoconstriction has been reported in chronically hypoxemic patients. Adaptive mechanisms in response to chronic hypoxemia apparently do not include the preservation of sympathetic function. There is, however, prompt improvement in vasoconstrictor responses when the hypoxemia is temporarily corrected. The abnormal sympathetic function is probably of central and/or peripheral origin and related to impairment of neural transmission rather than to a permanent autonomic neuropathic process (Heistad, Abboud, Mark, & Schmid, 1972). Some confirmatory evidence for these findings was also reported in a study on normal men who were inhaling carbon monoxide but arterial oxygen tension was kept normal. Impaired vasoconstrictor responses attributed to the hypoxia in spite of normal arterial oxygen tension were found (Heistad & Wheeler, 1972).

In tabes dorsalis, it has been shown that the Valsalva maneuver is not associated with a normal overshoot. In such patients, heating of the feet may cause normal vasodilatation of the hands and a gasp will produce vasoconstriction in the skin. From the results of these physiological studies, it was concluded that the efferents to the blood vessels remained normal and that the baroreceptor reflex pathway in tabes dorsalis was interrupted on the afferent side (Sharpey Schafer, 1956).

8.4.6.7 Peripheral nerve damage in central autonomic failure

Multiple system atrophy is classically considered a preganglionic autonomic disorder. Nevertheless, several reports have implicated the peripheral nerves as well. In one case, definite evidence of damage in sympathetic ganglia was found (Rajput & Rodzilsky, 1976). In one patient with an additional severe sensory neuropathy, the 9th and 10th cranial nerves were also involved, and a marked loss of fibers of the tractus solitarius was found (Oppenheimer, 1983). Kramer and colleagues performed forehead cooling, intradermal microdialysis of norepinephrine, assessment of orthostatic hypotension, and quantitative sensory testing in 20 patients with Parkinson's disease, 10 patients with multiple system atrophy, and 10 healthy subjects. The authors found evidence of peripheral sympathetic impairment with denervation hypersensitivity in Parkinson's disease but not in multiple system atrophy. Moreover, they reported evidence for small fiber neuropathy in both patient groups, including mechanical hyperalgesia (Krämer, Rebhorn, Geber, & Birklein, 2021). α-Synuclein deposits were described in epidermal and subepidermal nerve fibers in both Parkinson's disease and multiple system atrophy. The mechanism why neuropathy develops in neurodegenerative extrapyramidal diseases is unclear. Impairment of vitamin B12 and folate metabolism has been proposed as an explanation, or peripheral somatosensory nerve degeneration is an inherent part of Parkinson's disease and multiple system atrophy pathology. A study of 232 patients with MSA who underwent both postganglionic sympathetic sudomotor testing with quantitative sudomotor axon reflex testing (QSART) and thermoregulatory sweat test sought to determine the frequency of pre- and postganglionic sudomotor impairment in MSA (Coon et al., 2017). Postganglionic impairment was present in 59% of patients on initial testing. Utilizing QSART and thermoregulatory sweat test to determine the site of the lesion, central was evident in 47%, while 12% had only postganglionic impairment, and 41% had evidence of mixed (central and peripheral) involvement. As the disease progressed, the percentage of patients with evidence of postganglionic impairment increased, suggesting greater involvement of small fiber nerves or denervation atrophy from a primarily central process. In a prospective study of 43 patients with MSA and 57 patients with PD, there was greater postganglionic sympathetic sudomotor impairment in MSA patients (Provitera et al., 2022).

8.4.6.8 Autonomic dysfunction in Machado–Joseph disease

An inherited neurologic disorder mostly found among Portuguese and their descendants, but also reported from Japan, has been termed Machado–Joseph disease. There are three types: Type I is clinically recognizable by amyotrophic lateral sclerosis, parkinsonism, and dystonia. This type was originally reported in the Joseph family. Type 2 is clinically recognized by cerebellar signs, and Type 3 has, in addition to cerebellar signs, pronounced distal muscle atrophy. Some families from Japan have autosomal dominant inheritance, while others have bulging eyes and prominent autonomic deficits, including sleep apnea, urinary-sphincter disturbances, and low levels of homovanillic acid in the CSF (Kitamura, Kubuki, & Tsuruta, 1989). Studying 15 patients with genetically confirmed Machado–Joseph disease, Yeh et al. found that voiding problems and thermoregulatory disturbance were the most common symptoms. Ten (71%) of 14 patients had abnormal R–R interval variation with a significant reduction of the mean ratio. Prolonged latency or absence of sympathetic skin response to electrical stimulation was identified in 73% of patients and to magnetic stimulation, in 53%. R–R interval variation and sympathetic skin response showed a good correlation with the functional stage (Yeh et al., 2005).

8.4.6.9 Causalgia

Causalgia is characterized by hyperpathia, trophic changes, and autonomic phenomena (see Chapter 15). Some patients develop swelling of the legs, edema, and severe pain on movement after a considerable interval following back surgery. The edema is often responsive to diuretics, and there might not be trophic changes in the affected limb. On occasion, Raynaud phenomenon develops in the hands, but the exact mechanism of the excessive vasoconstriction in the skin of the hands in response to cold has not been determined. This could result, however, from an abnormal firing of the whole of the postganglionic sympathetic nervous system induced by the initial trauma to the back. Lumbar sympathectomy in such patients is also effective in removing the symptoms if it is carried out soon after the clinical onset of the disorder.

A case of causalgia complicating meningococcal meningitis has also been reported. This patient developed severe burning pain with paresthesia over the left foot and lower leg. The pain increased in severity and was associated with wasting and weakness in spite of the rapid improvement of his meningitis. No local cause for the extensive denervation and pain was found on full investigation, and the patient made a remarkable recovery after five guanethidine blocks, the technique is similar to that of a Biers block. Clinically, the patient was indistinguishable from one with trauma-associated causalgia. His symptoms included hyperpathia, burning spreading pain, and increased sweating and vasodilatation. Vasodilatation responded well to guanethidine blockade, which probably blocked the larger afferent input to the spinal cord (McLelland & Ellis, 1986).

On rare occasions, lumbar sympathectomy itself may lead to neuralgic pain. This usually occurs about two weeks after surgery. The pain is localized to the thighs and is deep and nagging in character and disappears spontaneously, as a rule, within 10 days or 2 weeks. A few patients, however, have severe pain which is not relieved by narcotics. A number of patients were markedly improved by the administration of diphenylhydantoin or carbamazepine (Raskin, Levinson, Hoffman, Pickett, & Fields, 1974). The most common surgical indication which led to sympathectomy in the patients who developed postsympathectomy neuralgia was vascular intermittent claudication and leg ulcers. The mechanism of postsympathectomy neuralgia has not been elucidated, but the effectiveness of drugs that alter the excitability of nerve membranes and depress polysynaptic responses suggests abnormal excitability in postsympathectomy neuralgia.

In some patients with reflex sympathetic dystrophy, pathologic examination of sympathetic ganglia shows giant sympathetic neurons—a degenerative change similar to the pathology in patients with diabetic autonomic neuropathy (Appenzeller & Richardson, 1966) (Fig. 8.29).

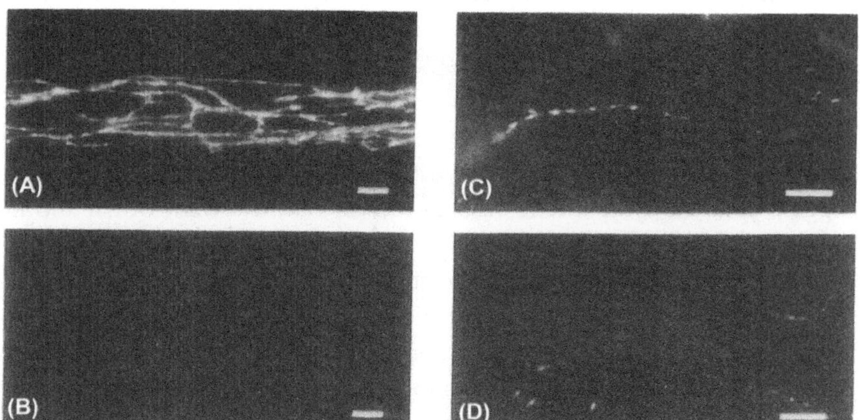

Figure 8.29 Giant sympathetic ganglion cell in surgically removed lumbar sympathetic ganglion for reflex sympathetic dystrophy in the foot from a 24-year-old man.

8.4.6.10 The postpolio syndrome

Postpolio patients frequently have lower temperatures in their affected limbs, regardless of ambient temperature. Often a severe burning pain is present. Hyperesthesia and color changes, ranging from cyanosis to reddish-violet or deep dark blue with mottling are also frequent. These color changes and discomfort persist for hours, and lengthy rewarming to relieve pain is often necessary. Many patients report that during these periods, the dexterity of the affected limb and its remaining muscle strength decreases. The cold-related impairment of limbs associated with postpolio syndrome increases with advancing age.

The hypothesis that these manifestations represent reflex sympathetic dystrophy is supported by the fact that histologic damage to the intermediolateral cell columns occurs after poliovirus infection. This damage has been blamed for a decrease in inhibition of sympathetic outflow to the affected limbs. One consequence of decreased sympathetic inhibition would be an increase in tonic sympathetic vasoconstrictor tone. Even mildly cold ambient temperatures might produce exaggerated vasospastic phenomena resulting in a marked decrease in limb temperature.

This interpretation of postpolio syndrome does not account for the pain and paresthesia of affected limbs, particularly in elderly postpolio subjects. Nor does it account for the fact that neither consistently lower skin temperatures nor decreases in muscle blood flow, are found when ambient temperatures are neutral. Inconsistent findings of temperature changes in affected limbs during the day and the permanent violet hue of the extremity suggest venous dilatation and not arterial constriction. These findings also argue against an increase in sympathetic tone due to a fall in inhibitory output from the intermediolateral cell columns.

Further study of postpolio subjects showed that equal decreases in blood flow with cooling of an affected limb occurred both in the affected limbs of patients with postpolio syndrome and in those of controls. This suggests that the sympathetic vasoconstrictor outflow was normal and not increased. Cortical activity, or thinking, which normally initiates and then increases sympathetic tone, in postpolio patients causes a decrease in sympathetic tone. Often in such patients, cortical activity also initiated a paradoxical vasomotor response, suggesting that stress-induced vasoconstriction is decreased in the postpolio syndrome. This may perhaps result from both pre- and postganglionic sympathetic neuron damage. Mean blood flow values in affected limbs at 30°C ambient temperature were lower. This was attributed to damaged sympathetic vasoconstrictor neurons, causing passive dilatation of the cutaneous venous (capacitance) vascular beds. This passive dilatation would account for venous congestion that can cause a decrease in arterial inflow to the skin. Together, venous dilatation and a decrease in arterial inflow will promote an increase in heat loss from the skin, which in turn results in the very cold limbs often found in such patients.

The cause of cold-induced pain of postpolio limbs remains to be explained, but it may be related to local hypersensitivity or a change in the peptide content of nerves or of perivascular nerve plexuses. The cold-associated decrease in conduction velocity has been implicated in the reduced dexterity of limbs exposed to cold (Bruno, Johnson, & Berman, 1985).

8.4.6.11 Navajo neuropathy

Navajo neuropathy was first reported by Appenzeller and Snyder (1976). It is a unique sensorimotor disease with corneal ulceration hitherto confined to one ethnic group. At least 36 cases have been identified and more patients are being recognized. The onset of the disease is between 5 days and 4.5 years of age (mean 1.2 years). The presenting symptom is usually infection (in 28%) corneal ulceration (in 20%), liver disease (in 24%), and failure to thrive (in 24%). All cases had weakness with sensory deficit (80%) and an abnormal sural nerve biopsy (12 out of 13) showed decreases in large and small myelinated fibers and relative preservation of unmyelinated fibers. Pneumonia, meningitis, and other sepsis are common. Eventually, 76% develop corneal ulceration, and liver disease causes morbidity and mortality in a large proportion of patients. The CSF protein may be elevated without cellular responses. Magnetic resonance imaging (5 out of 7) showed progressive cerebral and matutine cerebellar white matter lesions and cervical cord thinning (Figs. 8.30 and 8.31). The patients eventually become wheelchair-bound, and death has occurred between 6 months and 25 years after onset (mean 9.9 years), with the oldest living case being 35 years old (Fig. 8.32). Affected siblings have been identified, and the condition seems unrelated to Cree encephalitis or leukoencephalopathy, except that all these conditions seem restricted to Athapaskan North American Indians (Singleton et al., 1990).

Autonomic dysfunction, though present, is overshadowed by the severe and mutilating neuropathy. The innervation of blood vessels in the sural nerve suggests a depletion of noradrenergic perivascular nerve fibers and of norepinephrine in the vasa nervorum (Fig. 8.33).

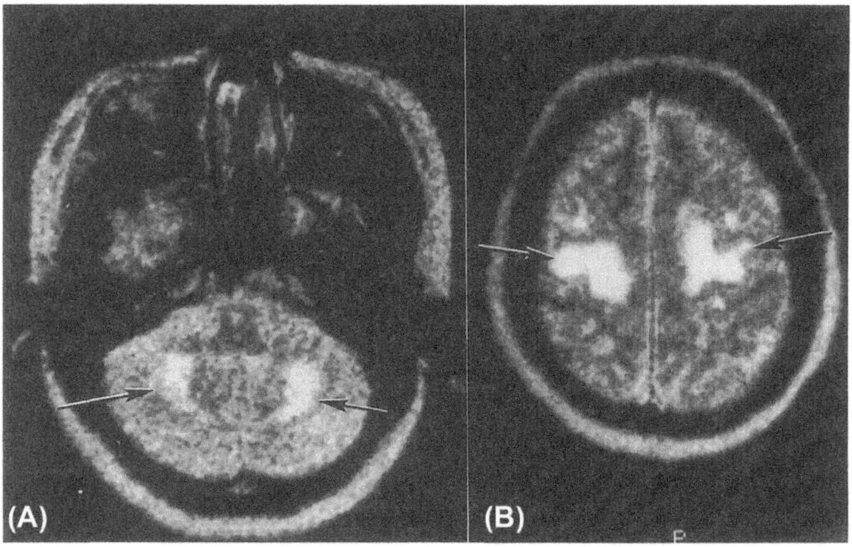

Figure 8.30 T_2 weighted coronal MRI (1.5 Tesla, TR-2000, TE-80). (A) The earliest observable lesion in CNS in this peripheral neuropathy. Abnormal increased signal in cerebellar white matter (arrows). (B) Same patient 1 year later. New biparietal white matter lesions (arrows). (C) Abnormal increased signal in cerebellar white matter (arrows). The matutine cerebellar butterfly in another patient (1.5 Tesla, TR-2500, TE-80).

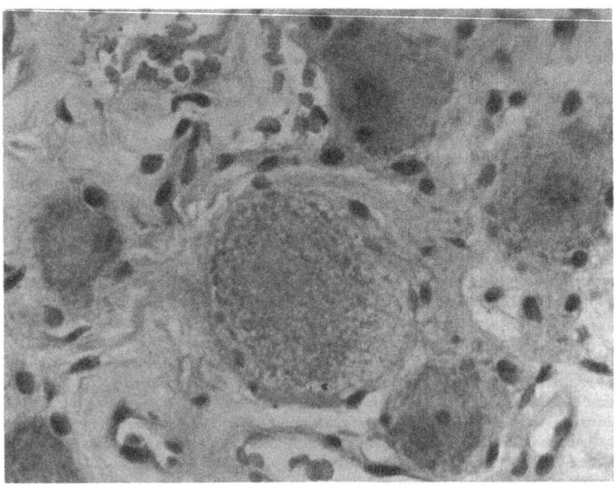

Figure 8.31 Progressive cerebral white matter lesions in Navajo neuropathy.

One infant with the disease became symptomatic with respiratory difficulties at two months of age. By 7 months, he was ventilator-dependent, paretic, and hypotonic, with barely obtainable tendon jerks. No peroneal

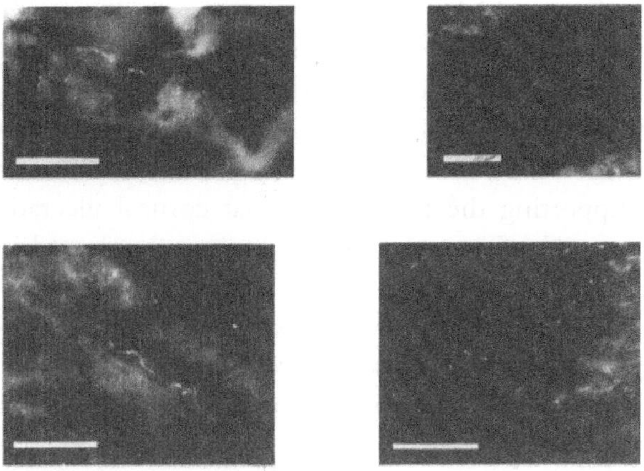

Figure 8.32 Oldest surviving patient with Navajo neuropathy, age 6 (upper left), age 15 (lower left), age 25 (upper right). Mutilation and severe atrophy of hands, age 25 (lower right).

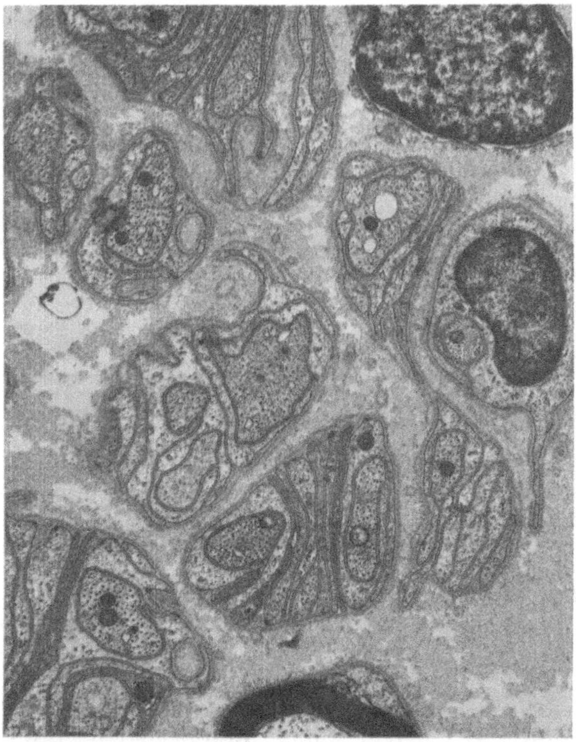

Figure 8.33 Catecholamine fluorescent varicose perivascular plexus of vasa nervorum in sural nerve biopsy from a patient with Navajo neuropathy. Note the decreased density of perivascular nerve fibers.

nerve conduction could be obtained. Sural nerve biopsy showed axonal degeneration with decrease in myelinated and preservation of unmyelinated fibers, similar to the findings in other cases of Navajo neuropathy. Corneal abnormalities were not as yet present. This case extends the clinical spectrum of this disease, implying intrauterine insult to the nervous system and supporting the proposition that corneal ulceration present in older children results from corneal anesthesia and repeated injury and not from corneal dystrophy.

The spectrum of the disease seems to be changing. More cases are being recognized in infancy presenting with liver failure after what appears to be infantile hepatitis. These patients usually have MRI abnormalities of the brain before they die at a very early age. The childhood variety presents with developmental delay and neuropathic signs that vary in severity. These patients usually go on to fulminant liver failure in preschool years. There are also mixed cases presenting with early hepatitis which usually resolves followed by a gradual neurologic function impairment. The fourth group of patients has clinically a pure neurologic disease (as originally described by Appenzeller and Snyder, (1976) with gradual progression and abnormal liver function tests.

The pathogenesis of this disease remains obscure but, because of the predominant occurrence on the western Navajo reservation, a founder effect is suspected with a gene frequency of 1/20. Genetic studies based on 8 DNA samples focused on 8 areas that have high homozygosity (Kurth, personal communication, 1996).

Livers from patients with Navajo neuropathy have become available. In four samples, there was little if any mRNA for Multidrug Resistant 3 protein or P-Glycoprotein-3 (Arias, personal communication, 1996). While the absence of these substances in mice produces liver disease, peripheral neuropathy is not found in the animal model. The relation of these intriguing findings in Navajo neuropathy patients to the pathogenesis of the disease requires further study.

8.4.6.12 Navajo neuropathic arthropathy

Eight patients with this condition in the Navajo pediatric population have been reported (Johnson & Johnsen, 1987). This disease is presently confined to Navajo children. It is familial and is characterized by hyperkeratosis of the palms and Charcot joints, unsuspected fractures with normal neurologic examination, and heat intolerance. Electromyography and nerve conduction velocities are normal. Sural nerve biopsies showed mainly a reduction of unmyelinated nerve fibers with fibrosis and a mild loss of small myelinated fibers. This finding contrasts with the biopsy findings in Navajo neuropathy, where the unmyelinated fiber population is relatively preserved, while the myelinated nerve fibers show a marked loss. In contrast to Navajo neuropathy, the development of children with Navajo neuropathic arthropathy is usually normal until the second decade of life, when painless fractures typically are present, primarily in weight-bearing bones. A progressive severe arthropathy, associated with recurrent joint swelling, develops mostly in the knees and ankles.

Fusion and fixation of the affected joints are required. Heat intolerance and lack of sweating without noted hyperpyrexia are present in all cases. In contrast to patients with Navajo neuropathy, patients with Navajo neuropathic arthropathy have neither mutilation nor clinical somatic neurologic deficits. In particular, they have no corneal ulceration (typically found in Navajo neuropathy); however, their pupillary reflexes are difficult to elicit.

Extensive autonomic testing of one patient showed sympathetic and parasympathetic denervation of postganglionic type, and, in addition, an abnormal Lewis hunting reaction, suggestive of a primary abnormality of blood vessels. The pupil cycle time was 1720 ms (upper limit of normal 1000 ms). In this disease, from which no recovery has yet been reported, extensive autonomic deficits occur in a setting of relative preservation of somatic neurologic function, including normal electrophysiologic tests. The central nervous system in these patients has not as yet been imaged, and the vasa nervorum and their innervation have not been studied.

8.4.6.13 Miscellaneous disorders associated with autonomic dysfunction

8.4.6.13.1 Hyperbradykininemia

Syncope and orthostatic lightheadedness have been described in association with an orthostatic fall in pulse pressure and rise in heart rate in five patients with erythema of the face and neck and purple discoloration of the legs after standing. High plasma bradykinin levels were found. The syndrome is sometimes familial. The high bradykinin levels were the result of impaired destruction of this substance in the circulation. Marked improvement in symptoms and blood pressures changes and a fall in pulse rate was achieved by treating patients with propranolol, fludrocortisone, or cyproheptadine (Streeten et al., 1972).

8.4.6.13.2 Dysautonomia in mitral valve prolapse

Many patients with mitral valve prolapse suffer from symptoms suggesting the possibility of abnormal autonomic activity not least of which is the often-demonstrated normal coronary arteriograms with ST-segment depression during submaximal treadmill exercise and normal thallium-201 uptake by the myocardium, associated with angina-like pain. There is also easy fatiguability, a decrease in effort tolerance, and vasomotor instability. A large number of patients with this condition are said to suffer from migraine. Inappropriate heart rate responses during general anesthesia, excessive bradycardia, and increased tendency to ventricular fibrillation during coronary angiography have all been considered manifestations of possible autonomic abnormalities. A study designed to assess the incidence of dysautonomia in mitral valve prolapse included heart rate and blood pressure responses to Valsalva maneuver and postural tilting in 44 untreated patients who had demonstrated mitral valve prolapse. The patients differed from controls in their widely oscillating heart rates during the upright posture, in exaggerated and prolonged bradycardia during the recovery phase after the Valsalva maneuver, and in greater respiratory variation in pulse intervals, which became particularly prominent during postural change (Coghlan; Phares et al., 1979). It is of interest that this study suggests that autonomic responses in patients with mitral valve prolapse are not blunted or absent, but the pattern of reflex regulation of the circulation is different, almost suggesting a hypersensitivity to autonomic stimulation. Syncope may be present in some patients with mitral valve prolapse. The molecular basis for these disorders is not understood but an abnormal guanine nucleotide regulatory protein may be important (Davies, Su, & Balasubramanyam, 1991).

8.4.6.13.3 Cystic fibrosis

Numerous studies have shown abnormalities in autonomic activity which contribute to the clinical picture in cystic fibrosis. Alpha-adrenergic regulation of the pupil is abnormal, beta-adrenergic responses are deficient in leukocytes, and the reactions to cholinergic agents are distorted in the submaxillary gland and bronchial smooth muscle. Because almost all types of autonomic function seem to be involved, it is possible that cystic fibrosis is, in part at least, the result of abnormal responsiveness of autonomic end organs. Several patients with cystic fibrosis had reduced mydriasis in response to darkness when compared to normal control pupils, reduced cardiovascular sensitivity to ß-adrenergic stimulation, increased sensitivity to inhaled methacholine, hypertrophy, and hyperplasia of the mucus-secreting apparatus in airways, and other abnormalities, which were similar to those found in patients with allergic asthma. These findings in two different conditions have led to the suggestion that α-adrenergic and perhaps cholinergic sensitivity contributes to the severity of the illness and its symptomatology but are not of primary pathogenetic importance in cystic fibrosis. Against this interpretation are the abnormal α-adrenergic and cholinergic pupillary sensitivity in asymptomatic obligate heterozygotes, which suggests that they are inherited, genetically determined parts of the disease, and not due to chronic illness, malnutrition, or the administration of a number of autonomically active drugs. Further studies are necessary to define the molecular basis of these deficits (Davis, Shelhamer, & Kaliner, 1980).

8.4.6.13.4 Weight loss and sympathetic nervous system activity

It has been advocated that in semistarvation or other therapies for obesity, low-calorie protein diets are advantageous because of their protein-sparing property over conventional isocaloric carbohydrate-mixed diets. The comparative effects of these two types of diet on body fluids and on the metabolism of electrolytes and sympathetic nervous system activity have been studied. Surprisingly, mixed diets are not less advantageous in comparison with hypocaloric protein diets as far as nitrogen balance is concerned. But low-calorie protein diets result in greater sodium depletion and a greater compromise in sympathetic nervous system activity. Such diets are also associated with orthostatic hypotension (DeHaven, Sherwin, et al., 1980).

The problem of obesity continues to bedevil a major portion of Western society, and because of its unquestionable deleterious effect on a number of diseases and longevity, attempts have been made at weight reduction. The importance of associated symptoms during hypocaloric periods in obese subjects cannot be overestimated since adherence to low-calorie diets in obese patients is often a problem. It is suggested that the increased natriuresis, which occurs in the ketogenic high-protein and low-carbohydrate diet results from the increased secretion of ketone bodies, and accounts for the greater weight loss with this type of diet. However, larger orthostatic decreases in blood pressure occur with protein than with mixed diets and all patients have symptomatic postural hypotension with this type of diet. Plasma levels of norepinephrine are reduced in the basal state and after two minutes of standing in those on protein diets. One would have expected, however, plasma norepinephrine levels to be exaggerated because of sodium depletion. This provides evidence of a decline in sympathetic activity on standing and could contribute to the orthostatic hypotension. In underfed human beings, there is a reduction in norepinephrine turnover. In rats overfed with sucrose, an increase in norepinephrine turnover occurs. In humans, therefore, the decrease in norepinephrine turnover is primarily a result of the dietary exclusion of carbohydrates, rather than a reduction in total calories taken.

Of related interest are the findings of hypothalamic dysfunction associated with simple weight loss and secondary amenorrhea, which are similar to those occurring in anorexia nervosa but are of a lesser degree (Vigersky; Andersen et al., 1977). Such patients have abnormal thermoregulation, which correlates closely with the percentage below the ideal body weight of each individual. Partial diabetes insipidus is present in about a quarter of patients with simple weight loss. Delayed peak plasma luteinizing hormone levels after the administration of luteinizing hormone-releasing factor also correlates with the percentage below ideal body weight. A delay in peak-plasma thyrotropin levels after thyrotropin-releasing factor administration occurs. No abnormalities in the levels of prolactin, pituitary, thyroid, or adrenal function were found. The evidence of hypothalamic disturbances with simple weight loss and amenorrhea might be compared with the amenorrhea of endurance training, but thermoregulation in amenorrheic athletes has not been investigated.

A study utilized a variety of methodologies to assess autonomic nervous system function in women with a current diagnosis of anorexia, a past diagnosis of anorexia who were weight-restored and healthy controls. The authors found that muscle sympathetic nerve activity was significantly decreased in both anorexia and weight-restored groups, as compared to controls. Participants with anorexia had a significantly lower standard deviation in heart rate, lower blood pressure variability, and decreased sudomotor function as compared to controls. Weight-restored participants demonstrated decreased baroreflex sensitivity in response to head-up tilt as compared to controls. Therefore, women with a current or previous diagnosis of anorexia have decreased sympathetic activity, which may reflect a physiological response to decreased energy intake (Jenkins et al., 2021).

8.4.6.13.5 Lesch–Nyhan syndrome

Lesch–Nyhan syndrome is a disorder of purine metabolism, which is characterized by hyperuricemia and excessive production of uric acid. It is an X-linked disorder and clinically manifests profound neurologic dysfunction including spasticity, choreoathetosis, self-mutilation, and mental retardation. There is an almost total absence of the enzyme hypoxanthine-guanine phosphoribosyltransferase (HGPRT). Some patients have lesser deficits in HGPRT; they have hyperuricemia, excessive production of uric acid and gout, but do not show self-mutilation although other neurologic dysfunctions are occasionally present.

The pathogenesis of behavioral and neurologic disturbances in patients with HGPRT deficiency has not been established. Nevertheless, because of self-mutilating behavior in laboratory animals after the administration of caffeine and large doses of amphetamine, the suggestion has been made that this part of the Lesch–Nyhan syndrome might be due to alterations in function of pathways in the central nervous system susceptible to these agents. A study of patients with Lesch–Nyhan syndrome and self-mutilating behavior has shown a unique pattern of adrenergic dysfunction. They have an elevated plasma dopamine-ß-hydroxylase, which catalyzes the formation of norepinephrine from dopamine. This enzyme is simultaneously released into the synaptic cleft with the neurotransmitter norepinephrine. The circulating dopamine-ß-hydroxylase levels serve as a quantitative index of adrenergic function. Though elevated levels of this enzyme were found in patients with the Lesch–Nyhan syndrome, none of them had evidence of autonomic manifestation of excessive adrenergic activity such as hypertension, tachycardia, or mydriasis. Clinical tests of adrenergic responsiveness such as the cold pressor test however showed the normally expected rise in blood pressure in controls and patients with HGPRT deficiency without self-mutilation, but absent rise in blood pressure in those with the Lesch–Nyhan syndrome who exhibited self-mutilation. This abnormal response is attributed to an absence of vasoconstriction though the exact mechanism of this has not been established. Whether the abnormally high plasma dopamine-ß-hydroxylase activity in patients with HGPRT deficiency and self-mutilation is responsible for their bizarre behavioral manifestation is not clear, but it is tempting to speculate that monoamine pathways might be involved centrally and appear to be impaired peripherally in these patients and that this somehow accounts for their peculiar destructive activity (Rockson, Stone, Weydenvander, & Kelley, 1974).

8.4.6.13.6 Canine dysautonomia

A cluster of this condition has been reported near Columbia, Missouri, and sporadically elsewhere. Primary dysautonomias are relatively rare in animals. Equine dysautonomia (Grass sickness) is the most common in animals and has been described in Great Britain since 1928; it was once widespread in Britain but is now largely confined to Scotland. The condition has been described in nondomestic equines and also in wild European hares. Recently cases of leporine dysautonomia in England have occurred on the same pastures as equine cases. Dysautonomia was first described in cats in the United Kingdom in 1982, and since throughout Europe, New Zealand, and the United States. The first case of canine dysautonomia was reported in England in 1983, and then in Scotland, Belgium, Norway, France, and the first American case was in 1991. Like in human acute pandysautonomia, the cause of the animal disease remains unknown. Experimentally, it is possible to induce characteristic lesions in the autonomic ganglia by injecting horses intraperitoneally with serum and plasma from affected horses. Serum from clinical equine cases injected into the parotid salivary glands of normal ponies caused chromatolysis in the ipsilateral cranial cervical ganglion.

Dysautonomia was diagnosed in 11 young, predominantly medium to large breed dogs from 1988 to 1995. Clinical signs caused by autonomic dysfunction of the urinary, alimentary, and ocular systems were most common. Dysuria, mydriasis, absence of pupillary light reflexes, decreased tear production, dry mucous membranes, weight loss, and decreased anal tone were present in over 75% of affected dogs. Ocular pharmacological testing demonstrated iris sphincter dysfunction in all dogs. Cholinergic and adrenergic receptor dysfunction was found in four dogs. All dogs died or were euthanized. Neuronal depletion, with gliosis and minimal inflammation, were noted histologically in the autonomic ganglia. The pelvic, ciliary, celiac, cranial cervical, and cranial and caudal mesenteric ganglia were affected in all animals (Randall, Longshore, & Davis, 1996). The clinical description of the dogs is remarkably similar to human acute pandysautonomia cases. In horses, recovery from the mildest form of the disease has been noted and an immunologic or infectious agent or both have been suspected in the pathogenesis of this veterinary disease.

References

Aguayo, A. J., Nair, C. P. V., & Bray, G. M. (1971). Peripheral nerve abnormalities in the Riley Day syndrome. *Arch. Neurol.*, *24*(2), 106–116.

Amenta, F., Mione, M. C., & Napoleone, P. (1983). The autonomic innervation of the vasa nervorum. *J. Neural Trans. (Vienna, Austria: 1996)*, *58*(3–4), 291–297.

Anand, P., Antia, N. H., & Blanck, M. A. e a (1983). Neuropeptides in leprosy: Evidence of a role for VIP in the reorganisation or regeneration of the nervous system. *Clin. Sci.*, *65*, 64P.

Anand, P., Llewelyn, J. G., Thomas, P. K., Gillon, K. R., Lisk, R., & Bloom, S. R. (1988). Water content, vasoactive intestinal polypeptide and substance P in intact and crushed sciatic nerves of normal and streptozotocin-diabetic rats. *J. Neurol. Sci.*, *83* (2–3), 167–177.

Andrade, C. (1952). A peculiar form of peripheral neuropathy: Familial atypical generalized amyloidosis with special involvement of peripheral nerves. *Brain, 75*, 408–427.

Appenzeller, O. (1963). *The neurogenic control of vasomotor function in the hand: A study of patients with lesions at various levels of the nervous system* (Thesis). London.

Appenzeller, O. (1983). Immune autonomic neuropathies: Experimental studies in autonomic function. In R. Bannister (Ed.), *Autonomic failure: A textbook of clinical disorders of the autonomic nervous system* (pp. 640–649). Oxford: Oxford University Press.

Appenzeller, O., Dhital, K. K., Cowen, T., & Burnstock, G. (1984). The nerves to blood vessels supplying blood to nerves: The innervation of vasa nervorum. *Brain Res., 304*(2), 383–386.

Appenzeller, O., & Kornfeld, M. (1973). Acute pandysautonomia clinical and morphologic study. *Arch. Neurol., 29*, 334–339.

Appenzeller, O., Lincoln, J., Zumwalt, R., et al. (1989). Vasa nervorum and nervi nervorum: Adrenergic and peptidergic innervation of normal human and neuropathic sural nerves. *Ann. Neurol., 26*(1), 185.

Appenzeller, O., & Lueker, R. (1972). Autonomic dysfunction in patients on chronic hemodialysis. *Neurology, 22*, 395.

Appenzeller, O., & Marshall, J. (1963). Vasomotor disturbances in Landry Guillain Barre syndrome. *Arch. Neurol., 9*, 368–372.

Appenzeller, O., Milner, P., Lincoln, J., et al. (1989). Nervi nervorum and perivascular nerve plexus of vasa nervorum: Effect of streptozotocin diabetes on NPY and VIP like nerve profiles. *Annals Neurol., 26*(2), 307.

Appenzeller, O., Parks, R., & MacGee, J. (1968). Peripheral neuropathy associated with chronic respiratory tract disease. *Am. J. Med., 44*, 873–880.

Appenzeller, O., & Richardson, E. P., Jr (1966). The sympathetic chain in patients with diabetic and alcoholic polyneuropathy. *Neurology (Minneapolis), 16*, 1205–1209.

Appenzeller, O., & True, C. W. (1967). Amyloidosis confined to the central and peripheral nervous systems. *J. Neuropath. Exp. Neurol., 26*, 174.

Appenzeller, O. M. K., & Snyder, R. (1976). Acromutilating paralyzing neuropathy with corneal ulceration in Navajo children. *Arch. Neurol., 33*(11), 733–738.

Axelrod, F. B. (2004). Familial dysautonomia. *Muscle Nerve, 29*(3), 352–363.

Banerjee, D., Roeker, L. E., Grogan, M., Swiecicki, P., Poterucha, J., Heimbach, J., ... Dispenzieri, A. (2017). Outcomes of patients with familial transthyretin amyloidosis after liver transplantation. *Prog. Trans. (Aliso Viejo, Calif.), 27*(3), 246–250.

Barraclough, M. A., & Sharpey-Schafer, E. P. (1963). Hypotention from absent circulatory reflexes. Effects of alcohol, barbiturates, psychotherapeutic drugs and other mechanisms. *Lancet, 1*, 1121–1126.

Bell, M. A., & Weddell, A. G. (1984). A descriptive study of the blood vessels of the sciatic nerve in the rat, man and other mammals. *Brain, 107*(Pt 3), 871–898.

Bichat, M. F. X. (1830). General Anatomy (Vol. 1, p. 168). London.

Birchfield, R. I. (1964). Postural hypotension in Wernicke's disease. *Am. J. Med., 36*, 404–414.

Blanco, R., De Tejada, I. S., Goldstein, I., et al. (1990). Dysfunctional penile cholinergic nerves in diabetic impotent men. *J. Urol., 144*, 278–280.

Blunt, J. M. (1957). Functional and clinical implications of the vascular anatomy of peripheral nerves. *Postgraduate Med. J., 33*, 68–72.

Britland, S. T., Young, R. J., Sharma, A. K., et al. (1990). Vagus nerve morphology in diabetic gastropathy. *Diabetic Med.*, 7, 780–787.

Brooks, A. P. (1980). Abnormal vascular reflexes in Charcot Marie Tooth disease. *J. Neurol. Neurosurg. Psychiat., 43*, 348–350.

Bruno, R. L., Johnson, J. C., & Berman, W. S. (1985). Motor and sensory function with changing ambient temperature in post polio subjects: Autonomic and electrophysiological correlates. In L. S. Halstead, & D. O. Wiechser (Eds.), *Late effects of poliomyelitis* (pp. 95–108). Miami, FL: Symposia Foundation.

Brunt, P. W. (1967). Unusual cause of Charcot joints in early adolescence (Riley Day syndrome). *Br. Med. J.*, *4*, 277–278.

Bulawa, C. E., Connelly, S., Devit, M., Wang, L., Weigel, C., Fleming, J. A., ... Labaudinière, R. (2012). Tafamidis, a potent and selective transthyretin kinetic stabilizer that inhibits the amyloid cascade. *Proc. Natl. Acad. Sci. U. S. A.*, *109*(24), 9629–9634.

Burnstock, G. (1977). Autonomic neuroeffector junctions—reflex vasodilatation of the skin. *J. Invest. Derm.*, *69*, 47–57.

Cable, W. J. L., Kolodny, E. H., et al. (1980). Fabry disease: A clinical demonstration of impaired autonomic function. *Neurology*, *30*, 1352.

Cardone, C., Paiusco, P. G., Marchetti, G., et al. (1990). Cough test to assess cardiovascular autonomic reflexes in diabetes. *Diabetes Care*, *13*, 719–724.

Carra, M., Brambilla, P., Meschi, F., et al. (1990). Pancreatic polypeptide secretion after insulin infusion and protein meal in juvenile type 1 diabetic subjects. *Acta Diabetologica Latina*, *27*, 165–171.

Charcot, J. M., & Marie, P. (1886). Sur une form particuliere datrophie musculaire progressive souvent familiale debutant par les pieds et les jambes et atteignant plus tard les mains. *Rev. Med.*, *6*, 97–138.

Chopra, J. S., & Hurwitz, L. J. (1967). Internodal length of sural nerve fibers in chronic occlusive vascular disease. *J. Neurol. Neurosurg. Psychiat.*, *30*, 207–214.

Christlieb, A. R., Choodappasetty, M., & Braaten, J. T. (1974). Decreased response of plasma renin activity to orthostatis in diabetic patients with orthostatic hypotension. *Diabetes*, *23*, 835–840.

Coelho, T., Maia, L. F., da Silva, A. M., Cruz, M. W., Planté-Bordeneuve, V., Suhr, O. B., ... Grogan, D. R. (2013). Long-term effects of tafamidis for the treatment of transthyretin familial amyloid polyneuropathy. *J. Neurol.*, *260*(11), 2802–2814.

Coelho, T., Maia, L. F., Martins da Silva, A., Waddington Cruz, M., Planté-Bordeneuve, V., Lozeron, P., ... Grogan, D. R. (2012). Tafamidis for transthyretin familial amyloid polyneuropathy: A randomized, controlled trial. *Neurology*, *79*(8), 785–792.

Coghlan, H. C., Phares, P., et al. (1979). Dysautonomia in mitral valve prolapse. *Am. J. Med.*, *67*, 236–244.

Cohen, P., & Solomon, N. H. (1955). Familial dysautonomia, case report with autopsy. *J. Pediatr.*, *46*, 663–670.

Coon, E. A., Fealey, R. D., Sletten, D. M., Mandrekar, J. N., Benarroch, E. E., Sandroni, P., ... Singer, W. (2017). Anhidrosis in multiple system atrophy involves pre- and postganglionic sudomotor dysfunction. *Mov. Disord.*, *32*(3), 397–404.

Cortese, A., Vita, G., Luigetti, M., Russo, M., Bisogni, G., Sabatelli, M., ... Pareyson, D. (2016). Monitoring effectiveness and safety of Tafamidis in transthyretin amyloidosis in Italy: A longitudinal multicenter study in a non-endemic area. *J. Neurol.*, *263*(5), 916–924.

Cowen, T., Haven, A. J., Wen Quin, C., et al. (1982). Development and ageing of perivascular adrenergic nerves in the rabbit. A quantitative fluorescence histochemical study using image analysis. *J. Auton. Nerv. Syst.*, *5*, 317–336.

Dalos, N. P., Borel, C., & Hanley, D. F. (1988). Cardiovascular autonomic dysfunction in Guillain Barre Syndrome: Therapeutic implications of Swan Ganz monitoring. *Arch. Neurol.*, *45*, 115–117.

Davies, A. L., Su, C. J., Balasubramanyam, A., et al. (1991). Abnormal guanine nucleotide regulatory protein in MVP dysautonomia: Evidence from reconstitution of GS. *J. Clin Endocrinol. Metabol.*, *72*, 867–875.

Davis, P. B., Shelhamer, J. R., & Kaliner, M. (1980). Abnormal adrenergic and cholinergic sensitivity in cystic fibrosis. *New England J. Med.*, *302*(26), 1453–1456.

Day, R., & Klingman, W. O. (1939). Effect of sleep on skin temperature reactions in case of acrocyanosis. *J. Clin. Investig.*, *18*, 271–276.

De Haven, J., Sherwin, R., et al. (1980). Nitrogen and sodium balance and sympathetic nervous system activity in obese subjects treated with low calorie protein or mixed diet. *N. Engl. J. Med.*, *302*, 477–482.

Denny Brown, D., & Brenner, C. (1944). Paralysis of nerve induced by direct pressure and by tournique. *Arch. Neurol. Psychiat. (Chic.)*, *51*, 1–26.

Dhital, K., Lincoln, J., Appenzeller, O., & Burnstock, G. (1986). Adrenergic innervation of vasa and nervi nervorum of optic, sciatic, vagus and sympathetic nerve trunks in normal and streptozotocin-diabetic rats. *Brain Res.*, *367*(1–2), 39–44.

Dhital, K. K., & Appenzeller, O. (1988). Innervation of vasa nervorum, Chapter 11. In G. Burnstock, & S. G. Griffith (Eds.), *Noradrenergic innervation of blood vessels* (pp. 191–211). Boca Raton, FL: CRC Press.

Di Carlo, V. (1984). Perivascular serotoninergic neurons: Somatodendritic contacts and axonic innervation of blood vessels. *Neurosci. Lett.*, *51*(3), 295–302.

Di SantAgnese, P. A., Darling, R. C., Perera, G. A., & Shea, E. (1953). Abnormal electrolyte composition of sweat in cyctic fibrosis of the pancreas; clinical significance and relationship to disease. *Pediatrics*, *12*, 549–563.

Dunnington, J. H. (1954). Congenital alacrimia in familial autonomic dysfunction. A.M.A. *Arch. Ophthalmol.*, *52*, 925–931.

Dyck, P. J., & Lambert, E. H. (1969). Dissociated sensation in amyloidosis compound action potential, quantitative histologic and teased fiber, and electron microscopic studies of sural nerve biopsies. *Arch. Neurol.*, *20*, 490–507.

Eames, R. A., & Lange, L. S. (1967). Clinical and pathological study of ischemic neuropathy. *J. Neurol. Neurosurg. Psychiat.*, *30*, 215–226.

Edelman, N. H., Richards, E. C., & Fishman, A. P. (1967). Abnormal ventilatory and circulatory responses to hypoxia in familial dysautonomia. *J. Clin. Investig.*, *46*, 1051.

Edvinsson, L. (1985). Functional role of perivascular peptides in the control of cerebral circulation. *Trends Neurosci.*, *8*, 126–131.

Edvinsson, L., Degueurce, A., Duverge, D., et al. (1983). Central serotoninergic nerves project to the pial vessels of the brain. *Nature (London)*, *306*, 55–57.

Edvinsson, L., Edman, R., Ottoson, A., & Uddman, R. (1985). Distribution, concentration and effects of neuropeptide Y (NPY), Substance P (SP) and vasoactive intestinal polypeptide (VIP) in human cerebral blood vessels. *J. Cereb. Blood Flow Metab.*, *5* (Suppl. 1), S545.

Engel, G. L., & Aring, C. D. (1945). Hypothalamic attacks with thalamic lesion. I. Physiologic and psychologic considerations. *Arch. Neurol. Psychiat. (Chicago)*, *54*, 37–43.

Esler, M., Dudley, F., Jennings, G., et al. (1992). Increased sympathetic nervous activity and the effects of its inhibition with clonidine in alcoholic cirrhosis. *Ann. Int. Med.*, *116*, 446–455.

Ewing, D. J. (1991). Clinical aspects of primary and secondary autonomic failure. *Curr. Opinion Neurol. Neurosurg.*, *4*, 539–544.

Ewing, D. J., Campbell, I. W., Burt, A. A., & Clarke, B. F. (1973). Vascular reflexes in diabetic autonomic neuropathy. *Lancet*, *2*, 1354–1356.

Ewing, D. J., Campbell, I. W., et al. (1980). Assessment of cardiovascular effects in diabetic autonomic neuropathy and prognostic implications. *Ann. Int. Med.*, *92*(Part 2), 308–311.

Ewing, D. J., & Neilson, J. M. M. (1990). QT interval length and diabetic neuropathy. *Diabeti. Med.*, 7, 23–26.

Fagerberg, S. E. (1959). Diabetic neuropathy: A clinical and histological study on the significance of vascular affections. *Acta Med. Scand.*, *164*(Suppl. 345).

Fanciulli, A., Kerer, K., Leys, F., Seppi, K., Kaufmann, H., Norcliffe-Kaufmann, L., & Wenning, G. K. (2020). Validation of the neurogenic orthostatic hypotension ratio with active standing. *Ann. Neurol.*, *88*(3), 643–645.

Feer, E. (1923). Eine eigenartige neurose des vegetativen systems beim Kleinkinde. *Ergebn. Inn. Med. Kinderheilk*, *24*, 100–122.

Figueroa, J. J., Dyck, P. J., Laughlin, R. S., Mercado, J. A., Massie, R., Sandroni, P., ... Low, P. A. (2012). Autonomic dysfunction in chronic inflammatory demyelinating polyradiculoneuropathy. *Neurology*, *78*(10), 702–708.

Filling Katz, M. R., Merrick, H. F., Fink, J. K., et al. (1989). Carbamazepine in Fabry's disease: Effective analgesia with dose dependent exacerbation of autonomic dysfunction. *Neurology*, *39*, 598–600.

Fiorucci, S., Bosso, R., Scionti, L., et al. (1990). Neurohumoral control of gall bladder motility in healthy subjects and diabetic patients with or without autonomic neuropathy. *Digest. Dis. Sci.*, *35*, 1089–1097.

Fox, J. L., & Kenmore, Pl (1967). The effect of ischemia on nerve conduction. *Exp. Neurol.*, *17*, 403–419.

Freedman, L. S., Ebstein, R. P., Goldstein, M., Axelrod, F. B., & Dancis, J. (1975). Serum dopamine b hydroxylase in familial dysautonomia. *J. Lab. Clin. Med.*, *85*, 1008–1012.

Freeman, R., Roberts, M. S., Friedman, L. S., et al. (1990). Autonomic function in human immunodeficiency virus infection. *Neurology*, *40*, 575–580.

Geltzer, A. I., Gluck, L., Talner, N. S., & Polesky, H. F. (1964). Familial dysautonomia studies in a newborn infant. *N. Engl. J. Med.*, *271*, 436–440.

Gibbons, C. H., & Freeman, R. (2010). Treatment-induced diabetic neuropathy: A reversible painful autonomic neuropathy. *Ann. Neurol.*, *67*(4), 534–541.

Gibson, S. J., McGregor, G., Bloom, S. R., et al. (1982). Local application of capsaicin to one sciatic nerve of the adult rat induces a marked depletion in the peptide content of the lumbar dorsal horn. *Neuroscience*, 7, 3153–3162.

Gilliatt, R. W., & Harrison, M. J. G. (1984). Nerve compression and entrapment in peripheral nerve disorders. In A. K. Ashbury, & R. W. Gilliatt (Eds.), *Butterworth-Heinemann international medical reviews* (pp. 243–286). London: Butterworth.

Gilmore, J. E., Allen, J. A., & Hayes, J. R. (1990). A comparison of peripheral vasoconstrictor responses and cardiovascular autonomic function tests in diabetic patients. *Diabetologia*, *33*, 350–356.

Goadby, H. K., & Downman, C. B. B. (1973). Peripheral vascular and sweat gland reflexes in diabetic neuropathy. *Clin. Sci. Mol. Med.*, *45*, 281–289.

Goldenberg, M., Pines, K. L., Baldwin, E. D. F., Greene, D. G., & Roh, C. E. (1948). Hemodynamic response of man to norepinephrine and epinephrine and its relation to problem of hypertension. *Am. J. Med.*, *5*, 792–806.

Gotelli, C. A., Astolfi, E., Cox, C., Cernichiari, E., & Clarkson, T. W. (1985). Early biochemical effects of an organic mercury fungicide on infants: Dose makes the poison. *Science (New York, N.Y.)*, *227*, 638–640.

Griffith, S. G., Lincoln, J., & Burnstock, G. (1982). Serotonin as a neurotransmitter in cerebral arteries. *Brain Res.*, *274*, 388–392.

Grover Johnson, N. M., Baumann, F. G., Imparato, A. M., et al. (1981). Abnormal innervation of lower limb epineurial arterioles in human diabetes. *Diabetologia*, *10*, 31–38.

Guy, R. J. C., Richards, F., Edmonds, M. E., et al. (1984). Diabetic autonomic neuropathy and iritis: An association suggesting an immunological cause. *Br. Med. J.*, *289*, 343–345.

Haimovich, H. (1950). Peripheral arterial embolism. *Angiology*, *1*, 20–45.

Haller, V. A. (1756). Icones anatomicae. A. Vanderhoeck, Gottingen.

Hanko, J., Sundler, F., Hardebo, J. E., & Owman, C. (1985). Dilatation of pial and peripheral arteries by calcitonin gene related peptides. *J. Cereb. Blood Flow Metab.*, *5* (Suppl. 1), S507.

Harris, D. J., Yang, B. I. Y., Wolf, B., & Snodgrass, P. J. (1980). Dysautonomia in an infant with secondary hyperammonemia due to propionyl coenzyme A carboxylase deficiency. *Pediatrics*, *65*(1), 107–110.

Heidbreder, E. K. S., & Heidland, A. (1985). Autonomic neuropathy in chronic renal insufficiency. *Nephron*, *41*(1), 50–56.

Heistad, D. D., Abboud, F. M., Mark, A. L., & Schmid, P. G. (1972). Impaired reflex vasoconstriction in chronically hypoxemic patients. *J. Clin. Investig.*, *51*, 331–337.

Heistad, D. D., & Wheeler, R. C. (1972). Effect of carbon monoxide on reflex vasoconstriction in man. *J. Appl. Physiol. (Bethesda, Md.: 1985)*, *32*, 7–11.

Hellstrand, P., & Jarhult, J. (1980). Effects of nine different gastrointestinal polypeptides on vascular smooth muscle in vitro. *Acta Physiol. Scand.*, *110*, 89–94.

Hepburn, D. A., Patrick, A. W., Eadington, D. W., et al. (1990). Unawareness of hypoglycemia in insulin treated diabetic patients: Prevalence and relationship to autonomic neuropathy. *Diabetic Med.*, 7, 711–717.

Hilsted, J., Richier, E., Madsbad, S., et al. (1987). Metabolic and cardiovascular responses to epinephrine in diabetic autonomic neuropathy. *N. Engl. J. Med.*, *317*, 421–426.

Horstink, M. W. I. M., Gabreels, F. J. M., Joosten, E. M. G., Gabreels Festen, A. W. M., Jaspar, H. H. J., Van Haelst, V. J. G., & Korten, J. J. (1974). Multiple mucosal neuromas, dysautonomia and abnormal intradermal histamine reaction. *Clin. Neurol. Neurosurg.*, *3/4*, 212–224.

Hromada, J. (1963). On the nerve supply of the connective tissue of some peripheral nervous system components. *Acta Anatomica*, *55*, 343–351.

Iosa, D., Dequattro, V., Lee, D. D. P., et al. (1990). Pathogenesis of cardiac neuromyopathy in Chagas disease and the role of the autonomic nervous system. *J. Auton. Nerv. Syst.*, *30*, S83–S88.

Isenflamm, J. F., & Doerffler, J. R. (1768). De vasis nervorum. Erlangen.

Ivy, H. K. (1961). Renal sodium loss and bronchogenic carcinoma: Associated autonomic neuropathy. *Arch. Int. Med.*, *108*, 47–55.

Janssens, J., Peeters, T. L., Vantrappen, G., et al. (1990). Improvement of gastric emptying in diabetic gastroparesis by erythromycin. *N. Engl. J. Med.*, *322*, 1028–1031.

Jenkins, Z. M., Castle, D. J., Eikelis, N., Phillipou, A., Lambert, G. W., & Lambert, E. A. (2021). Autonomic nervous system function in women with anorexia nervosa. *Clin. Auton. Res.*

Jenzer, G., Mumenthaler, M., Ludin, H. P., & Robert, F. (1975). Autonomic dysfunction in botulism B: A clinical report. *Neurology*, *25*, 150–153.

Johnson, P. C., & Johnsen, S. D. (1987). Hereditary neuropathies in Navajo children. *Neurology*, *37*, 255.

Johnson, R. H., McLellan, D. L., & Love, D. R. (1971). Orthostatic hypotension and the Holmes Adie syndrome. *J. Neurol. Neurosurg. Psychiatry*, *34*, 562–570.

Karagiorgou, Katerina, Dandoulaki, Maria, Mantegazza, Renato, Andreetta, Francesca, Furlan, Raffaello, Lindstrom, Jon, ... Tzartos, John (2022). Novel Cell-Based Assay for Alpha-3 Nicotinic Receptor Antibodies Detects Antibodies Exclusively in Autoimmune Autonomic Ganglionopathy. *Neurol. Neuroimmunol. Neuroinflamm.*, *9*(3). Available from https://doi.org/10.1212/NXI.0000000000001162.

Kennedy, W. R., Navarro, X., Goetz, F. C., et al. (1990). Effects of pancreatic transplantation on diabetic neuropathy. *N. Engl. J. Med.*, *322*, 1031–1037.

Kersh, E. S., Kronfield, S. J., Unger, A., Popper, R. W., Cantor, S., & Cohn, K. (1974). Autonomic insufficiency in uremia as a cause of hemodialysis induced hypotenson. *N. Engl. J. Med.*, *290*, 650–653.

Key, A., & Retzius, G. (1876). *Studien in der Anatomie des Nervensystems und des Bindegewebes*. Stockholm: Samson and Wallin.

Kitamura, J., Kubuki, Y., Tsuruta, K., et al. (1989). A new family with Joseph disease in Japan. *Arch. Neurol.*, *46*, 425–428.

Koay, S., Vichayanrat, E., Bremner, F., Panicker, J. N., Lang, B., Lunn, M. P., ... Iodice, V. (2021). Multimodal biomarkers quantify recovery in autoimmune autonomic ganglionopathy. *Ann. Neurol.*, *89*(4), 753–768.

Kobayashi, S., Kyoshima, K., Olschowka, J. A., & Jacobowitz, D. M. (1983). Vasoactive intestinal polypeptide immunoreactive and cholinergic nerves in the wholemount preparation of the major cerebral arteries of the rat. *Histochemistry*, *79*, 377–381.

Krämer, H. H., Rebhorn, C., Geber, C., & Birklein, F. (2021). Sympathetic and sensory nerve fiber function in multiple system atrophy and idiopathic Parkinson's disease. *J. Neurol.*, *268*(9), 3435–3443.

Krarup, T., Schwartz, T. W., et al. (1979). Impaired response of pancreatic polypeptide to hypoglycaemia: An early sign of autonomic neuropathy in diabetics. *Br. Med. J.*, *2*, 1544–1546.

Kroop, I. G. (1956). The production of tears in familial dysautonomia. *J. Pediat.*, *48*, 328–329.

Krucke, W. (1955). Erkränkungen des peripheren Nervensystems. In G. Doring, E. Herzog, W. Krucke, & H. Orthner (Eds.), *Handbuch der speziellen pathologischen Anatomie und Histologie* (Vol. XIII/V, pp. 1–203). Berlin: Springer.

Lamotte, G., & Sandroni, P. (2021). A practical approach to peripheral autonomic neuropathies. *Curr. Opin. Neurol.*, *34(5)*, 638–647.

Lamotte, G., Sandroni, P., Cutsforth-Gregory, J. K., Berini, S. E., Benarroch, E. E., Shouman, K., ... Coon, E. A. (2021). Clinical presentation and autonomic profile in Ross syndrome. *J. Neurol.*, *268*(10), 3852–3860.

Lang, J. (1962). Über das Bindegewebe und die Gefässe der Nerven. *Z. Anat. Entw. Ges.*, *123*, 61–69.

Lanpus, W. E., Collins, L. R., Bloodworth, R., & Wood, E. J. (1962). Vasoconstrictor responses in infancy and childhood. Society for Pediatric Research Meeting, Atlantic City, May 8–10 (1962). *Am. J. Dis. Child.*, *104*, 577–578.

Lanting, P., Bos, J. E., Aartsen, J., et al. (1990). Assessment of pupillary light reflex latency and darkness adapted pupil size in control subjects and in diabetic patients with and without cardiovascular autonomic neuropathy. *J. Neurol. Neurosurg. Psychiatry*, *53*, 912–914.

León-Velarde, F., & Arregui, A. (1994). Desadaptación a la vida en las grandes alturas. *Travaux de lInstitut Fran_ais dÉtudes Andines*, *85*, 80.

Levy, S. B., Lilley, J. J., et al. (1978). Baroreflex function in uremic and hypertensive man. *Am. J. Med. Sci.*, *276*, 57–66.

Lincoln, J., Milner, P., Appenzeller, O., et al. (1993). Innervation of normal human sural and optic nerves by noradrenalin and peptide containing nervi vasorum and nervorum: Effect od diabetes and alcoholism. *Brain Res.*, *632*, 48–56.

Linde, L. M. (1955). Dysautonomia (case report of a variant). *J. Pediatrics*, *46*, 453–455.

Linde, L. M., & Westover, J. L. (1962). Esophageal and gastric abnormalities in dysautonomia. *Pediatrics*, *29*, 303–306.

Lourie, R. S. (1943). Role of secretion of the parotid glands in normal children: Measurement of function of the autonomic nervous system. *Am. J. Dis. Child.*, *65*, 455–479.

Lundberg, J. M., Hökfelt, T., Schultzberg, M., Uvnäs-Wallensten, K., Köhler, C., & Said, S. I. (1979). Occurrence of vasoactive intestinal polypeptide (VIP)-like immunoreactivity in certain cholinergic neurons of the cat: Evidence from combined immunohistochemistry and acetylcholinesterase staining. *Neuroscience*, *4*(11), 1539–1559.

Lundborg, G. (1970). Ischemic nerve injury. Experimental studies on intraneural microvascular pathophysiology and nerve function in a limb subjected to temporary circulatory arrest. *Scand. J. Plast. Reconstruct. Surg. Suppl.*, *6*, 3–113.

Mai, T. H., Wu, J., Diedrich, A., Garland, E. M., & Robertson, D. (2014). Calcitonin gene-related peptide (CGRP) in autonomic cardiovascular regulation and vascular structure. *J. Am. Soc. Hypertension*, *8*(5), 286–296.

Malpas, S. C., & Maling, T. J. B. (1990). Heart rate variability and cardiac autonomic function in diabetes. *Diabetes*, *39*, 1177–1181.

Mancini, R. E. (1963). Connective tissue and serum proteins. *Int. Rev. Cytol.*, *14*, 193–222.

Marcarian, H. Q., & Smith, R. D. (1968). A quantitative study on the vasa nervorum in the ulnar nerve of cats. *Anatomical Record*, *161*(1), 105–109.

St. Martin, D. A. (1953). Familial autonomic dysfunction: Its occurrence in two siblings. *Clin. Proc. Child. Hosp. (Wash.)*, *9*, 81–85.

Mason, D. T., Kopin, I. J., & Braunwald, E. (1966). Abnormalities in reflex control of the circulation in familial dysautonomia. *Am. J. Med.*, *41*, 898–909.

Matsuyama, T., Matsumoto, M., Shiosaka, S., et al. (1984). Serotoninergic and peptidergic innervation of the cerebral arteries: An immunohistochemical and electron microscopic study. *Neurosci. Lett.* (Suppl. 17), S118.

McGregor, G. P., Gibson, S. J., Sabate, I. M., Blank, M. A., Christofides, N. D., Wall, P. D., ... Bloom, S. R. (1984). Effect of peripheral nerve section and nerve crush on spinal cord neuropeptides in the rat; increased VIP and PHI in the dorsal horn. *Neuroscience*, *13*(1), 207–216.

McLelland, J., & Ellis, S. J. (1986). Causalgia as a complication of meningococcal meningitis. *Br. Med. J.*, *292*(6537), 1710.

Mellick, R. S., & Cavanagh, J. B. (1968). Changes in blood vessel permeability during degeneration and regeneration in peripheral nerves. *Brain*, *91*(1), 141–160.

Miglietta, O. (1967). Electrophysiological studies in chronic occlusive peripheral vascular disease. *Arch. Phys. Med.*, *48*, 89–96.

Mittag, T. W., Mindel, J. S., & Green, J. P. (1974). Choline acetyltransferase in familial dysautonomia. *Ann. N. Y. Acad. Sci.*, *228*, 301–306.

Mittal, C. M., Wig, N., Mishra, S., & Deepak, K. K. (2004). Heart rate variability in human immunodeficiency virus-positive individuals. *Int. J. Cardiol.*, *94*(1), 1–6.

Moloshok, R. E., & Moseley, J. E. (1956). Familial dysautonomia: Pulmonary manifestations. *Pediatrics*, *17*, 327–333.

Moorhouse, J. A., Carter, S. A., & Doupe, J. (1966). Vascular responses in diabetic peripheral neuropathy. *Br. Med. J.*, *1*, 883–888.

Moskowitz, M. A. (1984). The neurobiology of vascular head pain. *Ann. Neurol.*, *16*, 157–168.

Navarro, X., Kennedy, W. R., Loewenson, R. B., et al. (1990). Influence of pancreas transplantation on cardiorespiratory reflexes, nerve conduction, and mortality in diabetes mellitus. *Diabetes*, *39*, 802–806.

Nobel, W., & Black, D. (1974). The microcirculation of peripheral nerves: Techniques for perfusion and microangiographic, macrophotographic, and photomicrographic recordings in animals. *J. Neurosurg.*, *41*(1), 83–91.

Nolano, M., Provitera, V., Perretti, A., Stancanelli, A., Saltalamacchia, A. M., Donadio, V., ... Santoro, L. (2006). Ross syndrome: A rare or a misknown disorder of thermoregulation? A skin innervation study on 12 subjects. *Brain*, *129*(Pt 8), 2119–2131.

Norcliffe-Kaufmann, L., Kaufmann, H., Palma, J. A., Shibao, C. A., Biaggioni, I., Peltier, A. C., ... Autonomic Disorders, C. (2018). Orthostatic heart rate changes in patients with autonomic failure caused by neurodegenerative synucleinopathies. *Ann. Neurol.*, *83*(3), 522–531.

Nordborg, C., Kristensson, K., Olsson, Y., & Sourander, P. (1973). Involvement of the autonomous nervous system in primary and secondary amyloidosis. *Acta Neurol. Scand.*, *49*, 31–38.

Olesen, S. P. (1985). A calcium dependent reversible permeability increase in microvessels in frog brain induced by serotonin. *J. Physiol. (Lond.)*, *361*, 103–111.

Olsson, Y. (1968). Mast cells in the nervous system. *Int. Rev. Cytol.*, *24*, 27–70.

Olsson, Y. (1972). The involvement of vasa nervorum in disease of peripheral nerves, Part 2 In P. J. Vinken, & G. W. Bruyn (Eds.), *Handbook of clinical neurology* (Vol. 12, pp. 644–664). Amsterdam: North Holland.

Olsson, Y., & Kristensson, K. (1979). Recent applications of tracer techniques to neuropathology with particular reference to vascular and axonal flow: A review. In T. S. Smith, & J. B. Cavanaugh (Eds.), *Recent advances in neuropathology* (pp. 1–25). Edinburgh: Churchill/Livingston.

Olsson, Y., & Reese, T. S. (1969). Inaccessibility of mouse sciatic nerve to exogenous proteins. *Anatomical Record*, *163*, 1319–1321.

Oppenheimer, D. (1983). Neuropathology of progressive autonomic failure. In R. Bannister (Ed.), *Autonomic failure: A textbook of clinical disorders of the autonomic nervous system* (pp. 267–283). Oxford: Oxford University Press.

Parratt, J. R., & West, G. B. (1957). 5-Hydroxytryptamine and tissue mast cells. *J. Physiol.*, *137*(2), 169–178.

Pearson, J. (1979). Familial dysautonomia (a brief review). *J. Auton. N. Syst*, *1*, 119–126.

Pearson, J., Axelrod, F., & Dancis, J. (1974). Current concepts of dysautonomia: Neuropatholgical defects. *Ann. N.Y. Acad. Sci.*, *228*, 288–300.

Pernow, B. (1983). Substance P. *Pharmacol. Rev.*, *35*, 85–141.

Potulska-Chromik, A., Zakrzewska-Pniewska, B., Szmidt-Sałkowska, E., Lewandowski, J., Siński, M., Przyjałkowski, W., & Kostera-Pruszczyk, A. (2013). Long lasting dysautonomia due to botulinum toxin B poisoning: Clinical-laboratory follow up and difficulties in initial diagnosis. *BMC Res. Notes*, *6*, 438.

Provitera, Vincenzo, Iodice, Valeria, Manganelli, Fiore, Mozzillo, Stefania, Caporaso, Giuseppe, Stancanelli, Annamaria, … Nolano, Maria (2022). Postganglionic sudomotor assessment in early stage of multiple system atrophy and Parkinson disease: a morpho-functional study. *Neurology*, *98*(12), e1282–e1291. Available from https://doi.org/10.1212/WNL.0000000000013300.35017309.

Puentes, D., Teijelo, D., Stiep, T. S., Mannava, S., & Margolesky, J. (2021). Takotsubo cardiomyopathy as a manifestation of dysautonomia in Guillain–Barré syndrome: A case series and review of the literature. *Cureus*, *13*(6), e16069.

Qureshi, A. I., Saleem, M. A., Ahrar, A., & Raja, F. (2017). Imaging of the vasa nervorum using contrast-enhanced ultrasound. *J. Neuroimaging*, *27*(6), 583–588.

Rajput, A. H., & Rodzilsky, B. (1976). Dysautonomia in Parkinsonism: A clinicopathological study. *J. Neurol. Neurosurg. Psychiat.*, *39*(11), 1092–1100.

Ramage, D. (1927). The blood supply to the peripheral nerves of the superior extremity. *J. Anat.*, *61*, 198–205.

Randall, C. A., Longshore, F. A., Davis, M. A., et al. (1996). Dysautonomiain dogs: A retrospective study. *J. Vet. Int. Med.*, *10*, 103–109.

Ranvier, L. (1878). *Lecons sur Lhistologie du Système Nerveux*. Paris: F. Savy.

Raskin, N. H., Levinson, S. A., Hoffman, P. M., Pickett, J. B. E., III, & Fields, H. L. (1974). Postsympathectomy neuralgia: Amelioration with diphenylhydantoin and carbamazepine. *Ann. J. Surg*, *128*(1), 75–78.

Rathmann, W., Ziegler, D., Jahnke, M., Haastert, B., & Gries, F. A. (1993). Mortality in diabetic patients with cardiovascular autonomic neuropathy. *Diabet. Med.*, *10*(9), 820–824.

Riley, C. M. (1952). Familial autonomic dysfunction. *JAMA*, *149*, 1532–1535.

Riley, C. M. (1957). Familial dysautonomia. *Advanc. Pediat.*, *9*, 157–190.

Riley, C. M., Freedman, A. M., & Langford, W. S. (1954). Further observations on familial dysautonomia. *Pediatrics*, *14*, 475–480.

Robertson, D., Hollister, A. S., Biaggioni, I., Netterville, J. L., Mosqueda-Garcia, R., & Robertson, R. M. (1993). The diagnosis and treatment of baroreflex failure. 1993. *N. Engl. J. Med.*, *329*, 1449–1455.

Rockson, S., Stone, R., Weydenvander, M., & Kelley, W. N. (1974). Lesch Nyhan syndrome: Evidence for abnormal adrenergic function. *Science (New York, N.Y.)*, *186*, 934–935.

Ross, A. T. (1958). Progressive selective sudomotor denervation; a case with coexisting Adie's syndrome. *Neurology*, *8*(11), 809–817.

Ruysch, F. (1701). Thesaurus Anatomicus Primus. Amsterdam.

Ryder, R. E. J., Kennedy, R. L., Newrick, P. G., et al. (1990). Autonomic denervation may be a prerequisite of diabetic neuropathic foot ulceration. *Diabet. Med.*, *7*, 726–730.

Ryder, R. E. J., Owens, D. R., Hayes, T. M., et al. (1990). Unawareness of hypoglycemia and inadequate hypoglycemic counterregulation: No causal relation with diabetic autonomic neuropathy. *Br. Med. J.*, *301*, 783–787.

Sampson, M. J., Wilson, S., Kargianis, P., et al. (1990). Progression of diabetic autonomic neuropathy over a decade in insulin dependent diabetics. *Quart. J. Med.*, *75*, 635–646.

Scatton, B., Duverger, D., LHeureux, R., & MacKenzie, E. T. (1985). Neurochemical studies on the nature of the serotoninergic innervation of the cerebral circulation. *J. Cereb. Blood Flow Metab.*, *5*(Suppl. 1), S572.

Schwartz, J. P., & Breakefield, X. O. (1980). Altered nerve growth factor in fibroblasts from patients with familial dysautonomia. *Proc. Natl. Acad. Sci. U. S. A.*, 77, 1154–1158.

Seibold, J. R., & Jageneau, A. H. M. (1984). Treatment of Raynauds phenomenon with ketanserin, a selective antagonist of the serotonin 2(5 HT2) receptor. *Arthr. Rheum.*, *27*, 139–146.

Senard, J. M., Barb., Estan, L., et al. (1990). Decreased high affinity state in platelet alpha2 adrenoceptors from diabetic patients with orthostatic hypotension. *J. Clin. Endocrinol. Metabol.*, *71*, 311–317.

Sharpey Schafer, E. P. (1956). Circulatory reflexes in chronic disease of the afferent nervous system. *J. Physiol. (London)*, *134*, 1–10.

Sharpey-Schafer, E. P., & Taylor, P. J. (1960). Absent circulatory reflexes in diabetic neuritis. *Lancet*, *1*, 559–562.

Singleton, R., Helgerson, S. D., Snyder, R. D., OConner, P. J., Nelson, S., Johnsen, S. D., & Allanson, J. E. (1990). Neuropathy in Navajo children: Clinical and epidemiologic features. *Neurology*, *40*(2), 363–367.

Smith, A. A., & Dancis, J. (1963). Physiologic studies in familial dysautonomia. *J. Pediat.*, *63*, 838–840.

Smith, A. A., & Dancis, J. (1964). Exaggerated response to infused norepinephrine in familial dysautonomia. *N. Engl. J. Med.*, *270*, 704–707.

Smith, A. A., & Dancis, J. (1967). Catecholamine release in familial dysautonomia. *N. Engl. J. Med.*, *277*, 61.

Smith, A. A., Hirsch, J. I., & Dancis, J. (1965). Response to infused metacholine in familial dysautonomia. *Pediatrics*, *36*, 225.

Spiegel, V. D. (1627). A. De humani corpus fabrica. Venice.

Streeten, D. P. H., Kerr, L. P., Kerry, C. B., Prior, J. C., & Dalakos, T. G. (1972). Hyperbradykininism: A new orthostatic syndrome. *Lancet*, *2*, 1048–1053.

Sunderland, S. (1945). Blood supply of peripheral nerves; practical considerations. *Arch. Neurol. Psychiatry*, *54*, 280–282.

Sunderland, S. (1965). The connective tissues of peripheral nerves. *Brain*, *88*(4), 841–854. Available from https://doi.org/10.1093/brain/88.4.841.

Sunderland, S. (1978). *The blood vessels of peripheral nerve trunks* Chapter 4 *Nerves and nerve injuries* (2nd ed., pp. 46–60). Edinburgh: Churchill/Livingston.
Swanson, A. G., Buchan, G. C., & Alvord, E. P., Jr (1963). Absence of Lissauers tract and small dorsal root axons in familial congenital universal insensitivity to pain. *Trans. Amer. Neurol. Ass.*, *88*, 99–103.
Thawani, S. P., Brannagan, T. H., 3rd, Lebwohl, B., Green, P. H., & Ludvigsson, J. F. (2015). Risk of Neuropathy Among 28,232 Patients With Biopsy-Verified Celiac Disease. *JAMA Neurol*, *72*(7), 806–811.
Thomas, P. K., & Lascelles, R. G. (1965). The pathology of diabetic neuropathy. *Quart. J. Med.*, *35*, 489–501.
Tuor, V. I., Kelly, P., Edvinsson, L., et al. (1985). Neuropeptide Y and the regulation of local cerebral blood flow: A quantitative autoradiographic study. *J. Cereb. Blood Flow Metab.* (Suppl. 1), S513.
Vanhoutte, P. M., Cohen, R. A., & van Neuten, J. M. (1984). Serotonin and arterial vessels. *J. Cardiovasc. Pharmacol.*, *6*, S421–S428.
Vigersky, R. A., Andersen, A. E., et al. (1977). Hypothalamic dysfunction in secondary amenorrhea associated with simple weight loss. *New Engl. J. Med.*, *297*, 1141–1145.
Vollmer, H. (1927). Periodische Pathothermia. Z. Kinderheilk. *43*, 88–102.
Waggener, J. D., Bunn, S. M., & Beggs, J. (1965). The diffusion of ferritin within the peripheral nerve sheath: An electron microscopy study. *J. Neuropathol. Exp. Neurol.*, *24*(3), 430–443.
Waksman, B. H. (1961). Experimental study of diphtheritic polyneuritis in the rabbit and guinea pig. III. The blood nerve barrier in the rabbit. *J. Neuropathol. Exp. Neurol.*, *20*, 35–77.
Weinshilboum, R. M., & Axelrod, J. (1971). Serum dopamine b hydroxylase decrease after chemical sympathectomy. *Science (New York, N.Y.)*, *173*, 931.
Weinshilboum, R. M., Kvetnansky, R., & Axelrod, J. (1971). Serum dopamine beta hydroxylase activity with forced immobilization. *Nature (New Biol.)*, *230*, 287.
Williams, T. J. (1982). Vasoactive intestinal polypeptide is more potent than prostaglandin E2 as a vasodilator and oedema potentiator in rabbit skin. *Br. J. Pharmacol.*, *77*, 505–509.
Wu, T. Y., Taylor, J. M., Kilfoyle, D. H., Smith, A. D., McGuinness, B. J., Simpson, M. P., … Roxburgh, R. H. (2014). Autonomic dysfunction is a major feature of cerebellar ataxia, neuropathy, vestibular areflexia 'CANVAS' syndrome. *Brain*, *137*(Pt 10), 2649–2656.
Yaksh, T. L., Abay, E. O., 2nd, & Go, V. L. (1982). Studies on the location and release of cholecystokinin and vasoactive intestinal peptide in rat and cat spinal cord. *Brain Res.*, *242*(2), 279–290. Available from https://doi.org/10.1016/0006-8993(82)90311-0.
Yeh, T. H., Lu, C. S., Chou, Y. H., Chong, C. C., Wu, T., Han, N. H., & Chen, R. S. (2005). Autonomic dysfunction in Machado-Joseph disease. *Arch. Neurol.*, *62*(4), 630–636.
Young, R. R. (1969). Pure pan dysautonomia with recovery. *Trans. Amer. Neurol. Ass.*, *94*, 355–357.
Zawadski, J. V., Furchgott, R. F., & Cherry, P. (1981). The obligatory role of endothelial cells in the relaxation of arterial smooth muscle by Substance P. *Federation Proceedings*, *40*, 689.
Zuckerkandl, O. (1885). Zwei Fälle von collateral Kreislauf. *Medsche J.C. Schr*, *15*, 273–282.

CHAPTER 9

Progressive autonomic failure

9.1 Progressive autonomic failure—autonomic dysfunction in synucleinopathies

9.1.1 Introduction

Pure autonomic failure (PAF), Parkinson's disease (PD), dementia with Lewy bodies (DLB), and multiple system atrophy (MSA) are distinct neurodegenerative clinical syndromes characterized by cytoplasmic deposition of the protein α-synuclein and are termed synucleinopathies. These disorders differ in the cellular location and pattern of α-synuclein deposition. Parkinson's disease and DLB are characterized by deposition of α-synuclein in Lewy bodies in brainstem neurons, whereas MSA features α-synuclein deposition in glial cytoplasmic inclusions (Fanciulli & Wenning, 2015; Poewe et al., 2017). Peripheral autonomic structures are more affected in the Lewy body disorders compared with in MSA, which has primarily a preganglionic central autonomic involvement.

Autopsy studies of PAF cases report Lewy bodies in sympathetic ganglia and Lewy neurites in distal sympathetic nerves but there is also pathological involvement of central structures (Hague, Lento, Morgello, Caro, & Kaufmann, 1997; van Ingelghem, van Zandijcke, & Lammens, 1994). Therefore, PAF could be considered as a prototypical Lewy body disorder manifesting with autonomic failure that may evolve into PD or DLB. On the other hand, if PAF is defined clinically by the presence of chronic neurogenic orthostatic hypotension without an identified secondary cause or clinical evidence of central neurodegeneration regardless of biomarkers suggesting Lewy body pathology (Kaufmann, 1996), then PAF can evolve into PD, DLB, or MSA (Coon, 2020; Kaufmann et al., 2017). Research laboratory biomarkers that have the potential to differentiate Lewy body disorders from MSA include the analysis of α-synuclein oligomers and neurofilament light chain in the cerebrospinal fluid (CSF) (Singer et al., 2020; Singer et al., 2021), cardiac sympathetic neuroimaging (Lenka, Lamotte, & Goldstein, 2021), and immunofluorescence microscopic methods to visualize α-synuclein in sympathetic noradrenergic nerves in skin biopsies (Donadio, 2019). Laboratory biomarkers may also help assess

DOI: https://doi.org/10.1016/B978-0-323-96011-3.00014-0

the prognosis in synucleinopathies. For example, plasma neurofilament light chains are a reliable biomarker for the disease severity of MSA and monitoring the progression of MSA, but not the progression of cognition (Zhang et al., 2021).

Although dysfunction of the autonomic nervous system is characteristic of the synucleinopathies, the frequency, degrees of severity, and patterns vary in the different disorders. For example, cardiovascular adrenergic failure characterized by neurogenic orthostatic hypotension is seen in all patients with PAF and in the majority of patients with MSA, whereas neurogenic orthostatic hypotension is seen in 30%–60% of patients with Lewy body disorders (Hiorth, Pedersen, Dalen, Tysnes, & Alves, 2019; Velseboer, de Haan, Wieling, Goldstein, & de Bie, 2011). Autonomic dysfunction in the synucleinopathies may occur at all disease stages and is not limited to the cardiovascular autonomic nervous system and involves the thermoregulatory, urogenital, and gastrointestinal systems (Coon, 2020).

9.1.2 Historical perspective

Pure autonomic failure was first described by Bradbury and Eggleston (1925). The authors suggested that because postural hypotension was the chief and most disabling symptom in autonomic failure, it was the primary feature of this disease. Less apparent neurologic disturbances associated with autonomic failure included disturbances of the bladder, sexual dysfunction, and heat intolerance due to abnormalities in sweating. In the first edition of his textbook, *Autonomic Failure*, Bannister used the abbreviation, PAF, to denote "progressive autonomic failure" (Bannister, 1983), which he viewed as including the Lewy body disease PD and the non-Lewy body disease MSA. In the second edition of his textbook, Bannister noted that idiopathic orthostatic hypotension can be "pure" in the sense of lacking clinical evidence of motor or cognitive deficits indicating central nervous system neurodegeneration (Bannister, 1988). The term pure autonomic failure is now accepted and used by most clinicians and researchers, even though the pathogenesis and the postmortem pathologic features are not well known. Furthermore, the clinical presentation is not purely limited to autonomic failure, and most patients with PAF exhibit symptoms of central nervous system involvement such as anosmia and rapid-eye-movement sleep behavior disorder (Coon, Singer, & Low, 2019).

Shy and Drager (1960) reported for the first time the syndrome that was later named after them in 1960. The syndrome, which they studied in two patients, consisted of the following symptoms: orthostatic hypotension, urinary and rectal incontinence, loss of sweating, iris atrophy, external ocular palsies, rigidity of limbs, distal tremor, loss of arm movements associated with walking, impotence, loss of rectal sphincter tone, atonic bladder, fasciculations, and wasting of distal muscles. They also found evidence of neuropathic lesions in the electromyogram, suggesting the involvement of the anterior horn cells, and a neuropathic pattern on muscle biopsy. They observed that the onset of the disease was usually after the fifth decade of life. In 1966, Johnson et al. were the first to note that degeneration of intermediolateral cell columns accompanies postural hypotension, and from their pathologic material, they suggested that olivopontocerebellar atrophy should be linked with the Shy-Drager syndrome (Johnson, Lee Gde, Oppenheimer, & Spalding, 1966). The term MSA was first proposed by Graham and Oppenheimer in 1969 and later developed as a label to encompass different previously described disorders such as olivopontocerebellar atrophy, Shy-Drager syndrome, and striatonigral degeneration (Graham & Oppenheimer, 1969).

Parkinson's disease was formally described by James Parkinson in the classic "An Essay on the Shaking Palsy" published in 1817 (Parkinson, 1817). In this seminal description, he describes nonmotor features of PD with severe constipation requiring mechanical removal of the feces, continuous drooling of saliva, and depression ("the unhappy sufferer"). Clinicians and researchers recognized autonomic failure accompanying otherwise clinically typical PD (Appenzeller & Goss, 1971). Importantly, although autonomic failure has been described in PD, even in its early stage, the UK PD Society Brain Bank criteria list early severe autonomic involvement as exclusionary for PD (Hughes, Daniel, Blankson, & Lees, 1993). An orthostatic decrease of blood pressure within 3 minutes of standing by at least 30 mmHg systolic or 15 mmHg diastolic is listed as a red flag for a diagnosis of PD in the revised MDS clinical diagnostic criteria for PD (Postuma et al., 2015).

9.1.3 Pathophysiology

The exact pathophysiological mechanisms that contribute to autonomic dysfunction in synucleinopathies are not well understood. In PAF, the prominent phenotype is of postganglionic efferent autonomic failure with impaired neuronal norepinephrine biosynthesis and decreased myocardial norepinephrine content (Coon, Singer, et al., 2019; Goldstein, Holmes, Sharabi, Brentzel, & Eisenhofer, 2003). A feature that is shared by Lewy body disorders is peripheral adrenoreceptors hypersensitivity secondary to chronically low circulating norepinephrine and upregulation of postsynaptic receptors. Plasma norepinephrine and 3,4-dihydroxyphenylglycol (DHPG) levels are especially low in PAF related to reduced turnover of norepinephrine stores in sympathetic nerves. The mechanisms of the peripheral noradrenergic lesion are not fully elucidated and the clinical correlation of cardiac noradrenergic deficiency in Lewy body disorders is not known (Lamotte & Benarroch, 2021). A study using a combined positron emission tomographic (PET) neuroimaging approach with ^{11}C-methylreboxetine, a selective ligand for the cell membrane norepinephrine transporter, and ^{18}F-dopamine to assess intraneuronal vesicular storage found that PAF entailed moderately decreased cardiac sympathetic innervation and a substantial vesicular storage defect in residual nerves (Goldstein, Isonaka, Holmes, Ding, & Sharabi, 2020). The results of this study suggest that the majority of residual cardiac postganglionic sympathetic nerves are dysfunctional in PAF, in other words, "sick-but-not-dead" (Goldstein et al., 2020; Goldstein, 2020b). Postganglionic sympathetic noradrenergic denervation/dysfunction is a feature shared by the Lewy body disorders (PD and DLB) and leads to impaired vasoconstriction upon standing, with accentuated venous pooling contributing to orthostatic hypotension. The loss of intermediolateral nucleus (IML) neurons in the spinal cord and dorsal motor nucleus of the vagus also contributes to orthostatic hypotension. All patients with orthostatic hypotension in the synucleinopathies show a reduction of neurons in the IML with the most severe being in MSA (Hatsuta et al., 2016). In addition to inadequate reflexive vasoconstriction characterizing cardiovascular adrenergic failure, patients with different forms of synucleinopathies are well known to have an inadequate reflexive increase in heart rate (Norcliffe-Kaufmann et al., 2018). Heart rate variability is impaired in patients with autonomic failure that affect preganglionic pathways (MSA), or predominantly postganglionic neurons (PAF, PD, DLB) (Coon, Singer, et al., 2019; Kaufmann & Goldstein, 2013; Rahman & Goldstein, 2014). Relative preservation of chronotropic function has been reported in patients with MSA compared to PD (Goldstein et al., 2000). In contrast with the predominant postganglionic efferent impairment in Lewy body disorders, involvement of central autonomic structures contributing to autonomic failure is implicated in MSA. The rostral ventrolateral medulla is critical for centrally mediated control of sympathetic vasomotor tone and baroreflex control of blood pressure and demonstrates severe neuronal loss and α-synuclein inclusions in patients with MSA (Benarroch, Smithson, Low, & Parisi, 1998; Coon et al., 2016), and to a lesser extent, in patients with Lewy body disorders (Benarroch et al., 2005).

The clinical relevance of α-synuclein aggregation in peripheral and central autonomic structures in different synucleinopathies is not known. Alpha-synuclein and Lewy bodies have been found in the enteric nervous system of patients with PD, which might then be transmitted in a prion-like manner to the central nervous system (Borghammer, 2018; Masuda-Suzukake et al., 2014; Prusiner et al., 2015; Van Den Berge et al., 2019). Misfolded α-synuclein can be taken up by neighboring neurons to form neuronal cytoplasmic and/or nuclear inclusions, leading to calls for MSA to be considered a "prion-like disease" (Prusiner et al., 2015). Arguing against MSA being termed a prion disease, there have been no reported cases of human-to-human transmission of MSA. A link between α-synuclein aggregation and autonomic failure may also be suggested by genetic synucleinopathies where the autonomic features are frequent and involve both the central and peripheral autonomic systems (Chelban, Vichayanrat, Schottlaende, Iodice, & Houlden, 2018). In these syndromes, the distribution and the degree of autonomic failure appear to be related to the spread of α-synuclein pathology, where genetic forms of PD that do not present with significant Lewy body pathology also do not show significant autonomic dysfunction and are mostly restricted to brain pathology (Chelban et al., 2018). A possible link between α-synuclein pathology and catecholaminergic neurodegeneration is further supported by the catechol aldehyde hypothesis and the potential role of the autotoxic intraneuronal metabolite of dopamine, 3,4-dihydroxyphenylacetaldehyde (DOPAL) (Goldstein, 2020a). Studies have found that DOPAL potently oligomerizes and quinonizes α-synuclein (Burke et al., 2008; Jinsmaa, Sharabi, Sullivan, Isonaka, & Goldstein, 2018); DOPAL-induced synuclein oligomers impede vesicular functions of catecholaminergic neurons (Goldstein, 2020a). Nevertheless, the exact role of α-synuclein oligomers in synucleinopathies is far from being elucidated. A study found that tissue norepinephrine contents in gastrointestinal organs were similar in patients with Lewy body disorders compared to controls suggesting that α-synuclein aggregation is not necessarily associated with sympathetic noradrenergic denervation (Lamotte, Holmes, Sullivan, Lenka, & Goldstein, 2020). The myocardium is the exception where intraneuronal α-synuclein deposition may play a pathophysiological role in Lewy body forms of neurogenic orthostatic hypotension (Isonaka et al., 2019). Distinct forms of α-synuclein may underlie the specific pathology in different synucleinopathies and a different pattern of aggregation of α-synuclein in CSF was demonstrated using protein misfolding cyclic amplification between MSA and the Lewy body synucleinopathies PD and DLB (Fig. 9.1) (Singer et al., 2020). Finally, other indirect evidence from human studies suggests that protein aggregation in sporadic cases may be protective and not capable of discriminating clinical disease subtypes (Espay et al., 2019).

Studying CSF levels of catecholamine metabolites may allow for identification of central catecholamine deficiency in synucleinopathies. A study measuring CSF levels of the dopamine metabolite 3,4-dihydroxyphenylacetic acid (DOPAC), norepinephrine metabolite DHPG, homovanillic acid (HVA, metabolite of DOPAC), and 3-methoxy-4-hydroxyphenylglycol (MHPG, metabolite of DHPG) in 36 PD, 37 MSA, and 19 PAF patients and in 38 controls demonstrated that the PD, MSA, and PAF groups had decreased CSF MHPG, DHPG and norepinephrine compared to the control group. CSF HVA and DOPAC were decreased in PD and MSA but

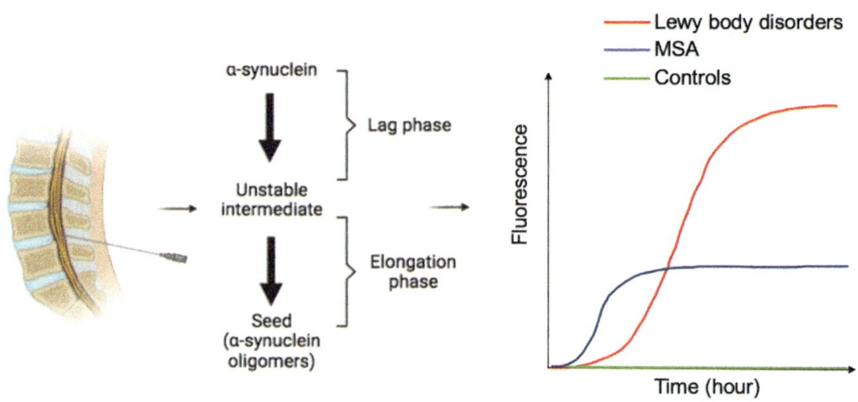

Figure 9.1 Illustrative example of the technique of protein misfolding cyclic amplification assay (PMCA). Samples of CSF from patients with Lewy body disorders (i.e., Parkinson's disease, dementia with Lewy bodies), patients with multiple system atrophy (MSA) or healthy control individuals are subjected to α-syn-protein misfolding cyclic amplification and the extent of aggregation was monitored by Thioflavin T fluorescence. The representative aggregation curves of α-syn in the presence of CSF from patients with Lewy body disorders differ from patients with MSA and healthy controls. *Adapted from Shahnawaz et al. Discriminating α-synuclein strains in Parkinson's disease and multiple system atrophy. Nature. 2020 Feb;578(7794):273–277.*

not in PAF. Therefore, the three synucleinopathies have in common in vivo evidence of central noradrenergic deficiency but differ in extents of central dopaminergic deficiency—prominent in PD and MSA, less apparent in PAF (Goldstein et al., 2021). Looking at postmortem data, the same group confirmed these findings with shared degeneration of locus coeruleus noradrenergic neurons in PD, PAF, and MSA and variable loss of substantia nigra dopaminergic neurons, which was much less pronounced in PAF (Goldstein et al., 2021).

9.1.4 Pathology

The central autonomic network is variably affected in the different synucleinopathies (Coon, Cutsforth-Gregory, & Benarroch, 2018) (Table 9.1). Different interconnected areas of the central autonomic network are involved in MSA and to a lesser extent in Lewy body disorders. These areas include the insular cortex, extended amygdala, hypothalamus, periaqueductal gray, ventrolateral medulla, medullary raphe, the dorsal motor nucleus of the vagus, the nucleus ambiguus, the Onuf nucleus, and sacral

Table 9.1 Key central and peripheral autonomic structures affected in synucleinopathies.

Structure	MSA	PD/DLB	PAF
Hypothalamus	++	+	−
Periaqueductal gray	++	+	−
Ventrolateral medulla	+++	+	−
Medullary raphe	++	+	−
Dorsal motor nucleus of vagus	++	+++	−
Nucleus ambiguus	++	+	−
Pontine micturition center	++	+	−
Sacral preganglionic neurons	+	+	−
Onuf nucleus	++	+	−
Enteric nervous system	?	++	?
Postganglionic sympathetic neurons	+	++	+++
Cholinergic sudomotor	+/−	+	++
Noradrenergic cardiac	+/−	+++	++
Noradrenergic vasomotor	−	+	++
Adrenal gland	−	+	?
Postganglionic parasympathetic neurons	?	++	?
Submandibular gland	?	++	?
Pelvic ganglia	?	+	?
Gonadal tissue	?	+	?

preganglionic neurons (Coon et al., 2018). The involvement of the pontine micturition center is specific to MSA. The insular cortex participates in autonomic control via descending outputs to the amygdala, hypothalamus, and brainstem (Cechetto, 2014). In PD, involvement of the insular cortex has been related to nonmotor symptoms, including autonomic dysfunction and arousal (Christopher et al., 2015). The presence of Lewy bodies in the left posterior insula has been shown to correlate with orthostatic hypotension in PD (Papapetropoulos & Mash, 2007). The anterior insula and anterior cingulate cortex are nodes of the so-called salience network involved in the processing of behaviorally salient information (Seeley, 2019). Activation of the salience network is associated with cardiovagal and sympathetic modulation related to behavioral arousal (Benarroch, 1993). The insula is involved in MSA and hypometabolism of the left insula on brain ^{18}F-fluorodeoxyglucose PET scan is associated with shorter survival in MSA (Grimaldi et al., 2021). Involvement of the hypothalamus has been shown in PD and MSA and may result in thermoregulatory dysfunction, sweating abnormalities, and circadian abnormalities (Langston & Forno, 1978; Swinn et al., 2003). In MSA, the loss of vasopressin neurons in the paraventricular nucleus of the hypothalamus may contribute to orthostatic hypotension (Deguchi et al., 2004; Puritz, Lightman, Wilcox, Forsling, & Bannister, 1983). The loss of C1-group neurons in the rostral ventrolateral medulla also likely contributes to orthostatic hypotension in MSA with the severity of neuronal loss in MSA exceeding that of PD and DLB (Benarroch et al., 1998; Benarroch et al., 2005; Coon et al., 2016). Sleep-related cardiovagal and respiratory dysregulation in MSA may be associated with the loss of GABAergic neurons of the lateral paragigantocellular nucleus in the ventrolateral medulla (Schmeichel et al., 2021). Experimentally, norepinephrine, serotonin, and dopamine have an important role in the control of breathing (Dempsey, Olson, & Skatrud). Experimental evidence also shows that dopamine can affect respiration. This action can be on central structures when the neurotransmitter stimulates breathing or peripheral structures through the carotid body chemoreceptor, where dopamine is inhibitory on respiration. Imbalance between these two effects on respiration may contribute to respiratory dysrhythmia. The loss of serotonergic neurons in the medullary raphe and involvement of the pre-Botzinger complex may contribute to respiratory dysfunction in MSA (Benarroch, Schmeichel, Low, & Parisi, 2004; Presti, Schmeichel, Low, Parisi, & Benarroch, 2014). The dorsal motor nucleus of the vagus is involved in the premotor stages of PD and also in MSA. Loss of cholinergic neurons in the dorsal motor nucleus of the vagus likely contributes to gastrointestinal symptoms in synucleinopathies (Benarroch, Schmeichel, Sandroni, Low, & Parisi, 2006). The dorsal motor nucleus of the vagus prominently controls the esophagus and stomach, leading to esophageal peristalsis, gastric accommodation, and emptying via vasovagal reflexes (Travagli, Hermann, Browning, & Rogers, 2006). In MSA and Lewy body disorders, there is involvement of cholinergic parasympathetic neurons innervating vagal preganglionic neurons located in the nucleus ambiguus that project to cardiac ganglion neurons to elicit bradycardia (Benarroch et al., 2006). Widespread involvement of structures involved in the control of micturition in MSA (i.e., pontine micturition center, Onuf nucleus, sacral preganglionic neurons innervating the detrusor) explains the early and severe bladder dysfunction in MSA (Benarroch, 2003).

Involvement of the enteric nervous system and ganglionic and postganglionic sympathetic and parasympathetic neurons is common in Lewy body disorders and considerably less so in MSA (Table 9.1). Peripheral autonomic involvement seems to occur early in the pathogenetic sequence leading to relatively morbid "body-first" Lewy body diseases (Cersosimo & Benarroch, 2012; Fereshtehnejad et al., 2015; Horsager et al., 2020). Several studies have reported α-synuclein aggregates in the enteric nervous system in PD, which might then be transmitted in a prion-like manner to the central nervous system (Borghammer, 2018; Prusiner et al., 2015). Alpha-synuclein pathology has been demonstrated in the submandibular gland, lower esophagus, stomach, small bowel, colon, and rectum; however, the presence of synuclein pathology in the enteric nervous system is not necessarily associated with neuronal loss (Dickson, 2012; Lamotte et al., 2020). The peripheral sympathetic pathways are variably involved in synucleinopathies. The loss of sympathetic fibers to blood vessels, sweat glands, and erector pili muscles, as shown in PD skin biopsy specimens, causes thermoregulatory abnormalities and impaired vasoconstriction (Isonaka et al., 2019; Isonaka, Sullivan, Jinsmaa, Corrales, & Goldstein, 2018). On the other hand, anhidrosis in MSA results from both preganglionic and postganglionic sudomotor denervation, with more postganglionic dysfunction seen later in the disease (Coon et al., 2017). Schwann cell cytoplasmic inclusions on skin biopsy may be a specific marker of MSA (Donadio et al., 2022). Cardiac noradrenergic denervation can be visualized with ^{123}I-metaiodobenzyl guanidine scintigraphy or 6−^{18}F-dopamine PET scan in patients with PD and DLB (Fig. 9.2). Most MSA patients have biomarkers of intact cardiac sympathetic innervation; however,

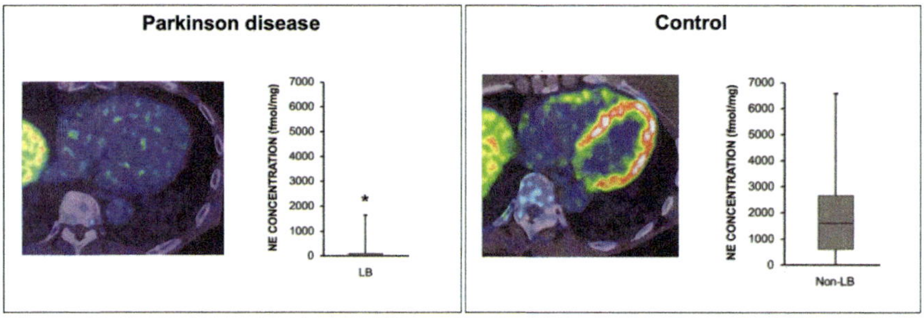

Figure 9.2 Illustrative examples of cardiac 18-fluorodopamine positron emission tomography scanning in a control subject (A) and a patient with Parkinson's disease (B). Analysis of tissue NE content demonstrates profound myocardial NE depletion in Lewy body disorders compared to controls (97% reduction). Images from David S. Goldstein, Autonomic Meidcine Section, National Institutes of Health, Besthesda, MD, USA (reproduced with permission).

rare cases of MSA with neuroimaging evidence of cardiac noradrenergic deficiency have been reported (Cook, Sullivan, Holmes, & Goldstein, 2014; Goldstein, Holmes, Cannon, Eisenhofer, & Kopin, 1997; Lamotte et al., 2020). The contribution of cardiac noradrenergic deficiency to the pathogenesis of cardiovascular autonomic dysfunction in Lewy body disorders, including orthostatic hypotension, is debated. For example, some studies found a correlation between cardiac sympathetic denervation and orthostatic hypotension (Goldstein, Holmes, Dendi, Bruce, & Li, 2002; Shibata, Morita, Shimizu, Takahashi, & Suzuki, 2009), but other studies did not confirm these findings (Haensch, Lerch, Jorg, & Isenmann, 2009; Matsui et al., 2006). The involvement of peripheral parasympathetic structures is best established in PD. The involvement of the submandibular gland in PD may explain the hyposialorrhea in some patients (Beach et al., 2010). Studies using ^{11}C-donepezil PET showed reduced levels of acetylcholine in the intestine, pancreas, and kidneys of patients with PD, suggesting possible parasympathetic denervation in these areas (Gjerloff et al., 2015).

9.1.5 Epidemiology

Parkinson's disease is the most common of the synucleinopathies with a mean incidence of 33.3 per 100,000 person-years, with incidence increasing with age, affecting 2%–3% of the population over the age of 65 (Savica, Grossardt, Bower, Ahlskog, & Rocca, 2016). Dementia with Lewy bodies is the second most common form of dementia, with an incidence of 3.5 cases per 100,000 person-years (Savica, Grossardt, Bower,

Ahlskog, & Rocca, 2013). The annual incidence rate of MSA is approximately 3 per 100,000 population for subjects above age 50 years, while the estimated point prevalence is 4.4 cases per 100,000 population (Bower, Maraganore, McDonnell, & Rocca, 1997). Available data on the clinical incidence and prevalence of MSA may be an underestimate as various pathological studies suggest that MSA might be more common accounting for 4%–5% of those with parkinsonism. The incidence of PAF is unknown (Coon, Singer, et al., 2019). The prevalence of autonomic dysfunction in the synucleinopathies varies between the different diseases. The most commonly reported symptoms of autonomic dysfunction in patients with PD include constipation (33%–48% of patients), urinary problems (37%–63% of patients), orthostatic lightheadedness or prior history of fainting (29%–36% of patients), dysphagia (30% of patients), and hypersialorrhea (33%–46% of patients) (Rodriguez-Blazquez et al., 2021). The prevalence of orthostatic hypotension in PD increases with age and disease duration; however, multiple studies have documented that orthostatic hypotension may appear early in the course of PD (Kim et al., 2016; Strano et al., 2016). The reported estimated prevalence of orthostatic hypotension in PD ranges from 30% to 65% (9, 10); however, symptomatic orthostatic hypotension may be much less common. Patients with DLB frequently have orthostatic symptoms, and neurogenic orthostatic hypotension is detected in 50% of patients (Thaisetthawatkul et al., 2004). Constipation and genitourinary failure are less commonly reported in DLB than in PD and MSA (Thaisetthawatkul et al., 2004). The prevalence of orthostatic hypotension in patients with MSA is 70 to 80%, while urinary symptoms (84%) and constipation (80%) are commonly reported (Coon et al., 2015; Low et al., 2015). Based on current consensus criteria, all patients with PAF have neurogenic orthostatic hypotension, and frequently report urinary symptoms (75%) and gastrointestinal symptoms (83%) (Coon, Singer, et al., 2019).

9.1.6 Clinical presentation

9.1.6.1 Neurogenic orthostatic hypotension

Orthostatic hypotension is defined as an orthostatic decrease of blood pressure within 3 minutes by at least 20 mmHg systolic or 10 mmHg diastolic (Cheshire et al., 2020). In the synucleinopathies, orthostatic hypotension is neurogenic in nature with a blunted heart-rate increase on standing and a ratio delta heart rate/delta systolic blood pressure < 0.5 bpm/mmHg (Norcliffe-Kaufmann et al., 2018). The severity of orthostatic hypotension varies in the different synucleinopathies. In comparison to patients with PD and MSA, the degree of blood pressure drop in DLB is typically moderate (Coon, 2020). Patients with PAF and MSA tend to show severe orthostatic hypotension that is frequently the presenting feature of the disease (Coon, Singer, et al., 2019; Fanciulli & Wenning, 2015). Symptoms of cerebral hypoperfusion in orthostatic hypotension include lightheadedness, dizziness, vision changes, fatigue, weakness, cognitive complaints, and occipital-cervical ("coat hanger") distribution pain. An upright mean systolic blood pressure below 75 mmHg has been associated with symptoms of orthostatic intolerance in PD and this cutoff has a sensitivity of 97% and a specificity of 98% for detecting symptomatic orthostatic hypotension (Palma et al., 2015). Factors that can exacerbate orthostatic hypotension due to redistribution of blood volume and vasodilation of skin vessels include eating (i.e., postprandial hypotension) and exposure to high ambient heat. Postexercise hypotension has been reported in patients with PD with symptoms of orthostatic intolerance (Low, Vichayanrat, Iodice, & Mathias, 2014); however, the effects of baroreflex-cardiovagal and baroreflex-sympathetic noradrenergic impairment during and following aerobic exercise in PD are poorly understood. Hypotension symptoms are often worse in the morning, related to nocturnal diuresis that is frequently augmented in patients with supine hypertension. Supine hypertension is present in approximately half of all patients with PAF, PD, and MSA (Berganzo et al., 2013; Fanciulli et al., 2016). Different types of orthostatic hypotension have been reported in synucleinopathies. Classic orthostatic hypotension is defined by a sustained systolic blood pressure fall ≥ 20 mmHg or diastolic ≥ 10 mmHg within 3 minutes (Cheshire et al., 2020), whereas initial or transient orthostatic hypotension is defined by systolic blood pressure fall ≥ 20 mmHg or diastolic ≥ 10 mmHg resolving within the first minute upon standing or upright tilt table testing (Fanciulli et al., 2020), and delayed orthostatic hypotension is defined by a fall in blood pressure fall in blood pressure meeting the orthostatic hypotension criteria beyond 3 minutes of orthostatic stress (Gibbons & Freeman, 2020). Patients with MSA and PAF typically present with classic orthostatic hypotension, whereas all three types of orthostatic hypotension have been reported in PD and can contribute to orthostatic intolerance and syncope (Fanciulli et al., 2020; Gibbons & Freeman, 2020).

9.1.6.2 Gastrointestinal dysfunction

Clinical features of gastrointestinal dysfunction in synucleinopathies include dysphagia and gastrointestinal dysmotility. Patients with PD and

DLB typically have mild dysphagia that occurs late in the disease course whereas patients with MSA may have early and severe dysphagia (Calandra-Buonaura et al., 2021; Suttrup & Warnecke, 2016). Esophageal abnormalities in patients with PD may include incomplete relaxation of the upper portion of the lower esophageal sphincter, esophageal spasms, and reduced esophageal peristalsis (Suttrup et al., 2017). Clinical manifestations include early satiety with anorexia, nausea, vomiting, abdominal fullness, and bloating. Symptoms may be exacerbated by the effect of levodopa on dopaminergic enteric neurons that additionally slows gastric motility and gastric emptying (Bestetti, Capozza, Lacerenza, Manfredi, & Mancini, 2017). Constipation is frequent in all synucleinopathies. Constipation results from enteric nervous system involvement of the gut and/or degeneration of the dorsal motor nucleus of the vagus leading to slowed colonic transit, rectal/anal dysfunction, or commonly both (Chung & Pfeiffer, 2021).

9.1.6.3 Genitourinary dysfunction

Urinary symptoms are frequent in the synucleinopathies and include increased urinary frequency, urinary urgency, nocturia, and urinary retention. Patients with Lewy body disorders typically have mild or moderate urinary symptoms that are more prominent later in the disease course (Sakakibara et al., 2005). In contrast, patients with MSA have severe urinary dysfunction that typically occurs early in the disease and may be a presenting feature with urge incontinence or urinary retention (Gilman et al., 2008; Low et al., 2015; Wenning et al., 2013). Sexual dysfunction is common in patients with synucleinopathies and typically manifests as erectile dysfunction, ejaculatory failure, and anorgasmia in males. Sexual dysfunction may be a premotor sign of disease in PD and MSA and is frequently reported by patients with PAF. Sexual dysfunction is also reported in female patients with PD and MSA, who report vaginal dryness, decreased libido, and anorgasmia (Coon, Nelson, et al., 2019).

9.1.6.4 Thermoregulatory dysfunction

Thermoregulatory disturbances are likely underrecognized in synucleinopathies. Classically, patients insidiously develop anhidrosis. However, patients may report hyperhidrosis of areas with preserved sweating (compensatory hyperhidrosis) or may report heat intolerance. The severity and pattern of anhidrosis vary in the different synucleinopathies with the lowest degree of anhidrosis typically seen in patients with PD, a moderate degree of anhidrosis seen in patients with DLB, and global anhidrosis typically seen in MSA (often with sparing of the hands) and PAF (Coon, 2020).

9.1.6.5 Survival and autonomic dysfunction in synucleinopathies

More severe autonomic dysfunction has an impact on morbidity and influences survival in synucleinopathies. The presence of orthostatic hypotension has been associated with more rapid disease progression, shorter survival time, cognitive impairment, and falls in PD (De Pablo-Fernandez et al., 2017; Fanciulli et al., 2020; Rascol et al., 2015). Autonomic dysfunction is also a predictor of survival in patients with DLB (Stubendorff, Aarsland, Minthon, & Londos, 2012). Early autonomic involvement is associated with shorter survival in MSA (Low et al., 2015). Specifically, early and severe urinary failure requiring catheterization is associated with a shortened disease course in MSA (Watanabe et al., 2002; Wenning et al., 2013). The prognosis of PAF is related to phenoconversion. This is an active area of research and isolated autonomic failure can be the initial presentation in MSA or Lewy body disorders (Coon et al., 2020). Patients with PAF who phenoconvert to MSA tend to demonstrate more severe autonomic dysfunction, typically with a central pattern on testing (Coon et al., 2020) (Table 9.2).

9.1.7 Differential diagnosis

Nonneurogenic causes of orthostatic hypotension such as hypovolemia, medications (antihypertensive and dopaminergic medications), anemia, and deconditioning often contribute to orthostatic hypotension in synucleinopathies. Other causes of neurogenic orthostatic hypotension may be superimposed such as diabetic autonomic neuropathy or amyloid autonomic neuropathy. Finally, other neurogenic orthostatic hypotension syndromes may fit the current clinical criteria for PAF. Such syndromes include autoimmune autonomic ganglionopathy where the acetylcholine receptor ganglionic antibody is detected in only about 50% of patients (Vernino et al., 2000), hyperadrenergic orthostatic hypotension with high upright plasma norepinephrine (Mar et al., 2015), autoimmunity-associated autonomic failure with sympathetic denervation (Goldstein et al., 2017), and autonomic failure due to CYB561 or CHRNA3 mutations (Shibao et al., 2020; Shibao et al., 2021). Laboratory biomarkers that have the potential to identify the different synucleinopathies include the analysis of α-synuclein oligomers and neurofilament light chain in the CSF (Singer et al., 2020; Singer et al., 2021), cardiac sympathetic

Table 9.2 Features associated with the conversion from pure autonomic failure to other synucleinopathies.

Conversion to multiple system atrophy
Subtle motor symptoms of parkinsonism and/or cerebellar dysfunction, frequent falls
REM sleep behavior disorder
Preserved olfaction
Normal supine norepinephrine levels ($>$ 100 pg/mL)
Preganglionic pattern of anhidrosis
Severe bladder dysfunction with urinary incontinence or retention
Preserved cardiac sympathetic innervation on $^{123-}$MIBG SPECT imaging
MSA pattern of α-synuclein aggregation (lower maximum-ThT-fluorescence) and higher Nfl levels in the cerebrospinal fluid with PMCA
Preserved heart rate increase to tilt
Putaminal/infratentorial abnormalities and increase putaminal/middle cerebellar peduncle diffusivity on MRI
Stridor
Conversion to Parkinson's disease or dementia with Lewy bodies
Subtle motor symptoms of parkinsonism
Visual hallucinations, illusions
REM sleep behavior disorder
Impaired olfaction
Longer duration of illness
Cardiac sympathetic denervation on $^{123-}$MIBG SPECT imaging
Lewy body pattern of α-synuclein aggregation (higher maximum-ThT-fluorescence) and lower Nfl levels in the cerebrospinal fluid with PMCA

MIBG, metaiodobenzylguanidine; *MSA*, multiple system atrophy; *Nfl*, neurofilament light chains; *PMCA*, protein misfolding cyclic amplification; *REM*, rapid-eye-movement; *SPECT*, single-photon emission computed tomography.

neuroimaging (Lenka et al., 2021), and immunofluorescence microscopic methods to visualize alpha-synuclein in sympathetic noradrenergic nerves in skin biopsies (Fig. 9.3) (Donadio, 2019). The most recent consensus criteria for the diagnosis of MSA were published in 2022 (Wenning et al., 2022) (Table 9.3). The 2022 consensus criteria distinguish three clinical categories: clinically established, clinically probable, and prodromal MSA (Wenning et al., 2022). All three clinical diagnostic categories need validation in future studies and possible prodromal MSA is a research category that should be refined with emerging data. Recognizing the usefulness of brain magnetic resonance imaging (MRI) in the diagnosis of MSA, MRI findings classically seen in these patients e.g., pontine and/or cerebellar atrophy, "hot cross bun" sign, putaminal hyperintense rim on T2

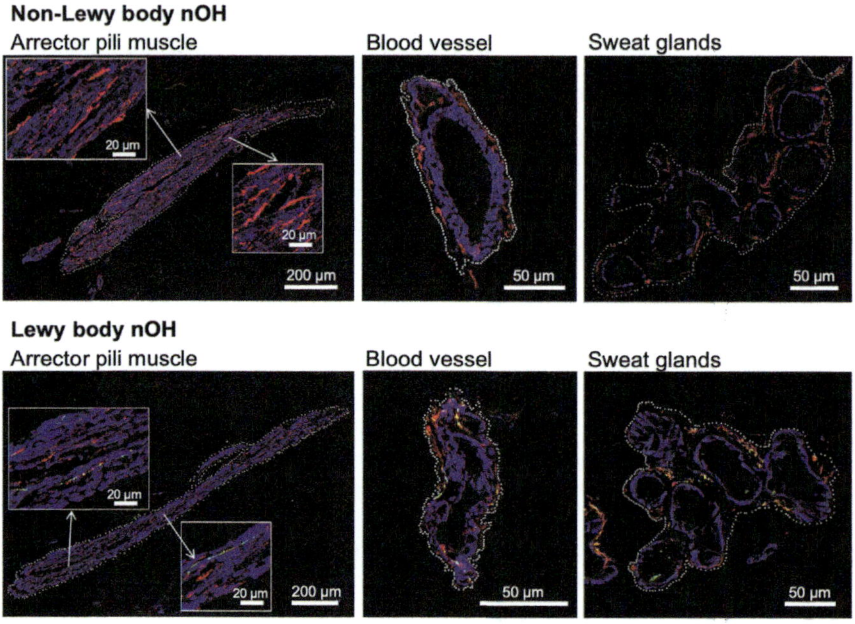

Figure 9.3 Immunofluorescence microscopic images of sympathetic noradrenergically innervated constituents in skin biopsies. Smooth muscle actin (blue) and tyrosine hydroxylase (red) are present in arrector pili muscles, blood vessels, and sweat glands. White dots indicate placement of regions of interest for quantifying α-synuclein (green) and colocalized α-synuclein with tyrosine hydroxylase (yellow). Upper panels show images from patients with non-Lewy body forms of neurogenic orthostatic hypotension and bottom panels images from patients with Lewy body forms of neurogenic orthostatic hypotension. Note α-synuclein deposition in tyrosine hydroxylase-positive fibers in the Lewy body patients. Images from Risa Isonaka, Autonomic Meidcine Section, National Institutes of Health, Besthesda, MD, USA (reproduced with permission).

sequences became required for a diagnosis of clinically established MSA. The most exciting but challenging part of the 2022 diagnostic criteria is the introduction of possible prodromal MSA as a research category. The essential features are either polysomnography-proven rapid eye movement sleep behavior disorder (RBD), or isolated autonomic failure, defined as at least one of urogenital failure with PVR >100 mL or urinary urge incontinence, or neurogenic OH within 10 minutes of standing or HUT, as well as the presence of subtle parkinsonian motor signs or cerebellar signs. This category has limited specificity, especially without other imaging or laboratory biomarkers, and it should only be used for research purposes at this time.

Table 9.3 Diagnostic criteria for clinically established, clinically probable, and possible prodromal multiple system atrophy.

Essential features	**A sporadic, progressive adult (> 30 years) onset disease**		
	Clinically established MSA	**Clinically probable MSA**	**Possible prodromal MSA**
Core clinical features	1. Autonomic dysfunction (at least one required) a. Unexplained voiding difficulties with PVR ≥ 100 mL b. Unexplained urinary urge incontinence c. Neurogenic OH (≥20/10) within 3 minutes of standing or head-up tilt test and at least one 2. Poorly L-dopa-responsive parkinsonism 3. Cerebellar syndrome	At least two of: 1. Autonomic dysfunction defined as (at least one is required): a. Unexplained voiding difficulties with PVR b. Unexplained urinary urge incontinence c. Neurogenic OH (≥20/10) within 10 minutes of standing or head-up tilt test 2. Parkinsonism 3. Cerebellar syndrome	At least one of the following: 1. RBD (polysomnography) 2. Neurogenic OH (≥20/10) within 10 minutes of standing or head-up tilt 3. Urogenital failure (ED in males < 60 years + at least one of unexplained voiding difficulties with PVR >100 mL or unexplained urinary urge incontinence) and at least one 4. Subtle parkinsonism 5. Subtle cerebellar signs
Supportive features[a]	At least two	At least one	NA
MRI marker	At least one	Not required	Not required
Exclusion criteria[b]	Absence	Absence	Absence

[a]Supportive clinical features: rapid progression within 3 years of motor onset, moderate to severe postural instability within 3 years of motor onset, craniocervical dystonia induced or exacerbated by L-dopa in the absence of limb dyskinesia, severe speech impairment within 3 years of motor onset, severe dysphagia within 3 years of motor onset, unexplained Babinski sign, jerky myoclonic postural or kinetic tremor, postural deformities, stridor, inspiratory sighs, cold discolored hands and feet, erectile dysfunction (below age of 60 years for clinically probable MSA), pathologic laughter or crying.

[b]Exclusion criteria: Substantial and persistent beneficial response to dopaminergic medications, unexplained anosmia on olfactory testing, fluctuating cognition with pronounced variation in attention and alertness and early decline in visuoperceptual abilities, recurrent visual hallucinations not induced by drugs within 3 years of disease onset, dementia according to DSM-V within 3 years of disease onset, downgaze supranuclear palsy or slowing of vertical saccades, brain MRI findings suggestive of an alternative diagnosis (e.g., PSP, multiple sclerosis, vascular parkinsonism, symptomatic cerebellar disease, etc.), documentation of an alternative condition (MSA look-alike, including genetic or symptomatic ataxia and parkinsonism) known to produce autonomic failure, ataxia, or parkinsonism and plausibly connected to the patient's symptoms.

9.1.8 Autonomic testing

9.1.8.1 Autonomic reflex screen

Standardized autonomic function testing with autonomic reflex screen tests postganglionic sympathetic sudomotor, cardiovagal, and cardiovascular adrenergic functions (Cheshire et al., 2020). Analysis of postganglionic sympathetic sudomotor function reveals variable patterns in the synucleinopathies (Fig. 9.4). A distal- or length-dependent impairment is frequently seen in Lewy body disorders. Although it is considered to be a predominantly central or preganglionic thermoregulatory disorder, postganglionic sympathetic sudomotor axon reflex testing can be abnormal in about 60% of patients with MSA (Coon et al., 2017). In a prospective study of 43 patients with MSA and 57 patients with PD, sweating output

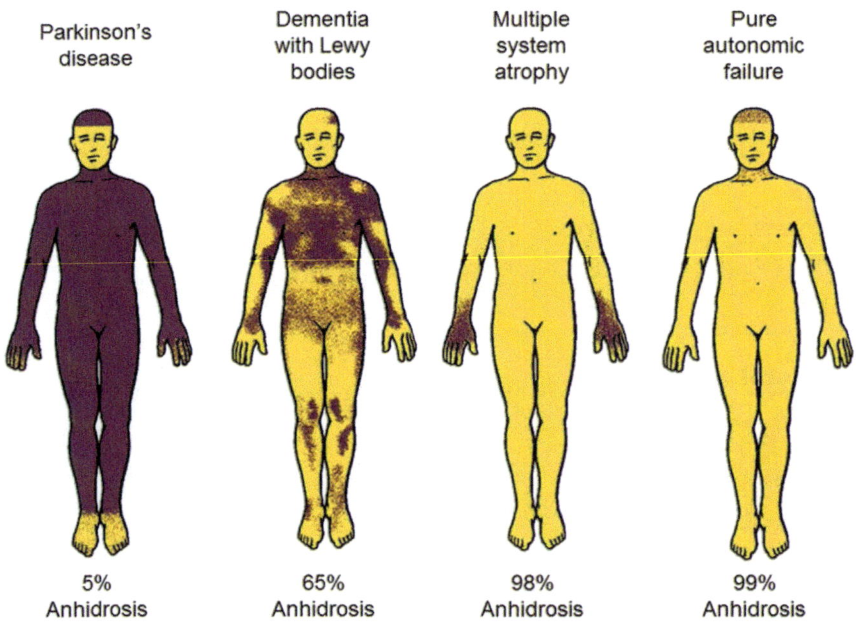

Figure 9.4 Patterns of anhidrosis on thermoregulatory sweat test in the synucleinopathies. Anhidrosis is demonstrated on thermoregulatory sweat test using an indicator powder that turns purple when exposed to sweat. Percentage of anhidrosis is calculated based on the number of yellow pixels compared with overall surface area. Varying degrees of anhidrosis are seen in the different synucleinopathies, with the lowest degree of anhidrosis typically seen in patients with Parkinson's disease, a moderate degree of anhidrosis seen in patients with dementia with Lewy bodies, and global anhidrosis typically seen in multiple system atrophy (often with sparing of the hands) and pure autonomic failure.

assessed by the dynamic sweat test and intraepidermal, pilomotor, and sudomotor nerve densities were lower in both groups compared to controls but with greater impairment in MSA (Provitera et al., 2022). This study suggests greater postganglionic sympathetic sudomotor impairment in MSA compared to PD. Possible explanations include transsynaptic postganglionic degeneration of sudomotor fibers or coexistence of Lewy body pathology in MSA. This study also highlights the limitations of sudomotor testing in the differential diagnosis between MSA and PD. Moreover, the value of sudomotor testing as a biomarker of Lewy body versus non-Lewy body pathology in patients with PAF is unknown. Cardiovagal impairment with abnormal heart rate responses to deep breathing, Valsalva ratio, and cardiovagal baroreflex sensitivity is frequently seen in synucleinopathies (Fig. 9.5). In patients with neurogenic orthostatic hypotension, the heart rate increase more in those with MSA compared to patients with Lewy body disorders suggesting relative preservation of baroreflex-mediated cardiovagal function in MSA; however,

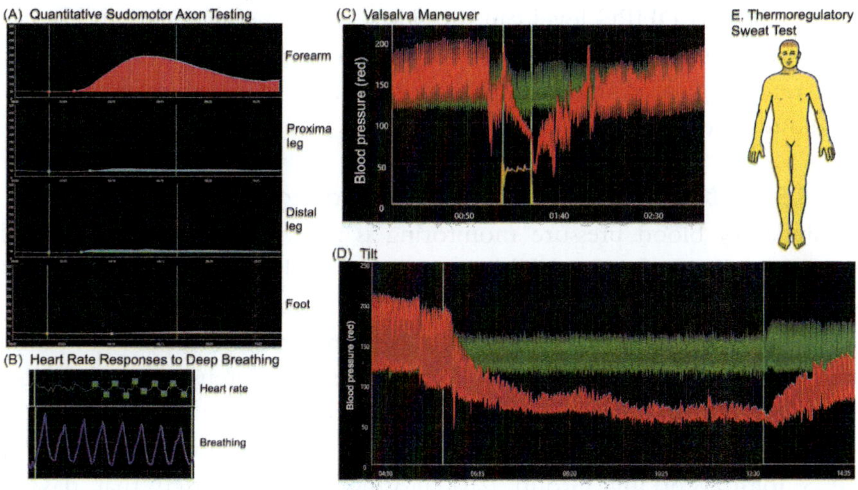

Figure 9.5 Autonomic reflex screen results in an illustrative case with pure autonomic failure. There is postganglionic sympathetic sudomotor impairment with reduced sweat volumes in the leg (Panel A). There is moderate to severe cardiovagal impairment with reduced heart rate responses to deep breathing (Panel B) as well as reduced Valsalva ratio (Panel C) and baroreflex sensitivity (Panel C). There is evidence of severe cardiovascular noradrenergic impairment indicated by the absence of late Phase II, prolonged blood pressure recovery time, and absent Phase IV during the Valsalva maneuver (Panel C) as well as evidence of neurogenic orthostatic hypotension during head-up tilt (Panel D). The thermoregulatory sweat test reveals global anhidrosis (Panel E).

there is considerable overlap in values between MSA and Lewy body synucleinopathies (Norcliffe-Kaufmann et al., 2018). There is evidence of noradrenergic impairment of the baroreflex producing characteristic abnormalities on beat-to-beat blood pressure associated with the Valsalva maneuver and orthostatic hypotension during head-up tilt (Cheshire et al., 2020). Although widespread autonomic failure on testing suggests a diagnosis of MSA, these abnormalities are not specific to MSA and can be seen in patients with PAF and in about 30% of patients with PD.

9.1.8.2 Fractionated plasma catecholamines

Analysis of plasma norepinephrine and its metabolites in the supine and upright positions is helpful in the diagnosis of synucleinopathies and risk stratification of patients with isolated orthostatic hypotension. Supine plasma norepinephrine levels under 100 pg/mL are common in PAF, whereas supine and standing plasma norepinephrine levels are usually normal in MSA but do not increase appropriately with standing (< 60% increment). Normal supine plasma norepinephrine levels can also be seen in PD. Plasma DHPG level can provide a neurochemical clue to the correct diagnosis. Plasma DHPG level is normal in MSA whereas it is low in PD reflecting decreased norepinephrine stores in sympathetic terminals (Goldstein et al., 2003).

9.1.8.3 Ambulatory blood pressure monitoring

Ambulatory blood pressure monitoring is a simple and widely available test that allows for "real-life" assessment while patients perform daily activities. Ambulatory blood pressure monitoring can determine diurnal and nocturnal blood pressure patterns that are characteristics of autonomic failure such as supine hypertension, postural hypotension, labile blood pressure, and nondipping of nocturnal blood pressure; it also allows for the evaluation of the impact of antihypertensive and pressor medications (JCS Joint Working Group, 2012). One study reported that daytime blood pressure variability was a good screening tool to identify patients with autonomic failure from controls (Lodhi et al., 2019). In our experience, parameters derived from ambulatory blood pressure monitoring, including standard deviations of blood pressure and heart rate, reflect autonomic function across different syndromes of orthostatic intolerance (unpublished data). Patients with synucleinopathies with autonomic failure have greater blood pressure variability and lesser heart rate variability over a 24-hour period. This approach may lead to early identification of patients with autonomic failure in the primary care setting where patients with orthostatic intolerance are often first evaluated.

9.1.8.4 Genitourinary testing and gastrointestinal motility studies

Urodynamic investigations in MSA reveal vesical detrusor overactivity, absence of detrusor-sphincter coordination, and atonic bladder (Sakakibara et al., 2019). Postvoid residual measurement greater than 100 mL should raise suspicion for the diagnosis of MSA (Yamamoto et al., 2017). Measurement of rectal sphincter tone and rectal manometry demonstrate loss of anal-sphincter relaxation and decreased anal tone in MSA, whereas electromyogram of the urethral and anal sphincters shows evidence of denervation and reinnervation (Fanciulli & Wenning, 2015). Objective evidence of esophageal dysmotility, delayed gastric emptying, and slow colonic transit can be seen in patients with synucleinopathies (Claus et al., 2018; Ivan et al., 2021). Esophageal abnormalities in patients with PD may include incomplete relaxation of the upper portion of the lower esophageal sphincter, esophageal spasms, and reduced esophageal peristalsis (Suttrup et al., 2017). The prognostic value of gastrointestinal motility studies in synucleinopathies requires further studies.

9.1.9 Treatment

9.1.9.1 Multidisciplinary approach

A multidisciplinary and individualized care model is recommended to optimally address the various symptoms of autonomic dysfunction in synucleinopathies. Optimal care may require the involvement of the patient's primary care physician and multiple specialties including autonomic disorders, urology, sleep medicine, physical therapy, speech therapy, occupational therapy, social work, gastroenterology, palliative care, among others (Coon et al., 2022).

9.1.9.2 Symptom management

Levodopa is a precursor of dopamine. In the treatment of PD, DLB, and MSA with parkinsonian features, levodopa is combined with an inhibitor of L-aromatic-acid-decarboxylase, such as carbidopa, which does not penetrate the blood-brain barrier. Carbidopa/levodopa remains the treatment of choice to treat the parkinsonian manifestations of PD and DLB but also MSA (Coon & Ahlskog, 2021; Poewe et al., 2017). In the parkinsonian subtype of MSA, transient response to levodopa is observed in 40 to 50% of patients with improvement of akinesia and rigidity (Martin et al.,

2020). The effect of levodopa on cardiovascular autonomic function in synucleinopathies is not fully understood. Orthostatic hypotension has been documented as a potential side effect of levodopa in different studies (Bouhaddi et al., 2004; Senard et al., 1997); however, other studies failed to show any clear direct relationship (Goldstein et al., 2005; Jost et al., 2020; Perez-Lloret et al., 2012). The lack of detailed investigation of baroreflex-cardiovagal and baroreflex-sympathetic noradrenergic functions and the fact that most studies did not test the same patients on and off levodopa are important limitations of prior studies (Bouhaddi et al., 2004; Goldstein et al., 2005; Jost et al., 2020; Perez-Lloret et al., 2012; Senard et al., 1997). In patients with parkinsonian features and orthostatic hypotension, it is recommended to titrate the dose of levodopa slowly to minimize and monitor potential side effects such as hypotension or dyskinesias. The use of midodrine or droxidopa with or without pyridostigmine can be necessary in cases of moderate to severe orthostatic hypotension.

Education of the patients and caregivers about autonomic dysfunction is crucial. They should be informed about the mechanisms of neurogenic orthostatic hypotension, blood pressure fluctuations, and the effect of diet and posture on symptoms. General recommendations include getting up slowly, urinating in sitting position for men, avoiding exposure to hot and humid environments, eating frequent small snack-like meals, and keeping a symptom and blood pressure diary (Palma & Kaufmann, 2020). It is important to eliminate medications that can contribute to orthostatic hypotension under the guidance of the patient's primary care physician. Nonpharmacological management of orthostatic hypotension includes increasing water intake to at least 2 L/day, increasing salt intake to 5–10 g daily, using compression garments such as compression stockings or an abdominal binder, and exercise regularly focusing on endurance exercise and strengthening of the core musculature (Chelimsky & Chelimsky, 2020).

Pharmacological management of neurogenic orthostatic hypotension includes sympathomimetic such as midodrine, the norepinephrine precursor droxidopa, and mineralocorticoid of fludrocortisone, which increases intravascular volume (Fig. 9.6) (Park, Okamoto, Shibao, & Biaggioni, 2020). Clinical trials are investigating norepinephrine transporter inhibitors in the symptomatic management of orthostatic hypotension in synucleinopathies. Negative results were reported in one randomized, double-blind, placebo-controlled study that investigated the pressor effect of ampreloxetine for four weeks compared to placebo in patients with symptomatic neurogenic

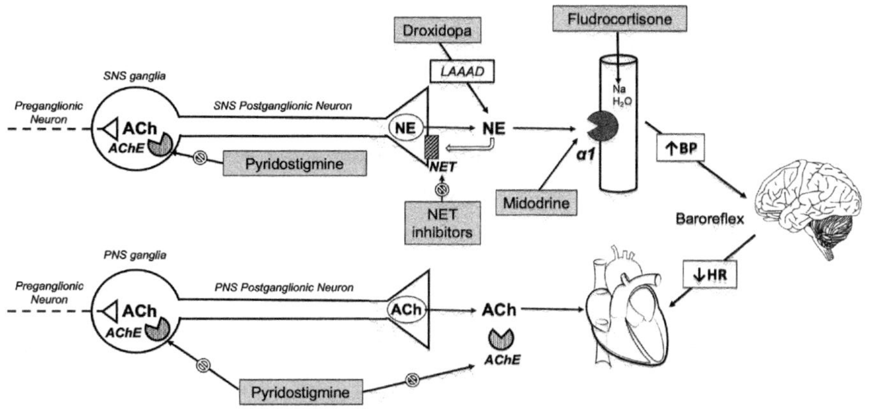

Figure 9.6 Pharmacological treatments of neurogenic orthostatic hypotension. Pyridostigmine inhibits acetylcholinesterase (AChE) and enhances both sympathetic ganglionic and parasympathetic ganglionic and postganglionic functions; fludrocortisone, a mineralocorticoid, promotes salt and water retention in blood vessels; midodrine, an alpha-1 (α1) adrenoreceptor agonist promoting vasoconstriction; droxidopa, a precursor of norepinephrine; and norepinephrine transporter (NET) inhibitors preventing the reuptake of norepinephrine (NE) from the synaptic cleft. *Ach*, acetylcholine; *AChE*, acetylcholinesterase; *BP*, blood pressure; *H_2O*, water; *HR*, heart rate; *LAAAD*, Aromatic L-amino acid decarboxylase; *Na*, sodium; *NE*, norepinephrine; *NET*, norepinephrine transporter; *PNS*, parasympathetic nervous system; *SNS*, sympathetic nervous system.

orthostatic hypotension (the primary outcome was the first item of the orthostatic hypotension symptom assessment). Exacerbation of supine hypertension is a potential side effect of therapies for orthostatic hypotension and patients should monitor supine blood pressure and raise the head of the bed with counseling to avoid total recumbence during the daytime. Short-acting antihypertensive agents may be considered at bedtime in cases of severe supine hypertension (Shibao & Biaggioni, 2020). Squair and colleagues implanted an epidural electrical stimulator in the thoracic spinal cord of a 48-year-old woman with MSA-P and orthostatic hypotension with frequent syncope (3x weekly) and who had failed midodrine and fludrocortisone (Squair et al., 2022). The neurostimulator was designed to deliver stimulation over the dorsal root entry zones of the lower thoracic spinal cord. It was implanted via T10-11 laminectomy under general anesthesia. Stimulation triggered systolic blood pressurepressor responses up to 20 mmHg, which increased linearly with the amplitude of stimulation. The stimulator included a three-axis accelerometer to detect body position and thus modulated its activity according to the patient's posture. Reported side effects included lower back discomfort and paresthesias if the stimulation continued beyond 5 minutes. The patient also underwent a period of autonomic rehabilitation for 3 months. At home, orthostatic intolerance was significantly improved and the patient reported almost no syncope when she was in an upright position and improved ambulation (Squair et al., 2022). Implantable neurostimulators may eventually become part of clinical autonomic practice for selected patients.

Patients with neurogenic bladder symptoms should be screened regularly for urinary tract infections. Detrusor overactivity can be treated with antimuscarinic agents; however, anticholinergic side effects must be monitored and botulinum toxin injections in the detrusor muscle may be beneficial in some patients, particularly early in the disease when there is no evidence of significant retention (Coon & Ahlskog, 2021). Clean intermittent self-catheterization is the first-line therapy for urinary retention with postvoid residual volumes above 100 mL. Unfortunately, lack of manual dexterity with parkinsonism or cerebellar appendicular ataxia may make this impossible and there is a risk of urethral ulceration long term such that chronic indwelling catheterization or suprapubic indwelling catheterization may be required in MSA. Nocturia can be alleviated with intranasal desmopressin at bedtime in patients with MSA.

A phosphodiesterase type 5 inhibitor such as sildenafil is efficacious in the treatment of erectile dysfunction in men; however, it can unmask or worsen orthostatic hypotension and should be used with caution. Alternative approaches include intracavernous injection of vasodilatory prostaglandins or penile implants (Bronner & Vodusek, 2011).

Constipation may be difficult to manage in some patients and lead to overflow fecal incontinence. Treatment is recommended with a high-fiber diet, bulk laxatives, or osmotic laxatives such as polyethylene glycol, suppositories, or enema (Seppi et al., 2019). Severe dysphagia and its complications should be anticipated by alteration of the diet. Surgical intervention with gastrostomy may be indicated in patients with PD, whereas it is rarely indicated in MSA given the overall prognosis of the disease. Involvement of palliative care medicine can be helpful for patients and caregivers.

The management of respiratory abnormalities in MSA is symptomatic and may be inadvisable in the later stages of the disease. Beyond general measures, such as avoidance of sedatives and alcohol or hypnotic drugs, which further depress respiratory neuron activity, noninvasive respiratory support with continuous positive airway pressure and invasive ventilation with tracheostomy to promote airway patency have been advocated with varying success (Cortelli et al., 2019). Of the two treatment modalities, tracheostomy may offer a survival benefit in patients with MSA (Nishida, Sakashita, Yamasaki, & Futamura, 2022).

Finally, anhidrosis can be associated with heat intolerance, and conservative management with cooling vests, spray fans, and adequate intake of water are recommended.

9.1.9.3 Disease-modifying approach

There is currently no available disease-modifying therapy to treat autonomic dysfunction in synucleinopathies. Many clinical trials have yielded negative results in Lewy body disorders and MSA. Because of its possible role in the pathogenesis of MSA, α-synuclein is currently one of the targets for the development of potential neuroprotective strategies. Because of the major impact on patients' quality of life, there is an urgent need for studies investigating symptomatic treatments for autonomic symptoms as well as nonpharmacological strategies until a disease-modifying approach becomes available.

9.1.10 Conclusion

Autonomic dysfunction is characteristic of the synucleinopathies. Peripheral or postganglionic involvement is prominent in the Lewy body disorders of PD, DLB, and PAF. In contrast, central or preganglionic involvement is characteristic of MSA. Nevertheless, central autonomic structures contributing to autonomic failure are also involved in PD and DLB and rare cases of MSA have evidence of postganglionic cardiac sympathetic noradrenergic deficiency. Early and severe autonomic failure is classically seen in MSA and PAF, whereas DLB tends to manifest an intermittent degree of autonomic failure, and PD manifests with a more mild-to-moderate severity later in the disease stage. Nevertheless, cardiovascular autonomic function tests performed in patients with synucleinopathies have limited value in discriminating between the different synuclein disorders because they can all feature prominent baroreflex-cardiovagal and cardiovascular noradrenergic failure (Leys et al., 2020). The severity of autonomic dysfunction has implications on prognosis and survival. Various forms of autonomic failure influence the quality of life in patients with synucleinopathies and treatment is recommended to improve autonomic symptoms. Further research will focus on the identification of subjects at risk for developing MSA or a Lewy body disorder. Further research is also needed to validate biomarkers of different synucleinopathies and the analysis of the pattern of protein misfolding cyclic amplification kinetics of alpha-synuclein in CSF or skin biopsies is promising in this regard.

9.2 Progressive autonomic failure—others

9.2.1 Human T-cell leukemia virus (HTLV) and autonomic failure

Chronic neurodegenerative disease associated with human T-cell leukemia virus (HTLV) type II has been recognized for the first time in 1992 amongst two Native American patients in New Mexico (Hjelle et al., 1992). Autonomic testing revealed prolonged pupil cycle time and lack of autonomic modulation of blood pressure (paradoxical increase with minor abdominal compression). These patients had spastic paraparesis without sphincter involvement and cerebellar signs resembling patients with MSA, and a myeloneuropathy. MRI scan showed cortical and cerebellar atrophy (Fig. 9.7) but the cervical and thoracic spinal cord were normal on MRI. Thus, autonomic failure may be associated with viral infection of the central nervous system. Another case series of 20 patients with human T-cell

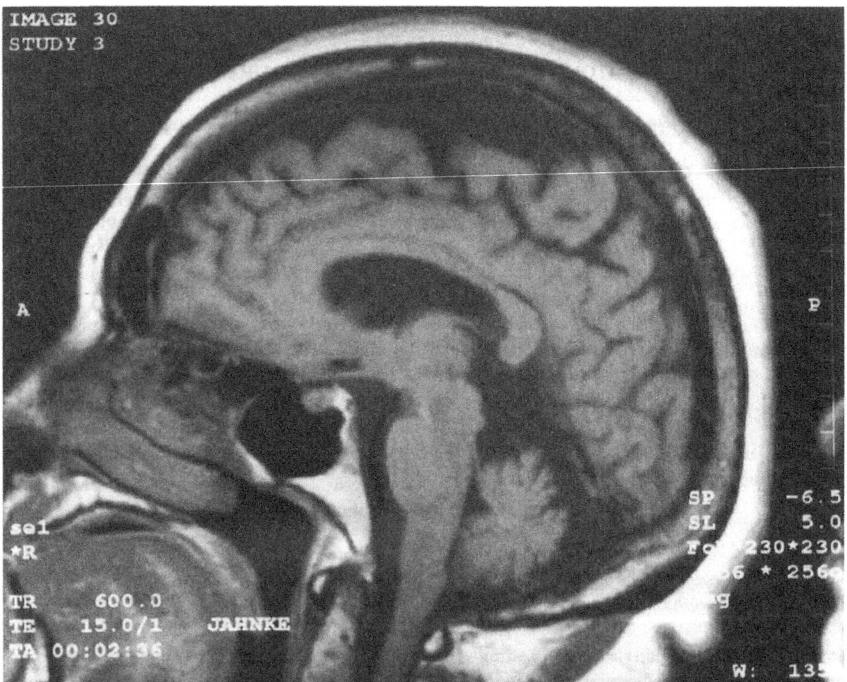

Figure 9.7 MRI showing severe cortical and cerebellar atrophy. The patient had clinically OPCA-MSA with autonomic failure, symptoms began at least 40 years before the scan was performed and pure HTLV-II infection was documented.

lymphotropic virus type I—associated myelopathy/tropical spastic paraparesis demonstrated the high frequency of autonomic dysfunction in these patients with prominent impairment of cardiovascular and sweat control (Alamy, Menezes, Leite, Nascimento, & Araújo, 2001). Postural hypotension is a common feature of human T-cell lymphotropic virus type I—associated myelopathy/tropical spastic paraparesis and should always be investigated and treated symptomatically. On pathology, there is involvement of the autonomic nervous system in the spinal cord with degeneration of neurons of the intermediolateral column and of Onuf's nucleus have been described (Iwasaki, 1990). Neuronal loss was more pronounced in the lateral and anterior columns of the spinal cord and inflammatory infiltrates were more intense in the lower thoracic cord and tended to decrease rostrally and caudally (Iwasaki, 1990). Kuriyama and colleagues demonstrated that the low to high-frequency ratio of heart rate variability was significantly lower in patients with human T-cell lymphotropic virus type I—associated myelopathy/tropical spastic paraparesis with thoracic cord atrophy (Kuriyama et al., 2009).

9.2.2 Genetic forms of autonomic failure

9.2.2.1 Dopamine-ß-hydroxylase (DBH) deficiency

Dopamine-ß-hydroxylase (DBH) converts dopamine to norepinephrine (Fig. 9.8). This deficiency is an autosomal recessive neurometabolic disorder. It is possible to recognize this rare disease in infants presenting with delayed eye opening, hypoglycemia, hypothermia and/or hypotension. A simple and definite assay of plasma norepinephrine and dopamine can clinch the diagnosis at this early stage and help management with appropriate therapy now available. DBH-deficiency leads to undetectable levels of norepinephrine and epinephrine and increased levels of dopamine in the central and autonomous nervous system and peripheral organs (Garland & Biaggioni, 1993).

The description of a 42-year-old man with a lifelong history of orthostatic intolerance is illustrative of the history (Biaggioni, Goldstein, Atkinson, & Robertson, 1990). This patient had a negative family history, bilateral ptosis, and frequent fainting spells, which during childhood were worsened by exercise and treated as epileptic. Though sexual maturation was normal, sexual function was characterized by retrograde ejaculation, and difficulty in maintaining an erection. Mental and physical development were otherwise entirely normal. Orthostatic symptoms became more incapacitating after heavy meals or alcohol. Hot environments

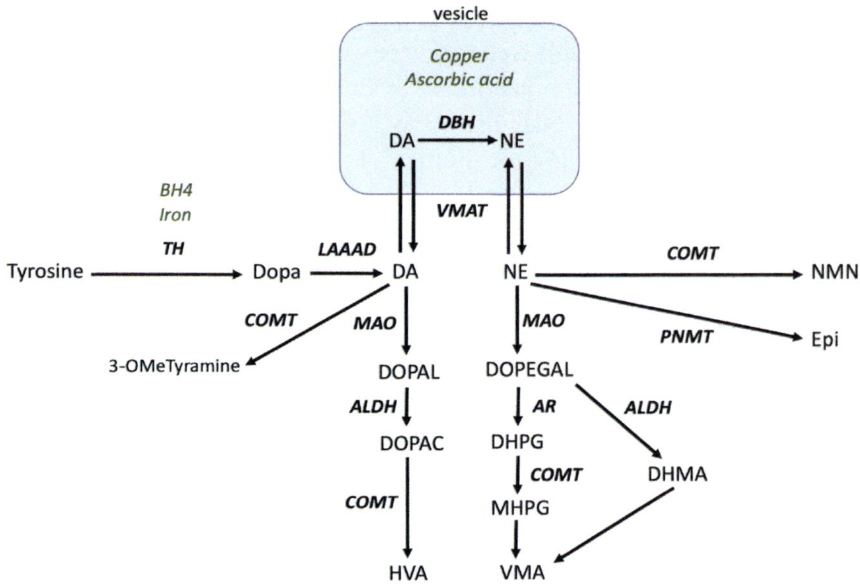

Figure 9.8 Catecholamine metabolism. *AR*, aldehyde reductase; *ALDH*, aldehyde dehydrogenase; *BH4*, tetrahydrobiopterin; *COMT*, catechol-O-methyltransferase; *DA*, dopamine; *DHPG*, dihydroxyphenylglycol; *DHMA*, 3,4-dihydroxymandelic acid; *Dopa*, 3,4-dihydroxyphénylalanine; *DOPAC*, 3,4-dihydroxyphenylacetic acid; *DOPAL*, 3,4-dihydroxyphenylacetaldehyde; *DOPEGAL*, 3,4-dihydroxyphenylglycolaldehyde; *HVA*, homovanillic acid; *DBH*, dopamine beta-hydroxylase; *LAAAD*, L-aromatic amino acid decarboxylase; *MAO*, monoamine oxidase; *MHPG*, 3-methoxy-4-hydroxyphenylglycol; *NMN*, normetanephrine; *PNMT*, phenyl ethanolamine N-methyltransferase; *TH*, tyrosine hydroxylase; *VMA*, vanillylmandelic acid; *VMAT*, vesicle monoamine transporter.

causing vasodilatation contributed to the worsening of the symptoms, as did micturition, also associated with vasodilatation. Orthostatic hypotension, documented at age 33, was treated unsuccessfully with fludrocortisone, pseudoephedrine, indomethacin, and tranylcypromine. The monoamine oxidase inhibitor produced paranoid thinking and tonic-clonic seizures with postictal symptoms, which occurred while assuming the upright posture. On neurologic examination, the only signs were postural hypotension and blepharoptosis. Pupils remained 2 mm in diameter, even in dim light, but were normally reactive, and a slight reduction in tendon reflexes was observed. The rest of the examination was entirely normal, including electroencephalography. Special autonomic tests showed normal sinus arrhythmia, indicating intact parasympathetic innervation of the heart, and an absent overshoot after the Valsalva maneuver, suggesting impaired reflex vasoconstriction. Beta-blockade did not affect resting heart rate, suggesting diminished sympathetic drive, but atropine caused a marked increase in heart rate, supporting the intactness of parasympathetic innervation to the heart. Tyramine had no effect on blood pressure, heart rate, or plasma norepinephrine levels, but increased plasma dopamine by 37%. CSF and plasma dopamine levels at rest were also markedly elevated. In addition, phenylephrine, 0.5% instillation into the conjunctival sacs, produced mydriasis, indicating noradrenergic hypersensitivity of the iris (Biaggioni et al., 1990).

These features are characteristic of a selective failure of noradrenergic function, with preservation of sympathetic cholinergic and parasympathetic function. Biochemically, the most important feature is the virtual absence of plasma and CSF norepinephrine and epinephrine and their metabolites, and a marked increase in plasma and CSF levels of dopa and dopamine and their metabolites (Garland & Biaggioni, 1993). Adrenoreceptors are markedly hypersensitive because of the absence of norepinephrine, and since the administration of tyramine increased plasma dopamine and not norepinephrine, the clinical syndrome and biochemical abnormalities are consistent with the proposition that instead of norepinephrine, dopamine is present in presynaptic sympathetic terminals. The virtual absence of 3,4-dihydroxylphenylglycol supports this interpretation. The administration of D,L-DOPS (droxidopa), a drug endogenously converted into norepinephrine, increases blood pressure and plasma norepinephrine levels. Of interest is also the possibility that the increased dopamine release from sympathetic terminals contributes to orthostatic hypotension since a reduction in dopamine production by metyrosine was significantly related to an increase in blood pressure (Biaggioni & Robertson, 1987).

The syndrome of congenital dopamine ß-hydroxylase deficiency is an experiment of nature that allows important deductions about the physiologic role of various catecholamines. The depletion of norepinephrine results in severe orthostatic hypotension. The impaired coinnervation by noradrenergic fibers of motor fibers may be responsible for the mild ptosis of eyelids and weakness of the facial musculature, including muscle hypotonia and decreased reflexes. The occurrence of spontaneous hypoglycemia with bursts of hyperinsulinism probably results from lack of epinephrine due to the failure of the adrenal medulla. The excessive production of dopamine can account for the reduced REM sleep periods and increase in slow-wave sleep. This also may underlie the hypoprolactinemia and the sodium wasting through the kidney in spite of low arterial pressures. The relative deficit of norepinephrine and epinephrine, when compared to the dopamine excess in the production of these clinical abnormalities is further supported by the evidence from substitution of the missing catecholamines by the administration of an alternate substrate for catecholamine synthesis that bypasses the dopamine ß-hydroxylase deficiency and the success of D,L-threodihydroxyphenylserine that can be decarboxylated to form norepinephrine in the treatment and maintenance of blood pressure of these rare patients.

9.2.2.2 Congenital absence of norepinephrine due to CYB561 mutations

Cytochrome b561 (CYB561) generates ascorbic acid, a cofactor in the enzymatic conversion of dopamine to norepinephrine by dopamine β-hydroxylase. Shibao et al. (2020) described 4 patients with lifelong orthostatic hypotension with CYB561 mutations. All four patients had normal birth and mental development. Three patients had episodes of hypoglycemia and all four patients had profound orthostatic hypotension. Ptosis was mild or questionable. One patient had retrograde ejaculation and three patients were treated for possible seizures in early childhood (unclear if these episodes were epileptic, hypoglycemic seizures, or the tonic-clonic movements that occur during hypotensive syncope). Droxidopa 100 mg three times a day improved orthostatic hypotension and increased plasma norepinephrine and DHPG levels (levels undetectable at baseline). Contrary to DBH deficiency, dopamine does not accumulate in CYB561 deficiency. The authors speculate that the presence of ascorbic acid within noradrenergic vesicles provides antioxidant capacity, and its absence in CYB561 deficiency leads to degradation of DOPA and dopamine, preventing their accumulation (Shibao et al., 2020).

9.2.2.3 Familial autonomic ganglionopathy caused by rare CHRNA3 genetic variants

The CHRNA3 subunit of the nicotinic acetylcholine receptor (nAChRα3) plays a crucial role in the autonomic function of multiple organs and is primarily located in the autonomic ganglia. Shibao et al. (2021) described a genetic disease with familial autonomic ganglionopathy caused by compound heterozygous and homozygous variants in the CHRNA3 gene that encodes the nAChRα3 subunit. The authors reported three individuals from two unrelated families with diffuse panautonomic failure characterized by severe neurogenic orthostatic hypotension, constipation, and miosis with poorly reactive pupils and peripheral radial transillumination defects in the dilator muscle area. The neurogenic orthostatic hypotension was not symptomatic until young adulthood in two patients, whereas the other patient had a much more severe condition with symptoms during infancy and her disease was also associated with intellectual impairment. All patients had impaired sympathetic vasoconstrictor activity on autonomic testing and parasympathetic nerves were also affected. Supine norepinephrine and DHPG levels were low and failed to increase while upright and there was evidence of hypersensitivity of α1 adrenoreceptors.

9.2.2.4 Other genetic disorders associated with autonomic failure

Autonomic failure has been associated with other genetic diseases including tetrahydrobiopterin deficiency, Menkes disease, norepinephrine transporter deficiency, aromatic l-amino acid decarboxylase deficiency, phenylketonuria, and monoamine oxidase deficiency. In Menkes disease, there is a mutation of the gene encoding a form of copper ATPase and this impairs incorporation of copper in DBH (Taneja & Robertson, 2003). Infants with Menkes disease have a loss of developmental milestones, hypotonia, seizures, and failure to thrive at or after 2–3 months of age. The diagnosis is suspected when infants exhibit typical neurological changes associated with characteristic changes of the hair (short, sparse, coarse, twisted, often lightly pigmented) and face (jowly appearance). Temperature instability and hypoglycemia in the neonatal period may be indicators of autonomic dysfunction.

9.2.2.5 Familial dysautonomia (see chapter on peripheral autonomic neuropathies)

9.2.2.5.1 Waardenburg's syndrome

This syndrome is a group of genetic conditions caused by congenital ectodermal germ-layer defects. It is inherited in an autosomal dominant form, with variable phenotypic expressivity (Ahmed jan, Mui, & Masood, 2021; Schweitzer & Clack, 1984). Several different gene mutations (insertion, deletion, frameshifts, missense, and nonsense mutations) in different genes that affect the function of the neural crest can cause Waardenburg syndrome. The syndrome is characterized by dystopia canthorum, broad nasal root, hypoplasia of the medial eyebrows, heterochromia iridis, white forelock and congenital deafness. Many additional features have been reported, which include meningocele, atresia of the esophagus, and

Hirschsprung's disease. In addition, a case with anal atresia, esophageal atresia, and tracheal esophageal fistula has also been described (Nutman, Steinherz, Sivan, & Goodman, 1986). The disease is found in many groups. An Inuit (Eskimo) boy with this disease had, in addition, vascular abnormalities, including left pulmonary artery stenosis, ocular ptosis, and unilateral duplication of the renal collecting system. On histologic examination, hypoganglionosis, hyperganglionosis, and ectopic ganglia in the lamina propria (neuronal colonic dysplasia) were found in the rectum. The hypopigmented skin was devoid of melanocytes, whereas the hyperpigmented skin patches contained melanin throughout the basal layer, but the melanocytes were not evenly distributed (Kaplan & de Chaderévian, 1988). Another patient with the syndrome had bilateral sectoral iris heterochromia and fundus bicolor. Complete iris heterochromia also is found in some patients. In three South African families of Afrikaner descent, the syndrome has been tracked back for 12 generations. The earliest known family member with this syndrome was born in 1842. Every case of Waardenburg syndrome does not present with all clinical features and four types have been identified based on genetic and clinical criteria (Ahmed jan et al., 2021).

This rare inherited disorder shows many features of autonomic dysfunction involving the eye, ear, and enteric nervous system. However, no functional autonomic tests have been performed on such patients, and the disease is still in the early descriptive state. Little pathogenetic knowledge exists, but the disorder result from a failure of migration of neural crest cells.

References

Ahmed jan, N., Mui, R. K., & Masood, S. (2021). *Waardenburg syndrome* Copyright © 2021, StatPearls Publishing LLC *StatPearls*. Treasure Island, FL: StatPearls Publishing.

Alamy, A. H., Menezes, F. B., Leite, A. C., Nascimento, O. M., & Araújo, A. Q. (2001). Dysautonomia in human T-cell lymphotrophic virus type I-associated myelopathy/tropical spastic paraparesis. *Annals of Neurology*, *50*(5), 681–685.

Appenzeller, O., & Goss, J. E. (1971). Autonomic deficits in Parkinson's syndrome. *Archives of Neurology*, *24*, 50–57.

Bannister, R. (1983). *Autonomic Failure*. New York, NY: Oxford University Press.

Bannister, R. (1988). *Autonomic Failure*. New York, NY: Oxford University Press.

Beach, T. G., Adler, C. H., Sue, L. I., Vedders, L., Lue, L., White Iii, C. L., & Walker, D. G. (2010). Multi-organ distribution of phosphorylated alpha-synuclein histopathology in subjects with Lewy body disorders. *Acta Neuropathologica*, *119*(6), 689–702.

Benarroch, E. E. (1993). The central autonomic network: functional organization, dysfunction, and perspective. *Mayo Clinic Proceedings. Mayo Clinic*, *68*(10), 988–1001.

Benarroch, E. E. (2003). Brainstem in multiple system atrophy: clinicopathological correlations. *Cellular and Molecular Neurobiology, 23*(4–5), 519–526.

Benarroch, E. E., Schmeichel, A. M., Low, P. A., Boeve, B. F., Sandroni, P., & Parisi, J. E. (2005). Involvement of medullary regions controlling sympathetic output in Lewy body disease. *Brain, 128*(Pt 2), 338–344.

Benarroch, E. E., Schmeichel, A. M., Low, P. A., & Parisi, J. E. (2004). Involvement of medullary serotonergic groups in multiple system atrophy. *Annals of Neurology, 55*(3), 418–422.

Benarroch, E. E., Schmeichel, A. M., Sandroni, P., Low, P. A., & Parisi, J. E. (2006). Involvement of vagal autonomic nuclei in multiple system atrophy and Lewy body disease. *Neurology, 66*(3), 378–383.

Benarroch, E. E., Smithson, I. L., Low, P. A., & Parisi, J. E. (1998). Depletion of catecholaminergic neurons of the rostral ventrolateral medulla in multiple systems atrophy with autonomic failure. *Annals of Neurology, 43*, 156–163.

Berganzo, K., Díez-Arrola, B., Tijero, B., Somme, J., Lezcano, E., Llorens, V., & Zarranz, J. J. (2013). Nocturnal hypertension and dysautonomia in patients with Parkinson's disease: are they related? *Journal of Neurology, 260*(7), 1752–1756.

Bestetti, A., Capozza, A., Lacerenza, M., Manfredi, L., & Mancini, F. (2017). Delayed gastric emptying in advanced Parkinson disease: correlation with therapeutic doses. *Clinical Nuclear Medicine, 42*(2), 83–87.

Biaggioni, I., Goldstein, D. S., Atkinson, T., & Robertson, D. (1990). Dopamine-beta-hydroxylase deficiency in humans. *Neurology, 40*, 370–373.

Biaggioni, I., & Robertson, D. (1987). Endogenous restoration of noradrenaline by precursor therapy in dopamine-beta-hydroxylase deficiency. *Lancet, 2*(8569), 1170–1172.

Borghammer, P. (2018). How does Parkinson's disease begin? Perspectives on neuroanatomical pathways, prions, and histology. *Movement Disorders: Official Journal of the Movement Disorder Society, 33*(1), 48–57.

Bouhaddi, M., Vuillier, F., Fortrat, J. O., Cappelle, S., Henriet, M. T., Rumbach, L., & Regnard, J. (2004). Impaired cardiovascular autonomic control in newly and long-term-treated patients with Parkinson's disease: involvement of L-dopa therapy. *Autonomic Neuroscience: Basic & Clinical, 116*(1–2), 30–38.

Bower, J. H., Maraganore, D. M., McDonnell, S. K., & Rocca, W. A. (1997). Incidence of progressive supranuclear palsy and multiple system atrophy in Olmsted County, Minnesota, 1976 to 1990. *Neurology, 49*(5), 1284–1288.

Bradbury, S., & Eggleston, C. (1925). Postural hypotension: a report of three cases. *American Heart Journal, 1*, 73–86.

Bronner, G., & Vodusek, D. B. (2011). Management of sexual dysfunction in Parkinson's disease. *Therapeutic Advances in Neurological Disorders, 4*(6), 375–383.

Burke, W. J., Kumar, V. B., Pandey, N., Panneton, W. M., Gan, Q., Franko, M. W., & Galvin, J. E. (2008). Aggregation of alpha-synuclein by DOPAL, the monoamine oxidase metabolite of dopamine. *Acta Neuropathologica, 115*(2), 193–203.

Calandra-Buonaura, G., Alfonsi, E., Vignatelli, L., Benarroch, E. E., Giannini, G., Iranzo, A., & Kaufmann, H. (2021). Dysphagia in multiple system atrophy consensus statement on diagnosis, prognosis and treatment. *Parkinsonism & Related Disorders, 86*, 124–132.

Cechetto, D. F. (2014). Cortical control of the autonomic nervous system. *Experimental Physiology, 99*(2), 326–331.

Cersosimo, M. G., & Benarroch, E. E. (2012). Autonomic involvement in Parkinson's disease: pathology, pathophysiology, clinical features and possible peripheral biomarkers. *Journal of the Neurological Sciences, 313*(1–2), 57–63.

Chelban, V., Vichayanrat, E., Schottlaende, L., Iodice, V., & Houlden, H. (2018). Autonomic dysfunction in genetic forms of synucleinopathies. *Movement Disorders, 33*(3), 359–371.

Chelimsky, G., & Chelimsky, T. (2020). Non-pharmacologic management of orthostatic hypotension. *Autonomic Neuroscience: Basic & Clinical, 229*, 102732.

Cheshire, W. P., Freeman, R., Gibbons, C. H., Cortelli, P., Wenning, G. K., Hilz, M. J., & Singer, W. (2020). Electrodiagnostic assessment of the autonomic nervous system: a consensus statement endorsed by the American Autonomic Society, American Academy of Neurology, and the International Federation of Clinical Neurophysiology. *Clinical Neurophysiology: Official Journal of the International Federation of Clinical Neurophysiology*.

Christopher, L., Duff-Canning, S., Koshimori, Y., Segura, B., Boileau, I., Chen, R., & Strafella, A. P. (2015). Salience network and parahippocampal dopamine dysfunction in memory-impaired Parkinson disease. *Annals of Neurology*, 77(2), 269–280.

Chung, K. A., & Pfeiffer, R. F. (2021). Gastrointestinal dysfunction in the synucleinopathies. *Clinical Autonomic Research: Official Journal of the Clinical Autonomic Research Society*, *31*(1), 77–99.

Claus, I., Suttrup, J., Muhle, P., Suntrup-Krueger, S., Siemer, M. L., Lenze, F., & Warnecke, T. (2018). Subtle esophageal motility alterations in parkinsonian syndromes: synucleinopathies vs. tauopathies. *Movement Disorders Clinical Practice*, *5*(4), 406–412.

Cook, G. A., Sullivan, P., Holmes, C., & Goldstein, D. S. (2014). Cardiac sympathetic denervation without Lewy bodies in a case of multiple system atrophy. *Parkinsonism & Related Disorders*, *20*(8), 926–928.

Coon, E. A. (2020). Autonomic dysfunction in the synucleinopathies. *Seminars in Neurology*, *40*(5), 492–501.

Coon, E. A., & Ahlskog, J. E. (2021). My treatment approach to multiple system atrophy. *Mayo Clinic Proceedings. Mayo Clinic*, *96*(3), 708–719.

Coon, E. A., Cutsforth-Gregory, J. K., & Benarroch, E. E. (2018). Neuropathology of autonomic dysfunction in synucleinopathies. *Movement Disorders: Official Journal of the Movement Disorder Society*, *33*(3), 349–358.

Coon, E. A., Fealey, R. D., Sletten, D. M., Mandrekar, J. N., Benarroch, E. E., Sandroni, P., & Singer, W. (2017). Anhidrosis in multiple system atrophy involves pre- and postganglionic sudomotor dysfunction. *Movement Disorders: Official Journal of the Movement Disorder Society*, *32*(3), 397–404.

Coon, E. A., Mandrekar, J. N., Berini, S. E., Benarroch, E. E., Sandroni, P., Low, P. A., & Singer, W. (2020). Predicting phenoconversion in pure autonomic failure. *Neurology*, *95*(7), e889–e897.

Coon, E. A., Nelson, R. M., Sletten, D. M., Suarez, M. D., Ahlskog, J. E., Benarroch, E. E., & Singer, W. (2019). Sex and gender influence symptom manifestation and survival in multiple system atrophy. *Autonomic Neuroscience: Basic & Clinical, 219*, 49–52.

Coon, E. A., Schmeichel, A. M., Parisi, J. E., Cykowski, M. D., Low, P. A., & Benarroch, E. E. (2016). Medullary neuronal loss is not associated with α-synuclein burden in multiple system atrophy. *Movement Disorders: Official Journal of the Movement Disorder Society*, *31*(12), 1802–1809.

Coon, E. A., Singer, W., & Low, P. A. (2019). Pure autonomic failure. *Mayo Clinic Proceedings. Mayo Clinic*.

Coon, E. A., Sletten, D. M., Suarez, M. D., Mandrekar, J. N., Ahlskog, J. E., Bower, J. H., & Singer, W. (2015). Clinical features and autonomic testing predict survival in multiple system atrophy. *Brain*, *138*(Pt 12), 3623–3631.

Coon, Elizabeth A., Golden, Elisabeth P., Bryarly, Meredith, McGregor, Tamara, Nguyen, Benjamin N., Moutvic, Margaret A., ... Vernino, Steven (2022). A call for multiple system atrophy centers of excellence. *Clin. Auton. Res.* Available from https://doi.org/10.1007/s10286-022-00866-1.

Cortelli, P., Calandra-Buonaura, G., Benarroch, E. E., Giannini, G., Iranzo, A., Low, P. A., & Meissner, W. G. (2019). Stridor in multiple system atrophy: consensus statement on diagnosis, prognosis, and treatment. *Neurology*, *93*(14), 630–639.

De Pablo-Fernandez, E., Tur, C., Revesz, T., Lees, A. J., Holton, J. L., & Warner, T. T. (2017). Association of autonomic dysfunction with disease progression and survival in Parkinson disease. *JAMA Neurology*, *74*(8), 970–976.

Deguchi, K., Sasaki, I., Touge, T., Tsukaguchi, M., Takeuchi, H., & Kuriyama, S. (2004). Abnormal baroreceptor-mediated vasopressin release as possible marker in early diagnosis of multiple system atrophy. *Journal of Neurology, Neurosurgery, and Psychiatry*, *75* (1), 110–115.

Dempsey, J.A., Olson Jr., E.B., & Skatrud, J.B. Hormones and neurochemicals in the regulation of breathing. In The Handbook of Physiology. The Respiratory System. Control of Breathing, A.P. Fishman and N.S. Cherniak (Eds.). American Physiological Society, Bethesda, MD, pp. 181–221.

Dickson, D. W. (2012). Parkinson's disease and parkinsonism: neuropathology. *Cold Spring Harbor Perspectives in Medicine*, *2*(8).

Donadio, V. (2019). Skin nerve alpha-synuclein deposits in Parkinson's disease and other synucleinopathies: a review. *Clinical Autonomic Research: Official Journal of the Clinical Autonomic Research Society*, *29*(6), 577–585.

Donadio, Vincenzo, Incensi, Alex, Rizzo, Giovanni, Westermark, Gunilla T., Devigili, Grazia, De Micco, Rosa, ... Liguori, Rocco (2022). Phosphorylated α-synuclein in skin Schwann cells: a new biomarker for multiple system atrophy. *Brain*. Available from https://doi.org/10.1093/brain/awac124, 35552610.

Espay, A. J., Vizcarra, J. A., Marsili, L., Lang, A. E., Simon, D. K., Merola, A., & Leverenz, J. B. (2019). Revisiting protein aggregation as pathogenic in sporadic Parkinson and Alzheimer diseases. *Neurology*, *92*(7), 329–337.

Fanciulli, A., Campese, N., Goebel, G., Ndayisaba, J. P., Eschlboeck, S., Kaindlstorfer, C., & Wenning, G. K. (2020). Association of transient orthostatic hypotension with falls and syncope in patients with Parkinson disease. *Neurology*, *95*(21), e2854–e2865.

Fanciulli, A., Göbel, G., Ndayisaba, J. P., Granata, R., Duerr, S., Strano, S., & Wenning, G. K. (2016). Supine hypertension in Parkinson's disease and multiple system atrophy. *Clinical Autonomic Research: Official Journal of the Clinical Autonomic Research Society*, *26* (2), 97–105.

Fanciulli, A., & Wenning, G. K. (2015). Multiple-system atrophy. *The New England Journal of Medicine*, *372*(3), 249–263.

Fereshtehnejad, S. M., Romenets, S. R., Anang, J. B., Latreille, V., Gagnon, J. F., & Postuma, R. B. (2015). New clinical subtypes of Parkinson disease and their longitudinal progression: a prospective cohort comparison with other phenotypes. *JAMA Neurology*, *72*(8), 863–873.

Garland, E. M., & Biaggioni, I. (1993). Dopamine beta-hydroxylase deficiency. In M. P. Adam, H. H. Ardinger, R. A. Pagon, S. E. Wallace, L. J. H. Bean, G. Mirzaa, & A. Amemiya (Eds.), *GeneReviews(®)*. Seattle, WA: University of Washington, Seattle.

Gibbons, C. H., & Freeman, R. (2020). Delayed orthostatic hypotension. *Autonomic Neuroscience: Basic & Clinical*, *229*, 102724.

Gilman, S., Wenning, G. K., Low, P. A., Brooks, D. J., Mathias, C. J., Trojanowski, J. Q., & Vidailhet, M. (2008). Second consensus statement on the diagnosis of multiple system atrophy. *Neurology*, *71*(9), 670–676.

Gjerloff, T., Fedorova, T., Knudsen, K., Munk, O. L., Nahimi, A., Jacobsen, S., & Borghammer, P. (2015). Imaging acetylcholinesterase density in peripheral organs in Parkinson's disease with 11C-donepezil PET. *Brain*, *138*(Pt 3), 653–663.

Goldstein, D. S. (2020a). The catecholaldehyde hypothesis: where MAO fits in. *Journal of Neural Transmission (Vienna)*, *127*(2), 169–177.

Goldstein, D. S. (2020b). The "sick-but-not-dead" phenomenon applied to catecholamine deficiency in neurodegenerative diseases. *Seminars in Neurology*, *40*(5), 502–514.

Goldstein, D. S., Eldadah, B. A., Holmes, C., Pechnik, S., Moak, J., Saleem, A., & Sharabi, Y. (2005). Neurocirculatory abnormalities in Parkinson disease with orthostatic hypotension: independence from levodopa treatment. *Hypertension, 46*(6), 1333–1339.

Goldstein, D. S., Holmes, C., Cannon, R. O., 3rd, Eisenhofer, G., & Kopin, I. J. (1997). Sympathetic cardioneuropathy in dysautonomias. *The New England Journal of Medicine, 336*(10), 696–702.

Goldstein, D. S., Holmes, C., Li, S. T., Bruce, S., Metman, L. V., & Cannon, R. O., 3rd (2000). Cardiac sympathetic denervation in Parkinson disease. *Annals of Internal Medicine, 133*(5), 338–347.

Goldstein, D. S., Holmes, C., Sharabi, Y., Brentzel, S., & Eisenhofer, G. (2003). Plasma levels of catechols and metanephrines in neurogenic orthostatic hypotension. *Neurology, 60*(8), 1327–1332.

Goldstein, D. S., Holmes, C., Sullivan, P., Donadio, V., Isonaka, R., Zhong, E., & Sharabi, Y. (2017). Autoimmunity-associated autonomic failure with sympathetic denervation. *Clinical Autonomic Research: Official Journal of the Clinical Autonomic Research Society, 27*(1), 57–62.

Goldstein, D. S., Holmes, C. S., Dendi, R., Bruce, S. R., & Li, S. T. (2002). Orthostatic hypotension from sympathetic denervation in Parkinson's disease. *Neurology, 58*(8), 1247–1255.

Goldstein, D. S., Isonaka, R., Holmes, C., Ding, Y. S., & Sharabi, Y. (2020). Cardiac sympathetic innervation and vesicular storage in pure autonomic failure. *Annals of Clinical and Translational Neurology*, 7(10), 1908–1918.

Goldstein, D. S., Sullivan, P., Holmes, C., Lamotte, G., Lenka, A., & Sharabi, Y. (2021). Differential abnormalities of cerebrospinal fluid dopaminergic vs noradrenergic indices in synucleinopathies. *Journal of Neurochemistry, 158*(2), 554–568.

Graham, J. G., & Oppenheimer, D. R. (1969). Orthostatic hypotension and nicotine sensitivity in a case of multiple system atrophy. *Journal of Neurology, Neurosurgery, and Psychiatry, 32*, 28–34.

Grimaldi, S., Boucekine, M., Witjas, T., Fluchere, F., Azulay, J. P., Guedj, E., & Eusebio, A. (2021). Early atypical signs and insula hypometabolism predict survival in multiple system atrophy. *Journal of Neurology, Neurosurgery, and Psychiatry.*

Haensch, C. A., Lerch, H., Jorg, J., & Isenmann, S. (2009). Cardiac denervation occurs independent of orthostatic hypotension and impaired heart rate variability in Parkinson's disease. *Parkinsonism & Related Disorders, 15*(2), 134–137.

Hague, K., Lento, P., Morgello, S., Caro, S., & Kaufmann, H. (1997). The distribution of Lewy bodies in pure autonomic failure: autopsy findings and review of the literature. *Acta Neuropathologica, 94*(2), 192–196.

Hatsuta, H., Takao, M., Nakano, Y., Nogami, A., Uchino, A., Sumikura, H., & Murayama, S. (2016). Reduction of small fibers of thoracic ventral roots and neurons of intermediolateral nucleus in Parkinson disease and dementia with Lewy bodies. *Journal of Parkinson's Disease, 6*(2), 325–334.

Hiorth, Y. H., Pedersen, K. F., Dalen, I., Tysnes, O. B., & Alves, G. (2019). Orthostatic hypotension in Parkinson disease: a 7-year prospective population-based study. *Neurology, 93*(16), e1526–e1534.

Hjelle, B., Appenzeller, O., Mills, R., Alexander, S., Torrez-Martinez, N., Jahnke, R., & Ross, G. (1992). Chronic neurodegenerative disease associated with HTLV-II infection. *Lancet, 339*(8794), 645–646.

Horsager, J., Andersen, K. B., Knudsen, K., Skjærbæk, C., Fedorova, T. D., Okkels, N., & Borghammer, P. (2020). Brain-first vs body-first Parkinson's disease: a multimodal imaging case-control study. *Brain, 143*(10), 3077–3088.

Hughes, A. J., Daniel, S. E., Blankson, S., & Lees, A. J. (1993). A clinicopathologic study of 100 cases of Parkinson's disease. *Archives of Neurology, 50*, 140–148.

Isonaka, R., Rosenberg, A. Z., Sullivan, P., Corrales, A., Holmes, C., Sharabi, Y., & Goldstein, D. S. (2019). Alpha-synuclein deposition within sympathetic noradrenergic neurons is associated with myocardial noradrenergic deficiency in neurogenic orthostatic hypotension. *Hypertension*, *73*(4), 910–918.

Isonaka, R., Sullivan, P., Jinsmaa, Y., Corrales, A., & Goldstein, D. S. (2018). Spectrum of abnormalities of sympathetic tyrosine hydroxylase and alpha-synuclein in chronic autonomic failure. *Clinical Autonomic Research: Official Journal of the Clinical Autonomic Research Society*, *28*(2), 223–230.

Ivan, I. F., Irincu, V. L., Diaconu, Ş., Falup-Pecurariu, O., CiopleiaŞ, B., & Falup-Pecurariu, C. (2021). Gastro-intestinal dysfunctions in Parkinson's disease (review). *Experimental and Therapeutic Medicine*, *22*(4), 1083.

Iwasaki, Y. (1990). Pathology of chronic myelopathy associated with HTLV-I infection (HAM/TSP). *Journal of the Neurological Sciences*, *96*(1), 103–123.

JCS Joint Working Group. (2012). Guidelines for the clinical use of 24 hour ambulatory blood pressure monitoring (ABPM) (JCS 2010): digest version. *Circulation Journal: Official Journal of the Japanese Circulation Society*, *76*(2), 508–519.

Jinsmaa, Y., Sharabi, Y., Sullivan, P., Isonaka, R., & Goldstein, D. S. (2018). 3,4-Dihydroxyphenylacetaldehyde-induced protein modifications and their mitigation by N-acetylcysteine. *The Journal of Pharmacology and Experimental Therapeutics*, *366*(1), 113–124.

Johnson, R. H., Lee Gde, J., Oppenheimer, D. R., & Spalding, J. M. (1966). Autonomic failure with orthostatic hypotension due to intermediolateral column degeneration. A report of two cases with autopsies. *The Quarterly Journal of Medicine*, *35*(138), 276–292.

Jost, W. H., Altmann, C., Fiesel, T., Becht, B., Ringwald, S., & Hoppe, T. (2020). Influence of levodopa on orthostatic hypotension in Parkinson's Disease. *Neurologia i Neurochirurgia Polska*, *54*(2), 200–203.

Kaplan, P., & de Chaderévian, J. P. (1988). Piebaldism-Waardenburg syndrome: histopathologic evidence for a neural crest syndrome. *American Journal of Medical Genetics*, *31*(3), 679–688.

Kaufmann, H. (1996). Consensus statement on the definition of orthostatic hypotension, pure autonomic failure and multiple system atrophy. *Clinical Autonomic Research: Official Journal of the Clinical Autonomic Research Society*, *6*, 125–126.

Kaufmann, H., & Goldstein, D. S. (2013). Autonomic dysfunction in Parkinson disease. *Handbook of Clinical Neurology*, *117*, 259–278.

Kaufmann, H., Norcliffe-Kaufmann, L., Palma, J. A., Biaggioni, I., Low, P. A., Singer, W., & Robertson, D. (2017). Natural history of pure autonomic failure: A United States prospective cohort. *Annals of Neurology*, *81*(2), 287–297.

Kim, J. S., Lee, S. H., Oh, Y. S., Park, J. W., An, J. Y., Park, S. K., & Lee, K. S. (2016). Cardiovascular autonomic dysfunction in mild and advanced Parkinson's disease. *Journal of Movement Disorders*, *9*(2), 97–103.

Kuriyama, N., Niwa, F., Watanabe, Y., Yamada, K., Tokuda, T., Mizuno, T., & Nakagawa, M. (2009). Evaluation of autonomic malfunction in HTLV-1 associated myelopathy (HAM). *Autonomic Neuroscience: Basic & Clinical*, *150*(1–2), 131–135.

Lamotte, G., & Benarroch, E. E. (2021). What is the clinical correlation of cardiac noradrenergic denervation in Parkinson disease? *Neurology*, *96*(16), 748.

Lamotte, G., Holmes, C., Sullivan, P., Lenka, A., & Goldstein, D. S. (2020). Cardioselective peripheral noradrenergic deficiency in Lewy body synucleinopathies. *Annals of Clinical and Translational Neurology*.

Langston, J. W., & Forno, L. S. (1978). The hypothalamus in Parkinson disease. *Annals of Neurology*, *3*(2), 129–133.

Lenka, A., Lamotte, G., & Goldstein, D. S. (2021). Cardiac (18)F-dopamine PET distinguishes PD with orthostatic hypotension from Parkinsonian MSA. *Movement Disorders Clinical Practice*, *8*(4), 582–586.

Leys, F., Fanciulli, A., Ndayisaba, J. P., Granata, R., Struhal, W., & Wenning, G. K. (2020). Cardiovascular autonomic function testing in multiple system atrophy and Parkinson's disease: an expert-based blinded evaluation. *Clinical Autonomic Research: Official Journal of the Clinical Autonomic Research Society*, *30*(3), 255–263.

Lodhi, H. A., Peri-Okonny, P. A., Schesing, K., Phelps, K., Ngo, C., Evans, H., & Vongpatanasin, W. (2019). Usefulness of blood pressure variability indices derived from 24-hour ambulatory blood pressure monitoring in detecting autonomic failure. *Journal of the American Heart Association*, *8*(7), e010161.

Low, D. A., Vichayanrat, E., Iodice, V., & Mathias, C. J. (2014). Exercise hemodynamics in Parkinson's disease and autonomic dysfunction. *Parkinsonism & Related Disorders*, *20* (5), 549–553.

Low, P. A., Reich, S. G., Jankovic, J., Shults, C. W., Stern, M. B., Novak, P., & Mandrekar, J. (2015). Natural history of multiple system atrophy in the USA: a prospective cohort study. *Lancet Neurology*, *14*(7), 710–719.

Mar, P. L., Shibao, C. A., Garland, E. M., Black, B. K., Biaggioni, I., Diedrich, A., & Raj, S. R. (2015). Neurogenic hyperadrenergic orthostatic hypotension: a newly recognized variant of orthostatic hypotension in older adults with elevated norepinephrine (noradrenaline). *Clinical Science (London)*, *129*(2), 107–116.

Martin, W. R. W., Miles, M., Zhong, Q., Hartlein, J., Racette, B. A., Norris, S. A., & Perlmutter, J. S. (2020). Is levodopa response a valid indicator of Parkinson's disease? *Movement Disorders: Official Journal of the Movement Disorder Society*.

Masuda-Suzukake, M., Nonaka, T., Hosokawa, M., Kubo, M., Shimozawa, A., Akiyama, H., & Hasegawa, M. (2014). Pathological alpha-synuclein propagates through neural networks. *Acta Neuropathologica Communications*, *2*, 88.

Matsui, H., Nishinaka, K., Oda, M., Komatsu, K., Kubori, T., & Udaka, F. (2006). Does cardiac metaiodobenzylguanidine (MIBG) uptake in Parkinson's disease correlate with major autonomic symptoms? *Parkinsonism & Related Disorders*, *12*(5), 284–288.

Nishida, Katsuya, Sakashita, Kento, Yamasaki, Hiroshi, & Futamura, Naonobu (2022). Impact of tracheostomy invasive ventilation on survival in Japanese patients with multiple system atrophy. *Parkinsonism Relat. Disord.*, *97*, 107–111. Available from https://doi.org/10.1016/j.parkreldis.2022.01.008.

Norcliffe-Kaufmann, L., Kaufmann, H., Palma, J. A., Shibao, C. A., Biaggioni, I., Peltier, A. C., & Autonomic Disorders, C. (2018). Orthostatic heart rate changes in patients with autonomic failure caused by neurodegenerative synucleinopathies. *Annals of Neurology*, *83*(3), 522–531.

Nutman, J., Steinherz, R., Sivan, Y., & Goodman, R. M. (1986). Possible Waardenburg syndrome with gastrointestinal anomalies. *Journal of Medical Genetics*, *23*(2), 175–178.

Palma, J. A., Gomez-Esteban, J. C., Norcliffe-Kaufmann, L., Martinez, J., Tijero, B., Berganzo, K., & Kaufmann, H. (2015). Orthostatic hypotension in Parkinson disease: how much you fall or how low you go? *Movement Disorders: Official Journal of the Movement Disorder Society*, *30*(5), 639–645.

Palma, J. A., & Kaufmann, H. (2020). Management of orthostatic hypotension. *Continuum (Minneap Minn)*, *26*(1), 154–177.

Papapetropoulos, S., & Mash, D. C. (2007). Insular pathology in Parkinson's disease patients with orthostatic hypotension. *Parkinsonism & Related Disorders*, *13*(5), 308–311.

Park, J. W., Okamoto, L. E., Shibao, C. A., & Biaggioni, I. (2020). Pharmacologic treatment of orthostatic hypotension. *Autonomic Neuroscience: Basic & Clinical*, *229*, 102721.

Parkinson, J. (1817). *An Essay on the Shaking Palsy*. London: Whittingham & Rowland.

Perez-Lloret, S., Rey, M. V., Fabre, N., Ory, F., Spampinato, U., Senard, J. M., & Rascol, O. (2012). Factors related to orthostatic hypotension in Parkinson's disease. *Parkinsonism & Related Disorders, 18*(5), 501–505.

Poewe, W., Seppi, K., Tanner, C. M., Halliday, G. M., Brundin, P., Volkmann, J., & Lang, A. E. (2017). Parkinson disease. *Nature Reviews Disease Primers, 3*, 17013.

Postuma, R. B., Berg, D., Stern, M., Poewe, W., Olanow, C. W., Oertel, W., & Deuschl, G. (2015). MDS clinical diagnostic criteria for Parkinson's disease. *Movement Disorders: Official Journal of the Movement Disorder Society, 30*(12), 1591–1601.

Presti, M. F., Schmeichel, A. M., Low, P. A., Parisi, J. E., & Benarroch, E. E. (2014). Degeneration of brainstem respiratory neurons in dementia with Lewy bodies. *Sleep, 37*(2), 373–378.

Provitera, V., Iodice, V., Manganelli, F., Mozzillo, S., Caporaso, G., Stancanelli, A., & Nolano, M. (2022). Postganglionic sudomotor assessment in early stage of multiple system atrophy and Parkinson disease: a morpho-functional study. *Neurology*.

Prusiner, S. B., Woerman, A. L., Mordes, D. A., Watts, J. C., Rampersaud, R., Berry, D. B., & Giles, K. (2015). Evidence for alpha-synuclein prions causing multiple system atrophy in humans with parkinsonism. *Proceedings of the National Academy of Sciences of the United States of America, 112*(38), E5308–E5317.

Puritz, R., Lightman, S. L., Wilcox, C. S., Forsling, M., & Bannister, R. (1983). Blood pressure and vasopressin in progressive autonomic failure. Response to postural stimulation, L-dopa and naloxone. *Brain, 106*(Pt 2), 503–511.

Rahman, F., & Goldstein, D. S. (2014). Quantitative indices of baroreflex-sympathoneural function: application to patients with chronic autonomic failure. *Clinical Autonomic Research: Official Journal of the Clinical Autonomic Research Society, 24*(3), 103–110.

Rascol, O., Perez-Lloret, S., Damier, P., Delval, A., Derkinderen, P., Destée, A., & Negre-Pages, L. (2015). Falls in ambulatory non-demented patients with Parkinson's disease. *Journal of Neural Transmission (Vienna), 122*(10), 1447–1455.

Rodriguez-Blazquez, C., Schrag, A., Rizos, A., Chaudhuri, K. R., Martinez-Martin, P., & Weintraub, D. (2021). Prevalence of non-motor symptoms and non-motor fluctuations in Parkinson's disease using the MDS-NMS. *Movement Disorders Clinical Practice, 8*(2), 231–239.

Sakakibara, R., Ito, T., Uchiyama, T., Asahina, M., Liu, Z., Yamamoto, T., & Hattori, T. (2005). Lower urinary tract function in dementia of Lewy body type. *Journal of Neurology, Neurosurgery, and Psychiatry, 76*(5), 729–732.

Sakakibara, R., Panicker, J., Simeoni, S., Uchiyama, T., Yamamoto, T., Tateno, F., & Aiba, Y. (2019). Bladder dysfunction as the initial presentation of multiple system atrophy: a prospective cohort study. *Clinical Autonomic Research: Official Journal of the Clinical Autonomic Research Society, 29*(6), 627–631.

Savica, R., Grossardt, B. R., Bower, J. H., Ahlskog, J. E., & Rocca, W. A. (2013). Incidence and pathology of synucleinopathies and tauopathies related to parkinsonism. *JAMA Neurology, 70*(7), 859–866.

Savica, R., Grossardt, B. R., Bower, J. H., Ahlskog, J. E., & Rocca, W. A. (2016). Time trends in the incidence of Parkinson disease. *JAMA Neurology, 73*(8), 981–989.

Schmeichel, A. M., Coon, E. A., Parisi, J. E., Singer, W., Low, P. A., & Benarroch, E. E. (2021). Loss of putative GABAergic neurons in the ventrolateral medulla in multiple system atrophy. *Sleep*.

Schweitzer, V. G., & Clack, T. D. (1984). Waardenburg's syndrome: a case report with CT scanning and cochleovestibular evaluation. *International Journal of Pediatric Otorhinolaryngology*, 7(3), 311–322.

Seeley, W. W. (2019). The salience network: a neural system for perceiving and responding to homeostatic demands. *The Journal of Neuroscience, 39*(50), 9878–9882.

Senard, J. M., Rai, S., Lapeyre-Mestre, M., Brefel, C., Rascol, O., Rascol, A., & Montastruc, J. L. (1997). Prevalence of orthostatic hypotension in Parkinson's disease. *Journal of Neurology, Neurosurgery, and Psychiatry*, *63*, 584–589.

Seppi, K., Ray Chaudhuri, K., Coelho, M., Fox, S. H., Katzenschlager, R., Perez Lloret, S., & Sampaio, C. (2019). Update on treatments for nonmotor symptoms of Parkinson's disease-an evidence-based medicine review. *Movement Disorders: Official Journal of the Movement Disorder Society*, *34*(2), 180–198.

Shibao, C. A., & Biaggioni, I. (2020). Management of orthostatic hypotension, postprandial hypotension, and supine hypertension. *Seminars in Neurology*, *40*(5), 515–522.

Shibao, C. A., Garland, E. M., Black, B. K., Mathias, C. J., Grant, M. B., Root, A. W., & Biaggioni, I. (2020). Congenital absence of norepinephrine due to CYB561 mutations. *Neurology*, *94*(2), e200–e204.

Shibao, C. A., Joos, K., Phillips, J. A., 3rd, Cogan, J., Newman, J. H., Hamid, R., & Biaggioni, I. (2021). Familial autonomic ganglionopathy caused by rare CHRNA3 genetic variants. *Neurology*, *97*(2), e145–e155.

Shibata, M., Morita, Y., Shimizu, T., Takahashi, K., & Suzuki, N. (2009). Cardiac parasympathetic dysfunction concurrent with cardiac sympathetic denervation in Parkinson's disease. *Journal of the Neurological Sciences*, *276*(1–2), 79–83.

Shy, G. M., & Drager, G. A. (1960). A neurological syndrome associated with orthostatic hypotension: a clinical-pathologic study. *Archives of Neurology*, *2*, 511–527.

Singer, W., Schmeichel, A. M., Shahnawaz, M., Schmelzer, J. D., Boeve, B. F., Sletten, D. M., & Low, P. A. (2020). Alpha-synuclein oligomers and neurofilament light chain in spinal fluid differentiate multiple system atrophy from Lewy body synucleinopathies. *Annals of Neurology*.

Singer, W., Schmeichel, A. M., Shahnawaz, M., Schmelzer, J. D., Sletten, D. M., Gehrking, T. L., & Low, P. A. (2021). Alpha-Synuclein Oligomers and Neurofilament Light Chain Predict Phenoconversion of Pure Autonomic Failure. *Annals of Neurology*, *89*(6), 1212–1220.

Squair, Jordan W, Berney, Maxime, Castro Jimenez, Mayte, Hankov, Nicolas, Demesmaeker, Robin, Amir, Suje, ... Bloch, Jocelyne (2022). Implanted system for orthostatic hypotension in multiple-system atrophy. *N. Engl. J. Med.*, *386*(14), 1339–1344. Available from https://doi.org/10.1056/NEJMoa2112809.

Strano, S., Fanciulli, A., Rizzo, M., Marinelli, P., Palange, P., Tiple, D., & Colosimo, C. (2016). Cardiovascular dysfunction in untreated Parkinson's disease: A multi-modality assessment. *Journal of the Neurological Sciences*, *370*, 251–255.

Stubendorff, K., Aarsland, D., Minthon, L., & Londos, E. (2012). The impact of autonomic dysfunction on survival in patients with dementia with Lewy bodies and Parkinson's disease with dementia. *PLoS One*, 7(10), e45451.

Suttrup, I., Suttrup, J., Suntrup-Krueger, S., Siemer, M. L., Bauer, J., Hamacher, C., & Warnecke, T. (2017). Esophageal dysfunction in different stages of Parkinson's disease. *Neurogastroenterology and Motility: the Official Journal of the European Gastrointestinal Motility Society*, *29*(1).

Suttrup, I., & Warnecke, T. (2016). Dysphagia in Parkinson's Disease. *Dysphagia*, *31*(1), 24–32.

Swinn, L., Schrag, A., Viswanathan, R., Bloem, B. R., Lees, A., & Quinn, N. (2003). Sweating dysfunction in Parkinson's disease. *Movement Disorders: Official Journal of the Movement Disorder Society*, *18*(12), 1459–1463.

Taneja, I., & Robertson, D. (2003). Genetic basis of autonomic dysfunction. *Seminars in Neurology*, *23*(4), 391–397.

Thaisetthawatkul, P., Boeve, B. F., Benarroch, E. E., Sandroni, P., Ferman, T. J., Petersen, R., & Low, P. A. (2004). Autonomic dysfunction in dementia with Lewy bodies. *Neurology*, *62*(10), 1804–1809.

Travagli, R. A., Hermann, G. E., Browning, K. N., & Rogers, R. C. (2006). Brainstem circuits regulating gastric function. *Annual Review of Physiology, 68*, 279–305.

Van Den Berge, N., Ferreira, N., Gram, H., Mikkelsen, T. W., Alstrup, A. K. O., Casadei, N., & Borghammer, P. (2019). Evidence for bidirectional and trans-synaptic parasympathetic and sympathetic propagation of alpha-synuclein in rats. *Acta Neuropathologica, 138*(4), 535–550.

van Ingelghem, E., van Zandijcke, M., & Lammens, M. (1994). Pure autonomic failure: a new case with clinical, biochemical, and necropsy data. *Journal of Neurology, Neurosurgery, and Psychiatry, 57*(6), 745–747.

Velseboer, D. C., de Haan, R. J., Wieling, W., Goldstein, D. S., & de Bie, R. M. (2011). Prevalence of orthostatic hypotension in Parkinson's disease: a systematic review and *meta*-analysis. *Parkinsonism & Related Disorders, 17*(10), 724–729.

Vernino, S., Low, P. A., Fealey, R. D., Stewart, J. D., Farrugia, G., & Lennon, V. A. (2000). Autoantibodies to ganglionic acetylcholine receptors in autoimmune autonomic neuropathies. *The New England Journal of Medicine, 343*(12), 847–855.

Watanabe, H., Saito, Y., Terao, S., Ando, T., Kachi, T., Mukai, E., & Sobue, G. (2002). Progression and prognosis in multiple system atrophy: an analysis of 230 Japanese patients. *Brain, 125*(Pt 5), 1070–1083.

Wenning, G. K., Geser, F., Krismer, F., Seppi, K., Duerr, S., Boesch, S., & European Multiple System Atrophy Study, G. (2013). The natural history of multiple system atrophy: a prospective European cohort study. *Lancet Neurology, 12*(3), 264–274.

Wenning, Gregor K., Stankovic, Iva, Vignatelli, Luca, Fanciulli, Alessandra, Calandra-Buonaura, Giovanna, Seppi, Klaus, ... Kaufmann, Horacio (2022). The Movement Disorder Society Criteria for the Diagnosis of Multiple System Atrophy. *Mov. Disord.* Available from https://doi.org/10.1002/mds.29005, 35445419.

Yamamoto, T., Asahina, M., Yamanaka, Y., Uchiyama, T., Hirano, S., Fuse, M., & Kuwabara, S. (2017). Postvoid residual predicts the diagnosis of multiple system atrophy in Parkinsonian syndrome. *Journal of the Neurological Sciences, 381*, 230–234.

Zhang, L., Cao, B., Hou, Y., Gu, X., Wei, Q., Ou, R., & Shang, H. (2021). Neurofilament light chain predicts disease severity and progression in multiple system atrophy. *Movement Disorders: Official Journal of the Movement Disorder Society.*

CHAPTER 10
Spinal cord injuries

10.1 Introduction

In this chapter, an attempt is made to summarize information about autonomic function in spinal cord dysfunction, particularly spinal cord injuries. Though most of the material presented here has been touched upon in other chapters, this scatter makes it rather difficult to apply all pertinent data to the individual patient with spinal cord injury. Because autonomic function in this situation is almost always affected and the extent of abnormalities is closely related to the success of the management of patients, it is thought useful to bring all pertinent information together.

10.1.1 Anatomy

A brief overview of the anatomy involved in spinal cord injury is necessary. The preganglionic neurons of the parasympathetic nervous system are in the brainstem and sacral cord. This part of the autonomic nervous system is sometimes called cranial-sacral division. In spinal cord trauma, the sacral division is usually disconnected from the influence of higher centers and on occasion directly injured. Most viscera are innervated by sympathetic and parasympathetic fibers although many structures have a predominantly single innervation for example some blood vessels and sweat glands.

The sympathetic preganglionic fibers arise in the intermedial lateral and intermedial medial cell columns of the spinal cord and join the ventral roots of T1 to L2. Variations in this arrangement are common, however, and preganglionic fibers from C7 and as low as L4 cord segments have been demonstrated. The preganglionic fibers synapse in the sympathetic chain and may pass through several sympathetic ganglia or up and down the chain before synapsing. Some go through the ganglia to synapse in collateral ganglia near the viscera they innervate. These fibers form the splanchnic nerves. The long postganglionic sympathetic fibers join peripheral nerves and are distributed to blood vessels, skin, and other structures or form visceral nerves such as the cardiac nerves. Other long postganglionic fibers form plexuses and these are distributed to the head and other structures.

Introduction to Clinical Aspects of the Autonomic Nervous System
DOI: https://doi.org/10.1016/B978-0-323-96011-3.00008-5

The preganglionic fibers of the parasympathetic nervous system arise in the visceral brainstem nuclei and the second to the fourth sacral segments. They are distributed through the third, seventh, ninth, and tenth cranial nerves and the bulbar accessory nerve to the head and neck, thorax, and abdominal viscera. The lower ganglia of the descending colon and pelvic organs receive their parasympathetic preganglionic fibers from the sacral segments.

10.1.2 Cardiovascular autonomic testing

The clinical symptomatology of autonomic dysfunction in spinal cord lesions is related to the anatomic arrangements and to functional interruption of the integrated activity of autonomic neurons situated above the lesion and uninhibited overactivity of neurons that remain intact below the lesion. For example, in cervical and high thoracic transection, the entire or a large part of the sympathetic outflow, together with the sacral parasympathetic outflow, is separated from cerebral control. Naturally, the type of disturbance seen depends upon whether the lesion is complete or incomplete in the cord and on the level of involvement. The longitudinal extent of the damage up and down the cord is also of importance. In some degenerative conditions in which autonomic neurons are affected predominantly the clinical manifestations of autonomic dysfunction may be accompanied by only minor symptoms and signs of motor or sensory involvement and autonomic dysreflexia is rare, the predominant symptom in such patients being the loss of visceral function.

In the management of paraplegia and tetraplegia from any cause, it is useful to determine the severity of vasomotor paralysis. If this is present, considerable delay in rehabilitation may occur. In patients with impaired baroreceptor function because of spinal cord injury, there is evidence of cardiovascular adrenergic failure on autonomic reflex screen with absent late phase II, prolonged blood pressure recovery time, and absent phase IV and reflex bradycardia. Some patients, particularly those with high spinal cord lesions, are unable to cooperate with the Valsalva maneuver. In such circumstances, the lungs can be passively inflated by an appropriate device that delivers an adequate volume at the desired pressure and the Valsalva maneuver becomes unnecessary. Clinical manifestations of defective baroreceptor function with spinal cord injury include orthostatic hypotension and, in some patients with high cord lesions, unconsciousness may occur within seconds when they are maintained in the upright position. In others, there may be a considerable fall in blood pressure but this is not always accompanied by symptoms of global cerebral ischemia. On occasion, the effect of orthostasis on the circulation in individual patients needs to be assessed in the autonomic laboratory. Because some paraplegic individuals cannot be tilted, the effect of gravity on the circulation can be simulated by exposing the body below the iliac crests to negative pressure for a short time.

Abnormal responses to Valsalva maneuver and orthostatic hypotension are both evidence of disordered circulatory noradrenergic reflex function and often occur together though only one of these abnormalities might be present in some patients. The reason for the dissociation of these two vasomotor responses in some patients with cervical cord lesions is not clear, but it might depend on a return of venous tone or a denervation supersensitivity of chronically denervated vessels, which develop increased responsiveness to stretch or to normal amounts of circulating catecholamines. Other possible explanations for this apparent discrepancy include the contribution of mechanisms independent of the baroreflex (e.g., hypovolemia) and that reflex pathways may only be partially damaged. Chronic loss of baroreflex function may not always lead to the expected consequences of symptomatic postural hypotension. In spinal cord injuries; therefore, direct assessment of the circulatory responses to tilting and sudden rises in blood pressure may be more informative than the measurement of the response to the Valsalva maneuver. In spinal cord injury patients, there is a progressive improvement in the blood pressure response to sudden tilts with continuous changes of posture from the horizontal to the vertical. The reason for this adaptive response or conditioning of the circulation to gravitational influences is not fully understood but may involve some of the mechanisms mentioned above. In addition, the lack of symptomatic cerebral ischemia might in part depend upon adaptive cerebral vasodilatation that insures adequate flow in the face of a lowered perfusion pressure.

Plasma renin concentration in patients with cervical cord lesions has been studied. In such patients, the range may be above normal and a further increase may occur on the assumption of the sitting or standing position. This may play a part in limiting the hypotension and in the adaptive responses of the circulation to repeated tilts (Mathias et al., 1976). Normally antidiuretic hormone is not released during rapid or gradual head-up tilt, but in tetraplegic individuals, a fourfold rise can be measured in the plasma. This may significantly contribute of the maintenance of blood pressure in such patients (Sved, McDowell, & Blessing, 1985).

Because of autonomic cardiovascular impairment and diminished plasma norepinephrine responses to orthostatic stress, there is an increased reliance on the renin-angiotensin-aldosterone system (RAAS) for maintenance of the orthostatic blood pressure, especially in individuals with injuries above T6 (Wecht et al., 2005). The chronic stimulation of the RAAS in patients with spinal cord injury may have deleterious consequences with stimulation of cell growth, inflammation, and vascular fibrosis (Duprez, 2006). Looking at pulse wave velocity—a marker of cardiovascular disease when increased—Katzelnick and colleagues demonstrated that increased arterial stiffness in individuals with spinal cord injury may be due to greater dependency on the RAAS to maintain hemodynamic stability during an orthostatic challenge (Katzelnick et al., 2021).

Reflex vasodilatation in hand vessels in response to heating the chest, a normal response, is absent in some patients with cervical cord or lower brainstem lesions. There is another vasodilatation that is induced in hand vessels after warm blood reaches central temperature-sensitive neurons. The relative importance in temperature regulation of these two vasodilator reflexes is uncertain. It is probable that skin receptor activity is responsible for vasomotor adjustments induced by rapid environment temperature swings, whereas the effect of blood temperature is to compensate for changes in heat production or to prevent overcompensation by skin reflexes. Since in humans the hypothalamic centers would still be normally responsive to changes in blood temperature, adjustments in vasomotor control could still be achieved but only above the level of the lesion and therefore they are usually inadequate. Because vasomotor reflex changes depending upon skin receptors may be abolished, marked swings in ambient temperature are poorly tolerated by spinal cord injured patients.

In patients with spinal cord lesions above the sympathetic outflow, reflex inspiratory vasoconstriction is preserved. This is a purely spinal reflex complete in the thoracic cord. Emotional stress causes reflex vasoconstriction and a rise in blood pressure; this is in part the result of blood from the splanchnic region and skin vessels shifting to blood vessels in muscles. Stress, therefore, causes a fall in skin blood flow and an increase in muscle blood flow. The observed change in blood pressure, however, depends on a balance between vasoconstriction in some vascular beds and vasodilatation in the blood vessels in muscles. This response is mediated by sympathetic cholinergic vasodilator fibers to blood vessels in the muscles and by the release of a hormonal substance, which most probably is epinephrine. The vasoconstriction in skin vessels is a purely neurologic mechanism independent of circulatory hormones and is due to sympathetic vasoconstrictor fiber activity, which has been initiated by the cortex. The skin vasoconstriction in response to emotional stress is absent in patients with cervical cord lesions. The importance of vasomotor changes induced by a single deep breath and by mental stress in normal temperature regulation is not fully known but some blood pressure elevation may perhaps be related to the vasoconstriction in response to gasping or that induced by rapid and excessive inflation of the chest in those with paralyzed respiratory muscle. In patients with tetraplegia, the diaphragm remains the only or the predominant active respiratory muscle and the rib cage is thought to behave passively depending very much upon diaphragmatic contractions, intrathoracic and intraabdominal pressure. The motion of the abdomen and rib cage was studied in earlier studies in patients with lower cervical cord lesions during normal breathing in the supine and sitting posture. In the supine position, the rib cage of patients moved paradoxically inward and the chest wall was deformed. During sitting, the paradoxical motion no longer occurred. During expiration, the inward recoil of the respiratory system is also evident in the deformed chest wall. The contractions of the diaphragm constrict the rib cage passively. Rib cage motion is therefore related to a balance of forces between muscle contractions, diaphragm dome shape, the geometry of the rib cage, and lung volume (Mortola & Sant'Ambrogio, 1978).

10.1.3 Thermoregulatory function following spinal cord injury

Sweating is important in thermoregulation and can be induced by a heat load applied to the body either by heating with a cradle or immersing a limb in warm water. In individuals with a spinal cord injury, the afferent input to the hypothalamus from below the lesion level is decreased. Sweating is induced also by the ingestion of fluids of normal temperature, particularly in dehydrated subjects. This sweating is produced regardless of the tonicity of the fluid but its intensity and duration are related to the tonicity of the fluid given. In patients with acute trauma to the spinal cord who may be dehydrated, the administration of fluids that induces sweating may complicate thermoregulation even further. Muscular exercise also induces sweating and this occurs within seconds after beginning exercise. This is a neurogenic reflex rather than a thermal stimulus and it can be produced when a limb is exercised with the circulation arrested.

This particular variety of sweating may be of importance during the early rehabilitative efforts of paraplegics where the muscular activity of normally innervated limbs becomes one of the main triggers of sweating. In some patients, there might be dissociation of sweating from other sympathetic functions because of the involvement of the intermedial lateral cell column in incomplete lesions of the spinal cord. One or other extremity may be cyanosed with extreme sweating. Such a limb is reminiscent of causalgia. Reflex sweating can also be elicited by distention of the bladder. This reflex is of spinal origin and has apparently no thermoregulatory function.

A key discriminating factor between tetraplegia and paraplegia is that of the available or recruitable muscle mass, in particular upper-limb function. Individuals with tetraplegia have affected upper-limb function and heat production was shown to be associated with due residual muscle mass in tetraplegic and paraplegic athletes (Griggs, Price, & Goosey-Tolfrey, 2015).

The level of the spinal lesion is also a determinant factor of thermoregulatory function following spinal cord injury and tetraplegic individuals with lesions above T1 are unable to control body temperature due to impairment in spinal sympathetic outflows. Patients with cervical cord injuries in hot climates need to be cared for in air-conditioned surroundings since they are like poikilotherms and disastrous rises in body temperature with consequent metabolic disturbances may occur. Conversely, hypothermia in cold climates also occurs particularly in the acute stages of the cervical cord lesions. In such patients because of vasomotor paralysis preventing thermoregulatory skin vasoconstriction, there is a loss of sensitivity to cold below the lesion. In hypothermia, muscle tone may be increased in patients with lesions of the cord but above the level of the lesion only. Other patients show generalized rigidity and neck stiffness during cold exposure. Shivering in this situation is characteristically absent even in muscles innervated from segments above the cord lesion. Though accidental hypothermia and its complications are not common in spinal cord disease, the treatment of hypothermia should be familiar to all who deal with cord injuries because some patients are brought in after prolonged exposure with an injury to the cord and may be hypothermic on admission. In patients with lesions below the sixth cervical and above T4 a constant temperature is usually maintained adequately when the air temperature is neutral about 27°C. This is similar to findings in patients with paraplegia with lesions below T8. In hot environments at 35°C–37°C, control subjects are capable of maintaining their body temperature due to sweat evaporation. Tetraplegic subjects, however, show no or only minimal sweating below the lesion, and a rapid rise in body temperature occurs. In cool environments at 18°C–20°C, there is also a difference between tetraplegic subjects, normal controls, and paraplegic subjects with lesions below T8. Tetraplegic individuals cool rapidly and their temperature falls within two hours. This is attributed to a lack of shivering, which could be observed only in muscles innervated from segments above the lesion. In an experiment on 21 patients with complete transections at levels between C5 and L4, 5, the arms were heated and vasodilatation in the feet was measured. No increase in blood flow occurred in patients with lesions between C5 and T10. This was in spite of a substantial increase in central body temperature and suggests that the expected release of vasoconstrictor tone with consequent vasodilatation in skin vessels of the feet in response to the warm blood acting on central thermoregulatory neurons did not occur. These findings are attributed to the interruption by the lesion of descending sympathetic fibers. Patients with lesions below L1 and controls showed normal vasodilator responses in the feet. The upper level of vasomotor outflow to the foot occurs therefore about L1 and it is somewhat higher than the level of sudomotor activity, which was examined by electrical stimulation of preganglionic sympathetic fibers. In lesions above T6, because of the interruption of the outflow from higher centers, vasomotor paralysis is so extensive that postural hypotension, initially at least, is the rule. Practically, this implies that patients should be moved in the horizontal position from the scene of an accident. Postural hypotension improves with time and the return of blood pressure maintenance in the face of gravitational influences can be accelerated and improved by carefully supervised conditioning of autonomic reflexes. To achieve this, the patient is placed on a tilt table 2–3 times daily and the head of the table is elevated to an angle that produces an asymptomatic fall in blood pressure preferably of 30–35 mm of mercury systolic. This position is maintained for 5 minutes at a time, the blood pressure is frequently checked during this interval because it may suddenly fall precipitously after having been well maintained for 2–3 minutes in the early days of conditioning. The mechanism by which conditioning works or why some measure of postural blood pressure control is eventually regained in those who complete lesions above T6 is not entirely clear.

The discussion of autonomic function in spinal cord lesions is incomplete without mentioning shivering. This consists of contraction of skeletal muscles normally first appearing in the face and then spreading to arms, trunks, and legs. In patients with spinal cord transection shivering occurs in normally innervated muscles above the lesion when the central temperature falls due to cooling of the legs. In this situation, there may be little change in superficial skin temperature and the increase in metabolism due to shivering is triggered by a fall in central temperature and not by receptors from normally innervated skin, the temperature of which remains unchanged. In humans, therefore, central structures are present, which can induce shivering when the central temperature of the blood falls, but the role of this shivering mechanism in normal subjects is not fully clarified because shivering with an intact nervous system is usually triggered first by a fall in skin temperature. During fever, shivering also occurs but in patients with cord lesions this is confined to muscles innervated from segments above the lesion. Therefore, this type of shivering must be initiated by the activity of pyrogen and central structures and not by a direct effect of the pyrogen on nerves or muscles. It is also noteworthy that in patients with hemiplegia for example thermoregulatory shivering may be confined at least during the early stages after a stroke to the uninvolved side of the body.

In the acute phase (<3 months) of spinal cord injury, the mean incidence of fever is most common in patients with neurologic lesion levels above T6 and of complete classification, and 4%–5% of fevers in this period are defined as "neurogenic" fever or fever of unknown origin (Unsal-Delialioglu et al., 2010). On the other hand, patients with spinal cord injury may not be able to mount a fever in the setting of inflammation or infection in the chronic phase of the injury. The pathophysiology of neurogenic fever is thought to be partially attributed to autonomic dysfunction postspinal cord injury.

10.1.4 Orthostatic hypotension following spinal cord injury: understanding clinical pathophysiology

In the acute phase after the injury (hours to days), there is a transient state of hypoexcitability of the isolated spinal cord ("spinal shock"). During spinal shock, vasodilatation combined with impaired autonomic reflexes often results in profound hypotension. Over time, the spinal shock resolves but inappropriate postural changes in blood pressure and heart rate often continue to occur. Orthostatic hypotension can persist for years, and often becomes worse with time (Frisbie & Steele, 1997). The prevalence of orthostatic hypotension is high (57% of patients on tilt, 25% of patients with symptoms) (Cariga, Ahmed, Mathias, & Gardner, 2002). Mechanisms contributing to orthostatic hypotension in patients with spinal cord injury include loss of descending supraspinal sympathetic excitatory and inhibitory control, generalized reduction in sympathetic activity with low plasma epinephrine and norepinephrine levels, atrophy of sympathetic preganglionic neurons, lack of skeletal muscle pumps, cardiovascular deconditioning, and altered salt and water balance (Claydon, Steeves, & Krassioukov, 2006; Krassioukov, Bunge, Pucket, & Bygrave, 1999).

When chronic tetraplegic individuals are placed on a tilt table, there is typically a fall in blood pressure and a rise in heart rate, which continues for about 2 minutes reaching a plateau. After they are returned to the horizontal, the heart rates returns to the resting level within 1 minute but a slight overshoot in blood pressure is observed over a period of a few minutes. In normal subjects under similar conditions, there is an immediate overshoot in blood pressure, which is rapidly compensated for by bradycardia. This was not seen in tetraplegic individuals. Forearm blood flow decreases to very low levels during head-up tilting in tetraplegic individuals also and venous constriction occurs. These responses are seen in the absence of muscle spasms and changes in bladder pressure, both of which profoundly affect cardiovascular function in such patients (Corbett, Frankel, & Harris, 1971c; Corbett, Frankel, & Harris, 1971d). In chronic tetraplegic individuals, profound degrees of hypotension can be seen without loss of consciousness. In tetraplegic individuals, the observed decrease in peripheral blood flow during tilting in the presence of falling blood pressure is attributed to active vasoconstriction, the afferent stimulus being probably in a low-pressure system of the intrathoracic or abdominal vascular beds and the vasoconstriction is presumably the result of cardiovascular reflexes, which are again complete in the isolated spinal cord. An appropriate increase in peripheral resistance when upright has been reported following spinal cord injury and the underlying mechanisms may include the rapid increase in plasma vasopressin upon postural change, spinal reflexes, and veno-arteriolar reflexes, or peripheral α-adrenoceptor hyperresponsiveness (Claydon et al., 2006).

Patients with spinal cord injury who demonstrate orthostatic hypotension also seem to have altered cerebral hemodynamics, which likely explains why many patients remain asymptomatic despite profound postural hypotension.

10.1.5 Hypertension following spinal cord injury

In earlier studies of tetraplegic subjects with chronic localized complete lesions of the cervical cord above the sympathetic outflow, it was found that both spontaneous and induced muscle spasms result in an increase in blood pressure and venous pressure (Corbett, Frankel, & Harris, 1971a). The heart rate decrease is presumably due to the activity of intact arterial baroreceptors, which have their afferent limb in the carotid sinus and aortic arc regions and through the vagus nerves, which conduct impulses to the cranial portion of the parasympathetic nervous system. In such patients, there was also a decrease in hand and calf blood flow. Hand blood flow reflects mainly flow through skin vessels whereas blood flow in the calf is predominantly through muscles. These changes occur within 2–3 seconds of a spasm and are independent of changes in bladder pressure. The increase in blood pressure is attributed solely to the vasoconstriction in skin and muscle vessels presumably from spinal sympathetic reflex connections to somatic afferents associated with muscular contraction.

Cardiovascular reflex responses to cutaneous and visceral stimuli in tetraplegic patients with cervical spinal cord lesions have also been studied (Corbett, Frankel, & Harris, 1971b). Pinprick and cold applied to the skin above the lesion and sudden loud noise produce no response. All these stimuli would be expected to induce cutaneous vasoconstriction in normal subjects. When pinprick and cold are applied below the lesion however an increase in blood pressure, bradycardia, venous constriction, and a decrease in hand blood flow is observed. These responses indicate active arterial vasoconstriction and because of the short time lapse between stimulus and response, these changes must be reflex in nature. The response to the application of cold below the lesion also indicates that some measure of reflex control of the temperature may be preserved in humans with chronic tetraplegic individuals. The effectiveness of this cold-induced vasoconstriction in actual thermoregulation remains, however, in doubt because of the short duration of vasoconstriction. The vasoconstrictor responses are sudden and do not last very long.

Animal experiments suggest a role of endorphins in the pathogenesis of endotoxic and hypovolemic shock and also in the regulation of blood pressure in spinal shock. Administration of naloxone to animals experimentally subjected to transection of the spinal cord improved hypotension, hypoventilation, and hypothermia. These beneficial effects were apparently mediated by the blockade of endorphin activity on opiate receptors in the central nervous system. These preliminary studies suggest that the sustained increase in mean arterial pressure after acute spinal shock obtained by the administration of naloxone might be therapeutically useful in treating spinal cord injury in humans and improves perfusion of the spinal cord and perhaps reduces ischemia (Holaday & Faden, 1980).

10.1.6 Respiratory function

Respiratory complications remain the most common cause of mortality following spinal cord injury. Determinants of the extent of respiratory dysfunction after spinal cord injury are the patterns and level of motor, sensory, and autonomic neurological impairment, with greater dysfunction seen at higher injury levels. During spinal shock, flaccid paralysis of the intercostal muscles creates an unstable chest wall. During inspiration, the negative intrathoracic pressure causes paradoxical inward depression of the ribs. This mechanical imbalance and disadvantage result in less efficient ventilation, increased work of breathing, and a tendency toward distal airway collapse and microatelectasis. Over time, the tendons, ligaments, and joints of the rib cage stiffen. The risk of respiratory failure is directly associated with injury level and will not be reviewed in detail in this chapter. Impairment in diaphragm function requiring endotracheal intubation and mechanical ventilation is seen with a complete injury above C5. At C5, diaphragm function is intact but intercostal and abdominal muscle paralysis results in decreased lung volumes, cough strength, and cough effectiveness. In cases of cervical lesions below C5 (C6—C8), patients are able to breathe independently and can augment inspiration and cough with accessory muscles if the lesion is below C7. Cough efficiency remains suboptimal in thoracic lesions, whereas respiratory function is essentially normal if the lesion is at T12 or below (Berlowitz, Wadsworth, & Ross, 2016).

10.1.7 Bladder function

In patients with spinal cord lesions, care is usually given to disordered bladder function. Immediately after complete transection of the spinal cord above the sacral segments, there is complete areflexia of the bladder but no satisfactory explanation for this phenomenon is available. Bladder sensation is absent but the patient may recognize bladder contraction by associated autonomic phenomena such as sweating, piloerection, and flushing. Some patients have a neurogenic bladder, which empties completely at frequent intervals. The capacity of the bladder is small, often no more than 100 mL. Emptying usually occurs without warning and control of micturition is very difficult. This type of bladder has been compared to the evolution of spasticity in the limbs. The final stage is a large capacity bladder and this more desirable state depends upon the prevention of changes in the bladder wall, which are associated with perpetual, though intermittently contracted state, in the early stages after injury. Residual urine may be present even though the bladder is of the spastic reflex neurogenic type. This is attributed to the occurrence of flexor spasms, which interrupt micturition due to the contraction of the pelvic floor muscles occurring simultaneously with the spasm. When extensor spasms are present, reflex micturition is more satisfactory and plantar flexion of the feet and extension of the knees often give warning of impending micturition.

Destruction of the sacral cord results in an autonomous bladder. Voluntary and reflex micturition is abolished but some bladder sensitivity remains particularly the sensation of fullness and that produced by irrigation of the urethra. Destruction of the cord above the sacral division produces the so-called cord bladder. This differs from the autonomous bladder because some spinal reflexes with central connections in the sacral cord remain intact. Voluntary initiation and the sensation of micturition are absent but reflex storage and voiding of urine are possible though residual urine is often present. Lesions of the conus medullaris interrupt the afferent and efferent parasympathetic pathways to the bladder. This results in a so-called decentralized, internuclear neurogenic, or autonomous neurogenic bladder. In this type of bladder voluntary and reflex micturition is absent. The remaining contractions of the bladder are small in amplitude although occasional large contractions occur. The bladder capacity is initially large but returns to normal sometime after injury. The intravesical pressure is high and micturition is frequent. This type of bladder is also seen in lesions of the cauda equina, sacral roots, pelvic nerves, or inferior hypogastric plexus.

10.1.8 Gastrointestinal function

The effect of interruption of extrinsic nerve supply to the intestine has unfortunately not been extensively studied in humans. In the early stage of spinal cord injury, there is vagal hyperactivity, which may contribute to acid hypersecretion, gastric ulcers, and hemorrhages. In patients with lesions in the cervical and upper thoracic spinal cord, activity under resting conditions is less and there are no periodic peristaltic waves that are usually seen in the sigmoid colon. On the other hand, lesions affecting the lumbar outflow are associated with greater than normal activity in the sigmoid colon and this indicates a normal inhibitory influence of the lumbar outflow on colonic motility. The increased peristaltic activity normally seen after meals, or the so-called gastrocolic reflex, is thought to depend upon the integrity of the thoracolumbar outflow but in paraplegic patients as in normal subjects colonic activity is normally increased after food.

The rectum is usually empty. The external anal sphincter is under voluntary control and is made up of striated muscles, which are continuously contracted. Defecation can normally be inhibited voluntarily. This is a complex act and the information about the fullness of the rectum must reach the brain. Severe acute injury of the spinal cord is associated with constipation and sometimes paralytic ileus, whereas in chronic lesions defecation becomes automatic without the subjective urge to defecate. A variety of autonomic responses may precede or accompany defecation and are associated with distention of the rectum and bladder. Patients with high spinal cord lesions may have impaired defecation because of respiratory muscle paralysis. Lesions in the lower spinal cord or cauda equina also interfere with the power to defecate but it can still be automatic. In paraplegic individuals, trigger zones in the sacral areas can be found for which stimulation initiates defecation. These trigger zones are of great help in the management of paraplegic patients and can be used to induce the act at suitable times. It is also important to consider the bowel habits of the patient prior to injury. The type of food eaten is also valuable in regulating bowel movements since the number of motions and their frequency can be influenced by the amount of fiber in the diet. A regular turning from side to side and taking advantage of the naturally occurring gastrocolic reflex may also facilitate bowel training. Initially after injury, particularly in tetraplegics, distention of the bowel by accumulated feces may push up the diaphragm and further affect respiration. In the later stages in lesions above T6, spasms of abdominal muscles and of the external rectal sphincter may interfere with the effective expulsion of feces. Low lesions involving the lumbosacral cord and roots cause a lower motor neuron type of disturbance with a flaccid patulous anus leading to fecal impaction and overflow incontinence. This type of dysfunction requires manual disimpaction at frequent intervals and these patients are most difficult to manage.

The bowel is hypertonic in patients with lesions affecting the corticospinal tract. In lesions of the sacral cord or cauda equine, which produce loss of all sensation including that of inflation of the rectum, the bowel is atonic, and major peristaltic waves are absent (Chung & Emmanuel, 2006). In patients who have had a lumbar sympathectomy, which interferes with the sympathetic nerve supply to the colon, no effect on defecation nor on colonic sensation has been found.

10.1.9 Sexual function

Impairment of sexual function is often a very distressing accompaniment of spinal cord injury. In men, psychogenic erections after appropriate erotic stimuli pass down the spinal cord to the second, third, and fourth sacral segments. Psychogenic erection, which is the most common initiating stimulus under normal circumstances, is abolished in patients with transection of the cord. On occasion, however, patients with complete lesions are capable of responding to erotic stimuli because some of the impulses to the sacral segments pass through the peripheral autonomic nervous system, which remains uninjured.

Reflex erection however is possible in men who have lesions above the sacral cord while women may maintain some sexual function and these have been discussed in Chapter 8 of this book. In the management of sexual dysfunction in patients with spinal cord injury, certain recurring questions usually arise that need compassionate understanding and lengthy explanation (Comarr & Vigue, 1978a,b; Eisenberg & Rustad, 1976).

10.1.10 Pupillomotor abnormalities

Abnormalities in pupillary function in spinal cord lesions have been reported. In complete lesions of the cervical cord, unilateral or bilateral Horner syndrome is often present. This syndrome is due to the interruption of oculopupillary fibers, which originate in the hypothalamus and descend in the anterolateral columns of the spinal cord. While the pupillary constriction is most marked if the seventh and eighth cervical and first thoracic roots have been interrupted or when the inferior cervical sympathetic ganglion has been removed, pupillary constriction also occurs in cord lesions alone. The pupil continues to react to light, accommodation, convergence and the lid and psychosensory reflexes are retained (see Chapter 13). The smallest diameter of the pupil occurs usually initially soon after the interruption of the sympathetic supply because of the overactivity of parasympathetic innervation of the pupil. But with time, there is progressive enlargement of the affected pupil though it usually remains somewhat smaller than the unaffected side. Eventually, the affected pupil may become equal to the normal one but the denervation can still be recognized by the use of dopamine drops, which will dilate the normal side and leave the affected side smaller. In some patients with cervical cord lesions, an excessive reflex pupillary dilatation can occur and this spasmodic dilatation of the pupil can cause severe photophobia. At the onset of the Horner syndrome, there is an initial rise in skin temperature on the affected side of the face, mainly due to interruption of vasomotor fibers and consequent vasodilatation. The loss of sweating may also contribute to the rise of skin temperature. In addition, in the acute stages of cervical cord lesions with bilateral Horner syndrome, there may be blockage of nasal passages due to vasomotor paralysis and marked vasodilatation of mucosal vessels. This may further affect respiration in those in whom intercostal muscles are already compromised by the lesion in the cord. In the later stages after cord injury, the complete Horner syndrome is not usually found and there may be dissociation of the various components of the syndrome, particularly a return of sweating and vasomotor tone. Ptosis may be observed but only with difficulty and if the patient is physically tired.

10.1.11 Trophic disorders

A puzzling feature regarding the paralysis of tetraplegic or paraplegic individuals is the early danger of pressure sore formation, which has not been fully explained. In contrast to spinal cord injury, the paralysis produced by anterior horn cell disease is remarkably free of this intractable complication. Patients with amyotrophic lateral sclerosis, for example, do not suffer from pressure sores and the simple explanation often proffered that this is because of their preserved sensation is not entirely satisfactory. Amyotrophic lateral sclerosis patients are equally exposed to pressure and a great deal of pressure sore prevention in spinal cord injury patients without sensation can be done by relieving pressure points. In tetraplegic individuals, osteoporosis, calciuria, and dystrophic cutaneous changes have been related to early metabolic disturbances in collagen degradation. Both bone and skin contain, in addition to collagen, glycosaminoglycans. The urinary excretion of these substances in tetraplegic individuals is larger but less oligosaccharides are excreted soon after injury while immediately after injury the excretion of oligosaccharides is increased. Several macromolecular components of connective tissue are, therefore, affected in tetraplegia almost immediately after injury and for weeks thereafter.

These changes cannot be entirely attributed to immobilization and other risk factors include nutritional deficiencies and imbalances, vascular and hemodynamic changes, physical risk factors, medications, psychological risk factors, and socioeconomic determinants (Byrne & Salzberg, 1996; Singh, DiFerrante, Strashun, & Gyorkey, 1977).

10.1.12 Phantom experiences following spinal cord injury

After amputation of a limb, many patients feel as though the missing limb is still present. This imaginary limb has been called a phantom limb and it may seem quite normal or feel distorted or painful. Patients with cord lesions without amputations may feel that their anesthetic limbs are either flexed or extended, but in fact, they may be in a different position altogether. Occasionally the lower limbs feel bloated to the patient and peculiar tingling sensations may be present. Contrary to phantom experience in amputees, paraplegic and tetraplegic phantoms are rarely if ever painful and usually not troublesome. The quality of phantom experience in cord lesions is often influenced or brought about by the actual position of the limbs and their posture immediately preceding the onset of paraplegia. In such patients, if subsequent amputation of a limb becomes necessary, the phantom experiences are not altered. In patients with cord injury with anesthesia below the waist, who prior to trauma had an amputation with phantom experiences, the cord lesion may cause phantoms of the unamputated leg similar to those experienced on the amputated side. Phantom experience without amputation is not confined to patients with cord lesions but is also seen in those with sensory loss due to parietal lobe lesions and in subjects undergoing spinal anesthesia.

10.1.13 Triple response of Lewis in patients with spinal cord injury

Transection of the spinal cord influences the triple response of Lewis. The flair component of the triple response in anesthetic dermatomes is often smaller or absent altogether. In addition, central nervous system lesions tend to modulate the flair also. In the examination of patients with cord lesions, the reduced flair of the triple response is helpful in defining the upper level of the lesion. If a heavy stroke of the skin is made beginning in areas above the suspected site of the lesion and extending well into anesthetic areas, a sudden decrease in the extent of the flair in dermatomes below the lesion can usually be seen. This decrease is as a rule one or two dermatomes above the upper limit of sensory loss detected on clinical examination. Occasionally in hyperpathic border zones between anesthetic and normal dermatomes, the flair response may be exaggerated. Also, during the initial examinations of patients with traumatic cord lesions, it may be found that the oiliness of the skin in dermatomes below the lesion is much decreased. It is generally believed that sebaceous glands, which are responsible for skin greasiness, are not influenced by denervation but other factors may contribute to the change in the texture of the skin in anesthetic dermatomes of paraplegics.

10.2 Autonomic dysreflexia

10.2.1 Introduction

Immediately after cord injury and the passing of the stage of flaccid paralysis, hyperreflexia (dysreflexia) appears not only in the somatic neuromuscular apparatus but also in autonomic neuroeffector systems (Sharif & Hou, 2017). Cutaneous or visceral stimulation below the level of the spinal cord injury initiates afferent impulses that trigger reflex sympathetic nervous system activity. The sympathetic response leads to diffuse vasoconstriction and a rise in blood pressure. In the setting of a spinal cord injury, the normal compensatory parasympathetic response cannot travel below the level of the spinal cord injury, and generalized vasoconstriction continues leading to systemic hypertension. The compensatory parasympathetic response leads to bradycardia and vasodilation, but only above the level of the spinal cord injury (Fig. 10.1). Such abnormal autonomic responses are common with thoracic (above T6) and cervical cord lesions and are at least initially observed to some extent in all cases with lesions above the sixth thoracic segment. The abnormal autonomic discharge is triggered by any cutaneous or other somatic stimuli but occurs often in response to bladder or rectal distention, sexual intercourse, or during labor. Percussion of an empty bladder unaccompanied by flexor spasms and cystometry can also be effective stimuli for autonomic dysreflexia (Wallin et al., 2013). At times, autonomic dysreflexia occurs in high spinal cord lesions without an obvious stimulus or manipulation of the body, and one is forced to the conclusion that peristaltic activity, sudden changes in environmental temperature, and other less obvious stimuli may lead to

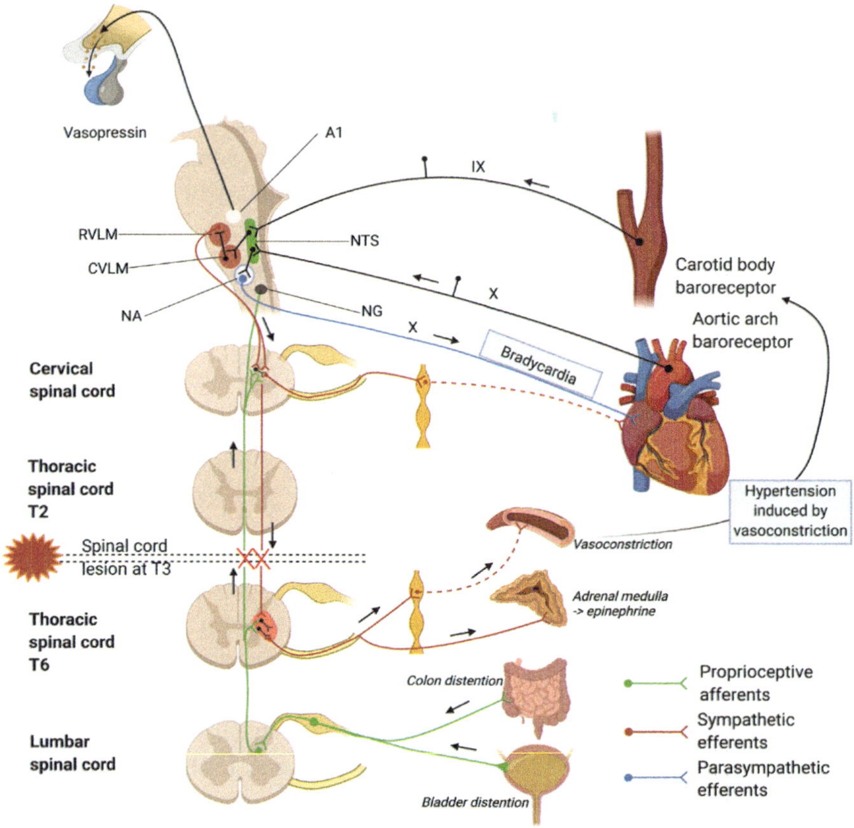

Figure 10.1 Neuronal pathways disrupted by complete spinal cord injury above the T6 spinal level are associated with the development of autonomic dysreflexia. Sensory afferent fibers (green) from the distended bladder or colon transmit stimuli to propriospinal neurons at corresponding spinal levels. These propriospinal neurons then relay the signal rostrally to activate sympathetic preganglionic neurons (red) and interneurons present in the thoracolumbar intermediolateral cell column to trigger sympathetic activation of the adrenal gland and norepinephrine-mediated peripheral vasoconstriction, which leads to hypertension. The increase in blood pressure is sensed by baroreceptors in the aortic arch and carotid sinuses, which relay the information via the nucleus tractus solitarius (NTS) in the medulla to the nucleus ambiguous (NA) that elicits a parasympathetic bradycardic response mediated by acetylcholine. Concomitant neuromodulation via caudal and rostral ventrolateral medulla (CVLM and RVLM) project to the intermediolateral cell column directly or via interneurons to maintain normal blood pressure. After spinal cord injury, the decentralized sympathetic preganglionic neurons elicit uninhibited vasoconstriction and hypertension, while the nucleus ambiguus signals baroreflex-mediated bradycardia. The lack of supraspinal regulation of sympathetic preganglionic neurons below the injury site maintains the hypertensive response until the noxious stimuli are removed, and maladaptive plasticity of both primary afferent fibers and propriospinal neurons is associated with the development of autonomic dysreflexia. *NG*, nucleus gracilis.

the autonomic discharge. Although it was previously thought that noxious inputs below the lesion could trigger autonomic dysreflexia, Brown and colleagues found that large-diameter somatic afferents, not small-diameter afferents, trigger the increases in vasoconstrictor drive that lead to autonomic dysreflexia (Brown, Burton, & Macefield, 2018).

Following spinal cord injury lesions, there is reactive sprouting of primary afferents impinging on interneurons, and exaggerated response of interneurons to somatic afferent stimulation (Rabchevsky, 2006). Reorganization of nociceptive circuits after spinal cord injury may result in changes in the effects of GABAergic inputs from inhibitory into excitatory on spinal interneurons (Shouman & Benarroch, 2019). Local vascular changes in the spinal cord may elicit ischemia in the setting of spinal cord injury and a paradoxical excess activity of serotonergic 5-HT1 on pericytes has been found (Li et al., 2017).

10.2.2 Clinical manifestation and autonomic testing

Autonomic dysreflexia consists of vasoconstriction in the paralyzed part of the body, piloerection, sweating and consequent lowering of the skin temperature in dermatomes below the lesion. The vasoconstriction in a large area of the skin prevents heat loss by convection and central temperature may rise leading to an attempted thermoregulatory activity with vasodilatation in dermatomes innervated from segments above the lesion; however, this is usually ineffective and fever often results. This vasodilatation does give rise to blotchy redness of the skin of the face and neck, and often a blocked nose.

The vasoconstriction in a large segment of resistance vessels leads to extreme hypertension and this is associated with bradycardia, which is related to the activity of baroreceptors in the great vessels and efferent discharges through the vagus to the heart. Cardiac arrhythmias are also common. During such episodes, feeling of anxiety, discomfort below the level of the lesion, shivering, goose pimples, and pounding headache occur and the pupils may dilate. An increase in circulating epinephrine or norepinephrine is also found and decreases in muscle blood flow and venous constriction is the rule. The cause of the pounding headache is not clear because this does not always occur in a definite temporal relationship to the sudden increase in blood pressure and studies on the release and of levels of vasoactive substances during such episodes are necessary to further elucidate the headache of autonomic hyperreflexia.

After high spinal cord transection, autonomic dysreflexia is prominent usually months or years after the injury. In some patients, the accompanying hypertension can be counteracted by head-up tilt. In three quadriplegic patients, however, hypotension was triggering or aggravating the symptoms of autonomic dysreflexia (Khurana, 1987). All three patients complained also of severe hyperhidrosis, which is a well-known complication of high spinal cord transection (Head & Riddoch, 1917).

The hypotension-induced autonomic dysreflexia remains unexplained, but possible afferent impulses from blood vessels and their perivascular innervation could produce a mass discharge of sympathetic neurons (Kewalramani, 1980). Alternatively, the spinal cord could be the source of sympathetic hyperactivity because of relative ischemia associated with tilting, but all these hypothetical explanations need to be supported by further study of tetraplegic patients. What is important is the responsiveness of the autonomic dysreflexia in these patients to fludrocortisone acetate, which improved posture-associated neurologic dysfunction in all.

Tilt table testing can demonstrate episodes of hypertension associated with the upright posture and/or muscle spasms and an overall tendency toward orthostatic hypotension.

Thermoregulatory sweating is preserved only in regions with intact sympathetic supply.

Measurement of plasma catecholamines during paroxysmal neurogenic hypertension associated with autonomic dysreflexia has shown consistent elevations in both norepinephrine and epinephrine levels, whereas circulating levels of catecholamines are chronically decreased in-between episodes, leading to lower baseline blood pressure. Infusion of norepinephrine to raise the blood pressure to comparable levels found in patients with distended bladders or during muscle contractions resulted in norepinephrine levels, which were 21 times higher than those spontaneously found during neurogenic hypertension. The neurogenic hypertension does not, therefore, seem to be secondary only to the rise in plasma norepinephrine per se, but reflects an increase in sympathetic nervous activity with $\alpha 1$-receptor hypersensitivity. Whether high epinephrine levels during autonomic dysreflexia can be related to the often accompanying headache or be pathogenetically important also in headache associated with coitus (see previous chapter) is not clear (Mathias, Christensen, Corbett, Frankel, & Spalding, 1976). Similar elevations in plasma dopamine ß-hydroxylase occurred some 3–5 minutes after peak blood pressure responses and in neurogenic hypertension. This indicates that in tetraplegic individuals, at least, dopamine ß-hydroxylase is released from sympathetic nerve terminals during sympathetic activity (Mathias, Smith, Frankel, & Spalding, 1976).

Autonomic dysreflexia in response to bladder filling or frequent uninhibited detrusor contractions is the result of exaggerated responses, which occur normally before and during bladder emptying. In subjects with intact spinal cords, these responses are modulated from higher centers. A rise in blood pressure is often found in those with full bladders and the blood pressure falls when the bladder is emptied, but these changes are compensated for by vasomotion. Sometimes precipitous falls in blood pressure in elderly otherwise intact individuals occur who because of their somewhat impaired vasomotor function may faint during micturition leading to so-called micturition syncope. Autonomic dysreflexia is usually milder and less frequent with rectal distention and occasionally it is the result of pressure sores and severe flexion contractures. At times paraplegic individuals have succumbed to cerebral hemorrhage during the height of the pressure rise and prompt reduction of systemic pressure is necessary (Pan et al., 2005).

10.2.3 Management of autonomic dysreflexia

The management of patients with autonomic dysreflexia includes careful manipulation, prevention of pressure sores, contractures, and uninhibited repeated bladder contractions. All these are important in the prophylaxis of abnormal autonomic discharges. Autonomic dysreflexia may give warning of an impending abdominal catastrophe such as a perforated viscus or it may occur during the last stages of labor in pregnant paraplegic women. In high cervical lesions, an intact phrenic nerve can mediate nociceptive impulses arising in the diaphragm to cord segments above the lesion. This occurs sometimes with perforation, distention, and other subdiaphragmatic emergencies. Pain in the shoulder may be a warning and should always alert the attending physician to carry out a careful examination of the abdomen. In chronic cord lesions above T5, the visceral sensation is not entirely absent. Afferent impulses due to spasm of the bowel, bladder distention, uterine contractions during labor, menstrual cramps or ureteric stone migration, and other visceral disorders may reach consciousness. Impulses apparently travel in extramedullary sympathetic pathways to reach the cord above the level of the lesion and eventually, though usually in a distorted fashion, consciousness.

Paroxysmal hypertension may be relieved by tilt-up or standing and vasodilator drugs (e.g., nitroglycerine, nifedipine) have been used with success. Agents such as prazosin (α1 adrenergic receptors blocker) or gabapentin (inhibiting presynaptic glutamate release) may be beneficial. Botulinum toxin injections may be utilized to relieve limb spasticity and detrusor muscle overactivity.

Vague discomfort and strange feelings in a part of the body below the lesion need careful consideration and appropriate investigation to elucidate the cause. Therefore, in most paraplegic individuals, some local and remote sensation of visceral activity remains and the loss of sensations in segments below the level of the lesion is not complete nor always permanent. For example, the sensation at the breaking point of breath-holding in normal subjects is relieved by breathing, a mixture that does not alter the hypoxia. It is assumed therefore that this sensation arises from stimuli in the contracting respiratory muscles (Noble, Frankel, Else, & Guz, 1971). Since paralysis of all respiratory muscles except the diaphragm and some accessory muscles of respiration does not abolish this sensation, it was attributed to afferent stimuli arising in the diaphragm. This was confirmed in a patient with complete cord transection at C3. This patient maintained adequate ventilation by the use of sternomastoid muscles. No unpleasant sensation of breath-holding reached consciousness in this patient. It is of interest that in this same patient in whom the chest wall and diaphragm was totally disconnected from the brain, the recognition of added resistance loads to respiration was preserved just like in patients with tracheostomies without neurologic lesions (Noble, Eisele, Frankel, Else, & Guz, 1971). The same vague sensation of resistance in the chest was reported by this patient as it occurred in controls. Since the only nerve connecting the chest with the brain was the vagus nerve, the sensation of added resistance to breathing must be related to afferent impulses traveling in this nerve.

Intolerable pain and hyperpathia may occur in dermatomal border zones between normal and lost cutaneous sensation. Tightness, burning sensations and intolerance to clothing, blotchy redness of the skin, piloerection, and cutaneous sweating may be observed in this zone. These phenomena are aggravated during episodes of autonomic hyperreflexia.

Epidural electrical stimulation of the spinal cord (T10–T11) with a closed-loop system is a promising treatment for spinal cord injury with autonomic dysreflexia. This "neuroprosthetic baroreflex" controlled hemodynamics in rodents, nonhuman primates, and humans, after both acute and chronic spinal cord injury (Squair et al., 2021). The neuroprosthetic baroreflex triggered well-controlled and reversible increases in blood pressure in one patient with tetraplegia. Future studies will investigate the hemodynamic responses to chronic stimulation and its effect on autonomic dysreflexia.

References

Berlowitz, D. J., Wadsworth, B., & Ross, J. (2016). Respiratory problems and management in people with spinal cord injury. *Breathe (Sheffield, England)*, *12*(4), 328–340.

Brown, R., Burton, A. R., & Macefield, V. G. (2018). Autonomic dysreflexia: somatosympathetic and viscerosympathetic vasoconstrictor responses to innocuous and noxious sensory stimulation below lesion in human spinal cord injury. *Autonomic Neuroscience: Basic & Clinical*, *209*, 71–78.

Byrne, D. W., & Salzberg, C. A. (1996). Major risk factors for pressure ulcers in the spinal cord disabled: a literature review. *Spinal Cord: the Official Journal of the International Medical Society of Paraplegia*, *34*(5), 255–263.

Cariga, P., Ahmed, S., Mathias, C. J., & Gardner, B. P. (2002). The prevalence and association of neck (coat-hanger) pain and orthostatic (postural) hypotension in human spinal cord injury. *Spinal Cord: the Official Journal of the International Medical Society of Paraplegia*, *40*(2), 77–82.

Chung, E. A., & Emmanuel, A. V. (2006). Gastrointestinal symptoms related to autonomic dysfunction following spinal cord injury. *Progress in Brain Research*, *152*, 317–333.

Claydon, V. E., Steeves, J. D., & Krassioukov, A. (2006). Orthostatic hypotension following spinal cord injury: understanding clinical pathophysiology. *Spinal Cord: The Official Journal of the International Medical Society of Paraplegia*, *44*(6), 341–351.

Comarr, A. E., & Vigue, M. (1978a). Sexual counseling among male and female patients with spinal cord and/or cauda equina injury. *American Journal of Physical Medicine*, *57* (3), 107–122.

Comarr, A. E., & Vigue, M. (1978b). Sexual counseling among male and female patients with spinal cord and/or cauda equina injury. Part II. Results of interview and neurological examinations of females. *American Journal of Physical Medicine*, *57*(5), 215–227.

Corbett, J. L., Frankel, H. L., & Harris, P. J. (1971a). Cardiovascular changes associated with skeletal muscle spasm in tetraplegic man. *The Journal of Physiology*, *215*(2), 381–393.

Corbett, J. L., Frankel, H. L., & Harris, P. J. (1971b). Cardiovascular reflex responses to cutaneous and visceral stimuli in spinal man. *The Journal of Physiology*, *215*(2), 395–409.

Corbett, J. L., Frankel, H. L., & Harris, P. J. (1971c). Cardiovascular reflexes in tetraplegia. *Paraplegia*, *9*(3), 113–122.

Corbett, J. L., Frankel, H. L., & Harris, P. J. (1971d). Cardiovascular responses to tilting in tetraplegic man. *The Journal of Physiology*, *215*(2), 411–431.

Duprez, D. A. (2006). Role of the renin-angiotensin-aldosterone system in vascular remodeling and inflammation: a clinical review. *Journal of Hypertension*, *24*(6), 983–991.

Eisenberg, M. G., & Rustad, L. C. (1976). Sex education and counseling program on a spinal cord injury service. *Archives of Physical Medicine and Rehabilitation*, *57*(3), 135–140.

Frisbie, J. H., & Steele, D. J. (1997). Postural hypotension and abnormalities of salt and water metabolism in myelopathy patients. *Spinal Cord: The Official Journal of the International Medical Society of Paraplegia*, *35*(5), 303–307.

Griggs, K. E., Price, M. J., & Goosey-Tolfrey, V. L. (2015). Cooling athletes with a spinal cord injury. *Sports Medicine (Auckland, N.Z.)*, *45*(1), 9–21.

Head, H., & Riddoch, G. (1917). The automatic bladder, excessive sweating and some other reflex conditions, in gross injuries of the spinal cord. *Brain*, *40*(2–3), 188–263.

Holaday, J. W., & Faden, A. I. (1980). Naloxone acts at central opiate receptors to reverse hypotension, hypothermia and hypoventilation in spinal shock. *Brain Research*, *189*(1), 295–300.

Katzelnick, C. G., Weir, J. P., Pinto Zipp, G., LaFountaine, M. F., Bauman, W. A., Dyson-Hudson, T. A., & Wecht, J. M. (2021). Increased pulse wave velocity in persons with spinal cord injury: the effect of the renin-angiotensin-aldosterone system. *American Journal of Physiology. Heart and Circulatory Physiology*, *320*(1), H272–H280.

Kewalramani, L. S. (1980). Autonomic dysreflexia in traumatic myelopathy. *American Journal of Physical Medicine*, *59*(1), 1–21.

Khurana, R. K. (1987). Orthostatic hypotension-induced autonomic dysreflexia. *Neurology*, *37*(7), 1221–1224.

Krassioukov, A. V., Bunge, R. P., Pucket, W. R., & Bygrave, M. A. (1999). The changes in human spinal sympathetic preganglionic neurons after spinal cord injury. *Spinal Cord: The Official Journal of the International Medical Society of Paraplegia*, *37*(1), 6–13.

Li, Y., Lucas-Osma, A. M., Black, S., Bandet, M. V., Stephens, M. J., Vavrek, R., & Bennett, D. J. (2017). Pericytes impair capillary blood flow and motor function after chronic spinal cord injury. *Nature Medicine*, *23*(6), 733–741.

Mathias, C. J., Christensen, N. J., Corbett, J. L., Frankel, H. L., & Spalding, J. M. (1976). Plasma catecholamines during paroxysmal neurogenic hypertension in quadriplegic man. *Circulation Research*, *39*(2), 204–208.

Mathias, C. J., Smith, A. D., Frankel, H. L., & Spalding, J. M. (1976). Dopamine beta-hydroxylase release during hypertension from sympathetic nervous overactivity in man. *Cardiovascular Research*, *10*(2), 176–181.

Mortola, J. P., & Sant'Ambrogio, G. (1978). Motion of the rib cage and the abdomen in tetraplegic patients. *Clinical Science and Molecular Medicine*, *54*(1), 25–32.

Noble, M. I., Eisele, J. R., Frankel, H. L., Else, W., & Guz, A. (1971). The role of the diaphragm in the sensation of holding the breath. *Clinical Science*, *41*(3), 275–283.

Noble, M. I., Frankel, H. L., Else, W., & Guz, A. (1971). The ability of man to detect added resistive loads to breathing. *Clinical Science*, *41*(3), 285–287.

Pan, S. L., Wang, Y. H., Lin, H. L., Chang, C. W., Wu, T. Y., & Hsieh, E. T. (2005). Intracerebral hemorrhage secondary to autonomic dysreflexia in a young person with incomplete C8 tetraplegia: a case report. *Archives of Physical Medicine and Rehabilitation*, *86*(3), 591–593.

Rabchevsky, A. G. (2006). Segmental organization of spinal reflexes mediating autonomic dysreflexia after spinal cord injury. *Progress in Brain Research*, *152*, 265–274.

Sharif, H., & Hou, S. (2017). Autonomic dysreflexia: a cardiovascular disorder following spinal cord injury. *Neural Regeneration Research*, *12*(9), 1390–1400.

Shouman, K., & Benarroch, E. E. (2019). Segmental spinal sympathetic machinery: implications for autonomic dysreflexia. *Neurology*, *93*(8), 339–345.

Singh, J., DiFerrante, N., Strashun, M. F., & Gyorkey, F. (1977). Urinary excretion of glycosaminoglycans in quadriplegia. *Proceedings of the Society for Experimental Biology and Medicine. Society for Experimental Biology and Medicine (New York, N.Y.)*, *156*(3), 488–490.

Squair, J. W., Gautier, M., Mahe, L., Soriano, J. E., Rowald, A., Bichat, A., & Phillips, A. A. (2021). Neuroprosthetic baroreflex controls haemodynamics after spinal cord injury. *Nature*, *590*(7845), 308–314.

Sved, A. F., McDowell, F. H., & Blessing, W. W. (1985). Release of antidiuretic hormone in quadriplegic subjects in response to head-up tilt. *Neurology*, *35*(1), 78–82. Available from https://doi.org/10.1212/wnl.35.1.78.

Unsal-Delialioglu, S., Kaya, K., Sahin-Onat, S., Kulakli, F., Culha, C., & Ozel, S. (2010). Fever during rehabilitation in patients with traumatic spinal cord injury: analysis of 392 cases from a national rehabilitation hospital in Turkey. *The Journal of Spinal Cord Medicine*, *33*(3), 243–248.

Wallin, B. G., Burton, A. R., Elam, M., Tamaddon, K., Millard, R., & Macefield, V. G. (2013). Viscerosympathetic reflexes in human spinal cord injury: relationships between detrusor pressure, blood pressure and skin blood flow during bladder distension. *Experimental Physiology*, *98*(6), 1081–1091.

Wecht, J. M., Radulovic, M., Weir, J. P., Lessey, J., Spungen, A. M., & Bauman, W. A. (2005). Partial angiotensin-converting enzyme inhibition during acute orthostatic stress in persons with tetraplegia. *The Journal of Spinal Cord Medicine*, *28*(2), 103–108.

CHAPTER 11

Autonomic adaptation to hypoxia: mountain medicine

The great classic of altitude physiology La Pression Barométrique was published by Paul Bert in 1878 (Bert, 1878). In this book, he described his experiments on the effects of barometric pressure changes on living beings. To his lasting credit and together with his collaborator Dr. Jourdanet, he concentrated on low barometric pressure first since for millions this is the natural condition of life. It is thus not only a fit subject in physiology and medicine but more generally of interest as a factor of human well-being (Dejours & Dejours, 1992). Bert also did many experiments on himself and on March 30, 1874, he exposed himself in his altitude chamber to a barometric pressure approximating that prevailing on top of Mount Everest thus anticipating more sophisticated human altitude chamber experiments by many years (Houston & Riley, 1947; Houston, Sutton, Cymerman, & Reeves, 1987). Historically, many noteworthy contributions to the understanding of the adaptation of humans to altitude have been made in the laboratory where conditions can be controlled. However, studies of altitude natives and of sojourners transiently exposed to the mountain environment allow comparisons that cannot be made in the laboratory and add different dimensions to the study of adaptation to hypoxia, albeit under less rigorously controlled conditions. Altitude is not synonymous with just a decrease in barometric pressure and consequent hypoxia, as achieved in an altitude chamber, but many additional ambient factors such as temperature, humidity, sensory stimulation, and radiation play important roles. Thus, both laboratory experiments and field studies have their place in defining adaptation to hypoxia.

In recent years, the role of the autonomic nervous system (ANS) in adaptation has become apparent. Many animal and a few human studies attest to the crucial importance of sympathetic activation that occurs on acute altitude exposure. Unfortunately, few altitude natives have been studied either in the laboratory or in the field.

The hypoxia of altitude has clinical significance at sea level. Many human diseases have hypoxia of tissues as their basic pathogenic mechanism.

Introduction to Clinical Aspects of the Autonomic Nervous System
DOI: https://doi.org/10.1016/B978-0-323-96011-3.00001-2

Thus, lessons from the mountain have relevance in the clinic and they may contribute to an understanding of the compensatory autonomic mechanisms that allow adaptation to aging and disease.

Paul Bert's 13 communications Experimental Research on the Influence the Changes in Barometric Pressure Exert on Life Processes, appeared in the Comptes Rendus de l'Academie des Sciences, beginning in 1871; they were sponsored by Claude Bernard, the father of autonomic physiology.

This chapter is confined to autonomic studies in normal humans in the laboratory and in the field. Relevant clinical studies on the influence of the changes in barometric pressure on life processes are reviewed.

11.1 Cardiovascular function at altitude

Well-controlled experiments on young healthy humans were conducted in an altitude chamber during Operation Everest II (Sutton et al., 1992). Although eight subjects were initially recruited for the ascent, only six made it to the top (234 Torr) where progressive exercise tests to exhaustion were carried out. The subjects spent 40 days and nights in the chamber. Maximal oxygen uptake decreased by an average of 72% of sea-level values at the summit. Progressive changes in electrocardiograms indicated increasing pulmonary artery hypertension. No changes indicative of myocardial ischemia nor arrhythmias were noted at rest in spite of profound hypoxemia. During exercise tests to exhaustion, performance progressively declined with increasing altitude and maximum heart rates decreased but ventilation increased. Perceived exertion (on the Borg scale) per unit of work was similar at all altitudes. At exhaustion, there was a progressive decline in blood lactate levels probably reflecting the decreasing quantity of work carried out at the summit. Muscle biopsies taken at the end of exercise also showed decreasing lactate values with increasing altitude (Green, Sutton, Young, Cymerman, & Houston, 1989).

Progressive decreases in left ventricular end-diastolic and end-systolic volumes and stroke volumes were noted on echocardiography. Indices of left ventricular systolic function were unchanged or even slightly enhanced by altitude reflecting slight increases in left ventricular contractility. Even exercise to exhaustion at the summit was not associated with electrocardiographic signs of myocardial ischemia.

Important changes in the right heart on electrocardiography were consistent with pulmonary hypertension and the echocardiograms showed moderate right ventricular enlargement and paradoxical septal wall motion

(Suarez, Alexander, & Houston, 1987). Right and left ventricular filling pressures remained normal with increasing altitude and the conclusion was reached that ventricular function remained normal and therefore it was not responsible for the declining exercise performance with increasing altitude.

Oxygen transport during exercise at altitude showed that major adaptation occurred in increases in ventilation, increases in hemoglobin, and in maximum oxygen extraction (Sutton et al., 1988).

There were no changes in cardiac output for a given oxygen uptake at all altitudes and no increases in cardiac output occurred (after acclimatization) to compensate for the marked arterial desaturation and decreased arterial oxygen content. Resting pulmonary artery pressures increased from 15 to 34 mmHg at the summit and pulmonary vascular resistance from 1.2 to 4.3 $\text{mmHg} \times 1 \times \text{min}^{-1}$; exercise at altitude considerably worsened pulmonary hypertension.

Reinhold Messner and Peter Habeler reached the summit of Mount Everest without supplemental oxygen on May 8, 1978, a feat that earlier altitude physiologists thought impossible. Operation Everest II gave some insights on how this historic achievement became possible but did not address the fundamental question of how the adaptive processes that allow such a feat are activated and the mechanisms that control such adaptation.

Field studies of exercising sojourners, well-adapted climbers, and Tibetan natives at the base camp of Mount Everest (5150 m), in general, confirmed the chamber studies but also allowed an assessment to be made of the effects of a rheologically active drug (pentoxifylline) on exercise performance and heart rate responses (Wood & Appenzeller, 1988). Sojourners taking the drug had a significantly better cardiac response to exercise as evidenced by higher maximal heart rates, yet their performance on a rowing ergometer was not significantly better and declined more with increasing altitude than in the controls. This was attributed to a greater degree of diffusion limitation to oxygen uptake because of the higher cardiac output in the medication group thus confirming laboratory data on limiting factors for exercise at altitude (Wagner et al., 1986). Tibetans, though never having exercised on a rowing ergometer, generated more power output than either the medication or control groups at the same altitude (White et al., 1987).

One distinguishing feature of successful extreme altitude climbers is their superior hypoxic ventilatory drive which allows the enormous hyperventilation seen at extreme altitude (>8000 m) and which, in turn, helps maintain oxygenation (West, 1988). It seems appropriate that cardiac output should also rise to increase oxygen delivery to tissues. However, this is not the case, as in adapted climbers cardiac output for a given amount of work is the same at sea level or extreme altitude. This paradox can only be explained by altitude-induced ANS impairment. Thus, the cardiac output does not keep up with ventilatory adaptation to improve oxygen delivery to tissues. In acute hypoxia, cardiac output is increased by an increase in heart rate but this is not maintained after acclimatization (Vogel & Harris, 1967). In chronic hypoxia, stroke volume falls and this is associated with tachycardia to maintain the cardiac output–work rate relationship although the fall in stroke volume is not due to the hypoxia-induced decrease in cardiac contractility (Groves et al., 1987). Studies of baroreceptor function with simultaneous power spectral analysis of heart rate, blood pressure, and microcirculation at altitude and in the field may provide answers to this paradox (Bernardi et al., 1995).

11.2 Cerebrovascular function at altitude

Headache is an integral and the most bothersome part of acute mountain sickness (Appenzeller, 1986). This headache has many clinical similarities to vascular headaches at sea level which are associated with an increase in cerebral blood flow. It was assumed, therefore, that the pathogenesis of this benign but annoying pain was related to the increase in cerebral blood flow on acute exposure to altitude (Hackett, 1988). This hypothesis is supported by the beneficial effects on altitude headache of sumatriptan (Bärtsch, Maggi, Kleger, Ballmer, & Baumgartner, 1994). Hypoxia-induced cerebral vasodilatation at altitude overrides the vasoconstrictor drive caused by the hyperventilatory hypocapnic alkalosis (Severinghaus, Chiodi, Eger, Brandstater, & Hornbein, 1966).

Using modern Doppler technology on Denali at an altitude of 4300 m, middle cerebral artery flow velocities were examined (Hackett, 1988). While climbers with headache alone had no different velocities from those without headache, persons with severe altitude illness had higher flow velocities indicating higher middle cerebral artery blood flows. These individuals were more severely hypoxemic and the administration of oxygen promptly decreased the blood flow and improved the headache. Using the administration of CO_2 to manipulate oxygenation by further driving ventilation, it was determined that the blood flow changes demonstrated in the middle cerebral artery were driven by the oxygen saturation in the blood and were not the primary cause of headache associated with high altitude (Hackett, 1988). While decreasing blood flow in the middle cerebral artery was clearly beneficial it remained only so while saturation was also improved.

The Valsalva maneuver has profound effects on the circulation and these effects are similar when measured by the Doppler technique in peripheral or intracranial arteries. The ANS has a modulating influence on the peripheral circulatory responses to the maneuver but the intracranial responses are mainly driven by changes in intracranial pressure induced by the maneuver (Greene, Appenzeller, Wood, & Appenzeller, 1991). A large number of subjects without headache, including high-altitude natives, were studied at various altitudes up to 5100 m (Appenzeller, Greene, Qualls, & Martignoni, 1994). The Valsalva ratios (that is the highest flow after the maneuver divided by the lowest flow during the maneuver) were independent of PCO_2. The ratios were inversely related to altitude and decreased with acclimatization. However, adjusting the flows for PCO_2 after acclimatization, sea level (40 Torr), and preValsalva (baseline) levels removed the differences in ratios attributed to altitude. Vertebral artery ratios were higher than middle cerebral artery ratios at all altitudes and there were no differences with age. Natives had significantly lower ratios than sojourners. Physical training had no effect on ratios in either artery before and after a 161 K trail race at an altitude of 3100 m. These studies support the idea that headache of altitude is primarily the result of altitude-induced changes in intracranial pressure but do not exclude a neurovascular component.

These results are consistent with earlier observations of middle cerebral artery blood flows at low and high altitude (Appenzeller, 1992). Natives had considerably lower flows than sojourners of the same age after correction for sea level PCO_2 and hematocrit of 45%. Aging was inversely related to flows at 5150 m. At 1500 m however, no relation of age to middle cerebral artery blood flow was found (Wood, Appenzeller, & Appenzeller, 1989).

Short-lasting baroreceptor manipulation by neck suction does not change blood pressure but increases RR intervals immediately after the application of 30 mmHg negative pressure to the neck (Sleight et al., 1995). The changes in flow velocities can be monitored by Doppler techniques in the renal and middle cerebral arteries. Six sojourners and six Indian yogi practitioners were studied at sea level upon arrival at 4970 m and 1 week later at the same altitude. As expected, the RR interval was significantly shortened on acute altitude exposure, the well-known altitude tachycardia. Flow velocities in the middle cerebral artery (MCA) increased but those in the renal artery (RA) were unaffected by altitude nor by PCO_2. Neck suction had no effect on flow velocities in the MCA but caused a significant fall in flow velocities in the RA. There were no differences between the responses of Indian yogis and sojourners. The degree of vasoconstriction in the renal arteries after neck suction at altitude was significantly less in older subjects but no age relationship could be recognized in the same subjects at sea level. Indian yogis in spite of different heart rate control have the same responses to baroreceptor manipulation in the MCA and RA at sea level and altitude (Figs. 11.1 and 11.2).

Molecular aspects of cerebrovascular function during hypoxia have been examined in animals (Harik, Behmand, & LaManna, 1994). The transport of glucose from the blood to the brain, essential for normal function, occurs primarily at the blood-brain barrier and is dependent on a number of glucose-transporting proteins found on brain microvessels. Chronic hypobaric hypoxia increases the density of brain microvessels as assessed by immunocytochemistry using an antibody to brain glucose

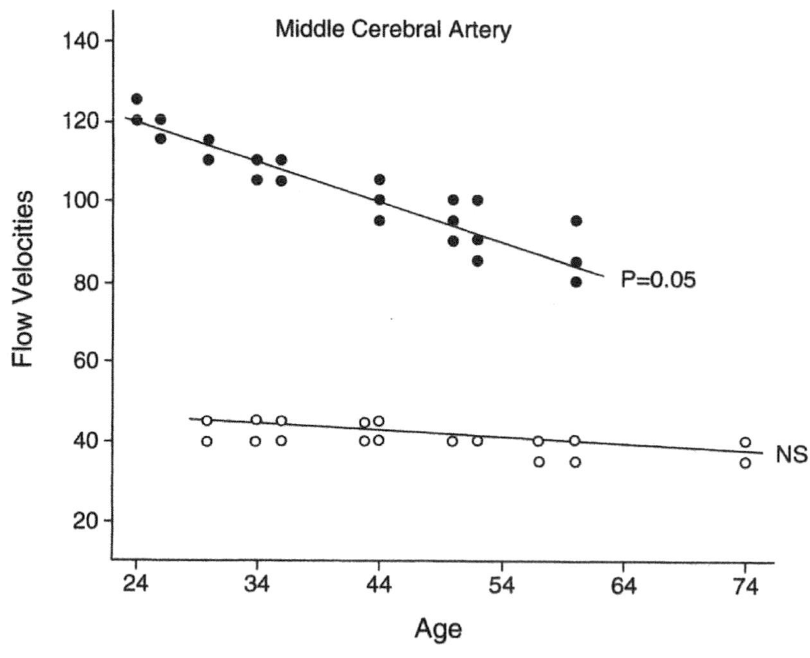

Figure 11.1 Middle cerebral artery flow velocities at sea level (open circles) and at 4970 m. (closed circles). There was a significant decrease in flow velocities in older subjects at altitude only.

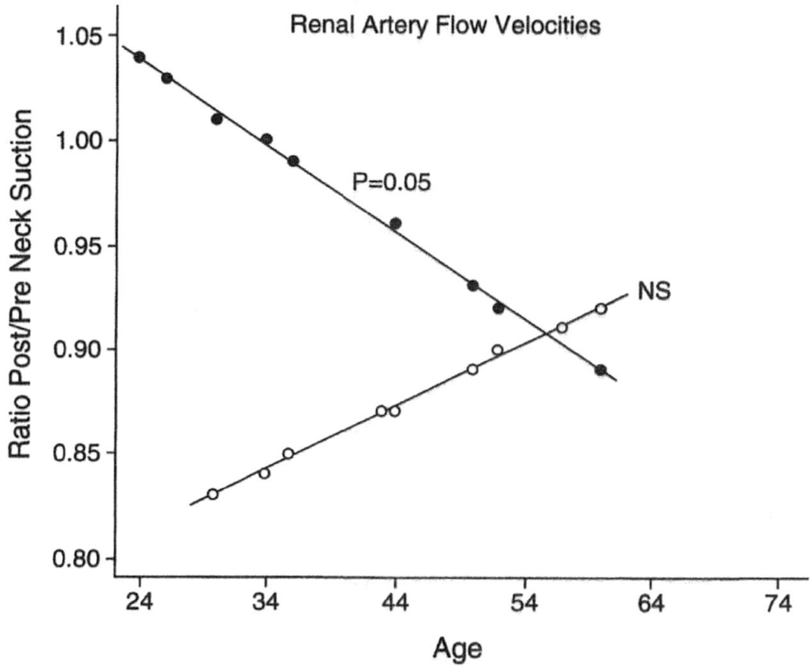

Figure 11.2 Ratios of flow velocities post-/preneck suction in the renal artery. A significant decrease in ratios with advancing age at altitude implies less vasoconstriction in older subjects at altitude.

transporter proteins and in isolated brain microvessels, the density of these transporter proteins is also increased. After 3 weeks of hypoxia, an increase in the transport of glucose at the blood–brain barrier in vivo was also found (Harik et al., 1994). The implications of these findings are that metabolic demands can change the microscopic, immunocytochemical, and membrane protein content in various brain regions differentially and adaptively to facilitate function even in adult animals. In cerebrovascular disease with regional hypoxia (stroke), however, such adaptive changes have yet to be demonstrated.

11.3 Oxygenation during exercise in Tibetan and Han-Chinese

The exercise capacity of Tibetans and Han-Chinese was examined at altitude. Differences in muscle and brain tissue oxygenation during exercise were found. Tibetans showed a blunted cerebrovascular response to hypocapnia during normoxic exercise. This, combined with a higher heart rate and cardiac output, enabled them to maintain higher cerebral tissue oxygenation at maximal effort than Han Chinese. Thus, Tibetans consistently outperformed their Han Chinese counterpart in both normoxic and hypoxic conditions, exhibiting a better energy economy at exercise exertion (Fan et al., 2021).

11.4 Sleep

Numerous studies have documented changes in sleep patterns on acute altitude exposure (Goldenberg, Richalet, Onnen, & Antezana, 1992; Nicholson, Smith, Stone, Bradwell, & Coote, 1988). Anecdotal reports include frequent awakenings, unsatisfactory sleep, lack of sleep, and frequent nightmares. Polysomnography has shown an increase in total sleep period, an increase in the number of awakenings, an increase in intrasleep wakefulness and the duration of stages 1 and 2 of sleep, and a decrease in slow-wave (SWS) and REM (rapid eye movement sleep) (Goldenberg et al., 1992).

In 1898, Mosso described periodic breathing at high altitude and many reports have since confirmed the presence of periodic breathing with apneic periods particularly during sleep (Goldenberg et al., 1992). Studies on Annapurna (4800 m) and on Mt. Sajama (6542 m) (Goldenberg et al., 1992) showed periodic breathing during sleep but its extent varied amongst individuals, and it was found during both non-REM and REM sleep. Periodic breathing at any altitude is thought to result from an instability of the ventilatory feedback control (Longobardo, Gothe, Goldman, & Cherniack, 1982). It is also believed that periodic breathing is an expression of high peripheral chemosensitivity. Difficulty in maintaining sleep and a shortening of SWS was also present. No correlation between periodic breathing and the number of arousals was found but a relationship with interval wakefulness duration was present. Of great interest was the tight correlation between periodic breathing and hypoxic ventilatory drive. The higher the hypoxic and hypercapnic chemosensitivity the more frequent and prolonged periodic breathing was present at 5360 m (Masuyama et al., 1989).

There also was a close negative correlation between desaturation during sleep and the frequency and duration of periodic breathing. Subjects with the severest form of mountain sickness had the shortest periods of periodic breathing during sleep and most profound falls in saturation during the night. Thus, periodic breathing on acute altitude exposure of sojourners seems to be protective and even though it disturbs the quality of sleep, it is important in maintaining oxygen saturation during the night. During SWS, there is normally hypoventilation so that the frequent arousal that is in part, dependent on periodic breathing, protects also from the hypoventilation that accompanies deepening SWS.

During REM sleep there is an activation of the autonomic nervous system (White et al., 1987). Thus, although there is a shortening of total REM time at altitude, the decreased SWS is more significant and this, together with autonomic activation (i.e. tachycardia, increase in cardiac output) and hyperventilation during REM, maintains better oxygenation during sleep.

Acclimatization is associated with a decrease in the severity and frequency of abnormalities in sleep and in periodic breathing during sleep but no change in the chemosensitivity has been found (Lahiri, 1984).

A decrease in the gain of peripheral chemoreceptors is associated with a decrease in periodicity of breathing and Himalayan natives are thought not to show periodic breathing during sleep, consistent with the decreased hypoxic ventilatory drive in such individuals (Lahiri & Data, 1992). A comparison of chemosensitivity and ventilation during sleep of Sherpa and Andean children in their normal habitat showed chemosensitivities consistent with hypoxic ventilatory drive in Sherpas but not in Andean natives whose responses were more like those of sea-level natives (Lahiri & Data, 1992). Thus, it seems that a central (or peripheral) autonomic mechanism can modulate the reflex chemosensitivity and this may be genetically determined since Andean altitude natives, unlike Himalayan natives, are thought to be maladapted to their high-altitude life (Monge, 1988).

High-altitude natives are born, live, and eventually die hypoxic. Thus their development and aging are influenced by hypoxia which most prominently affects oxygen-sensitive organs, particularly the dopaminergic system in the carotid bodies, superior cervical ganglia, and brain (Lahiri & Data, 1992). In the carotid body, the glomus cells of high-altitude natives show hypertrophy at autopsy, and yet during adult life they have a blunted hypoxic ventilatory drive (Heath & P., 1992). Clearly, the morphological changes which are initiated at birth persist but function is presumably modified by some central autonomic mechanism in the brain, perhaps related to aging of the nervous system because hypoxic ventilatory drive decreases with advancing age in sea level dwellers and presumably also in altitude natives.

11.5 Neuropeptides, altitude, and exercise

11.5.1 Opioids

Athletic events are held at ever-increasing altitudes (e.g., the Everest Marathon, reaching an altitude of 5450 m). Beta-endorphin-ß-lipotropin-like activity (BE-BL-like) in the blood increased significantly after a 46 K trail race reaching an altitude of 3200 m (Appenzeller, Standefer, Appenzeller, & R., 1980). On further study, it became apparent that BE-BL-like release did not occur until the finish of this long mountain race and levels did return to baseline within 3 hours after finishing the run. When the same runners were studied over a period of 3 years, significant falls in run-induced outpouring of BE-BL-like activity in the circulation were found (Appenzeller, Appenzeller, Standefer, Skipper, & Atkinson, 1984). This was not the result of better training because these very same runners had significantly higher myoglobin levels in the blood (a sign of muscle breakdown) with each succeeding year of participation. In earlier studies, it became apparent that aging was associated with a significant decrease in run-induced release of BE-BL-like activity in the same race (Appenzeller et al., 1980). Thus, the longitudinal decrease in BE-BL-like release was mostly a manifestation of aging. If men and women were analyzed separately, it was found that the average increase was entirely due to the marked increase in postrace BE-BL-like activity in men; the levels in women did not change after this 46 K run. Cortisol and melatonin levels increased significantly in both men and women (Strassman, Appenzeller, Lewy, Qualls, & Peake, 1989). In addition, 13 subjects received naltrexone, an orally active opioid antagonist just before the start of the race. These subjects were all veterans of the race on whom longitudinal performance data were available from previous years. Naltrexone did not change the run-induced increases in BE-BL-like activity, cortisol or melatonin and the drug did not affect performance, although most runners complained about their well-being and one developed nausea. The conclusion reached was that the release of these substances is a manifestation of stress and is not a specific response to altitude running. In support of this interpretation is a study of heatstroke patients carried out during the Mecca pilgrimages (Appenzeller, Khogali, & Carr, 1986). BE-BL-like activity, growth hormone, prolactin and cortisol were all significantly increased on admission and fell to normal levels after recovery but the most striking increase was in BE-BL-like activity (Appenzeller et al., 1986). A subsequent study on sheep with heat stroke showed that naloxone can be protective in this experimental model.

11.5.2 Endothelin (ET)

Exercise at moderate altitude is associated with significant increases in ET in circulation (Appenzeller & Wood, 1992). This occurs independently of the type of exercise (Fig. 11.3). There is a significant positive correlation between finishing times in races and ET. Since finishing times are positively related to age in most events, the exercise-induced outpouring of ET is also related to age. However, even at rest, ET levels of well-trained subjects are positively related to advancing age. In untrained individuals, no change in ET levels with advancing age could be detected (Lerman, Hildebrand, Aarhus, & Burnett, 1991). In these studies of athletes, no relation of ET to hypoxia nor changes in skin temperature could be found. Animal studies suggest, however, that hypoxia might be a stimulus for ET release (Lerman et al., 1991). In patients with high-altitude pulmonary edema while on oxygen therapy and after descent to an altitude of 3670 m, no increase in circulating ET was found. Nonexercising similarly aged subjects had no changes in ET levels at a similar altitude (3110 m.) even after 6 days of minor hypoxemia. Thus, most human studies of ET levels in the circulation do not support the findings in animal studies that have found that 12% oxygen in inspired air is associated with significant elevation of ET levels in the circulation.

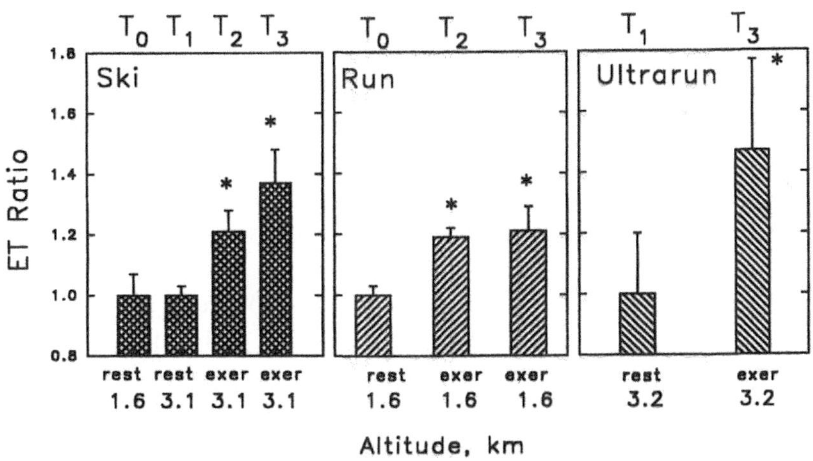

Figure 11.3 Endothelin-1 (ET) in the circulation before and after aerobic exercise. Significant increases in ET levels after each race but no significant differences in ET ratios. SKI = 5 km cross-country ski race ($n = 12$ mean age 38.5 ± 3.2); RUN = 10 km cross-country run ($n = 14$ mean age 45.0 ± 3.6) ULTRARUN = 161 km trail run ($n = 11$ mean age 48.6 ± 6.4) T_0 = rest at various times before race; T_1 = immediately before race; T_2 = immediately after finishing; T_3 = 20 minutes after finishing. *ET Ratio*, ET after exercise/ET before exercise; $^*P < .05$.

After acral cold exposure in patients with Raynaud phenomenon and after the cold pressor test in nonathletes, significant increases in ET have been reported (Fyhrquist et al., 1990; Zamora, O'Brien, Rutherford, & Weil, 1990). However, in cross-country skiers at an altitude of 3100 m significant falls in acral temperature occurred although no relation to ET levels was found (Appenzeller & Wood, 1992).

The application of ice to the neck of humans can selectively cool the brain and induce shivering in about 75% of subjects (Appenzeller, Martignoni, & Qualls, 1995). In a study of 19 sojourners and 6 natives at various altitudes in the Himalayas up to 5100 m, shivering was induced by the application of ice to the neck. In addition, 6 sojourners were studied at an altitude of 1500 m before and 3 minutes after the application of ice blocks to both malar regions of the face (2nd division of the trigeminal nerve). ET was measured in the circulation before and after cooling. There were no changes in ET levels after ice application to the neck. At 5100 m, ET levels were significantly higher than at other locations in sojourners and natives had significantly lower levels at this location. The application of ice to the cheeks induced significant increases in circulating ET. In sojourners who shivered after selective brain cooling, ET levels were significantly higher than in nonshiverers after 50 minutes of ice application to the neck at all altitudes. This study supports the proposition that selective brain cooling is not a stimulus for the release of ET but that neurogenic afferents are capable of increasing circulating ET levels probably accounting for the reported increases after the cold pressor test. Moreover muscle contraction, regardless of the means of activation such as induced through activity as during voluntary exercise, passive as with electrical stimulation of paraplegic muscles in humans, or during shivering, increases circulating ET levels (Robergs et al., 1993). Whether the increase in ET levels at 5100 m in sojourners is an expression of transient altitude-associated aging of neuroendothelial function remains to be determined (Figs. 11.4 and 11.5).

While it is probable that ET acts as a local hormone, measured increases in circulating ET levels are more likely the result of spillover into the circulation from ET release sites. Delays in the increases of ET levels after appropriate stimuli have been reported (Appenzeller & Wood, 1992; Zamora et al., 1990). Older subjects are slower in competition and had higher levels of ET 20 minutes after finishing a cross-country ski race at moderate altitude. Thus, it is likely that aging athletes had a more potent stimulus for ET release perhaps because of their longer exercise time.

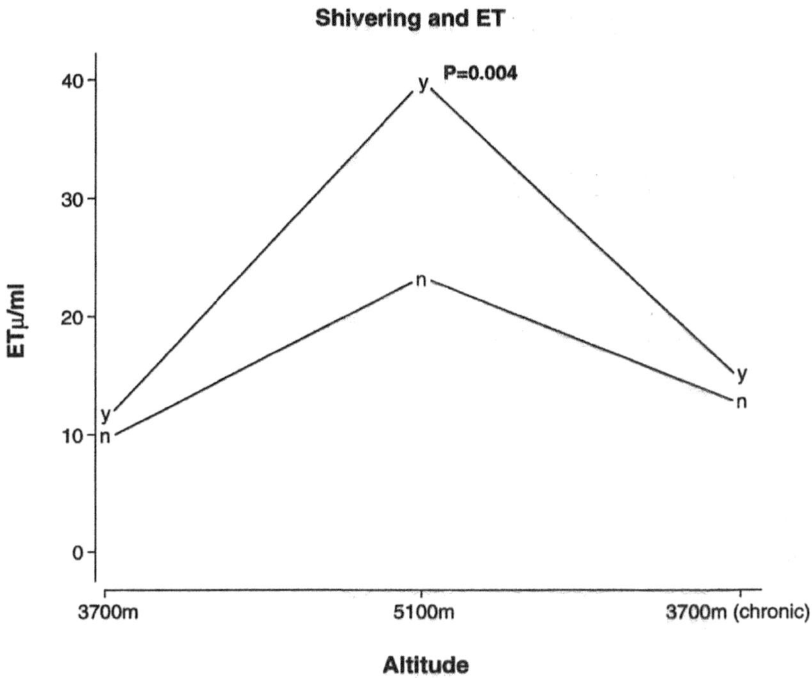

Figure 11.4 Circulating ET levels in shiverers (*y*) and nonshiverers (*n*) after 50 minutes of ice applied to the neck. Significantly higher levels were found at all altitudes in shiverers.

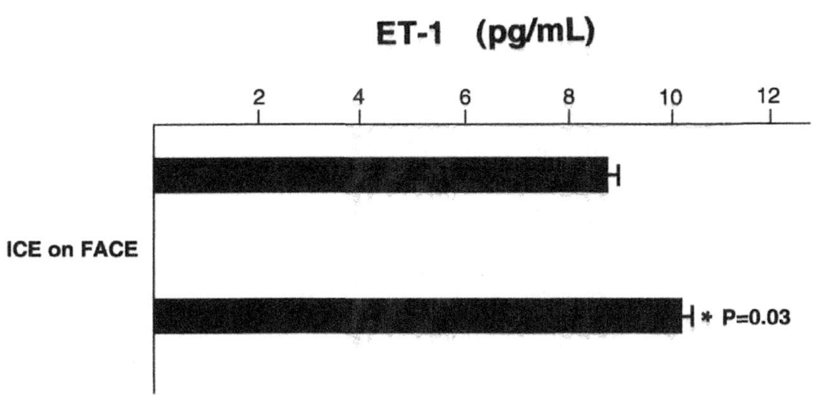

Figure 11.5 Circulating ET levels before and after the application of 2-inch diameter ice blocks to each cheek for 3 minutes. Note the significant elevation of ET after this short-term cooling without change in central or chest surface temperatures.

Endothelin increases the contractility of smooth and cardiac muscle and it has important effects on ion channels (Haber & Lee, 1994). These effects may also be crucial for the protection of the myocardium from the shortening of the action potential by catecholamines. Altitude at rest (Seals, Johnson, & Fregosi, 1991) and exercise at any altitude (Appenzeller & Schade, 1979) are associated with the release of catecholamines and the higher levels of ET may, as in patients with myocardial infarction, protect the myocardium from life-threatening ventricular arrhythmias associated with hypoxia.

11.5.3 Neuropeptide-Y (NPY-LI)

Laboratory studies have correlated the release of NPY-LI and norepinephrine (NE) from the levels found in the coronary sinus and arterial blood during cycle ergometer exercise while breathing air or 12% oxygen. Exercise caused the release of NPY-LI and NE but hypoxemia was an additional and preferential stimulus for the release of NPY-LI from the heart (Kaijser, Pernow, Berglund, Grubbström, & Lundberg, 1994).

11.6 Water and electrolytes

Altitude, in the absence of exercise, is associated with a loss of sodium and water by the kidney. Water and sodium diuresis has been found to be associated with a decrease in plasma and extracellular fluid volumes and increases in plasma aldosterone concentration and in plasma atrial natriuretic peptide (Milledge, 1992). Before symptoms of mountain sickness develop, subjects show pathological antidiuresis, retention of sodium, increased plasma and extracellular volumes and decreased plasma aldosterone levels when compared to altitude sickness-resistant individuals. No consistent relationship between plasma atrial natriuretic peptide and mountain sickness has been found. The pathological response of salt and water homeostasis to altitude is reminiscent of physiologic responses to exercise at sea level (Milledge, 1992). It is to be noted that both exercise and pathological responses of the salt and water homeostasis at altitude promote edema formation and may predispose to mountain sickness (Milledge, 1992).

Exercise at moderate altitude is associated with a significant increase in atrial natriuretic peptide and surprisingly, also a concomitant increase in arginine vasopressin (Solomon, Corwin, & Appenzeller, 1987). Under physiologic conditions at sea level, a see-saw relationship between changes in the plasma levels of these peptides is usually found in untrained individuals. Whether the simultaneous increases represent adaptive phenomena and account for the normally expanded plasma volumes of chronic exercisers is not known.

11.7 Autonomic reflexes and altitude

11.7.1 Pupillary function, age, and altitude

The pupil is smaller in infancy and widens progressively until adolescence. It becomes smaller again in old age. The magnitude of the pupillary responses parallels the size of the pupil. The smooth muscle of the iris is entirely innervated by the autonomic nervous system and is easily accessible as a test site for autonomic function. Pupil Cycle Time (PCT) is one simple way of assessing autonomic innervation of the iris. The normal PCT is 946 ± 120 milliseconds and the upper limit of normal is 1150 milliseconds. In sedentary individuals, there is an age-related slowing of PCT. In well-trained individuals, however, at low altitude, there is no relation of PCT to age but these same individuals show an acceleration of PCT at altitude which is positively related to age (Appenzeller, 1994). Thus, this part of autonomic innervation, though apparently rejuvenated (accelerated PCT) by altitude, in reality, reflects transient aging because of the age and altitude associated increase in circulating catecholamines which accelerate PCT.

11.7.2 Heart rate variability

The variation in heart rate can be analyzed by frequency domain analysis to assess the influence of sympathetic (low frequencies) and parasympathetic (high frequencies) drive to the heart. This is a noninvasive test based on simple electrocardiogram recording which requires, however, sophisticated computer-aided analysis. The total power of the spectrum and the high-frequency component was found to be lower in elderly healthy subjects. Patients with transplanted hearts also show a remarkable reduction in total power by as much as 96% (Korkushko, Shatilo, Plachinda Yu, & Shatilo, 1991).

Twenty-three competitors in a 100 mile (161 K) trail race at moderate altitude (average 3100 m) were studied before the race and 11 of these were also examined after the race. These American competitors were compared with 5 Tarahumara Indians (renowned for their ceremonial ultra-long distance runs) who participated in the same race. In this study of heart rate variability in ultralong distance runners, there was a significant decline in total power with advancing age before the race; after the race, there was significant aging but the younger runners (Tarahumaras) aged significantly more than the American competitors. Aging seems to lead to a disconnection of the heart, that is, a decrease in central control, a situation similar to that in heart transplant recipients. Thus, ultralong distance runners show the same age-related changes in heart rate variability as untrained individuals but long competition at altitude (hypoxia) leads to transient further aging.

A study of eight young professional mountain guides using 24-hour Holter monitors on acute and chronic exposure to an altitude of 5000 m showed an absence of the circadian reciprocity of high and low-frequency heart rate variation (found at sea level). There was no reduction in the low to high-frequency ratios at altitude but a significant decrease in the high-frequency component of the spectra was found. This was interpreted to show a decrease in vagal tone (Ponchia et al., 1994). Unfortunately, no respiratory signals were recorded. Respiration profoundly affects the high-frequency component of the power spectrum. Since profound changes in respiratory rate and tidal volume occur at altitude and Cheyne-Stokes respiration is prominent, the results may not reflect true, altitude-induced changes in sympathetic-parasympathetic balance controlling heart rate. Assuming that a bona fide reduction in the high-frequency component of the power spectrum did occur, it might also reflect transient, altitude-induced, aging of the ANS control of heart rate variability.

11.7.3 Swallow-induced tachycardia

Reflex tachycardia occurs on swallowing. Studies of this central autonomic (brain stem) reflex were carried out on well-trained sojourners and natives to assess the effects of age and ethnicity on this reflex at low and high altitude (Appenzeller, Qualls, & Martignoni, 1993).

Reflex tachycardia in response to swallowing occurred at 1500 m and 3700 m (acute altitude exposure). Continued altitude sojourn abolished the response. Highland natives, but not sojourners, had a response at 5100 m. Age was without effect in sojourners at 1500 m; however, it was inversely related to swallow-induced tachycardia at 3700 m. Latencies (as reflected by the most accelerated heartbeat after the swallow) were increased at altitude in all subjects.

This central autonomic reflex is abolished in sojourners after chronic altitude exposure and shows transient aging on acute altitude exposure, probably as the result of acute brain stem hypoxia. While high-altitude natives have normal swallow-induced tachycardia at 5100 m, altitude-adapted sojourners show no return of the reflex after 13 days at high altitude when many altitude-induced adaptations have occurred.

The reflex has not been studied in large populations to assess whether an age-related decay in reflex response occurs. However, the transient aging at altitude found in well-trained sojourners suggests that cross-sectional studies at sea level might reveal a decay in reflex responsiveness with advancing age.

The assessment (in real-time) of cognitive deficits associated with altitude exposure has been described (Lieberman, Protopapas, Reed, Youngs, & Kanki, 1994). In members of a Mount Everest Expedition, the study supported the idea that altitude exposure caused central nervous system dysfunction similar to that found in the aging brain and akin to the transient autonomic deficits at altitude attributed to aging.

11.7.4 Skin flare

When the skin is injured, the vascular reaction includes a flare due to spreading dilatation of small vessels. The flare is absent in denervated skin (Appenzeller, Wood, Qualls, Appenzeller, & Couzins, 1991). One chemical mediator of such neurogenic vasodilatation is substance-P (SP) and capsaicin applied to the skin induces C-fibers to release SP and to cause a flare (Appenzeller et al., 1991). Using capsaicin to induce the flare in 19 sojourners and 8 natives on acute and chronic altitude exposure, an altitude-induced qualitative and quantitative reduction in the flare was found and latencies of the vasodilatation were significantly increased by altitude. No ethnic differences in the response of the flare to altitude were noted. No effect of aging could be recognized in this group (Appenzeller et al., 1991). The reduction of the flare might result from a change in blood flow or blood vessel reactivity, or from a change in SP in skin sensory nerves (C-fibers). This reduction in the flare could also be a result of changes in rheologic properties of the blood induced by altitude (increased hematocrit) or altitude-associated skin edema (Heath & Williams, 1989). A similar observation, using a different technique to induce the flare, was made in one sojourner in South America (Heath & Williams, 1989).

11.8 Circulating catecholamines

Altitude is associated with a significant increase in circulating norepinephrine levels and resting as well as exercise-associated heart rates are increased but a definite decrease in maximum heart rates above 3000 m has repeatedly been found (Antezana, Richalet, Antezana, Spielvogel, & Kacimi, 1992). Thus, there is a progressive decrease in heart rate for the same plasma norepinephrine level with increasing altitude.

At 4350 m, a blunted response to infusion of isoproterenol, suggesting chronotropic refractoriness to sympathetic stimulation and a decreased ß adrenoceptor density on human lymphocytes, was found. It was therefore suggested that ß receptors are downregulated at altitude (Richalet et al., 1988). To investigate this phenomenon, Bolivian highlanders with normal or high hematocrits were compared with sea-level natives. The studies were done at sea level and at an altitude of 4800 m for sea-level natives acutely and after 21 days of acclimatization. The isoproterenol dose required to raise heart rate by 25 beats/minute was no different in the three groups when studied in their natural environment. However, the density of ß-adrenoceptors was significantly lower in the two high-altitude native groups than in sea-level natives. Thus, chronotropic responses are similar in natives of any altitude if examined in their usual environment but sea-level natives show a reduced response in hypobaric hypoxia most likely due to a reduction in ß-adrenoceptor density. It is also of interest that no adrenergic desensitization was found in Bolivian highlanders. The lower densities with normal responsiveness of ß-adrenoceptors of highlanders were thought to represent an adaptive phenomenon (Richalet, 1990). Of interest are the results in polycythemic high-altitude South American natives, most of whom are eventually destined to develop signs of chronic mountain sickness (Antezana et al., 1992). The decrease in the density of ß-adrenoceptors of polycythemic highlanders, when compared with sea-level natives, was less than in nonpolycythemic subjects. This could have implications for the right ventricular hypertrophy found in patients with chronic mountain sickness, a frequent sign of maladaptation in Andean natives (Antezana et al., 1992; Monge, 1988).

These findings might also help explain the frequency of acral frostbite in sojourners at high altitude even in relatively moderate ambient conditions and with adequate cold protection of the extremities. The blunted response of ß-adrenoceptors of sea-level natives in hypoxia could be present on blood vessels and not only on lymphocytes (Richalet et al., 1988).

Hypoxia may also potentiate exercise-induced sympathetic neural activity. By measuring sympathetic activity to leg muscle blood vessels in the peroneal nerve using microneurography, while performing handgrip exercise in normoxic and isocapnic hypoxia (inspired oxygen concentration 10%), it was shown that sympathetic bursts to blood vessels in the leg were significantly increased during hypoxia for the same amount of work whereas blood pressure remained the same during exercise in either hypoxia or normoxia (Seals et al., 1991).

Although there is activation of the sympathetic nervous system during very short acute hypoxia in the laboratory, there is no increase in blood pressure, no adrenergic vasoconstriction, and no increase in venous plasma norepinephrine. This apparent paradox was examined by assessing arterial norepinephrine kinetics and sympathetic nervous outflow to muscle using peroneal nerve microneurography before and during 30 minutes of mild hypoxemia (mean oxygen saturation 74%) in six healthy young men (Leuenberger et al., 1991). Forearm vascular resistance (muscle blood vessels) decreased but skin blood flow and resistance did not change during hypoxemia implying vasodilatation in muscle blood vessels. The failure of plasma norepinephrine to rise during this short mild hypoxic exposure was attributed to an increase in norepinephrine clearance (Leuenberger et al., 1991).

Altitude and exercise in the field are associated with dehydration, a significant factor in the early failures to reach the summit of Mount Everest. Eight men were examined during normoxia and hypoxic breathing of 10% or 12% oxygen in inspired air to assess the effects of mild central hypovolemia, induced by lower body negative suction, on muscle sympathetic activity in the peroneal nerve, norepinephrine levels, and vasoconstriction. Graded lower body negative suction was applied during normoxia and hypoxia (Rowell & Seals, 1990). No important deficit in sympathetic activity was found with graded hypovolemia either during normoxia or hypoxia as long as epinephrine levels remained unchanged. In two men, vasovagal syncope with associated bradycardia, decreases in muscle sympathetic nerve activity, hypotension, and increases in epinephrine levels were found during lower body negative suction (Leuenberger et al., 1991). Thus, in those early climbers rendered hypovolemic on Everest, failure to reach the summit cannot be attributed to an inadequate autonomic nervous system adaptation to the altitude unless they had increases in epinephrine levels and syncope.

A study to assess the effects of sympathetic blockade, using propranolol, on mountain sickness and well-being at 4300 m failed to show significant differences between placebo and medicated subjects (Fulco et al., 1989). This suggests that ß-adrenergic blockade does not impede the acclimatization nor does it change the feeling of well-being as assessed by a questionnaire, which was administered daily for 15 days after arrival at the high altitude (Fulco et al., 1989).

Prolonged hypoxemia depresses α-adrenergic function but the addition of a sympatho-excitatory stimulus may normalize hypoxemia-induced α-adrenergic depression. The activation of the sympathetic nervous pathways by hyperthermia during normoxia and hypoxia has been studied in the laboratory during a continuous cycle of normothermia-hyperthermia—normoxia and 10 minutes of hypoxia (Rowell, Brengelmann, Savage, & Freund, 1989). During hypoxia, increases in norepinephrine, epinephrine, plasma renin activity (ß-adrenergic), and heart rate were found. Hyperthermia was associated with normal increases in body temperature, sweat rates, and forearm blood flows both in normoxia and hypoxia (Wood & Appenzeller, 1988). This suggests that peripheral autonomic function remains unaltered by short-lived acute moderate hypoxia in a laboratory setting.

11.9 Thermoregulatory and vasomotor effects of high altitude

An important area of thermoregulation is the control of body temperature at altitude where serious frostbite and death from exposure are recognized hazards. Hypoxic stress, exercise, and changes in blood rheology due to erythrocytosis at altitude contribute to the complexity of autonomic thermoregulatory responses in sojourners and altitude natives. Vasomotor responses in skin and muscle can be examined noninvasively. They are of interest in thermoregulation since the skin and muscle blood flows contribute to heat conservation and heat loss. Excessive erythrocytosis (hemoglobin concentration [Hb] ≥21 g/dL in adult males) is associated with increased cardiovascular risk in highlander Andeans. Hyperviscosity, high hemoglobin, or both, actively contribute to acutely reversible impairments in flow-mediated dilation in male Adeans with excessive erythrocytosis (Tremblay et al., 2019).

A study of well-trained subjects in Tibet up to an altitude of 5150 m (Everest base camp) showed that vasoconstriction in skin vessels in response to a gasp was markedly increased at the Everest base camp as compared to the responses at sea level in Beijing (Appenzeller & Wood,

1988). Similarly, vasoconstriction in skin vessels in response to mental arithmetic was also increased by altitude. There was a close correlation between these two sympathetic vasoconstrictor responses. Clearly, accentuation of vasoconstriction in the hand can affect heat loss from the skin and promote frostbite at altitude (Appenzeller & Wood, 1988).

In muscle blood vessels, mental arithmetic produces reflex vasodilatation at sea level. At altitude, this is converted into a paradoxical vasoconstrictor response. After Valsalva maneuver, muscle blood vessels constrict but altitude had no effect on this reflex vasoconstriction (Appenzeller & Wood, 1988).

Altitude-induced erythrocytosis increases blood viscosity and may affect vasomotor function in skin and muscle during exercise and at rest (Appenzeller & Wood, 1988). Rheologic properties of the blood may also influence thermoregulation by affecting blood flow through heat exchange vessels. Pentoxifylline, a drug that renders red cells more deformable and therefore makes the blood less viscous, might improve autonomic nervous system function by improving perfusion and consequent oxygenation at altitude.

In a placebo-controlled double-blind study of pentoxifylline effects on vasomotor function at altitude, most sympathetically mediated vasoconstrictor responses, abnormal in placebo takers, were normalized by the drug. However, vasoconstriction in muscle blood vessels after Valsalva maneuver and the time required for reperfusion of blood vessels after complete circulatory arrest were not affected by the drug. Reperfusion was considerably prolonged by altitude in both placebo and medicated subjects (Appenzeller & Wood, 1988).

Fingertip temperature decreases significantly at altitude reflecting altitude-induced vasoconstriction in sojourners. After acclimatization, this vasoconstriction persists. Pentoxifylline significantly attenuated this response but Tibetans had less vasoconstriction than medicated subjects at the same altitude (Appenzeller & Wood, 1988).

Altitude is associated with a significant accentuation of normal sympathetically mediated vasoconstrictor responses in the skin whereas in muscle blood vessels vasomotor control becomes abnormal at altitude as shown by a paradoxical mental stress-induced vasoconstriction. Parasympathetic vasomotor function on the other hand remains unaffected by acute or chronic altitude exposure in sojourners (Appenzeller, 1991).

Pentoxifylline attenuated vasoconstrictor responses in cutaneous but not muscle vascular beds and protected normal reflex vasodilatation in muscle blood vessels. This rheologically active drug may therefore prove useful in the prevention of frostbite at altitude. Separate from its rheologic effects, pentoxifylline may also influence neurovascular function (Appenzeller & Wood, 1988).

The mechanisms involved in sympathetic activation during high-altitude exposure are not fully understood. Acute exposure to hypoxia causes chemoreflex activation of the sympathetic nervous system. After 4 weeks at an altitude of 5260 m, sympathetic nerve activity is highly elevated but administration of oxygen (to eliminate chemoreflex activation) and saline (to reduce cardiopulmonary baroreflex deactivation) has minimal to no effect on muscle sympathetic nerve activity. Exposure to high altitude is associated with a striking and long-lasting sympathetic overactivity that is not driven by chemoreflex activation by hypoxia and baroreflex deactivation by dehydration (Hansen & Sander, 2003).

Brachial artery vascular function was quantified at sea level, and after acclimatization to high altitude (4300 m; Cerro de Pasco, Peru), using forearm intraarterial infusions of vasoactive drugs (acetylcholine, endothelial-dependent vasodilation; sodium nitroprusside, endothelial-independent vasodilation) before and after forearm adrenergic blockade (phentolamine, nonselective α-adrenergic blockade; propranolol, nonselective β-adrenergic blockade). In lowlanders, 14 days of acclimatization to 4300 m resulted in a reduction in brachial artery endothelial-dependent function, which was abolished after administration of adrenergic blockade. High-altitude exposure elevated muscle sympathetic nervous activity and reduced endothelial-dependent function, which was fully restored after adrenergic blockade in lowlanders. Andeans highlanders with excessive erythrocytosis had elevated muscle sympathetic nervous activity and reduced endothelial-dependent function compared to Andean highlanders without erythrocytosis, but vascular function was normalized between groups after forearm adrenergic blockade. This study establishes the direct link between heightened neural sympathetic activity and reduced endothelial-dependent function at high altitude (Tymko et al., 2020).

11.10 Altitude hypoxia: exercise and cardiovascular responses

Improved performance after altitude training has persuaded many athletes to take to high altitude for training prior to competition at sea level. The circulatory effects of hypoxia at altitude have been analyzed (Heistad & Abboud, 1980). The local effects on the vasculature are inhibitory resulting in marked vasodilatation. Serious hypoxia, therefore, preferentially affects vessels that supply organs that are highly, metabolically active. This adjustment of the circulation and the redistribution of blood flow is important to organs with greater dependence on oxygen for proper function. Hypoxia induces chemoreceptor activity, which reflexively maintains arterial pressure and produces an important favorable redistribution of blood flow away from skeletal muscles and bowel. This vasoconstriction maintains arterial pressure. The cerebral vessels are little affected by chemoreceptor activity but vasodilatation occurs in coronary blood vessels so that proper cardiac function is insured. The vasoconstrictor effect on skeletal muscles in those training at high altitude might habituate muscle contraction if at all possible to relatively hypoxic conditions and thus enhance the training effects attributed to altitude exposure. This chemoreceptor activity is potentiated by a number of other known effects of hypoxia including acidosis, the release of circulating norepinephrine, and systemic hypotension. These factors act concurrently with hypoxia to increase chemoreceptor reflex drive and produce better adjustments of the circulation. Hypoxia may also suppress some chemoreceptor activity through dopamine release. Dopamine is present in glomus type 1 cells of the carotid body. It may be released by hypoxia to produce negative feedback in the chemoreceptor afferent nerves.

There is, in addition, a central pressor effect of hypoxia, which is more pronounced where there is also cerebral hypotension. This response is essential in the restoration of arterial pressure in order to maintain adequate cerebral perfusion. If systemic hypoxia is complicated by hypotension, then autonomic circulatory adjustments are to a large extent prevented. This failure of autonomic compensatory mechanisms occurs in patients with chronic hypoxia but it can be reversed by correcting arterial oxygen desaturation. How exercise, with its possible attendant hypoxia, may interfere with autonomic activity during athletic training at altitude has not been established.

High-altitude environment in the Himalayas was utilized as a model of hypobaric hypoxia by Verratti et al. The authors compared the delivery of oxygen to muscles in response to resistance exercise between Italian trekkers and native Nepalese porters. The porters had a greater muscle oxygen delivery than Italians, and this ability of porters was attributed to the long-range adaptive memory secondary to frequent exposures to hypoxia (Verratti et al., 2020).

Fluid overload and anemia are hazards further aggravated by altitude. Continuous arteriovenous hemofiltration has the potential to control fluid overload in flight at high altitude and can be useful for the evacuation of patients in acute renal failure, sometimes a complication of accidental hypothermia and multiple injuries that can occur at altitude.

Two young persons, one with multiple injuries and the other with severe preeclamptic toxemia, developed acute renal failure. Continuous arteriovenous filtration was used during aeromedical evacuation. With little sophisticated equipment, it successfully controlled fluid balance and fluid shifts in difficult confined conditions. The development of pulmonary edema, usually enhanced by a reduction in atmospheric pressure due to altitude, can best be managed by hemofiltration, rather than albumin administration to the edematous and anuric patient. It is to be emphasized that peritoneal dialysis is not a useful alternative to hemofiltration at altitude because hypoventilation due to the splinting of the diaphragm and possible catheter blockage with difficulty of fluid retrieval are serious hazards of altitude. The use of hemofiltration seems an advance in the safe transfer of casualties and allows flight much earlier than previously thought possible (Stevens, Bloodworth, & Rainford, 1986).

Hand-held Doppler shift measurement devices can be used for field studies at altitude in endurance-trained individuals. Doppler shifts in the common carotid and preauricular artery were recorded after exertion up to 17,000 feet A linear increase in velocities in both carotid and preauricular arteries and an accompanying decrease in pulsatility indices in these blood vessels with increasing altitude were found. These results imply the blood flow in the carotid and preauricular artery (a branch of the external carotid supplying the scalp) continues to increase in direct relationship to higher altitudes in endurance-trained individuals. While common carotid flow increases would be anticipated from the increasing hypoxia, the continued and parallel increase in flow in a blood vessel supplying a nonessential tissue such as the scalp is surprising. These findings may have some bearing on the occurrence of headache in acute mountain sickness and the precipitation of migrainous attacks in migraineurs at altitude. Similar longitudinal studies have not been carried out in untrained individuals where autonomic nervous system activity has not been reset or modified by athletic training at sea level or at increasing altitudes.

The reactivity of the middle cerebral artery (MCA) to CO_2 has been studied in sojourners and natives at altitude. Altitude increases cerebral blood flow and respiration rate and is associated with hypocapnia in sojourners. One feature of acute mountain sickness is pounding headache but high-altitude Himalayan natives do not suffer from acute mountain sickness, though chronic mountain sickness and headache are common in the Andes. The reactivity of cerebral blood vessels to CO_2 in the Himalayas might therefore shed light on the pathogenesis of headache at sea level. A study of natives and sojourners at 200, 1500, 3400, and at 4970 m after acclimatization was carried out (Anand, Greene, & Appenzeller, 1995). Breath-hold and voluntary hyperventilation were used to manipulate PCO_2. There were no differences in PCO_2 shifts induced by these maneuvers between natives and sojourners but flow velocities in sojourners were significantly enhanced or diminished by PCO_2 shifts in either direction when compared to the velocities in natives at the same altitudes. Even correction for PCO_2 did not change the significantly lower cerebrovascular reactivity of Himalayan natives. In these studies, age was negatively correlated with flow velocities and reactivity to CO_2 in all conditions and all subjects. Sojourners and Himalayan natives in their natural environments have normal flow velocities and cerebrovascular reactivity but when converting velocities to sea level PCO_2 (40 Torr) using an accepted formula, the Himalayan natives become abnormal just as sea-level natives are abnormal at altitude. These findings may reflect genetic adaptations of cerebrovascular reactivity to the environment and they have implications for the pathogenesis of headache at sea level and at altitude.

Whether the relative immunity of endurance-trained subjects to altitude headache reflects this increased CO_2 reactivity of the intracranial vasculature with increasing altitude remains to be determined by comparative measurements in those who travel to altitude but are not endurance-trained (Figs. 11.6–11.9).

Acute exposure to high altitude activates the sympathoadrenal axis but the effects of this activation on neural sympathovagal control of the heart, blood pressure, and particularly baroreceptors are unclear. To determine the effects of high-altitude exposure in the Himalayas on neurovascular control, sea-level natives and residents were compared with Himalayan natives who resided at sea level for at least a year before acute exposure to 4970 m together with the sea-level natives. The studies used noninvasive manipulation of baroreceptors by neck suction with simultaneous recordings of blood pressure oscillations, heart rate variability, and respiration. Power spectral analysis was then applied to the signals. In sea-level residents, acute altitude exposure caused, as expected, sympathetic activation. This, however, was only partly counteracted by the baroreceptors since blood pressure remained significantly elevated. High-altitude natives and

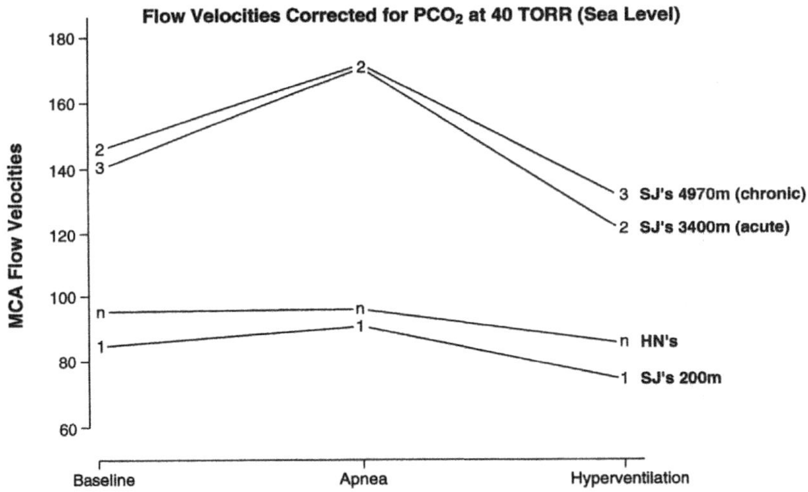

Figure 11.6 Flow velocities (FV) in the middle cerebral artery corrected for PCO_2 at sea level. Note the abnormal high FV in natives (HNs) (compare to Fig. 21.7) and the lack of difference in reactivity between sojourners (SJs) and HNs at 200 m altitude.

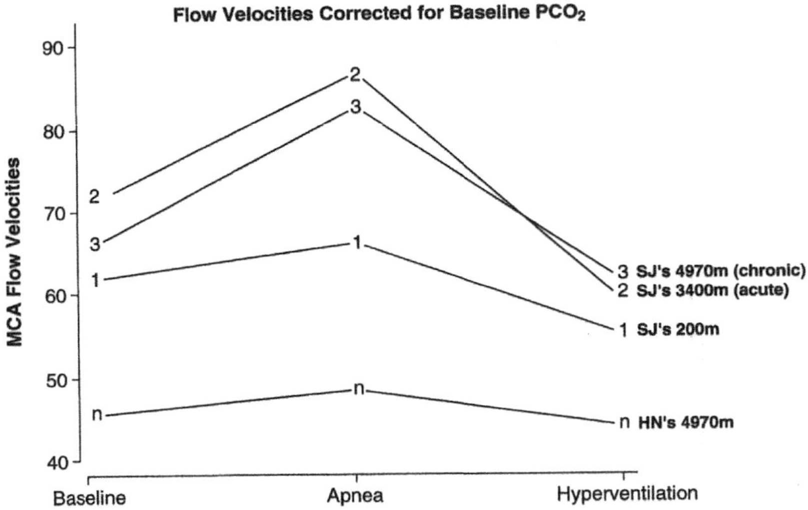

Figure 11.7 Flow velocities (FV) in the middle cerebral artery corrected for baseline (before apnea or hyperventilation) PCO_2. Note the difference in reactivity and FV in HNs and SJs in their natural habitat.

long-term sea-level residents also had sympathetic activation but to a lesser degree and their blood pressure was less elevated acutely than in sea-level natives indicating better baroreceptor modulation of blood pressure. After

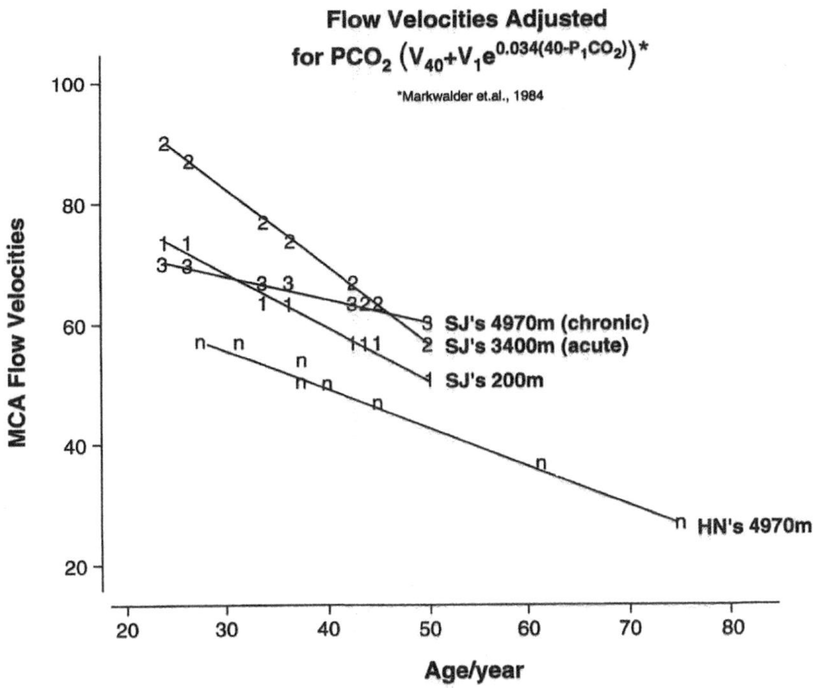

Figure 11.8 Flow velocities (FV) in the middle cerebral artery corrected for baseline (before apnea or hyperventilation) PCO_2. At various altitudes in SJs and HNs plotted against age. In all subjects and conditions, FV decreases with advancing age.

1 week at altitude, the two groups were restudied and no significant differences were found after chronic altitude exposure (Bernardi et al., 1995). This baroreceptor memory of Himalayan natives might be genetically determined. It would be interesting to compare Himalayan and Andean high-altitude natives since despite centuries of altitude residence, the latter are maladapted to their environment and many develop chronic mountain sickness which is extremely rare in Himalayan natives.

11.11 Pentoxifylline

11.11.1 Pentoxifylline—physiology and pharmacology

At altitude, or in patients with chronic obstructive pulmonary disease, congenital heart disease, or heat stroke, oxygen transport to the tissues is compromised. Systemic oxygen transport (SOT) is the product of flow (Q) and arterial oxygen content (CAo_2) ($SOT = Q \times CAo_2$). CAo_2 depends on

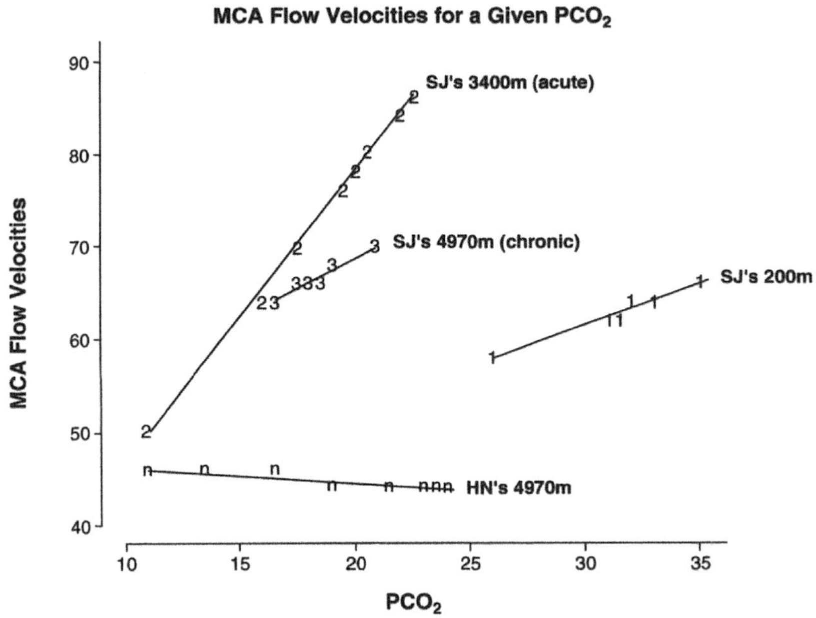

Figure 11.9 Flow velocities (FV) in the middle cerebral artery for a given PCO_2. Note the significant lower FV for any PCO_2 in HNs in their natural habitat.

hemoglobin and oxygen saturation. When blood becomes more viscous, due to platelet aggregation, increased plasma protein, or increased red cell mass, phlebotomy has been used to improve blood rheologic properties and blood flow. However, this has unpredictable effects on SOT; that is, increased flow with decreased CAo_2, increased flow with increased CAo_2 (due to increased PAo_2) or increased flow with no change in CAo_2 (decreased hemoglobin but increased oxygen saturation).

In contrast, theologically active drugs offer the possibility of reducing the viscosity of the blood without impairing systemic oxygen transport, a clear advantage in hypoxic altitude-induced erythrocytosis, and perhaps an advantage in heatstroke where circulatory collapse and reduced blood pressure compromise the microcirculation. Pentoxifylline (Trental®) is a theobromine derivative of methylxanthine. It was introduced in 1972 and used as a vasodilator and subsequently discovered to reduce blood viscosity (Aviado & Porter, 1984). Pentoxifylline is more soluble in water than other methylxanthines (e.g., theobromine, caffeine, and theophylline) and is readily absorbed from the gastrointestinal tract. Unlike aminophylline, pentoxifylline given in usual doses orally does not have vasodilator effects and does not influence heart rate, arterial blood pressure, or systemic vascular resistance. Thus, the action of the drug in humans is specific in improving flow properties, that is, a hemorheological activity.

With oral doses, though, no clear vasodilator effect has been observed, tissue oxygenation and blood flow in peripheral vascular tissues have been shown to improve in patients with peripheral vascular disease (Ehrly, 1982). Similar studies have not shown definite benefits in cerebral vascular disease.

Three mechanisms contribute to the rheologic effects of pentoxifylline; increased erythrocyte flexibility (Angelkort, 1979), slight reduction in fibrinogen concentration, and decreased platelet aggregation. One experiment demonstrated a vasodilator effect on the vascular smooth muscle that involves reduced production of thromboxane-A (Hoeffner, Farhys, & Vanhoutte, 1986), but a possible additional effect on the endothelium-derived relaxing factor (EDRF) has not been investigated.

Of special significance is that Sedova, Krasiukova, Kovaleva, Malykhina, and Ermil'chenko (1984) showed that pentoxifylline helped prevent thrombus formation in patients with polycythemia vera who are also prone to strokes (Sedova et al., 1984). Palareti et al. (1984) showed that members of an expedition to the Himalayas (above 7500 m) had increased viscosity and decreased red cell deformability at altitude (perhaps as a result of red cell ATP depletion) and treatment with pentoxifylline prevented these changes (Palareti et al., 1984).

11.11.2 Pentoxifylline—physiological effects at high altitude

There is significant interest in the physiology of humans at high altitude, sparked in part by the dramatic ascent of Mount Everest (8848 m) without supplemental oxygen by Messner and Habeler in 1978. At this altitude, exercise should not be possible because maximum O_2 uptake is predicted to be only slightly above the O_2 required for basal metabolic rate. At lower altitudes, exercise is possible but aerobic performance is limited by the diminished maximum O_2 uptake and, perhaps, increased rate of perceived exertion (RPE).

Running performance at any altitude is limited by the amount of oxygen delivered to the muscles used in running. This amount, termed systemic oxygen transport (SOT), is defined as the product of muscle blood flow (MBF) and arterial O_2 content (CAo_2):

$$SOT = MBF \times CAo_2.$$

Muscle blood flow is a function of the driving pressure (P) and resistance to flow (R):

$$MBF = P/R,$$

where R for laminar flow varies directly with blood viscosity and vessel length and varies inversely with vessel size. Blood viscosity is relatively ignored. However, experiments with dogs showed that cardiac output of denervated animals is strongly affected by changes in blood viscosity, falling or rising by 20% for a 10% change in viscosity (Dormandy, 1970). Blood viscosity depends on numerous factors including plasma viscosity, red blood cell deformability, shear rate, and hematocrit.

With exposure to high altitude, several factors alter the above relationships at rest and during exercise: At rest, arterial saturation with oxygen decreases as the partial pressure of O_2 in alveolar gas and arterial blood decreases. During exercise, arterial saturation may decrease an additional amount due to diffusion limitation in the lungs related to the reduced PO_2 difference and reduced equilibration time at the higher cardiac outputs associated with exercise. Erythropoiesis is stimulated and hemoconcentration occurs due to reduced plasma volume. The resulting elevation of hematocrit (Hct) may maintain a normal CAo_2 despite the reduced saturation. Thus, SOT may be maintained at normal values if MBF is normal (see equation). Blood viscosity rises with the increase in Hct and may also increase due to increased plasma viscosity secondary to elevated fibrinogen levels. This tends to reduce MBF as well as blood flow to other organs.

These considerations led to the hypothesis that exercise performance, known to diminish with altitude does so, in part, as a result of increased blood viscosity.

Autonomic function changes with altitude due to hypoxic effects on nervous system function in general and specifically due to reduced perfusion and Pentoxifylline, by rendering red blood cells more deformable and blood less viscous, might improve exercise performance and autonomic function.

A series of experiments were carried out on ten endurance-trained runners and 1 cyclist. The age and body mass of the subjects are listed in Table 11.1. In a double-blind, placebo-controlled, parallel protocol, five subjects took pentoxifylline in three daily oral doses of 400 mg each for 21 days during exposure to altitude (Figs. 11.10–11.14). Some tests on Chinese living in Lhasa (3600 m; $N = 37$), native Tibetans (yak herders) at

Table 11.1 Subjects in Tibet Research Run (1987).

Subjects	Age (years)	Body weight (kg)	Training
1	63	62.7	R(M, UM)
2	59	63.6	R(M, UM)
3	23	84.5	C
4	28	68.2	R (M)
5	27	52.3	R (M)
6	54	75	R (M)
7	29	63.6	R (M)
8	42	76.8	R (M)
9	47	59.1	R (M)
10	48	77.3	R (M)
11	45	75.0	R

Females are underlined before the run. Mean weight loss was 2.3 kg.
M, marathon participant; *UM*, ultramarathon participant; *C*, cyclist; *R*, running.

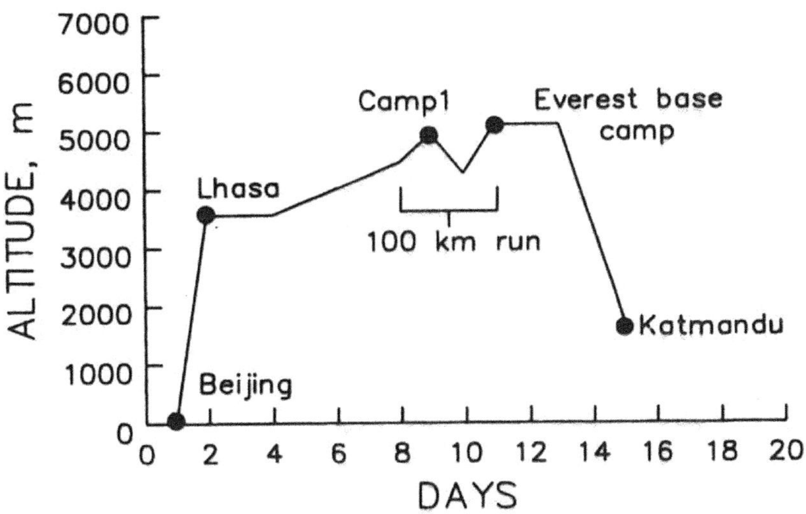

Figure 11.10 Altimetric graph of the Tibet Research Run, Sept. 21-Oct. 13, 1987 (from Wood, Appenzeller and Appenzeller (1989)).

4650 m ($N = 4$), and on members of an American climbing expedition ($N = 7$) who established base camp (5150 m) and higher camps (6360 m and 7800 m) prior to a Mount Everest summit attempt and had been at those altitudes for 5 weeks were also carried out.

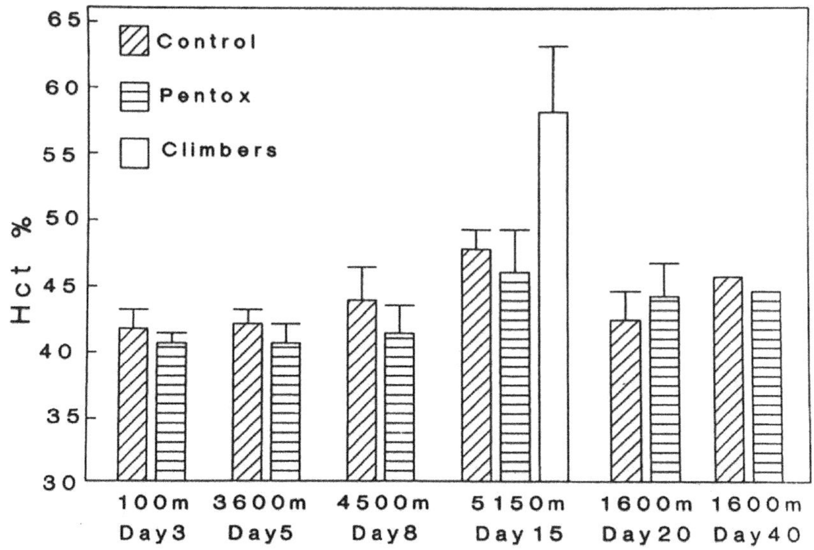

Figure 11.11 Packed red-cell volume (HCT) at various altitudes. Controls, subjects taking pentoxifylline (pentox), and American Everest team climbers (climbers). The HCT increase at altitude in pentox was significantly less ($P < .03$), but altitude had a significant effect in all subjects ($P < .0001$). *From Wood, Appenzeller and Appenzeller (1989).*

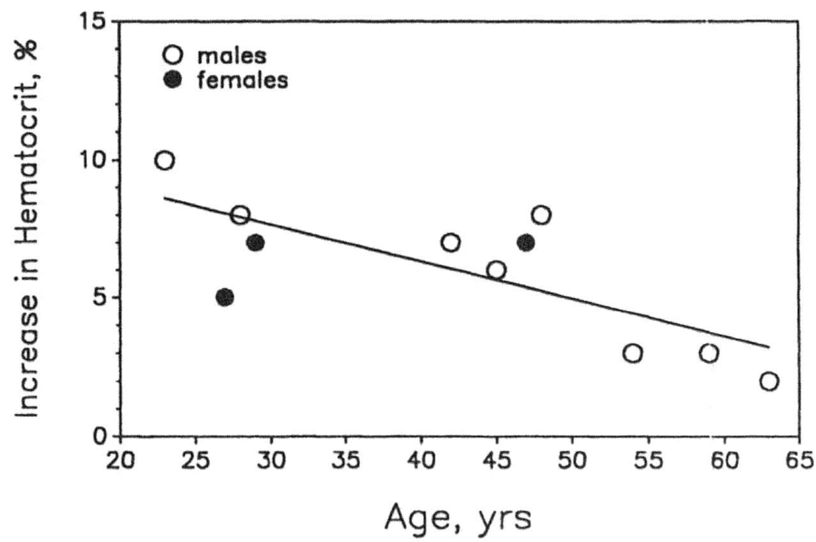

Figure 11.12 Relationship between age and increase in hematocrit. (Everest base camp, Beijing values) $P < .00001$ by Anova. *From Wood, Appenzeller and Appenzeller (1989).*

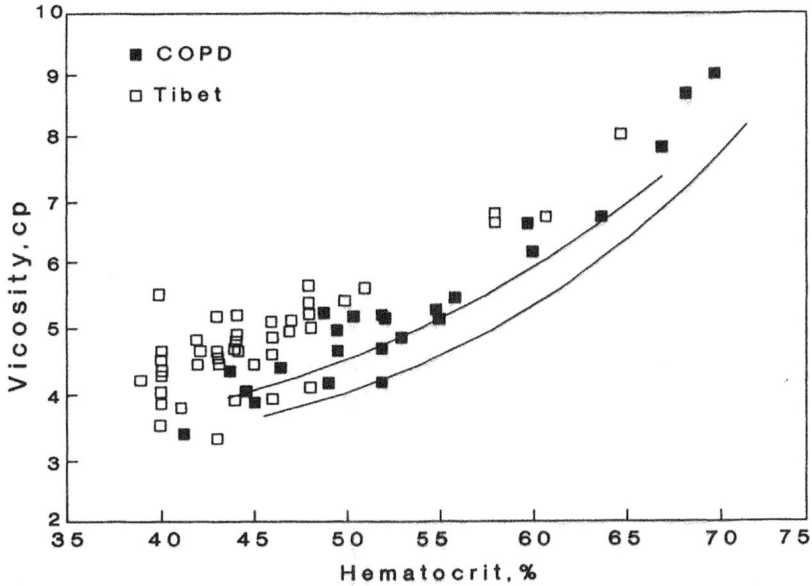

Figure 11.13 Whole-blood viscosity as a function of hematocrit in runners (Tibet) compared with the blood viscosity of patients with chronic lung disease (COPD). Blood viscosity increased significantly with altitude, but independently of hematocrit. *From Wood, Appenzeller and Appenzeller (1989).*

11.11.3 Pentoxifylline—neurologic and thermoregulatory effects at high altitude

The recent evolution in the understanding of the thermoregulatory system has made it a recurrent subject of physiologic investigation in a variety of settings. An important area of thermoregulation is control of body temperature under hypoxic stress and when blood rheology is altered due to hypoxia-induced erythrocytosis. With heightened interest in sporting activities at sea level and altitude, it is important to understand disturbances in thermoregulation induced by extreme environmental conditions and heat loads due to physical activity. Such conditions may cause serious and lasting disorders even in well-trained individuals.

Altitude may affect thermoregulation as follows:

1. Altitude-induced erythrocytosis leading to increased blood viscosity may affect vasomotor function in skin and muscle during exercise and at rest (Figs. 11.15–11.17).

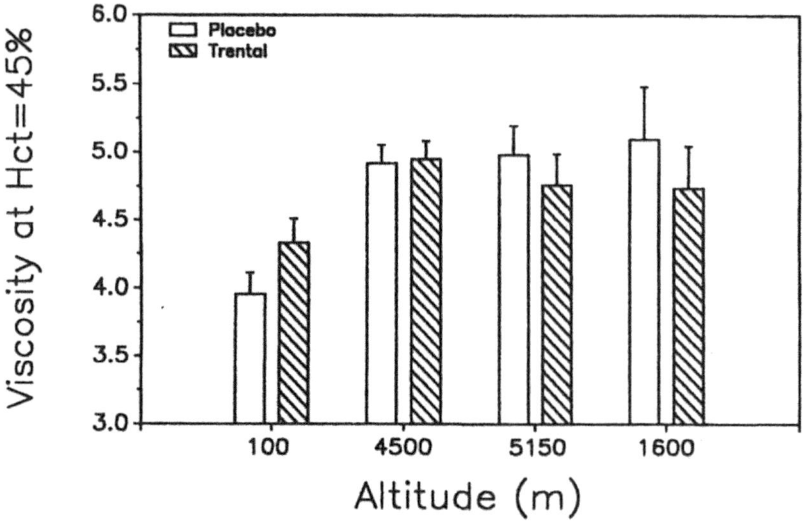

Figure 11.14 Whole-blood viscosity corrected to a hematocrit of 45%. At constant hematocrit blood viscosity increased with increasing altitude in all subjects ($P < .0001$). Subjects taking pentoxifylline had significantly less altitude-induced changes in viscosity ($P < .04$). *From Wood, Appenzeller and Appenzeller (1989).*

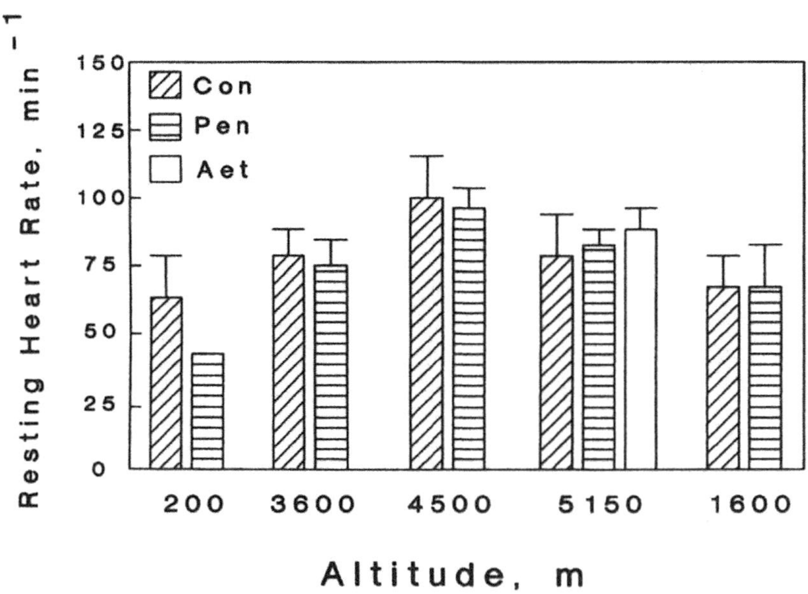

Figure 11.15 Resting heart rates at various altitudes. Altitude had a significant effect ($P < .02$). Pentoxifylline (PEN) takers had significantly lower heart rates than controls (CON) at 200 m, but not at any other altitude. *From Wood, Appenzeller and Appenzeller (1989).*

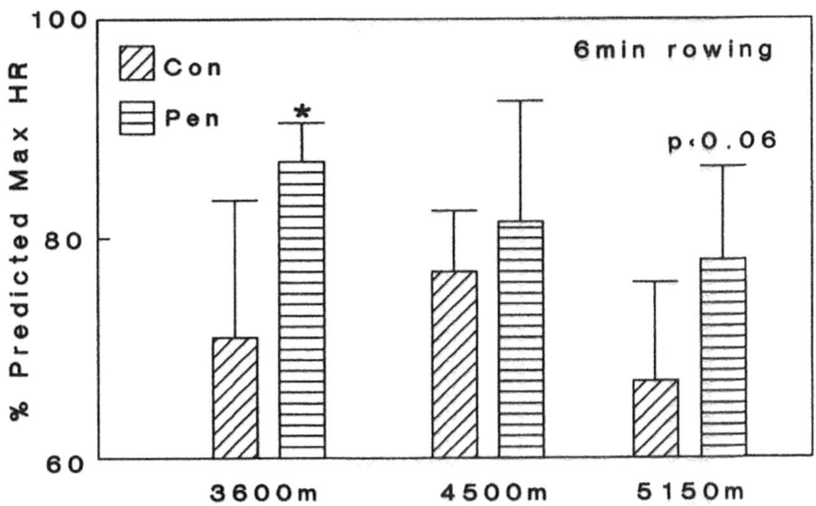

Figure 11.16 Percentage of predicted maximum heart rate after 6 min all-out rowing (predicted maximum heart rate at sea level is 220-age). Pentoxifylline (PEN)-taking was associated with significantly higher heart rates at altitude. *From Wood, Appenzeller and Appenzeller (1989).*

2. Autonomic function should deteriorate with altitude exposure (pari-passu with the deterioration in other nervous system function) due to hypoxic depression of neuronal activity and, perhaps more specifically due to changes in blood supply to this part of the nervous system.
3. Pentoxifylline, a substance that renders red blood cells more deformable, and therefore blood less viscous (see above), might improve neurogenic vasomotor control important in the maintenance of blood perfusion to brain muscles and skin and consequently in thermoregulation (Figs. 11.18–11.20).
4. Vascular tone in major vessels might be affected by altitude, exertion, and pentoxifylline intake and influence thermoregulation.
5. Parasympathetic function, evidenced by baroreceptor-initiated reflex slowing of the heart, will be influenced by altitude-induced changes in pulmonary blood volume or by direct effects on the afferent or efferent limb of the baroreflex arc.
6. Cognitive function is known to deteriorate at altitude. Ordinarily, the rate of perceived exertion (RPE) (a cognitive function) is related to heart rate, respiratory rate, and lactate production, all of which are markedly affected by altitude exposure and perhaps by pentoxifylline

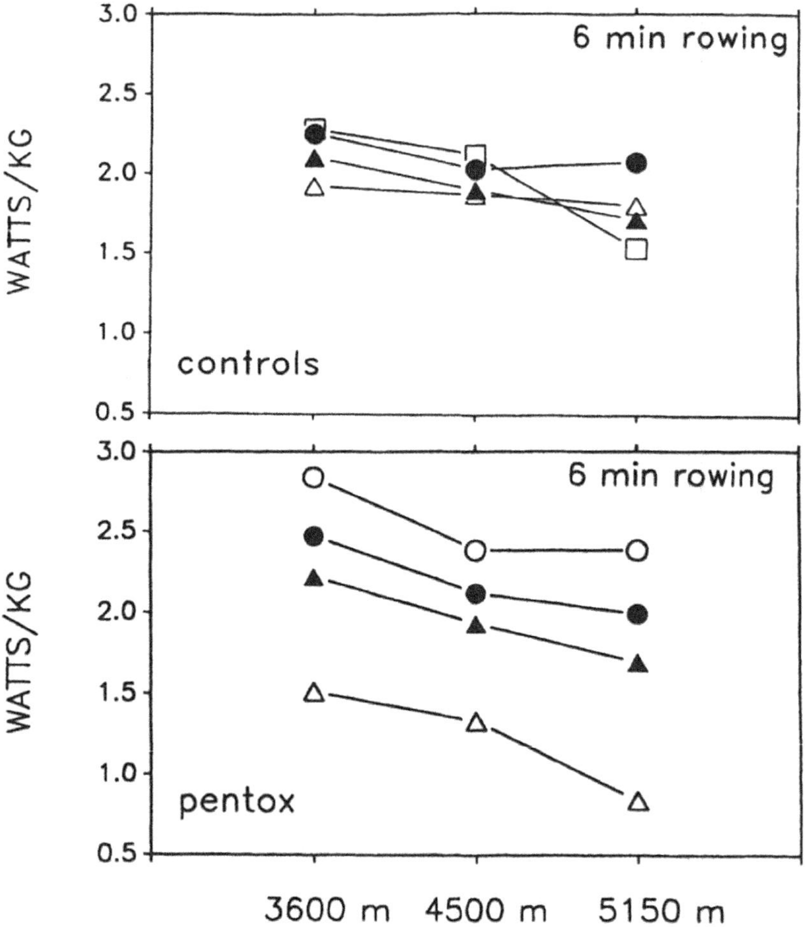

Figure 11.17 Power output (W/kg). Pentoxifylline had a negative effect on rowing performance. *From Wood, Appenzeller and Appenzeller (1989).*

secondarily through its putative effect on cardiac output (a product of stroke volume and heart rate). Therefore, RPE may affect all-out exercise performance at altitude.

11.11.4 Reflex vasoconstriction in the skin at altitude

Vasoconstriction in skin vessels after a gasp is shown in Fig. 11.21. Subjects on placebo had significantly larger vasoconstriction at basecamp than subjects on pentoxifylline, but no significant differences in the vasoconstrictor responses to gasps were found at sea level.

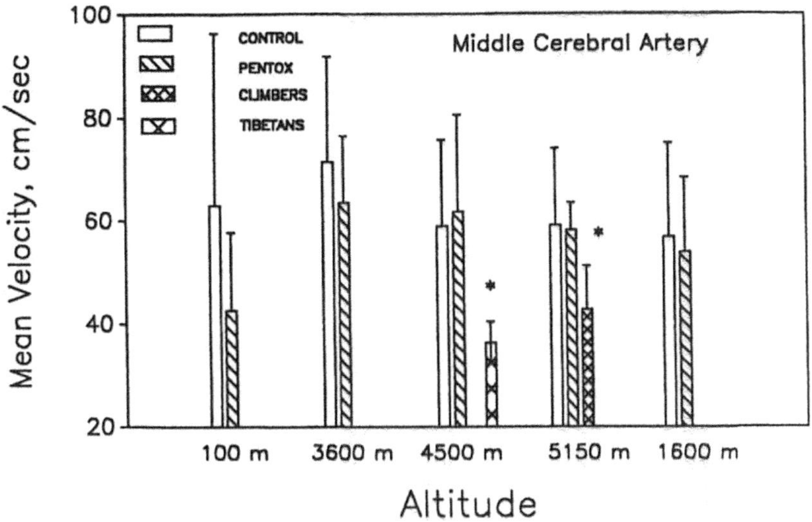

Figure 11.18 Doppler velocimetry. Mean velocity of climbers and Tibetans was significantly lower (*). *From Wood, S. C., Appenzeller, O., & Appenzeller, P. Transcranial measurement of cerebral blood flow in sojourners and natives at high altitudes. Ann. Sports Med. (1989) 4: 289–293.*

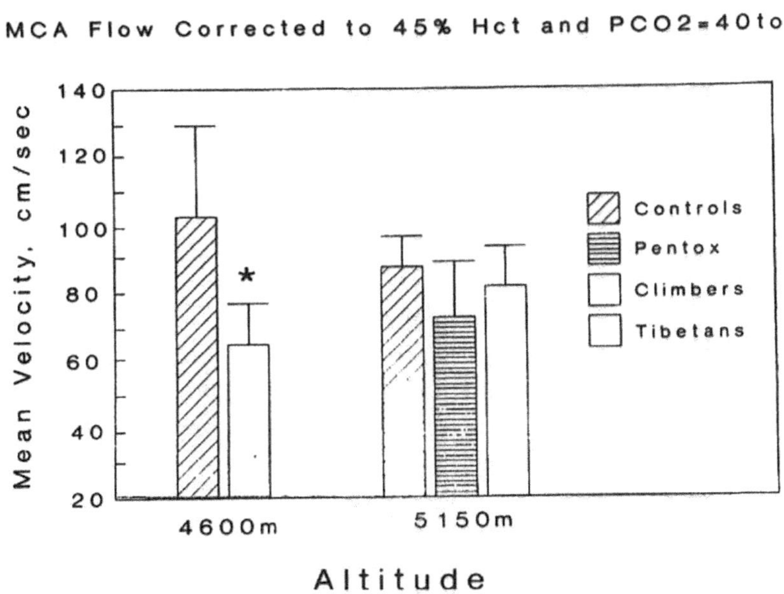

Figure 11.19 Doppler velocimetry. Middle cerebral artery (corrected to 45% hematocrit, $PCO_2 = 40$ torr). Tibetans have significantly lower velocities. Pentoxifylline had no measurable effect. *From Wood, S. C., Appenzeller, O., & Appenzeller, P. Transcranial measurement of cerebral blood flow in sojourners and natives at high altitudes. Ann. Sports Med. (1989) 4: 289–293.*

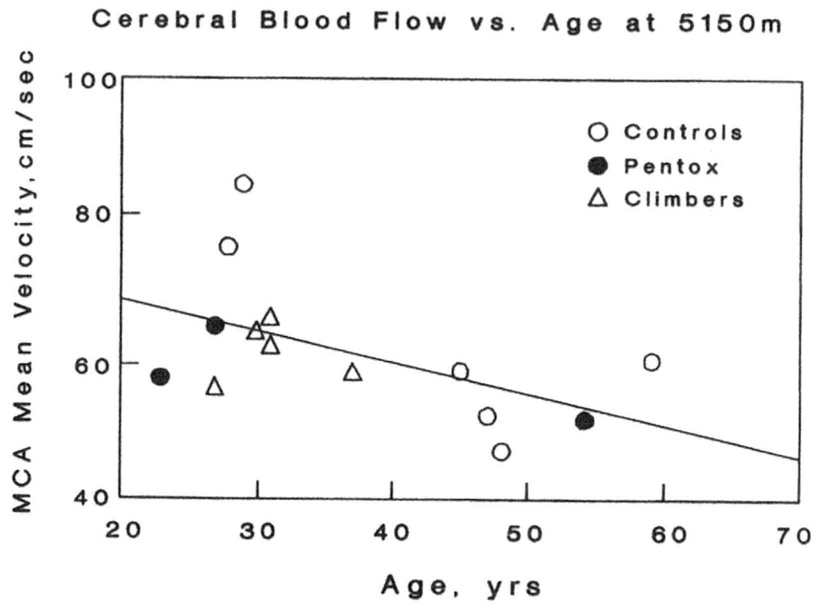

Figure 11.20 Doppler velocimetry. Middle cerebral artery corrected to 45% hematocrit and $Pco_2 = 40$ torr. A significant decline with advancing age only evident at 5150 m, but not at lower altitudes. *From Wood, S. C., Appenzeller, O., & Appenzeller, P. Transcranial measurement of cerebral blood flow in sojourners and natives at high altitudes. Ann. Sports Med. (1989) 4: 289–293.*

The response of skin vessels to mental arithmetic is shown in Fig. 11.22. Significant inhibition of the vasoconstrictor response by the drug is evident at Basecamp. A regression analysis of the responses to mental arithmetic and gasp in skin vessels is shown in Fig. 11.23. Pentoxifylline and placebo subjects show a significant correlation in vasoconstrictor responses to the two stimuli.

Fig. 11.24 shows the responses of the forearm muscle vasculature to mental stress. Placebo takers had paradoxical vasoconstriction at the base camp, whereas drug takers maintained normal vasodilatation throughout the study period. Vasoconstriction in muscle blood vessels after Valsalva maneuver is shown in Fig. 11.25. No consistent differences between the responses of drug or placebo takers were found. Regression analysis of the vasoconstrictor responses in skin (gasp) and muscle (after Valsalva maneuver) is shown in Fig. 11.26. A highly significant relation of these two adrenergically mediated constrictor responses was found in subjects taking the drug, but no significant correlation in those on placebo. Mental arithmetic induced skin vasoconstriction and muscle vasoconstriction after the Valsalva maneuver showed no significant relationship.

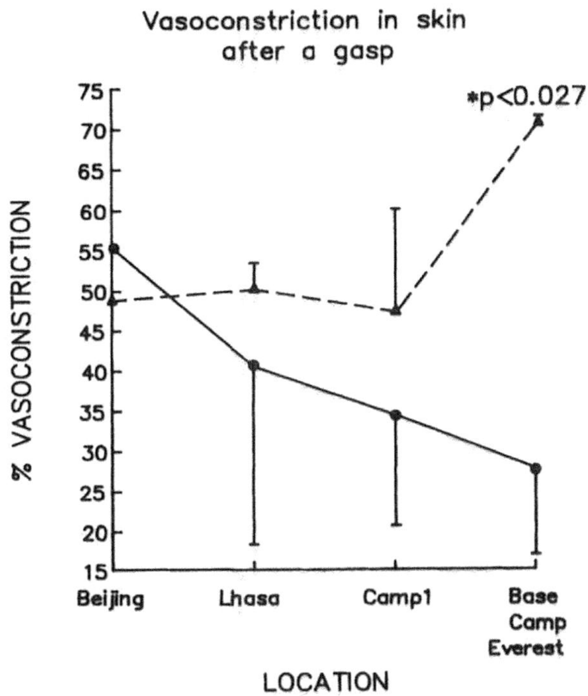

Figure 11.21 Reflex vasoconstriction in skin blood vessels after a gasp. Circles = subjects taking pentoxifylline. Triangles = subjects taking a placebo. A significant inhibition of altitude-induced vasoconstriction by the drug was found at 5150 m. *From Appenzeller and Wood (1988).*

11.11.5 Poststrain (Valsalva) bradycardia

The Valsalva maneuver is a complex stimulus and it is likely that several afferent mechanisms are influenced by the maneuver. Skeletal muscle vascular resistance is affected by cardiopulmonary receptors. Splanchnic vascular resistance and heart rate are modulated by carotid sinus and aortic arch baroreceptors. The situation is complicated because arterial baroreceptors have some influence on vasomotor outflow to skeletal muscles and to confound matters, decreases in central blood volume (which occur at altitude and during Valsalva maneuver) reduce activity in cardiopulmonary receptor afferents and may affect the sensitivity of the arterial baroreflexes controlling heart rate and blood pressure.

Forearm arteriolar resistance during Valsalva maneuver in normal subjects at sea level shows persistent vasoconstriction after the maneuver. The occurrence of post-Valsalva forearm vasodilatation has been taken as a sign of

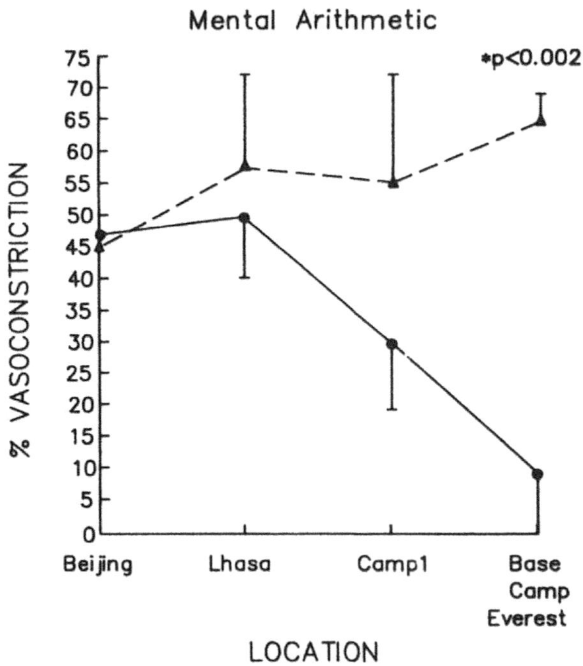

Figure 11.22 Reflex vasoconstriction in skin blood vessels induced by mental arithmetic. A significant inhibition of altitude-induced vasoconstriction by the drug was found at 5150 m. *From Appenzeller and Wood (1988).*

vasomotor abnormality in patients with autonomic dysfunction. The poststrain (Valsalva) bradycardia, therefore, most likely reflects parasympathetic activation (see Bennett & Gardiner, 1988). The poststrain (Valsalva) bradycardia is shown in Fig. 11.27. The bradycardia was accentuated in subjects on pentoxifylline with increasing altitude, but this did not reach statistical significance.

11.11.6 Reflow after circulatory arrest

In peripheral arterial disease, the pulse reappearance time can give a measure of arterial patency. Normally, the pulse reappearance time is nearly instantaneous and certainly less than one second (Fronek, Coel, & Bernstein, 1977). In patients with peripheral arterial disease, the pulse reappearance time is greatly prolonged.

Reflow after circulatory arrest by a cuff applied to the upper arm is shown in Fig. 11.28. No significant effect of the drug on pulse reappearance time was seen. However, an increasing delay with increasing altitude was found in all subjects tested.

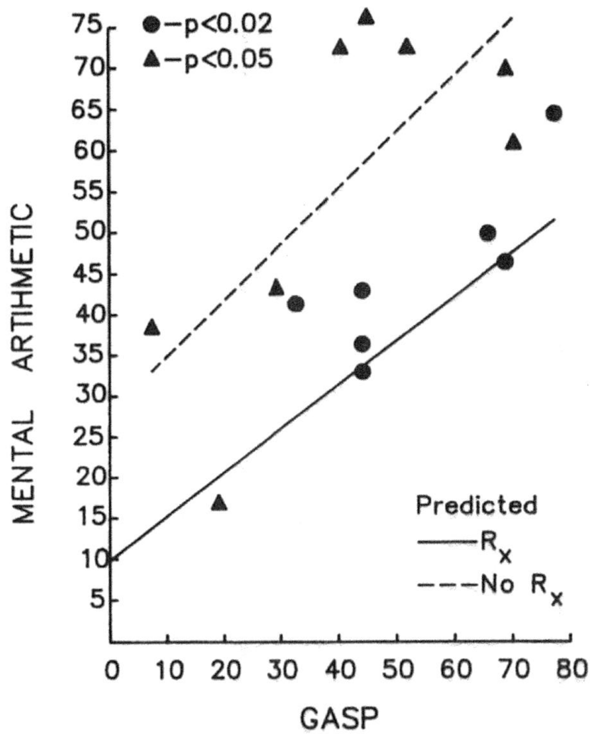

Figure 11.23 Subjects taking pentoxifylline (circles) had a significant correlation between sympathetically mediated vasoconstrictor responses in skin vessels (gasp and mental arithmetic). This correlation was less significant in subjects taking a placebo (triangles) *From Appenzeller and Wood (1988).*

11.11.7 Temperature difference—fingertips to palms

A satisfactory measurement of a natural invisible infrared emission of the human body and translation into a visible representation has been achieved by the use of thermography. Liquid crystal thermography can be used to obtain quantitative data on temperature of the palms and fingers. The amount of blood flowing through the skin determines skin temperature. Thermography is capable of giving a topographic representation of the amount of blood present in the skin. The color scheme on the polaroid print is produced by liquid crystals. Changes in temperature have been recorded in peripheral neuropathies, sympathetic dystrophy, and in patients with autonomic disturbances from other causes. Abnormalities in temperature of the skin produced by muscle or other injuries have been graphically displayed in the affected limb (Hobbins, 1988). Fig. 11.29

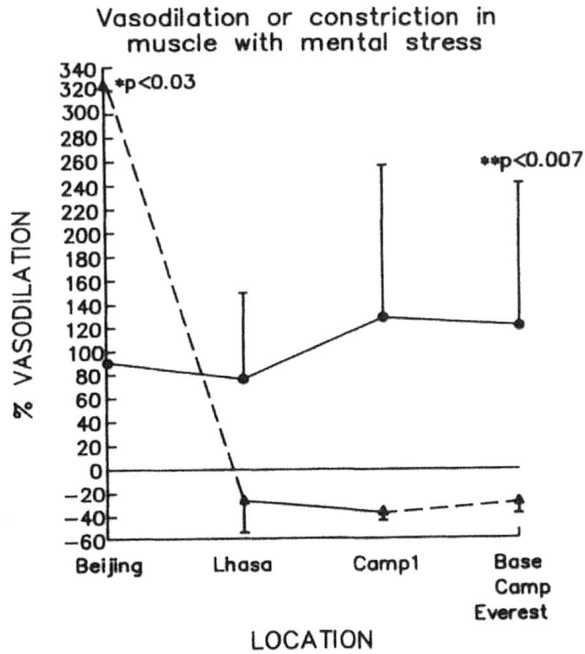

Figure 11.24 Mental stress caused normal reflex vasodilatation in muscle blood vessels at all altitudes in pentoxifylline takers, but not in placebo groups. *From Appenzeller and Wood (1988).*

shows the temperature difference between fingertips and palms in degrees centigrade. In an altitude study, a significant increase in the difference (vasoconstriction in fingertips) was seen in subjects receiving placebo at the base camp, but the difference in temperature between fingers and palms in those on active drug was not changed by increasing altitude and was similar to that of climbers, though still somewhat larger than that of native Tibetans tested at Camp 1.

Neurogenic vasomotor control in skin blood vessels with increasing exposure to altitude showed the vasoconstrictor responses to be preserved in the skin, but vasomotor function to muscle blood vessels becomes deranged with increasing altitude in normal individuals. This is evidenced by a conversion of normal muscle vasodilator response with mental stress to a significant vasoconstrictor response. There also is no relationship between vasoconstrictor responses in skin and in muscle to appropriate stimuli with increasing altitude. Parasympathetic function is shown to remain unaffected by altitude exposure as evidenced by poststrain bradycardia.

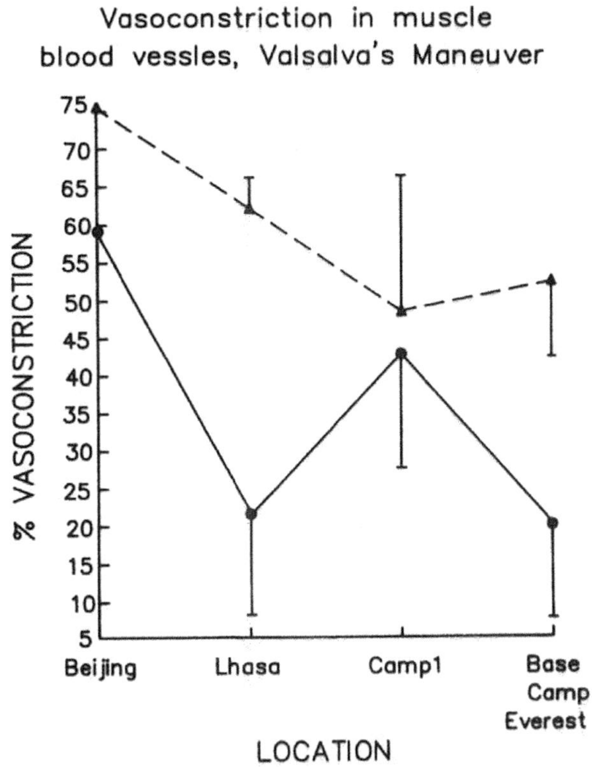

Figure 11.25 Reflex vasoconstriction in muscle blood vessels. No significant difference at any altitude between pentoxifylline and placebo groups. *From Appenzeller and Wood (1988).*

Frostbite to fingertips and toes in climbers are frequent. This may be related to the increasing temperature difference of fingertips when compared to palms with increasing altitude. Moreover, the time taken for the blood to recirculate to the muscle after complete arterial occlusion is considerably prolonged at very high altitude, suggesting that large-vessel vasoconstriction may contribute to poor perfusion of digits and thus promote frostbite at high altitude.

Pentoxifylline attenuates sympathetic vasoconstrictor responses in cutaneous, but not muscle, vascular beds, and the drug preserves muscle reflex vasodilatation at high altitude. The effect of the drug on parasympathetic function is less dramatic. An increased blood volume in the lung, or increased pulmonary tissue rigidity (having a similar effect) enhancing afferent stimulation, complicates the hemodynamics of the Valsalva maneuver and may account for the observed increase in poststrain bradycardia in drug takers at altitude.

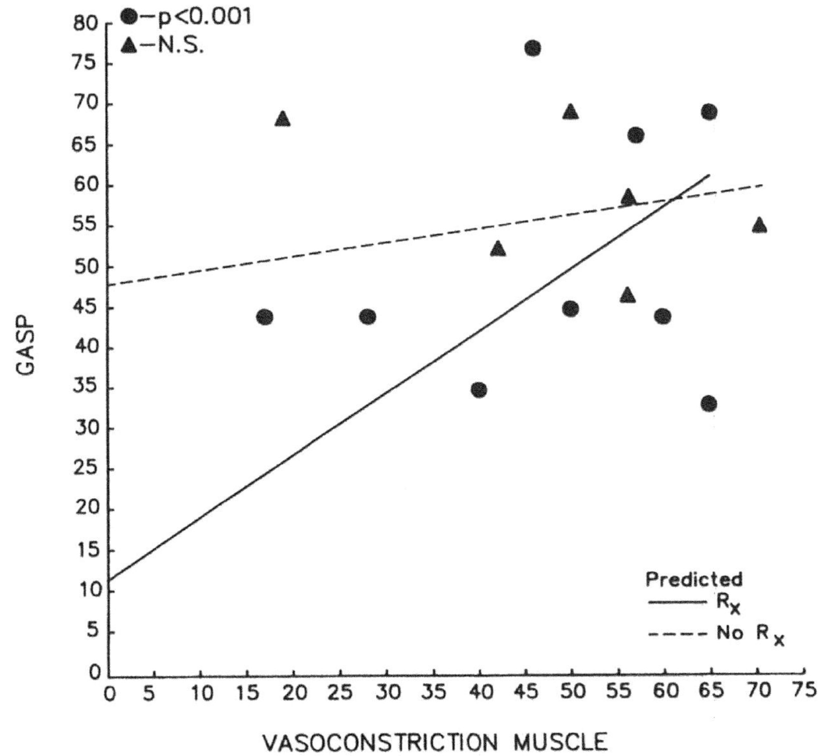

Figure 11.26 Vasoconstriction in skin blood vessels (gasp) and muscle blood vessels (Valsalva maneuver). A highly significant relationship between these sympathetic vasoconstrictor responses was found in the pentoxifylline group only, but not in the placebo group. *From Appenzeller and Wood (1988).*

The relative absence of fingertip to palm temperature difference in subjects on pentoxifylline at altitude and the considerable increase in temperature difference in those on placebo at base camp suggests that the drug preserves blood flow to fingertips at normal volume. However, the fingertips of altitude natives remained considerably warmer even when compared to study subjects taking pentoxifylline.

Responses to cold exposure vary in different races and individuals. Exposure to cold in indigenous Andean people shows that they maintain substantially warmer extremity temperatures than subjects of Caucasian descent (Little, Thomas, & Larrick, 1973). A similar genetically determined vasomotor response to cold may be evident in Tibetans whose fingertip temperature is consistently warmer when compared to sojourners.

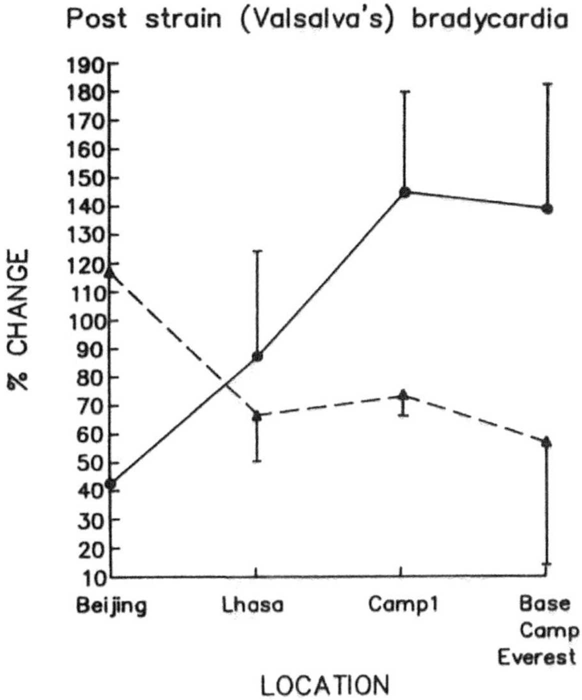

Figure 11.27 Parasympathetic activation (post-Valsalva bradycardia) was more pronounced in the pentoxifylline group at the highest altitude. *From Appenzeller and Wood (1988).*

The delay in reflow with increasing altitude (vasoconstriction in large vessels) might reflect a direct influence of hypoxia on blood vessel walls or on neurogenic control of several vascular beds (mainly resistance vessels). This large vessel function was unaffected by pentoxifylline, though clear evidence of the effect of the drug on blood viscosity and red cell deformability was obtained. Chamber studies at 4000, 5000, and 6000 m simulated acute altitude have shown significant increases in muscle sympathetic activity most likely stimulated by chemoreceptor activation. Surprisingly, there were large interindividual differences above 5000 m, which have been attributed to the direct effects on some central mechanism that has been independently influenced by the severe hypoxia (Saito et al., 1988). Unfortunately, this study was carried out on young subjects and it did not address the possible effects of acclimatization on muscle sympathetic activity.

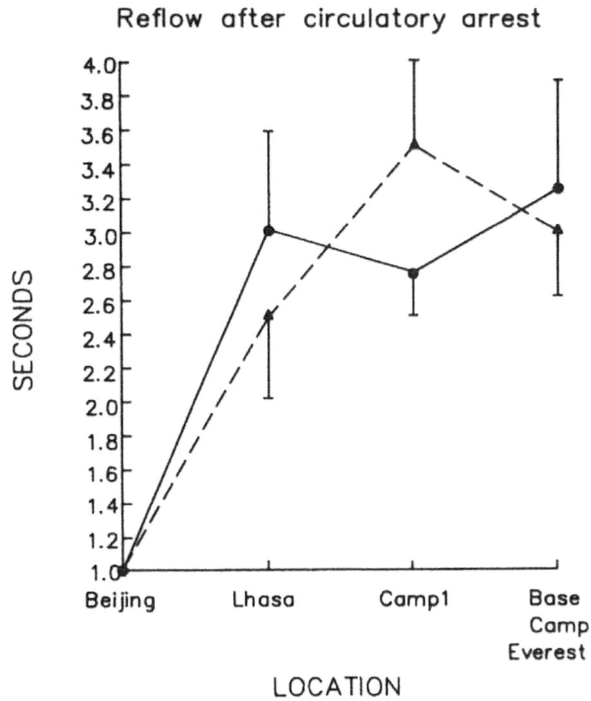

Figure 11.28 After arresting the circulation to the arm, the pulse reappearance time (after release) increased with increasing altitude. There was no effect of pentoxifylline on this delay in reperfusion. *From Appenzeller and Wood (1988).*

11.12 Clinical effects of altitude exposure

11.12.1 The pulmonary circulation

The asynchrony between the intermittent entry of air into blind-ended tubes susceptible to obstruction while breathing and the continuous, normally unobstructed flow of blood into the lung underlies the response of the pulmonary vasculature to hypoxia. At sea level, obstruction of air entry into part of the lung renders the blood vessels supplying the now hypoxic tissue useless for gas exchange. Hypoxia-induced pulmonary vasoconstriction shunts the blood away from the hypoxic portions of the lung to other parts with adequate ventilation, thus maintaining normal oxygenation of the blood (Heath & Williams, 1989). At altitude, the hypoxia-induced vasoconstriction becomes detrimental to blood oxygenation because all alveolar spaces are filled with hypoxic air due to the diminished barometric pressure. This altitude-induced pulmonary arterial

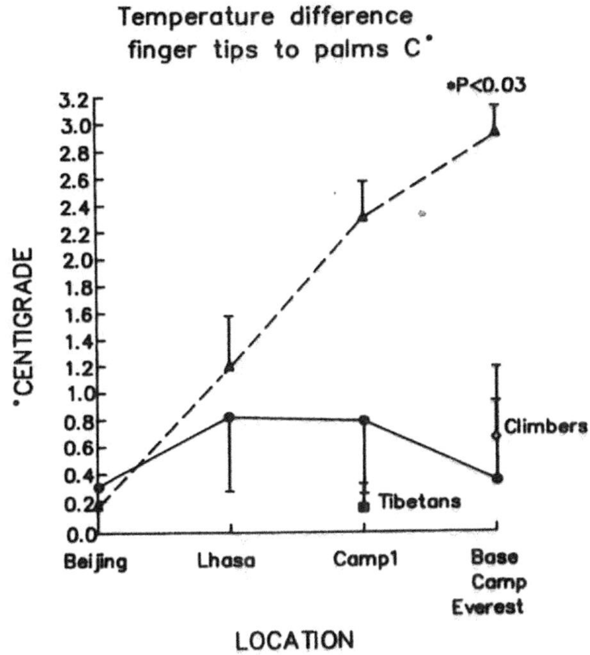

Figure 11.29 Temperature difference between fingertips and palms (a measure of skin blood vessel constriction) is significantly less in the pentoxifylline group and similar to that found in climbers and natives at high altitude. *From Appenzeller and Wood (1988).*

vasoconstriction underlies the increase in pulmonary artery pressure, which in some individuals becomes excessive and culminates in high-altitude pulmonary edema (HAPE). Maladaptation to altitude evidenced by persistent hypoxic pulmonary arterial vasoconstriction in humans and animals is associated with muscularization of pulmonary arteries (Heath & Williams, 1989). A direct relationship between sympathetic nerve activity and pulmonary artery pressure measured at low and high altitudes was demonstrated in one study suggesting that sympathetic overactivation may contribute to HAPE (Duplain et al., 1999).

The mechanism of hypoxic pulmonary vasoconstriction is controversial. Some support for the idea for a role of K^+ channels has come from the finding that hypoxia inhibits an outward K^+ current in isolated pulmonary artery smooth muscle cells causing depolarization of the resting membrane potential (Jensen, Micco, Czartolomna, Latham, & Voelkel, 1992). This is assumed to allow Ca^{2+} entry with consequent constriction (Anand, 1994).

The idea that hypoxic pulmonary vasoconstriction is the result of a local reflex is favored by those who point to the wide distribution of pulmonary endocrine cells which are arranged as innervated groups ideally suited for monitoring intrapulmonary gases. These clusters are known as neuroepithelial bodies (Lauweryns & Peuskens, 1972). The cells of neuroepithelial bodies have extensive innervation with afferent and efferent nerves and are modulated by the vagus (Gosney, 1994). There is direct evidence that neuroepithelial bodies act as chemoreceptors specifically reacting to hypoxia (Gosney, 1994). The result of hypoxia activation of neuroepithelial bodies is not clear, nevertheless, one assumption is that they induce localized reflex pulmonary vasoconstriction (Gosney, 1994). Thus changes in Ca^{2+} entry into pulmonary smooth muscle cells may be more suited to slower pulmonary vasoreactivity, whereas local reflexes mediated by neuroepithelial bodies may be more effective in short-latency responses to hypoxia.

11.12.2 J receptors

The juxta pulmonary capillary receptors (J receptors) have important functions in sensory perceptions in the upper chest and throat and if these are intensified, in dry cough as well. Moreover, they are also important in the breathlessness after exercise but probably not in the breathlessness induced by a tight-fitting vest in normal individuals (Paintal, 1995). J receptors are found near alveoli and are synonymous with pulmonary C fibers. The main stimulus for J receptor firing is an increase in pulmonary capillary pressure and pulmonary interstitial volume. Clearly then, they are important in conditions associated with increases in interstitial lung volume in normal individuals. These conditions are altitude exposure and after exercise. They may also be involved in the breathlessness, after mild exertion, in untrained elderly subjects (Cook & Shaper, 1989) because with less efficient left ventricular contractions an increase in left atrial pressure would result in an increase in pulmonary capillary pressure with consequent stimulation of J receptors.

J receptors are ordinarily dormant. When J receptors are activated, they produce the J reflex through central nervous system connections, which consists of an inhibition of all somatic muscle activity. Therefore, they may be instrumental in the postexercise collapse of athletes but this has not been investigated.

11.12.3 High-altitude pulmonary edema

Surveys of 350 high-altitude pulmonary edema (HAPE) subjects showed that 80% had breathlessness, pressure, choking feeling, and 50% had a dry cough. These sensations were located in the neck and upper sternal region. In humans, purposeful stimulation of J receptors by the injection of lobeline produces the same sensations and dry cough (Raj, Singh, Anand, & Paintal, 1995). There is also evidence that the breathlessness after exhaustive exercise and with even moderate exercise at altitude is J receptor-dependent (Paintal, 1995). The relationship between J receptors and baroreceptors has not been investigated. But coordination in activities between these receptor types under physiologic conditions and dysfunction in diseases of the autonomic nervous system is likely.

The pathogenesis of HAPE remains, however, controversial. Thus, mean pulmonary artery pressures in patients with HAPE range from 22 to 63 mmHg when measured during right heart catheterization. Using echo-Doppler estimates of systolic pulmonary artery pressures in patients with HAPE an average of 53 mmHg was found, only slightly higher than in healthy subjects exposed to the same normobaric or hypobaric hypoxia (Naeije, 1992). Subjects who have had an episode of HAPE are prone to recurrence of the condition when exposed to appropriate altitude and are said to have an exaggerated hypoxic pulmonary artery vasoconstriction at sea level.

However, echo-Doppler studies of pulmonary hemodynamics cannot reliably distinguish HAPE-prone subjects from normal individuals at sea level (Naeije, 1992). The noncardiac pulmonary edema of HAPE has been attributed to excessive and uneven reflex hypoxic pulmonary artery vasoconstriction (Naeije, 1992) but studies of pressures in the pulmonary artery have not fully supported this hypothesis. A different, mechanistic approach attributes HAPE to a stress failure of pulmonary capillaries with leakage of fluid and blood constituents into the alveolar spaces. This hypothesis is supported by animal experiments that show electron microscopic breaks in the pulmonary capillaries when the pressure in these vessels is raised to 40 mmHg (West, Tsukimoto, Mathieu-Costello, & Prediletto, 1991).

A landmark study attempted to assess the role of the autonomic nervous system in adaptation to altitude. A comparison with patients who had recently recovered from well-documented HAPE was also carried out (Mathew et al., 1985). Twenty young males were studied at sea level, 3 days after arrival at altitude, and at weekly intervals for 3 weeks at an altitude of 3500 m and again after return to sea level. At the same altitude, 10 acclimatized lowlanders, 10 high-altitude natives, and 6 patients who recently recovered from HAPE were also compared. Another phase of this study examined, at sea level, in a similar fashion, 15 subjects who had stayed at altitude without ill effects and compared the results with those from 10 subjects who had recovered from HAPE or acute mountain sickness. To assess autonomic function, a number of noninvasive tests were carried out which unfortunately did not reach present levels of sophistication to allow definitive conclusions. Nevertheless, the large number of subjects and well-matched controls in this study increases the likelihood that the results are meaningful. Sympathetic activation, evidenced by vasoconstriction and tachycardia, was found on acute altitude exposure of lowlanders. This activation subsided after acclimatization. High-altitude natives have a preponderance of parasympathetic tone affecting cardiovascular function. It is not clear, however, how much of this is attributable to the physically active lifestyle of the natives or to the lifelong residence at 3500 m. The results on HAPE patients showed sympathetic activation comparable to that of healthy lowlanders on acute altitude exposure and comparisons of recovered HAPE and acute mountain sickness subjects with altitude adapted subjects at sea level showed no significant differences. Unfortunately, this study did not include measurements of hormones, circulating catecholamines, or peptides to allow full assessment of autonomic balance. Pulmonary artery pressures and baroreceptor function were not assessed.

Rational therapy for HAPE, an often lethal complication of high-altitude sojourn, is now available. Since HAPE is associated with high pulmonary artery pressures, measures taken to reduce the pressure by descent, recompression if descent is impossible, oxygen administration, and drugs that relax pulmonary artery constriction such as calcium channel blockers have proven effective (Oelz et al., 1992).

In six subjects with radiographic evidence of HAPE, the calcium channel blocker nifedipine decreased pulmonary artery pressure, reduced alveolar-arterial oxygen gradient, and led to improved oxygenation, clinical well-being, and radiographic evidence of clearing of pulmonary edema. This improvement occurred in spite of continued exercise and without oxygen supplementation at an altitude of 4000 m (Oelz et al., 1992). In subjects with a history of documented HAPE, the prophylactic administration of the drug prevented the occurrence of HAPE on acute re-exposure to an altitude of 4559 m (Oelz et al., 1992). Subjects receiving the drug had lower pulmonary artery pressures, better oxygenation, and lower scores on a clinical scale of mountain sickness than subjects receiving placebo at the same altitude. Thus, nifedipine has a place not only in the acute treatment of HAPE but also as a means of prophylaxis in susceptible individuals. The use of nitrous oxide gas as a precursor for the gaseous neurotransmitter nitric oxide that dilates pulmonary vessels locally has been reported successful in ameliorating the severe hypoxemia of HAPE.

The hypothesis that the sympathetic nervous system is involved in the genesis of HAPE is supported by an investigation of 16 patients and 10 controls using various methods to lower pulmonary artery pressures (Hackett, Roach, Hartig, Greene, & Levine, 1992). The percent reduction in pulmonary vascular resistance and pulmonary artery pressure were, respectively, 46 and 33 for oxygen administration; 30 and 29 for nifedipine; 29 and 25 for hydralazine; 57 and 42 for phentolamine; and 72 and 52 for oxygen combined with phentolamine (Hackett et al., 1992). Thus, the marked effectiveness of α-adrenergic blockade in modifying the HAPE-associated pulmonary hypertension implies the presence of heightened sympathetic activity in patients with this condition. The exact mechanism by which such blockade brings about clinical improvement is, however, not clear.

In eight patients with HAPE, brain computed tomography (CT) and urinary and plasma catecholamines were assessed (Lataste, 1992). Catecholamines in the urine and plasma were significantly elevated during the illness when compared to the levels found after recovery and brain CTs were consistent with brain edema in all patients with HAPE. This supports the notion that HAPE and high-altitude cerebral edema (HACE) may coexist in patients and that HACE becomes symptomatic only late during the course of HAPE and perhaps only preterminal in such patients.

Nonpharmacologic treatment of HAPE and other types of mountain sickness is also possible. Some controlled studies using a portable hyperbaric chamber have been published indicating beneficial short-term effects after one hour of compression (Bärtsch, 1992). Rest alone is also beneficial since exercise at altitude increases tissue hypoxia, promotes edema formation by sodium and water retention (see above), and increases pulmonary artery pressure. Expiratory positive airway pressure has also been successful in short-term uncontrolled trials (Bärtsch, 1992) but the long-term effects of all these modalities have not been carefully evaluated. The most prudent and effective treatment of mountain sickness syndromes remains immediate descent.

11.12.4 Systemic blood pressure

After chronic altitude exposure, systemic blood pressure increases and there is sympathoadrenal activation (Mazzeo et al., 1991). This activation of the autonomic nervous system is the result of the neural release of norepinephrine and the adrenal release of epinephrine. The time course of the neurotransmitter release has been examined (Mazzeo et al., 1991). Early during altitude exposure, there is little change in epinephrine concentration in the blood when compared to sea level and epinephrine levels decline with acclimatization. Norepinephrine was unaltered on arrival at altitude but rose with acclimatization pari-passu with increases in blood pressure. From the time course of the changes in blood pressure and norepinephrine, it was concluded that altitude-associated hypertension after acclimatization was most consistent with an increase in α-adrenergic vasoconstriction (Reeves, Mazzeo, Wolfel, & Young, 1992). These studies were done at a fixed altitude of 4300 m and the findings may not reflect the responses of the blood pressure nor of the sympathetic nervous system at different altitudes. Other studies have shown extreme vasoconstriction in the microcirculation and hypertension on acute altitude exposure to 5000 m which persisted unchanged after acclimatization.

11.12.5 Saliva

A frequent complaint at high altitude is cough and a dry mouth. The effects of altitude on salivary gland function, predominantly under parasympathetic control, have been examined (Pilardeau et al., 1990). Twelve sea-level natives were studied before and after 2 days at 4350 m at rest and after maximal exercise. There were no differences in saliva flow rates, potassium levels nor in α-amylase concentrations after exercise in normoxia or hypoxia. After acute altitude exposure mean flow rates increased at rest and after exercise and potassium concentrations decreased in both basal and postexercise states. This seems to be the only suggestion that the parasympathetic nervous system innervating the salivary glands is activated by acute altitude exposure (Pilardeau et al., 1990).

11.12.6 Hypoxia at sea level

The value of altitude studies of normal subjects is reflected in patients with chronic hypoxia at sea level. Thirty-two patients with pulmonary tuberculosis were compared with 30 normal subjects to assess sympathetic function. Blood gases and spirometry were used to assess pulmonary function. Plasma norepinephrine and cyclic AMP concentration were directly related to hypoxemia and spirometric indices of pulmonary function (Watanabe, Ogawa, Ban, & Satake, 1981). This suggests that autonomic function is similarly affected by hypoxia at sea level caused by disease and that due to hypobaria.

11.12.7 Chronic mountain sickness (CMS)

Long-term altitude residents and natives may develop Monge's disease, first described by Professor Carlos Monge-M in 1928 at Cerro de Pasco. This disease is, with few exceptions, confined to indigenous people of the High Andes. Clinical autonomic symptoms include an increased prevalence of vascular headaches of the migrainous type and clubbing of the fingers.

The oxygen theory of the causation of migraine has been extensively reviewed by Amery (1987). This hypothesis suggests that hypoxia of the brain is the source of the migraine attack and that the clinical and other manifestations of migraine are the result of hypoxia-induced alterations of oxidative pathways of neuronal metabolism. Thus, any pathogenic factor that is capable of inducing regional or global hypoxia of the brain such as that found in hypobaric hypoxia may induce a migraine attack even in those who do not normally suffer from migraine. Epidemiologic support for this theory comes from a study in high-altitude natives of Cerro de Pasco where the prevalence of migraine is highly correlated with CMS severity in men (Arregui et al., 1994). CMS is thought to result from excessive erythremia which is taken as evidence of maladaptation to high-altitude life in South America (Monge, 1988). The ventilatory drive declines with advancing age in all individuals and hypoventilation, the cause of the accentuated erythremia in CMS, is the principal limiting condition for human life at very high altitudes in South America (Monge, 1988). In patients with CMS, the prevalence of migraine increases with advancing age but the correlation with ventilatory drive in these patients has yet to be examined. Thus, accelerated aging of the ventilatory drive in Andean native men leading to progressive hypoventilation might be the cause of the erythremia and of the high prevalence of migraine in these patients. It should be noted that at low altitudes migraine is more common in women and its prevalence decreases with advancing age, unlike the observations from Cerro de Pasco.

Chronic mountain sickness and its accompanying headache is practically unknown in Himalayan natives living at comparable altitudes. This suggests that genetic factors may play a role in the genesis of altitude migraine and in the cerebral vasomotor responses associated with migraine at any altitude. Himalayan natives also do not have the accelerated and age-related increase in erythrocytosis found in Andean populations. The cerebrovascular reactivity to CO_2 in South American high-altitude natives has not been compared with Himalayan natives or with that of sojourners. It is likely that differences in cerebrovascular reactivity will be uncovered to help account for the high prevalence of migraine in the Andes.

In chronic mountain sickness, systemic oxidative-inflammatory-nitrosamine stress, defined by an increase in free radical formation and a decrease in vascular nitric oxide bioavailability, is associated with blunted cerebral perfusion and vasoreactivity to hypercapnia, impaired cognition, and depression (Bailey et al., 2019).

Heatstroke is potentially lethal. A study by Sohal, Sun, Colcolough, and Burch (1968) pointed to heat-induced damage of endothelial cells that triggers disseminated intravascular coagulation (Sohal et al., 1968). Damage to pulmonary endothelial cells may contribute to the pulmonary edema common to both acute mountain sickness and heat stroke. In both diseases, there is a significant impairment of normal oxygen transport to tissues. In heatstroke, there is disruption due to increased O_2 demand (Q_{10} effect) combined with decreased circulatory transport. In mountain sickness, there is decreased external supply of O_2 and a fall in cardiac output, leading to decreased transport. Table 11.2 summarizes the clinical comparison between heat stroke and mountain sickness.

Table 11.2 Clinical manifestations of heat stroke and mountain sickness.

Heat stroke	Mountain sickness
Nervous system	
Dizziness, weakness, confusion, drowsiness, nausea, anorexia, headache, disorientation, apathy irritability, coma, stupor, delirium, RPE increased, seizures, cerebellar dysfunction, constricted pupils, brain edema, petechiae, degenerative ischemic changes in neurons	Dizziness, weakness, confusion, drowsiness, nausea, anorexia, headache, disorientation, apathy or irritability, coma, stupor, delirium, RPE increased, constricted pupils, phantom companion, brain edema, degenerative ischemic changes in neurons
Cardiovascular	
Sinus tachycardia, peripheral circulatory failure, flattened or inverted t-wave, circulatory	Sinus tachycardia, fall in cardiac output (hypoxic myocardium), proportion of CO to organ does

(*Continued*)

Table 11.2 (Continued)

Heat stroke	Mountain sickness
collapse, cutaneous vasodilatation or constriction, reduced total peripheral resistance, increased O_2 requirements	not equate to proportion of O_2 that organ receives, coronary artery flow decreased, cutaneous vasoconstriction, maximum heart rate decreased
Pulmonary	
Tachypnea, respiratory alkalosis, pulmonary edema	Tachypnea, respiratory alkalosis, pulmonary edema
Hematologic	
Thrombocytopenia, hemorrhagic diathesis, parenchymal organ petechiae, lung, renal, myocardial bleeding, endothelial cell damage, platelet aggregation, capillary microthrombi, hypoprothrombinemia, hypofibrinogenemia, diffuse intravascular coagulation	Thrombocytopenia, platelet aggregation? Petechiae, capillary microthrombi, hypercoagulability, hypoprothrombinemia, hypofibrinogenemia, capillary microthrombi (lungs) increased hematocrit
Water and electrolytes	
Low Na, normal K, hyperglycemia?	Low Na, normal K, polyuria, hyperglycemia not studied
Gastrointestinal	
Diarrhea	Anorexia, flatulence, explosive stools bulky and greasy Fat malabsorption
Renal	
Proteinuria, abundant granular casts and glomerular casts, red cell casts, acute renal failure, acute oliguric failure	Proteinuria, enlarged glomeruli
Endocrine	
Stress hormones increased, opioids specifically increased	Stress hormones all increased, opioids increased, 17-ochs-11-noradrenalin both increased, aldosterone falls, thyroid hormones? Vasopressin and atrial natriuretic factor increased

RPE, rate of perceived exertion; *CO*, cardiac output.

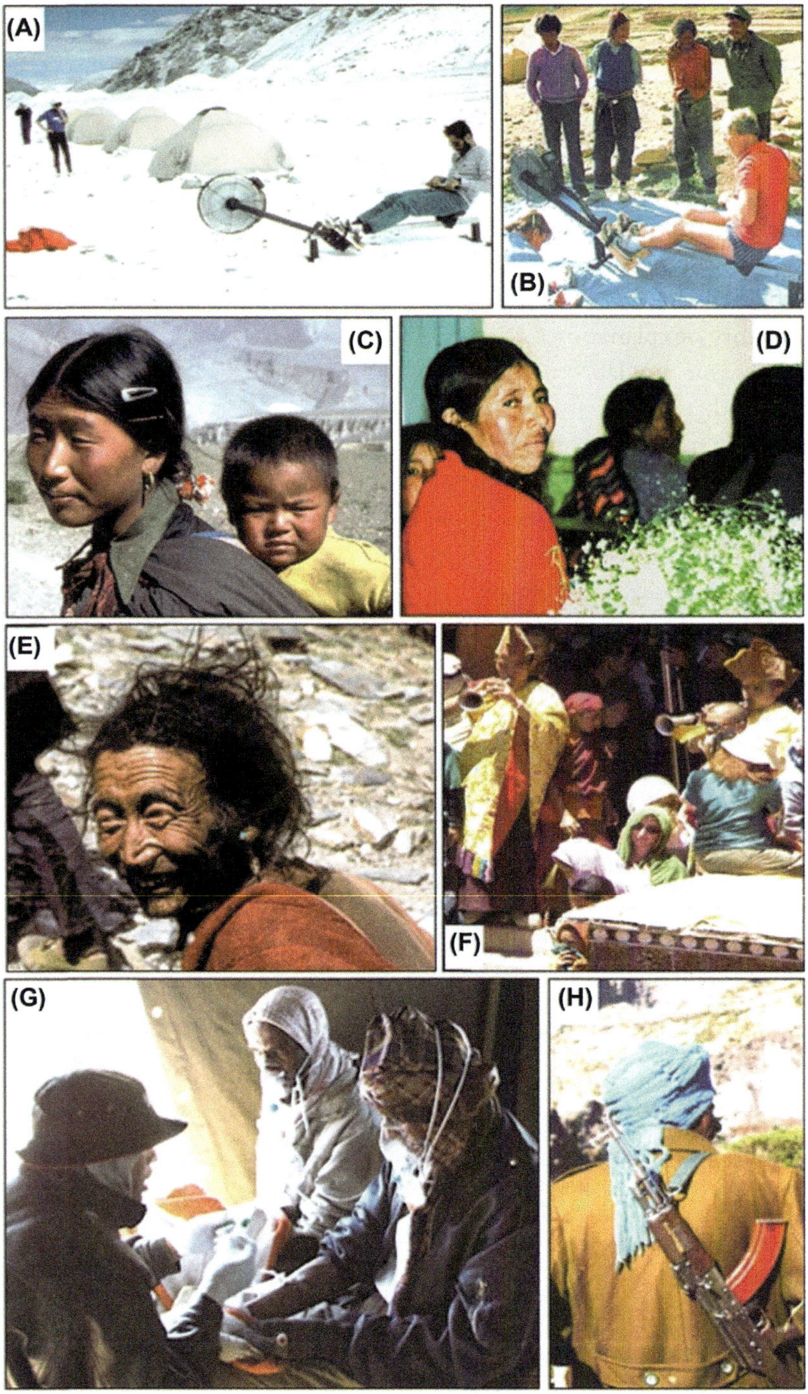

Figure 11.30 Freestyle neurology in action in the Himalayas, Peru, and Ethiopia. (A) Everest base camp in Tibet at 5500 m, where the rowing machine provided a standard exercise test. (B) On the way to the base camp at a lower altitude ~ 3800 m. Tibetan villagers watched the rowing machine in action before they in turn were tested at 5500 m. (C) on a trail in Tibet. (D) A wedding in Cuzco, Peru. (E) A Tibetan yak herder. (F) At a monastery festival in Tibet, a tourist attraction. (G) Simien mountains, Ethiopia, the only research site requiring an armed guard. (H). Our armed guard needed to prevent marauding locals from pilfering our equipment at night.

Clubbing of the fingers is mostly associated with disease of the lung and heart where hypoxia is often present even at sea level. It is also found in chronic mountain sickness. At sea level, the lesion associated with clubbing is an inflammatory periostitis and arthritis and this is often very painful. The pathology of clubbing in CMS is not known and there is no satisfactory explanation of this disorder, except that occasionally it regresses, in lowlanders, within days after section of the vagus. It is thought that vagotomy interrupts an abnormal pulmonary vascular reflex leading to a decrease in the peripheral circulation that accompanies the improvement in the clubbing. The hypothesis that vagotomy interrupts reflex hypoxic pulmonary artery vasoconstriction in patients with clubbing has not been disproved.

11.12.8 Altitude neuropathy

A frequent complaint amongst high-altitude South American natives is burning-hands-burning-feet, which has been attributed to peripheral neuropathy. A survey from the mining town of Cerro de Pasco (Peru), at an altitude of 4338 m showed that 39% of respondents had this complaint, whereas only 12% had this symptom in a comparable survey from Lima, Peru, situated at sea level. This presumed predominantly sensory neuropathy has been tentatively attributed to chronic hypoxia and maladaptation to life at very high altitudes (León Velarde & Arregui, 1994). It may be similar to the neuropathy of chronic respiratory disease with severe hypoxia seen at sea level (Appenzeller, Parks, & MacGee, 1968).

11.13 Freestyle neurology

The term "*freestyle Neurology*" was used for the first time in 2021 to denote advances in autonomic neuroscience gleaned at altitude in the mountains of Peru, Ethiopia, and the Himalayas. Because altitude sojourn causes autonomic nervous system activation, local people are readily available to serve as controls (O. Appenzeller, 2021) (Fig. 11.30).

References

Amery, W. K. (1987). The oxygen theory of migraine. In J. N. Blau (Ed.), *Migraine* (pp. 403–431). London: Chapman and Hall Medical.

Anand, I. S. (1994). Hypoxia and the pulmonary circulation. *Thorax*, *49*(S19), S24.

Anand, I., Greene, E. R., & Appenzeller, P. (1995). Middle cerebral artery (MCA) flow velocities (FV) and cerebrovascular reactivity to CO2 at altitude: effects of age and ethnicity. *Clin. Aut. Res.*, *5*, 330.

Angelkort, B. (1979). Thrombocytic function, plasmatic blood coagulation and fibrinolysis in chronic arterial occlusive disease. Study with pentoxifilline. *Die Medizinische Welt*, *30*(34), 1239–1243.

Antezana, A. M., Richalet, J. P., Antezana, G., Spielvogel, H., & Kacimi, R. (1992). Adrenergic system in high altitude residents. *Int. J. Sports Med.*, *13*(Suppl 1), S96–S100.

Appenzeller, O. (1986). Barogenic headache. In C. F. Rose (Ed.), *Handbook of clinical neurology (revised series) 4 headache* (pp. 395–404). Amsterdam: Elsevier.

Appenzeller, O. (1991). Aging, stress, and neurovegetative control. In A. R. Genazzani, G. Nappi, F. Patraglia, & E. Martignoni (Eds.), *Stress and related disorders* (pp. 281–293). Carnforth, Lancs, UK: Parthenon Publishing Co.

Appenzeller, O. (1992). Neurovascular headache: The autonomic nervous system, peptides and hypoxia. *Funct. Neurol.*, 7(Suppl. 3), 83–93.

Appenzeller, O. (1994). Aging stress and autonomic control. In M. L. Albert, & J. E. Knoefel (Eds.), *Clinical neurology of aging* (2nd. Edition, pp. 651–673). New York and Oxford: Oxford University Press.

Appenzeller, O. (2021). Reflections on the past three decades of high-altitude exercise research and the autonomic nervous system. *Clin. Auton. Res.*, *31*(1), 31–33.

Appenzeller, O., Appenzeller, J., Standefer, J., Skipper, B., & Atkinson, R. (1984). Opioids and endurance training; Longitudinal study. *Ann. Sports Med.*, *2*, 22–25.

Appenzeller, O., Greene, R., Qualls, C., & Martignoni, E. (1994). Valsalvas maneuver and intracranial circulation: effects of altitude and Ag. *Neurology*, *44*, A220.

Appenzeller, O., Khogali, M., & Carr, D. B. e a (1986). Makka Hajj: Heatstroke and endocrine responses. *Ann. Sports Med*, *3*, 30–32.

Appenzeller, O., Martignoni, E., & Qualls, C. (1995). Shivering thermogenesis and endothelin (ET): Effects of altitude and ethnicity. *Clin. Autonomic Res.*, *5*, 329.

Appenzeller, O., Parks, R. D., & MacGee, J. (1968). Peripheral neuropathy in chronic disease of the respiratory tract. *Am. J. Med.*, *44*(6), 873–880.

Appenzeller, O., Qualls, C., & Martignoni, E. (1993). Swallow induced tachycardia: Effects of age, altitude, and ethnicity on this brainstem autonomic reflex. *Neurology*, *43*, 757S.

Appenzeller, O., & Schade, D. (1979). Neurology of endurance training. III. *Sympathetic activity during a marathon race. Neurology*, *29*, 540.

Appenzeller, O., Standefer, J., Appenzeller, J., & R., A. (1980). Neurology of endurance training: V endorphins. *Neurology*, *30*, 418–419.

Appenzeller, O., & Wood, S. C. (1988). Pentoxifylline: Thermoregulatory and vasomotor effects at high altitudes. *Ann. Sports Med*, *4*, 279–285.

Appenzeller, O., & Wood, S. C. (1992). Peptides and exercise at high and low altitudes. *Int. J. Sports Med.*, *13*(Suppl 1), S135–S140.

Appenzeller, P., Wood, S., Qualls, C., Appenzeller, O., & Couzins, W. (1991). Vasomotor function at altitude: Effects of a rheologically active drug. In M. Yoshikawa, M. Uono, H. Tanabe, & S. Isikawa (Eds.), *New trends in autonomic nervous system research* (951, pp. 100–103). Amsterdam: Excepta Medica, Int. Congress Series.

Arregui, A., León-Velarde, F., Cabrera, J., Paredes, S., Vizcarra, D., & Umeres, H. (1994). Migraine, polycythemia and chronic mountain sickness. *Cephalalgia: Int. J. Headache*, *14*(5), 339–341.

Aviado, D. M., & Porter, J. M. (1984). Pentoxifylline: a new drug for the treatment of intermittent claudication. Mechanism of action, pharmacokinetics, clinical efficacy and adverse effects. *Pharmacotherapy*, *4*(6), 297–307.

Bailey, D. M., Brugniaux, J. V., Filipponi, T., Marley, C. J., Stacey, B., Soria, R., ... Sartori, C. (2019). Exaggerated systemic oxidative-inflammatory-nitrosative stress in chronic mountain sickness is associated with cognitive decline and depression. *J. Physiol.*, *597*(2), 611–629.

Bärtsch, P. (1992). Treatment of high altitude diseases without drugs. *Int. J. Sports Med.*, *13*(Suppl 1), S71–S74.

Bärtsch, P., Maggi, S., Kleger, G. R., Ballmer, P. E., & Baumgartner, R. W. (1994). Sumatriptan for high-altitude headache. *Lancet*, *344*(8934), 1445.

Bennett, T., & Gardiner, S. M. (1988). Cardiovascular pathophysiology in diabetes mellitus: A reappraisal. In R. Bannister (Ed.), *Autonomic Failure* (2nd ed., pp. 654–666). Oxford: Oxford University Press.

Bernardi, L., Bianchini, B., Spadacini, G., Leuzzi, S., Valle, F., Marchesi, E., et al. (1995). Demonstrable cardiac reinnervation after human heart transplantation by carotid baroreflex modulation of RR interval. *Circulation*, *92*(10), 2895–2903.

Bert, P. (1878). *La pression barometrique. Recherches de physiologie experimentale* (p. p1168) Paris: Masson.

Cook, D. G., & Shaper, A. G. (1989). Breathlessness, angina pectoris and coronary artery disease. *Am. J. Cardiol.*, *63*(13), 921–924.

Dejours, P., & Dejours, S. (1992). The effects of barometric pressure according to Paul Bert: The question today. *Int. J. Sports Med.*, *13*(Suppl 1), S1–S5.

Dormandy, J. A. (1970). Clinical significance of blood viscosity. *Ann. R. Coll. Surg. Engl.*, *47*(4), 211–228.

Duplain, H., Vollenweider, L., Delabays, A., Nicod, P., Bärtsch, P., & Scherrer, U. (1999). Augmented sympathetic activation during short-term hypoxia and high-altitude exposure in subjects susceptible to high-altitude pulmonary edema. *Circulation*, *99*(13), 1713–1718.

Ehrly, A. M. (1982). Effects of orally administered pentoxifylline on muscular oxygen pressure in patients with intermittent claudication. *IRCS Med. Sci.*, *10*, 401–402.

Fan, J. L., Wu, T. Y., Lovering, A. T., Nan, L., Bang, W. L., & Kayser, B. (2021). Differential brain and muscle tissue oxygenation responses to exercise in tibetans compared to Han Chinese. *Front. Physiol.*, *12*, 617954.

Fronek, A., Coel, M., & Bernstein, E. F. (1977). The pulse-reappearance time: An index of overall blood flow impairment in the ischemic extremity. *Surgery*, *81*(4), 376–381.

Fulco, C. S., Rock, P. B., Reeves, J. T., Trad, L. A., Young, P. M., & Cymerman, A. (1989). Effects of propranolol on acute mountain sickness (AMS) and well-being at 4,300 meters of altitude. *Aviat. Space Environ. Med.*, *60*(7), 679–683.

Fyhrquist, F., Saijonmaa, O., Metsärinne, K., Tikkanen, I., Rosenlöf, K., & Tikkanen, T. (1990). Raised plasma endothelin-I concentration following cold pressor test. *Biochem. Biophys. Res. Commun.*, *169*(1), 217–221.

Goldenberg, F., Richalet, J. P., Onnen, I., & Antezana, A. M. (1992). Sleep apneas and high altitude newcomers. *Int. J. Sports Med.*, *13*(Suppl 1), S34–S36.

Gosney, J. R. (1994). The endocrine lung and its response to hypoxia. *Thorax*, *49*(Suppl), S25–S26, Suppl.

Green, H. J., Sutton, J., Young, P., Cymerman, A., & Houston, C. S. (1989). Operation Everest II: muscle energetics during maximal exhaustive exercise. *J. Appl. Physiol. (1985)*, *66*(1), 142–150.

Greene, R., Appenzeller, P., Wood, S. C., & Appenzeller, O. (1991). Middle cerebral and extracranial vertebral artery flow velocities during Valsalvas maneuver; hypoxic and rheologic influences. *Neurology*, *41*, 1–418.

Groves, B. M., Reeves, J. T., Sutton, J. R., Wagner, P. D., Cymerman, A., Malconian, M. K., ... Houston, C. S. (1987). Operation Everest II: Elevated high-altitude pulmonary resistance unresponsive to oxygen. *J. Appl. Physiol. (1985)*, *63*(2), 521–530.

Haber, E., & Lee, M. E. (1994). Cardiac physiology. *Endothelin to the rescue? Nature*, *370* (6487), 252–253.

Hackett, P. H. (1988). Medical research on Mount Mckinley. *Ann. Sports Med, 4*, 232–244.

Hackett, P. H., Roach, R. C., Hartig, G. S., Greene, E. R., & Levine, B. D. (1992). The effect of vasodilators on pulmonary hemodynamics in high altitude pulmonary edema: a comparison. *Int. J. Sports Med., 13*(Suppl 1), S68–S71.

Hansen, J., & Sander, M. (2003). Sympathetic neural overactivity in healthy humans after prolonged exposure to hypobaric hypoxia. *J. Physiol., 546*(Pt 3), 921–929.

Harik, S. I., Behmand, R. A., & LaManna, J. C. (1994). Hypoxia increases glucose transport at blood-brain barrier in rats. *J. Appl. Physiol. (1985)*, 77(2), 896–901.

Heath, D., & P., S. (1992). *Diseases of the human carotid body*. London: Springer Verlag.

Heath, D., & Williams, D. R. (1989). *High altitude medicine and pathology*. London: Butterworths.

Heistad, D. D., & Abboud, F. M. (1980). Dickinson W. Richards lecture: circulatory adjustments to hypoxia. *Circulation, 61*(3), 463–470.

Hobbins, W. B. (1988). Thermography in sports medicine. In O. Appenzeller (Ed.), *Sports medicine: training, fitness, injuries* (pp. 395–402). Baltimore Munich: Urban and Schwarzenberg.

Hoeffner, U., Farhys, L. L., & Vanhoutte, P. M. (1986). Pharmacology of pentoxifylline in isolated canine arteries and veins. *Feder. Proc., 45*, 670.

Houston, C. S., & Riley, R. L. (1947). Respiratory and circulatory changes during acclimatization to high altitude. *Am. J. Physiol., 149*(3), 565–588.

Houston, C. S., Sutton, J. R., Cymerman, A., & Reeves, J. T. (1987). Operation Everest II: Man at extreme altitude. *J. Appl. Physiol. (1985), 63*(2), 877–882.

Jensen, K. S., Micco, A. J., Czartolomna, J., Latham, L., & Voelkel, N. F. (1992). Rapid onset of hypoxic vasoconstriction in isolated lungs. *J. Appl. Physiol. (1985), 72*(5), 2018–2023.

Kaijser, L., Pernow, J., Berglund, B., Grubbström, J., & Lundberg, J. M. (1994). Neuropeptide Y release from human heart is enhanced during prolonged exercise in hypoxia. *J. Appl. Physiol. (1985), 76*(3), 1346–1349.

Korkushko, O. V., Shatilo, V. B., Plachinda Yu, I., & Shatilo, T. V. (1991). Autonomic control of cardiac chronotropic function in man as a function of age: Assessment by power spectral analysis of heart rate variability. *J. Auton. Nerv. Syst., 32*(3), 191–198.

Lahiri, S. (1984). Respiratory control in Andean and Himalayan high altitude natives. In J. B. West, & S. Lahiri (Eds.), *High altitude and man* (147, p. 162). Bethesda: American Physiological Society.

Lahiri, S., & Data, P. G. (1992). Chemosensitivity and regulation of ventilation during sleep at high altitudes. *Int. J. Sports Med., 13*(Suppl 1), S31–S33.

Lataste, X. (1992). The blood-brain barrier in hypoxia. *Int. J. Sports Med., 13*(Suppl 1), S45–S47.

Lauweryns, J. M., & Peuskens, J. C. (1972). Neuro-epithelial bodies (neuroreceptor or secretory organs?) in human infant bronchial and bronchiolar epithelium. *Anat. Rec., 172*(3), 471–481.

León Velarde, F., & Arregui, A. (1994). Desadaptación a la vida en las grandes alturas. *Travaux de lInstitut Francais dÉtudes Andines*, 85.

Lerman, A., Hildebrand, F. L., Jr., Aarhus, L. L., & Burnett, J. C., Jr. (1991). Endothelin has biological actions at pathophysiological concentrations. *Circulation, 83*(5), 1808–1814.

Leuenberger, U., Gleeson, K., Wroblewski, K., Prophet, S., Zelis, R., Zwilliach, C., & Sinoway, L. (1991). Norepinephrine clearance is increased during acute hypoxemia in humans. *Am. J. Physiol., 261*, H1659–H1664.

Lieberman, P., Protopapas, A., Reed, E., Youngs, J. W., & Kanki, B. G. (1994). Cognitive defects at altitude. *Nature, 372*(6504), 325.

Little, M. A., Thomas, R. B., & Larrick, J. W. (1973). Skin temperature and cold pressor responses of Andean Indians during hand immersion in water at 4 degrees C. *Human Biol.: Int. Rec. Res.*, *45*(4), 643–662.

Longobardo, G. S., Gothe, B., Goldman, M. D., & Cherniack, N. S. (1982). Sleep apnea considered as a control system instability. *Respir. Physiol.*, *50*(3), 311–333.

Masuyama, S., Kohchiyama, S., Shinozaki, T., Okita, S., Kunitomo, F., Tojima, H., ... Honda, Y. (1989). Periodic breathing at high altitude and ventilatory responses to O2 and CO2. *Jap. J. Physiol.*, *39*(4), 523–535.

Mathew, L., Purkayastha, S. S., Jayashankar, A., Radhakrishnan, U., Sen Gupta, J., & Nayar, H. S. (1985). Responses of the autonomic nervous system in altitude adapted and high altitude pulmonary oedema subjects. *Int. J. Biometeorol.*, *29*(2), 131–143.

Mazzeo, R. S., Bender, P. R., Brooks, G. A., Butterfield, G. E., Groves, B. M., Sutton, J. R., ... Reeves, J. T. (1991). Arterial catecholamine responses during exercise with acute and chronic high-altitude exposure. *Am. J. Physiol.*, *261*(4 Pt 1), E419–E424.

Milledge, J. S. (1992). Salt and water control at altitude. *Int. J. Sports Med.*, *13*(Suppl 1), S61–S63.

Monge, C. C. (1988). Medical Research in the Andes. *Ann. Sports Med*, *4*, 245–254.

Naeije, R. (1992). Pulmonary circulation in hypoxia. *Int. J. Sports Med.*, *13*(Suppl 1), S27–S30.

Nicholson, A. N., Smith, P. A., Stone, B. M., Bradwell, A. R., & Coote, J. H. (1988). Altitude insomnia: studies during an expedition to the Himalayas. *Sleep*, *11*(4), 354–361.

Oelz, O., Maggiorini, M., Ritter, M., Noti, C., Waber, U., Vock, P., & Bärtsch, P. (1992). Prevention and treatment of high altitude pulmonary edema by a calcium channel blocker. *Int. J. Sports Med.*, *13*(Suppl 1), S65–S68.

Paintal, A. S. (1995). Some recent advances in studies on J receptors. *Adv. Exp. Med. Biol.*, *381*, 15–25.

Palareti, G., Coccheri, S., Poggi, M., Tricarico, M. G., Magelli, M., & Cavazzuti, F. (1984). Changes in the rheologic properties of blood after a high altitude expedition. *Angiology*, *35*(7), 451–458.

Pilardeau, P., Richalet, J. P., Bouissou, P., Vaysse, J., Larmignat, P., & Boom, A. (1990). Saliva flow and composition in humans exposed to acute altitude hypoxia. *Euro. J. Appl. Physiol. Occupat. Physiol.*, *59*(6), 450–453.

Ponchia, A., Noventa, D., Bertaglia, M., Carretta, R., Zaccaria, M., Miraglia, G., ... Buja, G. (1994). Cardiovascular neural regulation during and after prolonged high altitude exposure. *Euro. Heart J.*, *15*(11), 1463–1469.

Raj, H., Singh, V. K., Anand, A., & Paintal, A. S. (1995). Sensory origin of lobeline-induced sensations: a correlative study in man and cat. *J. Physiol.*, *482*(Pt 1), 235–246.

Reeves, J. T., Mazzeo, R. S., Wolfel, E. E., & Young, A. J. (1992). Increased arterial pressure after acclimatization to 4300 m: possible role of norepinephrine. *Int. J. Sports Med.*, *13*(Suppl 1), S18–S21.

Richalet, J. P. (1990). Heart and adrenergic system in hypoxia. In J. R. Sutton, G. Coates, & J. E. Remmers (Eds.), *Hypoxia the adaptations* (pp. 231–240). Toronto: P.C. Dekker.

Richalet, J. P., Larmignat, P., Rathat, C., Kéromès, A., Baud, P., & Lhoste, F. (1988). Decreased cardiac response to isoproterenol infusion in acute and chronic hypoxia. *J. Appl. Physiol. (1985)*, *65*(5), 1957–1961.

Robergs, R. A., Appenzeller, O., Qualls, C., Aisenbrey, J., Krauss, J., Kopriva, L., & DePaepe, J. (1993). Increased endothelin and creatine kinase after electrical stimulation of paraplegic muscle. *J. Appl. Physiol. (1985)*, *75*(6), 2400–2405.

Rowell, L. B., Brengelmann, G. L., Savage, M. V., & Freund, P. R. (1989). Does acute hypoxemia blunt sympathetic activity in hyperthermia? *J. Appl. Physiol. (1985)*, *66*(1), 28–33.

Rowell, L. B., & Seals, D. R. (1990). Sympathetic activity during graded central hypovolemia in hypoxemic humans. *Am. J. Physiol.*, *259*(4 Pt 2), H1197–H1206.

Saito, M., Mano, T., Iwase, S., Koga, K., Abe, H., & Yamazaki, Y. (1988). Responses in muscle sympathetic activity to acute hypoxia in humans. *J. Appl. Physiol. (1985)*, *65*(4), 1548–1552.

Seals, D. R., Johnson, D. G., & Fregosi, R. F. (1991). Hypoxia potentiates exercise-induced sympathetic neural activation in humans. *J. Appl. Physiol. (Beth., Md.: 1985)*, *71*(3), 1032–1040.

Sedova, G. T., Krasiukova, L. I., Kovaleva, L. G., Malykhina, L. S., & Ermil'chenko, G. V. (1984). Use of trental as an agent for preventing thrombus formation in patients with polycythemia vera. *Gematol. i Transfuziol.*, *29*(8), 19–22.

Severinghaus, J. W., Chiodi, H., Eger, E. I., 2nd, Brandstater, B., & Hornbein, T. F. (1966). Cerebral blood flow in man at high altitude. Role of cerebrospinal fluid pH in normalization of flow in chronic hypocapnia. *Circ. Res.*, *19*(2), 274–282.

Sleight, P., La Rovere, M. T., Mortara, A., Pinna, G., Maestri, R., Leuzzi, S., ... Bernardi, L. (1995). Physiology and pathophysiology of heart rate and blood pressure variability in humans: Is power spectral analysis largely an index of baroreflex gain? *Clin. Sci. (Lond., Engl.: 1979)*, *88*(1), 103–109.

Sohal, R. S., Sun, S. C., Colcolough, H. L., & Burch, G. E. (1968). Heat stroke. An electron microscopic study of endothelial cell damage and disseminated intravascular coagulation. *Arch. Int. Med.*, *122*(1), 43–47.

Solomon, S., Corwin, E., & Appenzeller, O. (1987). Atrial natriuretic factor and vasopressin in endurance trained subjects. *Ann. Sports Med.*, *3*, 163–166.

Stevens, P. E., Bloodworth, L. L., & Rainford, D. J. (1986). High altitude haemofiltration. *Br. Med. J. (Clin. Res. Ed.)*, *292*(6532), 1354.

Strassman, R. J., Appenzeller, O., Lewy, A. J., Qualls, C. R., & Peake, G. T. (1989). Increase in plasma melatonin, beta-endorphin, and cortisol after a 28.5-mile mountain race: Relationship to performance and lack of effect of naltrexone. *J. Clin. Endocrinol. Metab.*, *69*(3), 540–545.

Suarez, J., Alexander, J. K., & Houston, C. S. (1987). Enhanced left ventricular systolic performance at high altitude during Operation Everest II. *Am. J. Cardiol.*, *60*(1), 137–142.

Sutton, J. R., Reeves, J. T., Groves, B. M., Wagner, P. D., Alexander, J. K., Hultgren, H. N., ... Houston, C. S. (1992). Oxygen transport and cardiovascular function at extreme altitude: Lessons from Operation Everest II. *Int. J. Sports Med.*, *13*(Suppl 1), S13–S18.

Sutton, J. R., Reeves, J. T., Wagner, P. D., Groves, B. M., Cymerman, A., Malconian, M. K., ... Houston, C. S. (1988). Operation Everest II: Oxygen transport during exercise at extreme simulated altitude. *J. Appl. Physiol. (1985)*, *64*(4), 1309–1321.

Tremblay, J. C., Hoiland, R. L., Howe, C. A., Coombs, G. B., Vizcardo-Galindo, G. A., Figueroa-Mujíca, R. J., ... Ainslie, P. N. (2019). Global REACH 2018: High blood viscosity and hemoglobin concentration contribute to reduced flow-mediated dilation in high-altitude excessive erythrocytosis. *Hypertension*, *73*(6), 1327–1335.

Tymko, M. M., Lawley, J. S., Ainslie, P. N., Hansen, A. B., Hofstaetter, F., Rainer, S., ... Hearon, C. M., Jr. (2020). Global reach 2018 heightened α-adrenergic signaling impairs endothelial function during chronic exposure to hypobaric hypoxia. *Circ. Res.*, *127*(2), e1–e13.

Verratti, V., Bondi, D., Mulliri, G., Ghiani, G., Crisafulli, A., Pietrangelo, A., ... Cerretelli, P. (2020). Muscle oxygen delivery in the forearm and in the vastus lateralis muscles in response to resistance exercise: A comparison between Nepalese Porters and Italian Trekkers. *Front. Physiol., November 10*(11), 607616.

Vogel, J. A., & Harris, C. W. (1967). Cardiopulmonary responses of resting man during early exposure to high altitude. *J. Appl. Physiol. (Beth., Md.: 1985)*, *22*(6), 1124–1128.

Wagner, P. D., Gale, G. E., Moon, R. E., Torre-Bueno, J. R., Stolp, B. W., & Saltzman, H. A. (1986). Pulmonary gas exchange in humans exercising at sea level and simulated altitude. *J. Appl. Physiol. (1985)*, *61*(1), 260–270.

Watanabe, E., Ogawa, K., Ban, M., & Satake, T. (1981). Sympathetic nervous systems in chronic hypoxic states from pulmonary tuberculosis: a clinical study on plasma norepinephrine and cyclic AMP levels. *Jap. J. Med.*, *20*(3), 180–187.

West, J. B. (1988). Physiology at the summit of Mount Everest. *Ann. Sports Med*, *4*, 224–231.

West, J. B., Tsukimoto, K., Mathieu-Costello, O., & Prediletto, R. (1991). Stress failure in pulmonary capillaries. *J. Appl. Physiol. (1985)*, *70*(4), 1731–1742.

White, D. P., Gleeson, K., Pickett, C. K., Rannels, A. M., Cymerman, A., & Weil, J. V. (1987). Altitude acclimatization: influence on periodic breathing and chemoresponsiveness during sleep. *J. Appl. Physiol. (1985)*, *63*(1), 401–412.

Wood, S. C., & Appenzeller, O. (1988). Effects of pentoxifylline on blood rheological properties and exercise performance of endurance-trained athletes at high altitude. *Ann Sports Med*, *4*, 263–269.

Wood, S. C., Appenzeller, O., & Appenzeller, P. (1989). Transcranial measurement of cerebral blood flow in sojourners and natives at high altitudes. *Ann. Sports Med*, *4*, 289–293.

Zamora, M. R., O'Brien, R. F., Rutherford, R. B., & Weil, J. V. (1990). Serum endothelin-1 concentrations and cold provocation in primary Raynaud's phenomenon. *Lancet*, *336*(8724), 1144–1147.

CHAPTER 12

The autonomic nervous system in space exploration

12.1 Introduction

Weightlessness offers a unique opportunity to study the autonomic nervous system without the influence of gravity which many autonomic actions continuously oppose on Earth. While more than 60 years have elapsed since the first manned space mission in 1961 (a 108-minute orbital flight by the Soviet cosmonaut Yuri Gagarin), much remains to be learned regarding the effects of weightlessness on humans. Weightlessness (or microgravity) exists as a result of a gravitational force being counterbalanced by opposing centrifugal forces created by an orbital flight pattern (the spacecraft and crew are in a free-fall). For instance, while orbiting Earth at altitudes of 300 km and speeds of 7−8 km/second, spacecraft crews are subjected to gravitational forces of about 10^{-7} g (Nicogossian & Robbin, 1994; Pavy-Le-Traon et al., 1994). The most prominent autonomic changes present during microgravity reflect acute and chronic body fluid shifts and losses, changes in electrolyte concentrations, cardiovascular deconditioning, muscle and bone deconditioning, and space motion sickness (or space [mal] adaptation syndrome). These result mostly from the combined effects of weightlessness and confinement in relatively small spaces (such as a spacecraft) (Atkov & Benedenko, 1992). Other changes in the central nervous system resulting from spaceflight include vestibular, proprioceptive, sensory perception, and behavior alterations (Newberg, 1994). Some effects of weightlessness may interfere with the quality of sleep (Stampi, 1994).

Much effort has been devoted to studying the physiologic adaptations to weightlessness and to developing measures that counteract the effects of space travel (Bondar, 1994; Nicogossian & Nachtwey, 1989). Acute orthostatic intolerance on returning to 1-g conditions (Earth's gravity) represents an important problem encountered by humans once they are adapted to microgravity. The risk of symptomatic hypotension is of particular concern during space shuttle reentry when crews, who are required to manually pilot or be able to evacuate the spacecraft in case of an

Introduction to Clinical Aspects of the Autonomic Nervous System
DOI: https://doi.org/10.1016/B978-0-323-96011-3.00005-X

emergency, are subjected to gravitational forces along the head to toe axis of up to 1.2 g for 17–20 minutes. Other features present during readaptation to gravity are ataxia and postural disturbances, perceptual illusions, visual disturbances, and fatigue and weakness. Of particular significance to this chapter are the autonomic responses to weightlessness and their similarities to those resulting from deconditioning produced on Earth by bed rest (Kakurin, Lobachik, Mikhailov, & Senkevich, 1976).

Space travel expanded its frontiers in recent years. Future travelers to Mars will be exposed to numerous environmental and physiological challenges for up to 3 years, and there are vigorous research investments to help elucidate and mitigate the consequences of these long-duration exposures.

12.2 Fluid shifts during spaceflight

The acute changes in conditions to which space-bound humans are subjected to in rapid sequence through count-down, launch, and space flight, activate complex autonomic reflex responses producing a relatively rapid cardiovascular adaptation to the new environment. The nature of prelaunch conditions (sitting in a head-down position with hips flexed and legs elevated above the head for several hours), enduring 1.5–3 g forces during take-off and the 8- to 9-minute journey into orbit, and the abrupt insertion into weightlessness or microgravity (10^{-2}–10^{-7} g) produces dramatic fluid shifts that stress autonomic homeostatic mechanisms. There is general agreement that many of the clinical features that accompany insertion into microgravity are largely the result of the acute centripetal redistribution of fluids from the limbs (1.5–2.0 L) and lower body that takes place once the hydrostatic force generated by gravity is removed.

The redistribution of fluids to the upper body is apparent from the symptoms reported by astronauts, the changes in physical appearance, and many echocardiographic changes noted during the early periods of adaptation to weightlessness. Skylab astronaut E. Gibson reported that: *Early in the flight we experienced a sensation of head fullness. This is caused by a shift of the body fluids to the upper part of the body when one first enters into zero-g. One notices that the eyes turn red...the eye sockets themselves become a little puffy, the face a little rounder and a little redder, veins in the neck and forehead become distended and one's sinuses feel congested. These conditions did not change significantly in-flight, they just tapered off... We always felt a lot better for about half an hour to 2 hours after we exercised on the bicycle....also after eating...* (Gibson, 1977). In addition, as the thoracic volume increases, the waist becomes smaller, the calf volume decreases (the puffy head, bird legs syndrome), and the body mass center rises toward the chest (Thornton, Hoffler, & Rummel, 1977). In later phases of the adaptation, once most adjustments to the acute changes in volume distribution have taken place (see below), several factors appear to act together to maintain the visible engorgement of the upper body (Parazynski et al., 1991). These include loss of the head to heart hydrostatic pressure gradient, an increased microcirculatory blood flow in the head tissues (where precapillary blood flow control is less developed than the lower extremities), and a reduction in plasma colloid oncotic pressure. On Earth, head-down tilt experiments to simulate microgravity have shown an early increase in microvascular flow at the head and leg, possibly related to sympathetic withdrawal mediated through baroreflexes activated by centripetal fluid shifts (Stout, Watenpaugh, Breit, & Hargens, 1995).

12.3 Changes in blood volume with microgravity

12.3.1 Centralization of blood volume in microgravity

Early on during microgravity, the acute intra- and extravascular fluid shifts to the central circulation produce increases in cardiac size and output (Buckey et al., 1993; Bungo et al., 1986; Bungo, Goldwater, Popp, & Sandler, 1987). Engorgement of the central circulation is sensed by mechanoreceptors, which activate autonomic off-loading and volume regulating reflexes (Fig. 12.1). As a result, there is vasodilatation and pooling of blood in the viscera and renal fluid and salt loss. While most adaptations to increased central volume load occur within the first 6–10 hours of spaceflight (Moore & Thornton, 1987; Thornton et al., 1977), it is presumed that some may be very rapid and may take place, at least in part, while astronauts are in the pretakeoff head-down position. This, together with the effects of the expansion in chest size that occurs on exposure to microgravity, may explain the finding of direct and estimated central venous pressures that are lower during the early part of spaceflight than during supine 1 g conditions (Buckey et al., 1993; Kirsch et al., 1984). On Earth, assuming a seated head-down position from a standing posture had resulted in an immediate and transient rise in central venous pressure (CVP); CVP values in this position, however, were not different from supine values and continued to decrease thereafter (Charles, Bungo, & Ammerman, 1987) and central venous pressure becomes surprisingly reduced. These findings suggest that the adaptation of vascular beds to circulatory changes is rapid and may

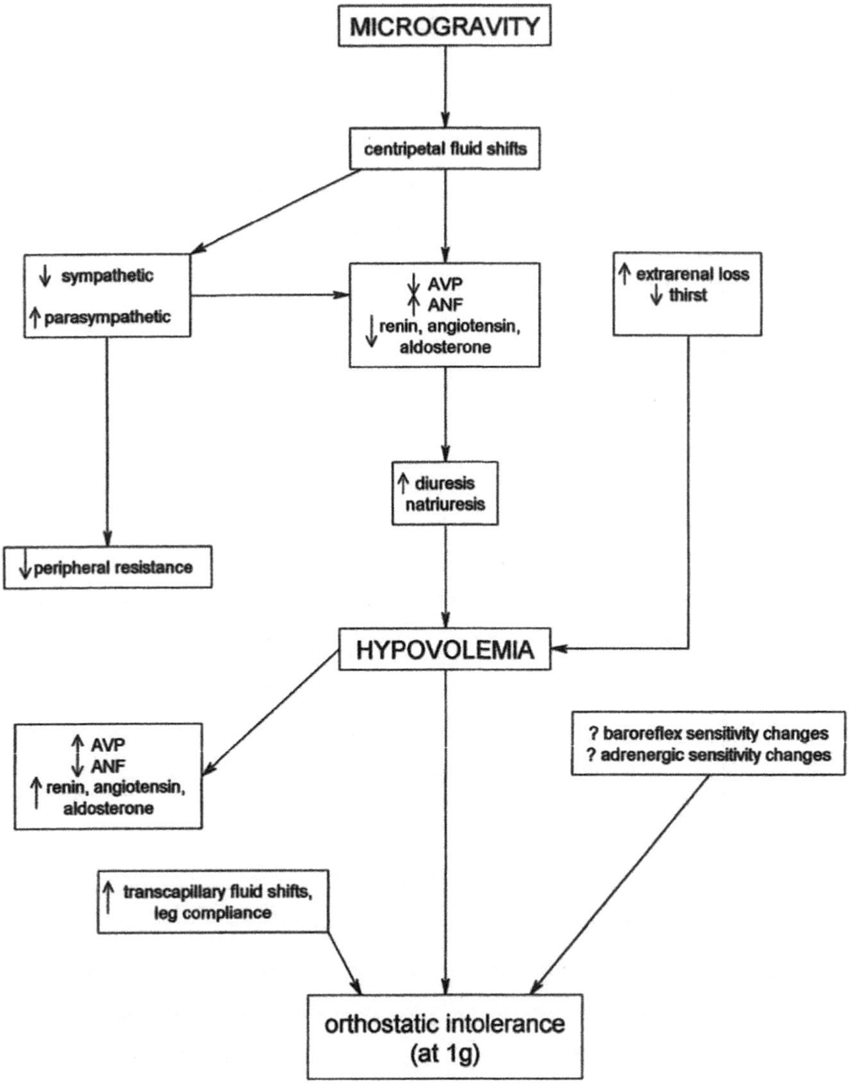

Figure 12.1 Cardiovascular adaptations to microgravity leading to orthostatic intolerance on return to Earth (1 g).

occur in part during the supine head-down position before lift-off (e.g., before microgravity conditions are reached). Some of the autonomic mechanisms responsible for the vascular adaptation to microgravity may involve vasodilatation produced by local endothelial factors such as nitric oxide. Increased plasma concentrations of vasopressin (AVP) found in some subjects early on during spaceflight instead of the predicted acute decrease (Greenleaf, Sciaraffa, Shvartz, Keil, & Brock, 1981), may have been triggered by the presence of nausea due to space motion sickness (Whitson, Cintron, Pietrzyk, Scotto, & Loeppky, 1994).

12.3.2 Decreased circulating volume

The acute autonomic cardiovascular adaptations to microgravity are supplemented by slower acting long-term volume-regulating responses that alter renal handling of water and electrolytes. Decreased circulating volume during spaceflight was first believed to be a result of the diuretic response to the centralization of blood volume in microgravity (see above); however, this has not been supported experimentally. Reductions in plasma volume during spaceflight are the result of decreased fluid intake, fluid shifts from the intravascular to interstitial space, as well as reductions in red cell mass. The increased urinary flow contributes to the reduction in plasma volume (the Gauer–Henry reflex) (Gauer & Henry, 1963). Increased insensible water losses into the dry spacecraft atmosphere, fluid shifts from the intravascular to interstitial space, and vomiting and decreased fluid intake if space motion sickness symptoms are prominent also contribute toward the decrease in circulatory volume. Decreased circulating blood volume and depressed cardiac output decrease carotid and aortic mechanoreceptor stretch, leading to sympathetic efferent stimulation.

There is rapid weight loss reflecting the decrease in circulatory volume (Leach, Inners, & Charles, 1991), and cardiac size and stroke volume indexes fall below preflight values after 24 hours of spaceflight (Bungo et al., 1986). After the first day of exposure to microgravity, the acute and early phases of adaptation are followed by a period of gradual adaptation (or resetting) to the new conditions, which eventually leads to a new steady-state situation with decreased plasma volume. Total body water measured during flight decreased by 3.4% (Leach et al., 1991), and plasma volumes measured shortly after prolonged space flight were reduced by as much as 13% when compared to preflight values (Johnson, Hoffler, & Nicogossian, 1977; Leach & Johnson, 1984). Consistent with this, circulatory stresses during space flight produce heart rate and blood pressure responses that mimic those found with hypovolemia and dehydration on Earth. For example, exaggerated heart rate increases, similar to those obtained with hypovolemia on Earth, are present during pooling of fluids in the lower body produced by application of negative pressure to the body below the umbilicus (lower body negative pressure, LBNP, see Chapter 23 and Fig. 12.2) during spaceflight (Johnson et al., 1977). The early inflight reduction in extracellular fluid volume may partly explain elevated plasma levels of noradrenalin, renin, and aldosterone found in four astronauts during adaptation to microgravity (Norsk et al., 1995).

Decreases in circulatory volumes during chronic exposure to microgravity are accompanied by decreases in red blood cell mass with anemia (spaceflight anemia). A report (Udden, Driscoll, Pickett, Leach-Huntoon, & Alfrey, 1995) attributed an 11% fall in red blood cell mass found during a 9-day spaceflight to ineffective erythropoiesis with low serum erythropoietin levels. The transient increase in hematocrit present during the first few days of microgravity due to plasma volume contraction may inhibit erythropoietin production. This results in decreased replacement of red blood cells while their destruction continues at a normal rate. Parallel reductions in circulatory volume and red cell mass explain the lack of significant pre- and inflight changes in hematocrit.

Cardiac natriuretic peptide deficiency could exacerbate central hypervolemia in astronauts and cosmonauts. Prolonged head-down bed rest at −6 degrees elicits sustained increases in circulating atrial natriuretic peptide concentrations (Maillet et al., 2000), whereas individual astronauts have shown reduced natriuretic peptide signaling in space (Rössler, Noskov, László, Polyakow, & Hinghofer-Szalkay, 2001). Studying natriuretic peptides in astronauts comparing lower-sodium and higher-sodium diets in space and on Earth, Frings-Meuthen and colleagues demonstrated

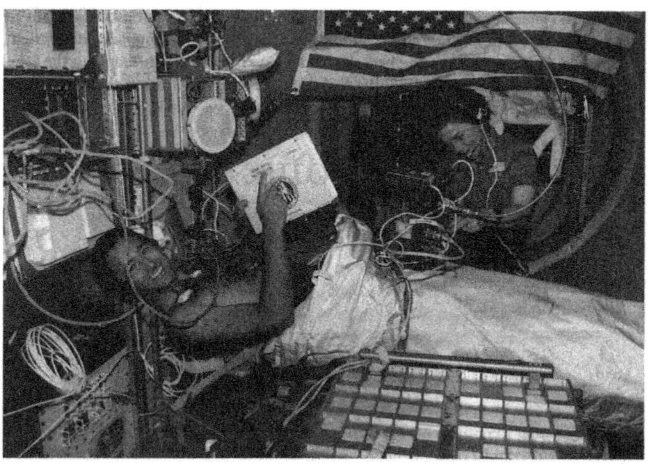

Figure 12.2 Lower body negative pressure used in Space. *Courtesy NASA.*

that cardiac natriuretic peptide concentrations respond to changes in sodium intake in space, however, they are reset to lower levels. The resetting of cardiac natriuretic peptide response at lower levels in space than on Earth may be caused by reductions in thoracic blood volume In contrast, serum aldosterone concentrations were similarly regulated in space and on Earth (Frings-Meuthen et al., 2020). Further studies should investigate whether resetting of natriuretic peptides reflects true deficiency or an appropriate response of changes in blood volume distribution in space.

12.3.3 Metabolic and inflammatory factors and vascular changes

Long-duration spaceflight (> 140 days) triggers a hyperinflammatory and aging immune phenotype that appears to be related to inflight chronic stress (Buchheim et al., 2019). Biomarkers of oxidative stress and inflammation have been found to increase inflight, but mostly restore within 1-week postflight (Lee et al., 2020). A 6-month flight increased insulin resistance index and glycated albumin level but did not alter the lipid profile, and inconsistently affected markers of inflammatory and oxidative stress (Hughson et al., 2016). These metabolic and inflammatory changes associated with an increase in vascular hydrostatic pressure and a reduction in shear stress may cause vascular remodeling and endothelial dysfunction contributing to autonomic dysregulation (Navasiolava et al., 2020).

12.4 Autonomic function testing

Reports on results of autonomic testing performed in space are scarce. The effects of microgravity on arterial blood pressure and heart rate remain unclear. For short flight duration, analysis of both heart rate and systolic blood pressure has shown them to be increased or unchanged (Mandsager, Robertson, & Diedrich, 2015). Data for long-duration flights (up to 6 months in duration) demonstrates largely unchanged heart rate and unchanged or even reduced blood pressure compared to supine preflight values (Verheyden, Liu, Beckers, & Aubert, 2009). Importantly, hydrostatic pressure, which influences vascular morphology and function, should be taken into account for vascular remodeling related to microgravity (Zhang, 2001).

Noninvasive measurement of sympathetic activity during spaceflight has been performed using the low-frequency power (LF), as well as the ratio of low-frequency and high-frequency power (LF/HF) of heart rate variability.

Although there is a trend toward an increase of the LF/HF ratio during short- or long-duration spaceflight, the results are not statistically significant (Baevsky et al., 2007; Hughson et al., 2012; Mandsager et al., 2015).

Frequency domain analysis (power spectrum analysis) of heart rate variability (see Chapter 23) did not show major changes during prolonged spaceflight. In a study on eight cosmonauts during long-term missions (flight lengths of 162–196 days), mean heart rate and heart rate variability in the low- and high-frequency bands did not change during spaceflight, although individual responses were different and correlated with preflight values (Baevsky et al., 2007). In the same study, heart rate responses to metronomic breathing at six cycles per minute decreased in six and increased in one cosmonaut during flight and the most pronounced changes in hemodynamic parameters occurred after landing suggesting functional adaptation (Baevsky et al., 2007).

In-flight sympathetic activity has also been indirectly measured using postflight platelet norepinephrine and epinephrine plasma levels. In a study of five cosmonauts, postflight platelet norepinephrine and epinephrine plasma levels were significantly increased compared to preflight values, consistent with activation of the sympathetic nervous system during the short-duration spaceflight (Christensen, Heer, Ivanova, & Norsk, 2005).

On the other hand, reduction of systemic vascular resistance was reported during short-duration flight, leading to the interpretation that exposure to microgravity results in vasorelaxation (Norsk et al., 2006); however, when the authors compared in-flight measurements to supine preflight controls, SVR was increased during flight, consistent with in-flight sympathetic activation.

Several studies have further evaluated vagal baroreflex function in space. Unfortunately, these results have been conflicting, demonstrating impaired (Eckberg et al., 2010), unchanged (Beckers, Verheyden, Liu, & Aubert, 2009), or even increased (Di Rienzo et al., 2008) baroreflex function during short-duration spaceflight. In long-duration studies, baroreflex sensitivity was unchanged compared to preflight values (Hughson et al., 2012; Verheyden, Liu, Beckers, & Aubert, 2010).

12.5 Deconditioning and miscellaneous autonomic manifestations

Physical deconditioning occurs during spaceflight resulting in significant muscle atrophy (Fitts et al., 2010) and decreased exercise capacity

(Hargens & Richardson, 2009). Deconditioning likely contributes to the impairment in vagal baroreflex function during spaceflight.

Various autonomic manifestations such as decreased bowel transit time and bowel sounds after a few days in space, Horner syndrome, hyperhidrosis, and changes in skin conductance, reported in some cosmonauts during prolonged spaceflight, await explanation (Atkov & Benedenko, 1992). The physiologic adaptations to microgravity can be divided into early (hours to days) and late (days to weeks) phases.

12.6 Altered central processing and brain changes in space

Central processing of baroreceptor afferents may be altered with exposure to microgravity and relative epinephrine, norepinephrine, and vasopressin release in response to orthostatic stress is abnormal following short-duration spaceflight (Meck et al., 2004).

Magnetic resonance imaging (MRI) has been used to assess the structure and function of the brains of crewmembers before and after spaceflight. Structural changes within the brain after spaceflight have been reported including narrowing of the central sulcus, an upward shift of the brain, twisting of the cerebral aqueduct, and increased ventricular volume in long-duration missions (+10.7% increase in total ventricular volumes) (Roberts et al., 2017). On 7-month follow up-scans, the ventricular volumes had decreased from the initial postflight measurements but remained significantly elevated from preflight measurements and these changes may persist for up to 1 year after return on Earth (Kramer et al., 2020).

Spaceflight also affects the gray and white matters of the brain. Kramer and colleagues reported a significant increase in the global white matter volume of long-duration astronauts immediately after spaceflight (5.5%) that persisted out to 12 months (4.6%), with no significant changes to the global gray matter on postflight scans (Kramer et al., 2020). Several studies have reported focal gray-matter changes in the brain following spaceflight. Decreases in gray-matter volume have been reported in the orbitofrontal and temporal poles following spaceflight (Van Ombergen et al., 2018), whereas gray matter volumes may be increased in other parts of the brain such as the sensorimotor and motor areas, specifically the supplementary motor area, and the precentral and postcentral gyrus with resolution at 6-months postflight (Hupfeld et al., 2020; Koppelmans, Bloomberg, Mulavara, & Seidler, 2016). The majority of these gray-matter changes are likely related to due to volume shift.

Functional imaging studies revealed decreased connectivity within the right insula, which is involved in vestibular processing and cognitive control, and altered connectivity between the motor cortex and cerebellum and between the cerebellum and regions involved with proprioception, visual, motor, and somatosensory functions (Pechenkova et al., 2019).

12.7 Simulated microgravity

The effects of exposure to microgravity on autonomic cardiovascular functions have been studied extensively during the simulation of microgravity conditions on Earth. Indeed, many manifestations noted during spaceflight and on return to 1 g had been predicted many years before space flight had ever occurred (Taylor, Erickson, Henschel, & Keys, 1945). The techniques used to simulate the effects of microgravity on Earth rely on gravitational unloading by redirecting or removing the weight load on the body to reduce fluid pooling. For example, by placing the body in the horizontal position, the load shifts from the vertical to the horizontal axis, and by immersing the body in water, the weight load is removed in all body positions. Bed rest and water immersion are frequently used to simulate many of the physiological responses produced by microgravity. Actual weightlessness can be transiently simulated by free-fall parabolic flight trajectories in specially adapted aircraft (KC-135). The experience gained from numerous Earth-based paradigms simulating microgravity has been applied toward the development of effective countermeasures aimed at reducing the effects of adaptation to microgravity.

Head-down bed rest and spaceflight also induce similar neuroimaging changes, including altered connectivity of motor, somatosensory and vestibular areas of the brain (Cassady et al., 2016).

In a study simulating prolonged microgravity exposure, parasympathetic cardiac activity was reported to decrease following 15-day bed rest (Crandall, Engelke, Pawelczyk, Raven, & Convertino, 1994). After exposure to short-term simulated microgravity (20 hours of 6-degree head-down tilt plus furosemide), another study reported reduced heart rate power in the respiratory frequency region and a concomitant increase in the low-frequency power in arterial pressure, heart rate, and radial flow (Patwardhan et al., 1995). While these findings suggest that exposure to microgravity results in decreased parasympathetic and increased sympathetic oscillations, interpretation of spectral analysis results is complicated by the confounding effects of detraining and the lack of control for ventilatory parameters in some studies. Additional evidence, based on the finding of decreased noradrenalin during space flight and bed rest experiments, indicates that sympathetic tone may be reduced (Leach, Altchuler, & Cintron-Trevino, 1983).

On Earth after bed rest, abnormalities appear more to reflect reduced plasma volumes rather than changes in reflex sensitivity. Following 20 days of horizontal bed rest, blood pressures decreased comparatively more during Valsalva and heart rates increased more in response to LBNP challenges and there was no change in the cold pressor response. In this study, although carotid vagal baroreflex sensitivities in a supine position were diminished, the sitting values were unchanged following bed rest (Haruna, Suzuki, Kawakubo, & Gunji, 1994). Splanchnic blood flow is affected by bed rest. Following 2 months of head-down bed rest, baseline portal flow diminishes by 19% ± 13%, related to a decrease in blood volume, and splanchnic vasoconstrictive response to orthostatic stress is also decreased (Arbeille, Kerbeci, Mattar, Shoemaker, & Hughson, 2008).

12.8 Postflight autonomic changes

Orthostatic intolerance upon return to Earth's gravity may persist for several days. For example, cosmonauts experienced significant orthostatic dizziness and weakness, felt extremely heavy and as if being pressed down from above, and preferred to remain recumbent after the 16-day Soyuz 9 mission (1970) (Atkov & Benedenko, 1992). Symptomatic orthostatic intolerance has been particularly severe in the Russian cosmonauts returning to Earth after prolonged spaceflight. In the US program, postflight orthostatic intolerance was present in 15%–20% of crew members despite countermeasures that included fluid and salt loading just before landing (Bungo, Charles, & Johnson, 1985). Postflight, there is resting and upright tachycardia (Mulvagh, Charles, Riddle, Rehbein, & Bungo, 1991) and a propensity to syncope during standing, head-up tilt, and LBNP (Charles & Lathers, 1991). Hypotension during postflight syncope or near-syncope is frequently accompanied by bradycardia (e.g., neurally mediated hypotension and bradycardia or a vasovagal response). Orthostatic intolerance after return to Earth is characterized primarily by a large fall in stroke volume and an augmented tachycardia. Inability to augment the total peripheral resistances above preflight upright levels and smaller increases of plasma norepinephrine concentrations during or immediately following standing have been documented (Fritsch-Yelle, Whitson, Bondar, & Brown, 1996). In addition to the well-documented hypovolemia, venous return may also be reduced as a consequence of the decreased venous pumping produced by muscles deconditioned during spaceflight. Decreased muscle activity and deconditioned muscles result in lower venous transmural pressures, which in turn promote transcapillary filtration of fluids into the extravascular spaces, further enhancing orthostatic intolerance (Hildebrandt et al., 1994). While most postflight autonomic cardiovascular manifestations largely reflect a reduced plasma volume (Bungo & Johnson, 1983), reduced baroreceptor sensitivity may be contributory (Fritsch, Charles, Bennett, Jones, & Eckberg, 1992) and exaggerated adrenergic responses (Robertson, Convertino, & Vernikos, 1994). The phenomena of acute postflight diuresis may also result from a not yet acclimated autonomic nervous system responding to the acute fluid and salt loading used as countermeasures to combat orthostatic intolerance. Measuring efferent muscle sympathetic nerve activity in both supine and upright positions, before and after astronauts' adaptation to microgravity, Levine and colleagues demonstrated normal muscle sympathetic nerve responses to upright tilt after 16 days of spaceflight (Levine et al., 2002). In this study, the linear correlation between left ventricular stroke volume and muscle sympathetic nerve activity suggested that sympathetic responses were appropriate for the hemodynamic challenge of upright tilt and were unaffected by spaceflight (Levine et al., 2002).

12.9 Space motion sickness and disruption of vestibular-autonomic reflexes

In addition to the changes in appearance, most astronauts complain of a syndrome (space motion sickness) of malaise, anorexia, nausea, vomiting, motion sensitivity, headaches, and in some instances, lethargy, that begins within the first few minutes of being exposed to microgravity and resolves gradually over 48–72 hours (Homick & J.M., 1989; Oman, Lichtenberg, & Money, 1990; Thornton, Moore, Pool, & Vanderploeg, 1987). Spaceflight alters the gravity-dependent sensory function of the vestibular organs of the inner ear, specifically the gravity-sensitive otolith organs. Abnormal otolith-ocular reflexes and spatial perception have been demonstrated during adaptation to microgravity as well as readaptation after return to Earth (Clément, Reschke, & Wood, 2005). Impairment in vestibulo-autonomic reflexes may contribute to autonomic manifestations during and after spaceflights, including orthostatic hypotension and orthostatic intolerance after return to Earth. Alterations in visual perception, which occur during space flight, may further disrupt vestibulo-autonomic function. Similarly, circadian rhythm disturbances during space missions may impair the vestibulo-hypothalamic pathway.

There are also striking similarities with acute mountain sickness (see Chapter 21). While the precise role of the acute fluid shift as a cause of space motion sickness remains to be defined, a hypothetical increase in CSF and intracranial pressure has been implicated but not yet confirmed as accurate noninvasive methods of measuring intracranial pressure in space are lacking (Keil, McKeever, Skidmore, Hines, & Severs, 1992; Simanonok & Charles, 1994). On Earth, head-down tilt simulation of the headward fluid shifts produced by microgravity increases tympanic membrane displacement, a measure that correlates with intracranial pressure (Reid, Cottingham, & Marchbanks, 1993). Contrary to expectations, in six young men after 6 days of −6-degree head-down tilt, there was no change in the middle cerebral or intracranial vertebral artery Valsalva ratios (calculated as flow velocities after a Valsalva maneuver divided by flow velocities during the maneuver and corrected for PCO_2) (O. Appenzeller: personal communication, 1996). Since these ratios reflect mainly intracranial pressure, support for an increase in intracranial pressure in microgravity is presently lacking. Overactivity of the parasympathetic nervous system may be responsible for causing nausea and vomiting, the main features of space motion sickness. The proposed parasympathetic overactivity may in turn be triggered by vestibular overstimulation elicited by exposure to microgravity.

12.10 Bed rest deconditioning and orthostatic intolerance

Prolonged bed rest (beyond the physiological nocturnal supine rest) results in important physiological adaptations (Table 12.1). While the main benefit derived from prolonged bed rest appears to be a reduction in energy consumption (which may be beneficial in the setting of certain acute illnesses), the disadvantages are multiple. Prolonged bed rest results in orthostatic intolerance (which may be disabling), impaired insulin-glucose tolerance, bone demineralization, muscle atrophy, weakness, joint stiffness, and motion restriction, increased urinary calculi, constipation, bedsores, deep vein thrombosis, pressure palsies, and other complications. The adaptations to bed rest have been grouped under the term bed rest deconditioning (Greenleaf, 1984).

The cardiovascular adaptations (cardiovascular deconditioning), part of the widespread adaptation to bed rest, are frequently manifested as orthostatic intolerance with tachycardia, hypotension, and a propensity to presyncope and syncope while in the upright posture. Lightheadedness, apprehension, and weakness can be severe and may be associated with fatigue. While bed rest deconditioning represents a disabling condition that appears to afflict many subjects who are confined to bed rest, it remains relatively unrecognized as a cause of reduced orthostatic tolerance.

The adaptative responses to bed rest and fluid shifts and the resulting orthostatic intolerance are similar to those encountered with microgravity. Assuming the horizontal posture produces a redistribution of fluids to the upper body activating autonomic off-loading and volume regulating reflexes. Increased pooling of blood in organs occurs, and renal fluid and salt losses increase. During the first days of bed rest, the plasma volume losses are paralleled by interstitial fluid losses. However, after several days of bed rest, the extracellular fluid volume returns to baseline while the plasma volume remains decreased. Red cell mass decreases in a linear manner reaching a 300–600 mL loss by 30 days (Miller, Johnson, & Lamb, 1965).

Table 12.1 Some physiological adaptations resulting from horizontal and head-down bed rest and exposure to microgravity.

Physiological adaptation	Bed rest (horizontal)	Bed rest (head-down tilt)	Microgravity
Fluid shifts to the upper body	+	+	+
Swelling of the upper body soft tissues	–[a]	+	+
Increased urinary flow	+	+	+
Hypovolemia/dehydration	+	+	+
Changes in baroreflex sensitivity	+	+	+
Orthostatic intolerance	+	+	+
Motion sickness	–	–[b]	+
Skeletal muscle deconditioning	+	+	+
Loss of bone density	+	+	+
Decreased RBC mass	+	+	+

+ , present; –, absent.
[a]Conjunctival injection and retinal vessel engorgement may be present after several days.
[b]Some symptoms may be present with 4- to 12-degree head-down tilt.

The enhanced urinary flow produced during the first days of bed rest is accompanied by increased electrolyte (an osmotic) losses, which gradually return to normal rates with prolonged bed rest. Significant losses of calcium (urinary and fecal), phosphorus, and magnesium are present after the first few days of bed rest due to bone disuse osteopenia. The overall fluid loss produced by bed rest is accompanied by a relative increase in body fat as lean body mass is reduced by dehydration. With bed rest, the heart rate increases (10–30 beats/minute during the first 10 days), and left ventricular size and blood pressure decrease. With prolonged (20 days) bed rest, the resting heart rate had increased by 10 beats/minute, systolic and diastolic pressure decreased (by 20 and 15 mmHg), left ventricular size decreased, and cardiac output decreased (by 0.8 L/minute). These changes resulted in reduced tolerance to LBNP (used here as a measure of orthostatic tolerance) (Takenaka et al., 1994). In another study, orthostatic intolerance with syncope was present after only 24 hours of supine bed rest in 8 of 11 healthy volunteers (Hildebrandt et al., 1994) suggesting that bed rest has a very powerful effect on the cardiovascular system. The authors suggested that significant transcapillary filtration of fluids into the legs may in part explain orthostatic intolerance following bed rest and exposure to microgravity.

Prolonged bed rest with the head slightly tilted down (i.e., a 6-degree head-down tilt) enhances the effects of bed rest and more closely simulates those of microgravity (Kakurin et al., 1976). In this posture, fluids move rapidly from the lower body to the upper body and the head and produce symptoms similar to those described by astronauts including the puffy face, bird legs syndrome (Hargens et al., 1983). The lower limbs lose about 1 L and the plasma volume increases by 6%–7% within the first 2 hours. Adaptations to the acute fluid shifts promote reductions in plasma volume that are similar to those present during microgravity (Blomqvist, Gaffney, & Nixon, 1983). With head-down bed rest, the plasma volume decreases progressively to reach a 10% drop after 24 hours (Nixon et al., 1979; Volicer, Jean-Charles, & Chobanian, 1976). Without exercise, the plasma volume continues to fall linearly (Greenleaf et al., 1977). With exercise (isotonic or isometric), hypovolemia reaches an equilibrium level of −10% (35–450 mL) by the fourth day.

Multiple factors are involved in orthostatic intolerance following bed-rest deconditioning. Vascular volume depletion plays an important role as procedures that restore plasma volume (such as fluid and salt replacement, periods of sitting, standing, and LBNP during periods of bed rest) tend to improve symptoms. Other factors are also involved as in some instances full restoration of vascular volume to normal prebed rest levels fails to prevent orthostatic intolerance. Because the changes in plasma volume and capillary filtration fail to fully account for orthostatic intolerance, impaired vasoconstriction and altered adrenoceptor function also appear to be important factors. Along these lines, bed rest studies have shown reduced catecholamines (Leach, Hulley, Rambaut, & Dietlein, 1973; Robertson et al., 1994) consistent with a decrease in sympathetic tone. In turn, adaptation to a reduced release of noradrenalin may lead to an increase in adrenoceptor sensitivity. Results of pharmacologic experiments (reviewed by Greenleaf, 1984) suggest that orthostatic intolerance following bed rest results in part, from impaired vasomotor responses due to a predominance of ß-adrenoceptor vasodilatation over α-adrenoceptor vasoconstriction. Thus, a combination of adrenoceptor supersensitivity (with more pronounced ß-adrenoceptor supersensitivity) and hypovolemia elicited by bed rest (or microgravity) may explain the tachycardia, orthostatic hypotension, and neurally mediated syncope that can occur in response to the upright posture (Robertson et al., 1994). There is a reduction in heart volume (cardiac atrophy) after periods of bed rest deconditioning (Levine, Zuckerman, & Pawelczyk, 1997). Baroreflex-mediated adjustments in heart rate and muscle sympathetic nerve activity are also impaired after prolonged bed rest (Barbic et al., 2019).

12.11 Effects of sex and gender on adaptation to space

Female astronauts are more susceptible to orthostatic intolerance after spaceflight than male astronauts (Waters, Ziegler, & Meck, 2002). Underlying mechanisms may include lower vascular resistance responses and a smaller stroke volume secondary to a smaller and less compliant left ventricle (Fu et al., 2004; Fu, Witkowski, Okazaki, & Levine, 2005). A greater reduction in plasma volume was also reported in female astronauts (Waters et al., 2002).

Cardiovascular alterations after simulated microgravity exposure with 6-degree head-down bed rest appear to be similar between sexes. Cardiac atrophy occurs in women similar to men following sedentary 60 days of head-down bed rest (Dorfman et al., 2007). A similar reduction in blood volume of about 9% in men and women has been reported after 7 days of head-down bed rest; however, other studies reported conflicting results with less reduction of plasma volume in women than men after 13 days of head-down bed rest in one study (Fortney, Turner, Steinmann, Driscoll, &

Alfrey, 1994) and a higher decrease in plasma volume after 3 days of head-down bed rest in women in another study (Vernikos, Dallman, Keil, O'Hara, & Convertino, 1993). These inconsistent findings may be attributable to differences in the study protocol or the time of study with the menstrual cycle in women (Grenon et al., 2006). Further studies should investigate the effect of sex and gender on adaptation to space.

12.12 Countermeasures to combat orthostatic intolerance

The best countermeasure to combat the effects of microgravity should be able to help ease the subjects into and out of microgravity. One approach is to create an environment where centrifugal forces generate artificial gravity. The benefits of this approach are limited by the need to design and construct new spacecraft with a diameter large enough to significantly reduce Coriolis stimulation of the vestibular apparatus (which can cause symptomatic motion sickness and affect crew performance) (Nicogossian & Robbin, 1994). Ideally, countermeasures should provide normal regional fluid distribution and high loads on the musculoskeletal system (Hargens, 1994).

Countermeasures currently in use to prevent orthostatic intolerance in astronauts returning to Earth rely on increasing intravascular volume and muscle tone (to increase venous return). These include a combination of periods of in-flight LBNP (Fig. 12.2) (either using conventional chambers or Chibis type suits, which allow sitting or standing during negative pressure), exercising, fluid and salt loading, and the use of antigravity suits that prevent the redistribution of fluids to the lower limbs (anti-G-suits). In-flight LBNP, in addition to serving as a measure of orthostatic tolerance, also improves postflight orthostatic tolerance (Fortney, 1991) as the fluid shifts it produces decreased glomerular filtration and promotes fluid and salt retention (Gilbert, Bricker, Springfield, Stevens, & Warren, 1966). Exercise, using an exercycle or a rowing machine, is performed routinely in an attempt to maintain cardiovascular and musculoskeletal conditioning through its effects of volume expansion and reinforcement of the blood pumping action of muscles. Along these lines, penguin suits (incorporating elastic elements that continuously oppose movements) have been widely used by Russian cosmonauts to maintain muscle tone. Before reentry to 1 g, crews rehydrate by drinking saline solutions and wear anti-G-suits that when inflated compress the legs and lower abdomen to increase venous return. Operant conditioning techniques, shown to increase supine and upright blood pressure on Earth, may be used in the future as adjunct countermeasures to combat hypotension during critical periods of reentry and once astronauts have reached 1 g conditions (Cowings et al., 1994). Resistive exercise and vibration could provide additional benefits (Scott, Weber, & Green, 2019). Irradiation factors become other major contributors to endothelial dysfunction and further research exploring the benefit of antioxidants, nutraceuticals, and radiation shielding is necessary (Hughson, Helm, & Durante, 2018).

Using ambulatory beat-to-beat blood pressure, Fu and colleagues demonstrated that prolonged spaceflight did not cause orthostatic hypotension or intolerance during routine postflight activities and had minimal impact on systolic blood pressure variability and distribution. The authors concluded that countermeasures on the International Space Station, including exercise training in flight and volume resuscitation on return, are sufficient to prevent clinically meaningful orthostatic intolerance after 6 months in space (Fu et al., 2019).

While these countermeasures may be considered too cumbersome and costly to apply on Earth to combat the effects of bed rest deconditioning, it is clear that the best measure is to prevent prolonged bed rest in the first place. Simple measures such as sitting, graded exercise, periodic upright tilting, and sleeping with the head of the bed elevated are very effective in reducing the incidence of orthostatic intolerance.

12.13 Brain and blood changes in humans during prolonged isolation

Exposure to environmental monotony and social isolation has harmful effects on the brain by reducing the generation of new neurons in the dentate gyrus of the hippocampus in animals. Similar stressors produce analogous changes in humans demonstrated in participants of a polar expedition (Stahn et al., 2019). Participants of future trips into space, where there is unavoidable prolonged social isolation and environmental monotony, will likely also demonstrate the same shrinkage of brain structures. Thus, future space travelers, when eventually arriving at their destination after years of isolation, may come without enough "brains" to carry out their mission. Thus, suggesting that humans cannot live permanently on other planets—we are just made to live on Earth. There were also changes in blood biomarkers of brain injury and degeneration after long-duration spaceflight. Blood samples were taken from cosmonauts before and after missions (mean duration, 169 days) on the International Space Station from 2016 to

2020. There was an increase in NfL (neurofilaments) and GFAP (glial fibrillary acidic protein). This suggests an axonal disintegration process along with astrocytic activation during the postflight course of a long-duration mission. The increases of both Aβ proteins over the entire postflight phase depicted an accumulative association of the cephalad fluid shift with the interstitial tissue. The findings could reflect coherent reparatory processes of intracranial pressure associations from cephalad fluid shift with the brain with subsequent restoration of the blood−brain barrier integrity (Zu Eulenburg et al., 2021). Astronauts also develop spaceflight-associated neuro-ocular syndrome and structural abnormalities such as choroid thickening and globe flattening may persist for years after returning to Earth. Loss of hydrostatic gradients and a headward shift in vascular and cerebrospinal fluid are hypothesized to be a primary physiological stress responsible for ocular remodeling. Hearon and colleagues demonstarted that nightly LBNP reinstated a footward fluid shift and mitigated the increase in choroid area and volume (Hearon et al., 2022).

References

Arbeille, P., Kerbeci, P., Mattar, L., Shoemaker, J. K., & Hughson, R. (2008). Insufficient flow reduction during LBNP in both splanchnic and lower limb areas is associated with orthostatic intolerance after bedrest. *Am. J. Physiol. Heart Circ. Physiol.*, *295*(5), H1846−1854.

Atkov, O. Y., & Benedenko, V. S. (1992). *Adaptation of the human organisms systems in hypokinesia and weightlessness. Hypokinesia and weightlessness: Clinical and physiological aspects* (pp. 1−94). Madison, CT: International Universities Press.

Baevsky, R. M., Baranov, V. M., Funtova, I. I., Diedrich, A., Pashenko, A. V., Chernikova, A. G., ... Tank, J. (2007). Autonomic cardiovascular and respiratory control during prolonged spaceflights aboard the International Space Station. *J. Appl. Physiol. (Bethesda, Md.: 1985)*, *103*(1), 156−161.

Barbic, F., Heusser, K., Minonzio, M., Shiffer, D., Cairo, B., Tank, J., ... Furlan, R. (2019). Effects of prolonged head-down bed rest on cardiac and vascular baroreceptor modulation and orthostatic tolerance in healthy individuals. *Front. Physiol.*, *10*, 1061.

Beckers, F., Verheyden, B., Liu, J., & Aubert, A. (2009). Cardiovascular autonomic control after short-duration spaceflights. *Acta Astronaut.*, *65*, 804−812.

Blomqvist, C. G., Gaffney, F. A., & Nixon, J. V. (1983). Cardiovascular responses to head down tilt in young and middle aged men. *Physiologist*, *26*, S81−S82.

Bondar, R. L. (1994). 28th Annual Harry G. Armstrong Lecture. Space qualified humans: the high five. *Aviat., Space, Environ. Med.*, *65*(2), 161−169.

Buchheim, J. I., Matzel, S., Rykova, M., Vassilieva, G., Ponomarev, S., Nichiporuk, I., ... Choukèr, A. (2019). Stress related shift toward inflammaging in cosmonauts after long-duration space flight. *Front. Physiol.*, *10*, 85.

Buckey, J. C., Gaffney, F. A., Lane, L. D., Levine, B. D., Watenpaugh, D. E., & Blomqvist, C. G. (1993). Central venous pressure in space. *N. Engl. J. Med.*, *328*(25), 1853−1854.

Bungo, M. W., Charles, J. B., & Johnson, P. C. (1985). Cardiovascular deconditioning during space flight and the use of saline as a countermeasure to orthostatic intolerance. *Aviat., Space Environ. Med.*, *56*, 985–990.

Bungo, M. W., Charles, J. B., Riddle, J., Roesch, J., Wolf, D. A., & Seddon, R. (1986). Echocardiographic investigation of the hemodynamics of weightlessness. *J. Am. Coll. Cardiol.*, 7, 192A.

Bungo, M. W., Goldwater, D. J., Popp, R. L., & Sandler, H. (1987). Echocardiographic evaluation of space shuttle crewmembers. *J. Appl. Physiol. (Bethesda, Md.: 1985)*, *62*(1), 278–283.

Bungo, M. W., & Johnson, P. C., Jr. (1983). Cardiovascular examinations and observations of deconditioning during the space shuttle orbital flight test program. *Aviat. Space Environ. Med.*, *54*(11), 1001–1004.

Cassady, K., Koppelmans, V., Reuter-Lorenz, P., De Dios, Y., Gadd, N., Wood, S., ... Seidler, R. (2016). Effects of a spaceflight analog environment on brain connectivity and behavior. *Neuroimage*, *141*, 18–30.

Charles, J. B., Bungo, M. W., Ammerman, B., et al. (1987). Hemodynamic alterations during the space shuttle pre launch posture. *Aviat. Space Environ. Med.*, *58*, 491.

Charles, J. B., & Lathers, C. M. (1991). Cardiovascular adaptation to spaceflight. *J. Clin. Pharmacol.*, *31*(10), 1010–1023.

Christensen, N. J., Heer, M., Ivanova, K., & Norsk, P. (2005). Sympathetic nervous activity decreases during head-down bed rest but not during microgravity. *J. Appl. Physiol. (Bethesda, Md.: 1985)*, *99*(4), 1552–1557.

Clément, G., Reschke, M., & Wood, S. (2005). Neurovestibular and sensorimotor studies in space and Earth benefits. *Curr. Pharmaceut. Biotechnol.*, *6*(4), 267–283.

Cowings, P. S., Toscano, W. B., Miller, N. E., Pickering, T. G., Shapiro, D., Stevenson, J., ... Knapp, J. (1994). Autogenic-feedback training: A potential treatment for orthostatic intolerance in aerospace crews. *J. Clin. Pharmacol.*, *34*(6), 599–608.

Crandall, C. G., Engelke, K. A., Pawelczyk, J. A., Raven, P. B., & Convertino, V. A. (1994). Power spectral and time based analysis of heart rate variability following 15 days head-down bed rest. *Aviat. Space Environ. Med.*, *65*(12), 1105–1109.

Di Rienzo, M., Castiglioni, P., Iellamo, F., Volterrani, M., Pagani, M., Mancia, G., ... Parati, G. (2008). Dynamic adaptation of cardiac baroreflex sensitivity to prolonged exposure to microgravity: Data from a 16-day spaceflight. *J. Appl. Physiol. (Bethesda, Md.: 1985)*, *105*(5), 1569–1575.

Dorfman, T. A., Levine, B. D., Tillery, T., Peshock, R. M., Hastings, J. L., Schneider, S. M., ... Hargens, A. R. (2007). Cardiac atrophy in women following bed rest. *J. Appl. Physiol. (Bethesda, Md.: 1985)*, *103*(1), 8–16.

Eckberg, D. L., Halliwill, J. R., Beightol, L. A., Brown, T. E., Taylor, J. A., & Goble, R. (2010). Human vagal baroreflex mechanisms in space. *J. Physiol.*, *588*(Pt 7), 1129–1138.

Fitts, R. H., Trappe, S. W., Costill, D. L., Gallagher, P. M., Creer, A. C., Colloton, P. A., ... Riley, D. A. (2010). Prolonged space flight-induced alterations in the structure and function of human skeletal muscle fibres. *J. Physiol.*, *588*(Pt 18), 3567–3592.

Fortney, S. M. (1991). Development of lower body negative pressure as a countermeasure for orthostatic intolerance. *J. Clin. Pharmacol.*, *31*(10), 888–892.

Fortney, S. M., Turner, C., Steinmann, L., Driscoll, T., & Alfrey, C. (1994). Blood volume responses of men and women to bed rest. *J. Clin. Pharmacol.*, *34*(5), 434–439.

Frings-Meuthen, P., Luchitskaya, E., Jordan, J., Tank, J., Lichtinghagen, R., Smith, S. M., & Heer, M. (2020). Natriuretic peptide resetting in astronauts. *Circulation*, *141*(19), 1593–1595.

Fritsch, J. M., Charles, J. B., Bennett, B. S., Jones, M. M., & Eckberg, D. L. (1992). Short-duration spaceflight impairs human carotid baroreceptor-cardiac reflex responses. *J. Appl. Physiol. (Bethesda, Md.: 1985)*, *73*(2), 664–671.

Fritsch-Yelle, J. M., Whitson, P. A., Bondar, R. L., & Brown, T. E. (1996). Subnormal norepinephrine release relates to presyncope in astronauts after spaceflight. *J. Appl. Physiol. (Bethesda, Md.: 1985)*, *81*(5), 2134–2141.

Fu, Q., Arbab-Zadeh, A., Perhonen, M. A., Zhang, R., Zuckerman, J. H., & Levine, B. D. (2004). Hemodynamics of orthostatic intolerance: Implications for gender differences. *Am. J. Physiol. Heart Circ. Physiol.*, *286*(1), H449–H457.

Fu, Q., Shibata, S., Hastings, J. L., Platts, S. H., Hamilton, D. M., Bungo, M. W., ... Levine, B. D. (2019). Impact of prolonged spaceflight on orthostatic tolerance during ambulation and blood pressure profiles in astronauts. *Circulation*, *140*(9), 729–738.

Fu, Q., Witkowski, S., Okazaki, K., & Levine, B. D. (2005). Effects of gender and hypovolemia on sympathetic neural responses to orthostatic stress. *Am. J. Physiol. Regul. Integr. Comparat. Physiol.*, *289*(1), R109–116.

Gauer, O. H., & Henry, J. P. (1963). Circulatory basis of fluid volume control. *Phys. Rev.*, *43*, 423–481.

Gibson, E. G. (1977). Skylab 4 crew observations. In R. S. Johnston, & L. F. Dietlein (Eds.), *Biomedical results from Skylab. (NASA SP 377)* (pp. 23–24). Washington, D.C.: U.S. Government Printing Office.

Gilbert, C. A., Bricker, L. A., Springfield, W. T., Jr., Stevens, P. M., & Warren, B. H. (1966). Sodium and water excretion and renal hemodynamics during lower body negative pressure. *J. Appl. Physiol. (Bethesda, Md.: 1985)*, *21*(6), 1699–1704.

Greenleaf, J. E. (1984). Physiological responses to prolonged bed rest and fluid immersion in humans. *J. Appl. Physiol. Respir. Environ. Exerc. Physiol.*, *57*(3), 619–633.

Greenleaf, J. E., Bernauer, E. M., Young, H. L., Morse, J. T., Staley, R. W., Juhos, L. T., & Van Beaumont, W. (1977). Fluid and electrolyte shifts during bed rest with isometric and isotonic exercise. *J. Appl. Physiol. Respir. Environ. Exerc. Physiol.*, *42*(1), 59–66.

Greenleaf, J. E., Sciaraffa, D., Shvartz, E., Keil, L. C., & Brock, P. J. (1981). Exercise training hypotension: Implications for plasma volume, renin, and vasopressin. *J. Appl. Physiol. Respir. Environ. Exerc. Physiol.*, *51*(2), 298–305.

Grenon, S. M., Xiao, X., Hurwitz, S., Sheynberg, N., Kim, C., Seely, E. W., ... Williams, G. H. (2006). Why is orthostatic tolerance lower in women than in men? Renal and cardiovascular responses to simulated microgravity and the role of midodrine. *J. Investig. Med.*, *54*(4), 180–190.

Hargens, A. R. (1994). Recent bed rest results and countermeasure development at NASA. *Acta Physiol. Scand. Suppl.*, *616*, 103–114.

Hargens, A. R., & Richardson, S. (2009). Cardiovascular adaptations, fluid shifts, and countermeasures related to space flight. *Respir. Physiol. Neurobiol.*, *169*(Suppl 1)), S30–33.

Hargens, A. R., Tipton, C. M., Gollnick, P. D., Mubarak, S. J., Tucker, B. J., & Akeson, W. H. (1983). Fluid shifts and muscle function in humans during acute simulated weightlessness. *J. Appl. Physiol. Respir. Environ. Exerc. Physiol.*, *54*(4), 1003–1009.

Haruna, Y., Suzuki, Y., Kawakubo, K., & Gunji, A. (1994). Orthostatic tolerance and autonomous nervous functions before and after 20-days bed rest. *Acta Physiol. Scand. Suppl.*, *616*, 71–81.

Hearon, Christopher M, Jr., Dias, Katrin A, Babu, Gautam, Marshall, John E T, Leidner, James, Peters, Kirsten, ... Levine, Benjamin D (2022). Effect of Nightly Lower Body Negative Pressure on Choroid Engorgement in a Model of Spaceflight-Associated Neuro-ocular Syndrome: A Randomized Crossover Trial. *JAMA Ophthalmol*, *140*(1), 59–65. Available from https://doi.org/10.1001/jamaophthalmol.2021.5200.

Hildebrandt, W., Gunga, H. C., Herrmann, J., Röcker, L., Kirsch, K., & Stegemann, J. (1994). Enhanced slow caudad fluid shifts in orthostatic intolerance after 24-h bedrest. *Euro. J. Appl. Physiol. Occup. Physiol.*, *69*(1), 61–70.

Homick, J. L., & J.M., V. (1989). The neurovestibular system. In A. E. Nicogossian, C. L. Huntoon, & S. L. Pool (Eds.), *Space physiology and medicine* (2nd ed., pp. 154–166). Philadelphia, Lea and Feibiger.

Hughson, R. L., Helm, A., & Durante, M. (2018). Heart in space: effect of the extraterrestrial environment on the cardiovascular system. *Nat. Rev. Cardiol.*, *15*(3), 167–180.

Hughson, R. L., Robertson, A. D., Arbeille, P., Shoemaker, J. K., Rush, J. W., Fraser, K. S., & Greaves, D. K. (2016). Increased postflight carotid artery stiffness and inflight insulin resistance resulting from 6-mo spaceflight in male and female astronauts. *Am. J. Physiol. Heart Circ. Physiol.*, *310*(5), H628–638.

Hughson, R. L., Shoemaker, J. K., Blaber, A. P., Arbeille, P., Greaves, D. K., Pereira-Junior, P. P., & Xu, D. (2012). Cardiovascular regulation during long-duration spaceflights to the International Space Station. *J. Appl. Physiol. (Bethesda, Md.: 1985)*, *112* (5), 719–727.

Hupfeld, K. E., McGregor, H. R., Lee, J. K., Beltran, N. E., Kofman, I. S., De Dios, Y. E., ... Seidler, R. D. (2020). The impact of 6 and 12 months in space on human brain structure and intracranial fluid shifts. *Cereb. Cortex Commun.*, *1*(1), tgaa023.

Johnson, R. L., Hoffler, G. W., Nicogossian, A. E., et al. (1977). Lower body negative pressure: third manned Skylab mission. In R. S. Johnston, & L. F. Dietlein (Eds.), *Biomedical results from Skylab. (NASA SP 377)* (pp. 284–312). Washington, DC: U.S. Government Printing Office.

Kakurin, L. I., Lobachik, V. I., Mikhailov, V. M., & Senkevich, Y. A. (1976). Antiorthostatic hypokinesia as a method of weightlessness simulation. *Aviat. Space Environ. Med.*, *47*(10), 1083–1086.

Keil, L. C., McKeever, K. H., Skidmore, M. G., Hines, J., & Severs, W. B. (1992). The effect of head-down tilt and water immersion on intracranial pressure in nonhuman primates. *Aviat. Space Environ. Med.*, *63*(3), 181–185.

Kirsch, K. A., Röcker, L., Gauer, O. H., Krause, R., Leach, C., Wicke, H. J., & Landry, R. (1984). Venous pressure in man during weightlessness. *Science (New York, N.Y.)*, *225*(4658), 218–219.

Koppelmans, V., Bloomberg, J. J., Mulavara, A. P., & Seidler, R. D. (2016). Brain structural plasticity with spaceflight. *NPJ Micrograv.*, *2*, 2.

Kramer, L. A., Hasan, K. M., Stenger, M. B., Sargsyan, A., Laurie, S. S., Otto, C., ... Macias, B. R. (2020). Intracranial effects of microgravity: A prospective longitudinal MRI study. *Radiology*, *295*(3), 640–648.

Leach, C. S., Altchuler, S. I., & Cintron-Trevino, N. M. (1983). The endocrine and metabolic responses to space flight. *Med. Sci. Sports Exer.*, *15*(5), 432–440.

Leach, C. S., Hulley, S. B., Rambaut, P. C., & Dietlein, L. F. (1973). The effect of bedrest on adrenal function. *Space Life Sci.*, *4*(3), 415–423.

Leach, C. S., Inners, L. D., & Charles, J. B. (1991). Changes in total body water during spaceflight. *J. Clin. Pharmacol.*, *31*(10), 1001–1006.

Leach, C. S., & Johnson, P. C. (1984). Influence of spaceflight on erythrokinetics in man. *Science*, *225*(4658), 216–218. Available from https://doi.org/10.1126/science.6729477.

Lee, S. M. C., Ribeiro, L. C., Martin, D. S., Zwart, S. R., Feiveson, A. H., Laurie, S. S., ... Stenger, M. B. (2020). Arterial structure and function during and after long-duration spaceflight. *J. Appl. Physiol. (Bethesda, Md.: 1985)*, *129*(1), 108–123.

Levine, B. D., Pawelczyk, J. A., Ertl, A. C., Cox, J. F., Zuckerman, J. H., Diedrich, A., ... Blomqvist, C. G. (2002). Human muscle sympathetic neural and haemodynamic responses to tilt following spaceflight. *J. Physiol.*, *538*(Pt 1), 331–340.

Levine, B. D., Zuckerman, J. H., & Pawelczyk, J. A. (1997). Cardiac atrophy after bedrest deconditioning: a nonneural mechanism for orthostatic intolerance. *Circulation*, *96* (2), 517–525.

Maillet, A., Zaouali-Ajina, M., Vorobiev, D., Blanc, S., Pastouchkova, L., Reushkina, G., ... Gauquelin-Koch, G. (2000). Orthostatic tolerance and hormonal changes in women during 120 days of head-down bed rest. *Aviat. Space Environ. Med.*, *71*(7), 706–714.

Mandsager, K. T., Robertson, D., & Diedrich, A. (2015). The function of the autonomic nervous system during spaceflight. *Clin. Autonom. Res.*, *25*(3), 141–151.

Meck, J. V., Waters, W. W., Ziegler, M. G., deBlock, H. F., Mills, P. J., Robertson, D., & Huang, P. L. (2004). Mechanisms of postspaceflight orthostatic hypotension: Low alpha1-adrenergic receptor responses before flight and central autonomic dysregulation postflight. *Am. J. Physiol. Heart Circ. Physiol.*, *286*(4), H1486–1495.

Miller, P. B., Johnson, R. L., & Lamb, L. E. (1965). Effects of moderate physical exercise during four weeks of bed rest on circulatory functions in man. *Aerosp. Med.*, *36*(11), 1077–1082.

Moore, T. P., & Thornton, W. E. (1987). Space shuttle inflight and postflight fluid shifts measured by leg volume changes. *Aviat. Space Environ. Med.*, *58*(9 Pt 2), A91–96.

Mulvagh, S. L., Charles, J. B., Riddle, J. M., Rehbein, T. L., & Bungo, M. W. (1991). Echocardiographic evaluation of the cardiovascular effects of short-duration spaceflight. *J. Clin. Pharmacol.*, *31*(10), 1024–1026.

Navasiolava, N., Yuan, M., Murphy, R., Robin, A., Coupé, M., Wang, L., ... Custaud, M. A. (2020). Vascular and microvascular dysfunction induced by microgravity and its analogs in humans: Mechanisms and countermeasures. *Front. Physiol.*, *11*, 952.

Newberg, A. B. (1994). Changes in the central nervous system and their clinical correlates during long-term spaceflight. *Aviat. Space Environ. Med.*, *65*(6), 562–572.

Nicogossian, A.E., & Nachtwey, D.S. (1989). Orbital flight. In: Space physiology and medicine (2nd ed.). A.E. Nicogossian, C.L. Huntoon and S.L. Pool (Eds.). Lea and Feibiger, Philadelphia, 47–58.

Nicogossian, A.E., & Robbin, D.E.I. (1994). Characteristics of the space environment. Space physiology and medicine (3rd ed.). A.E. Nicogossian, C.L. Huntoon and S.L. Pool (Eds.). Lea and Feibiger, Philadelphia, 50–62.

Nixon, J. V., Murray, R. G., Bryant, C., Johnson, R. L., Jr., Mitchell, J. H., Holland, O. B., ... Blomqvist, C. G. (1979). Early cardiovascular adaptation to simulated zero gravity. *J. Appl. Physiol. Respir. Environ. Exerc. Physiol.*, *46*(3), 541–548.

Norsk, P., Damgaard, M., Petersen, L., Gybel, M., Pump, B., Gabrielsen, A., & Christensen, N. J. (2006). Vasorelaxation in space. *Hypertension*, *47*(1), 69–73.

Norsk, P., Drummer, C., Röcker, L., Strollo, F., Christensen, N. J., Warberg, J., et al. (1995). Renal and endocrine responses in humans to isotonic saline infusion during microgravity. *J. Appl. Physiol. (Bethesda, Md.: 1985)*, *78*(6), 2253–2259.

Oman, C.M., Lichtenberg, B.K., & Money, K.E. (1990). Space motion sickness monitoring experiment: Spacelab 1. In: Motion and space sickness. G.H. Crampton (Ed.). Boca Raton, CRC Press, 217–246.

Parazynski, S. E., Hargens, A. R., Tucker, B., Aratow, M., Styf, J., & Crenshaw, A. (1991). Transcapillary fluid shifts in tissues of the head and neck during and after simulated microgravity. *J. Appl. Physiol. (Bethesda, Md.: 1985)*, *71*(6), 2469–2475.

Patwardhan, A. R., Evans, J. M., Berk, M., Grande, K. J., Charles, J. B., & Knapp, C. F. (1995). Spectral indices of cardiovascular adaptations to short-term simulated microgravity exposure. *Integr. Physiol. Behav. Sci.*, *30*(3), 201–214.

Pavy-Le-Traon, A., Vasseur, P., Maillet, A., Güell, A., Bes, A., & Gharib, C. (1994). Orthostatic hypotension after space flight. *La Presse Medicale*, *23*(22), 1031–1036.

Pechenkova, E., Nosikova, I., Rumshiskaya, A., Litvinova, L., Rukavishnikov, I., Mershina, E., ... Kozlovskaya, I. (2019). Alterations of functional brain connectivity after long-duration spaceflight as revealed by fMRI. *Front. Physiol.*, *10*, 761.

Reid, A., Cottingham, C. A., & Marchbanks, R. J. (1993). The prevalence of perilymphatic hypertension in subjects with tinnitus: a pilot study. *Scand. Audiol.*, *22*(1), 61–63.

Roberts, D. R., Albrecht, M. H., Collins, H. R., Asemani, D., Chatterjee, A. R., Spampinato, M. V., ... Antonucci, M. U. (2017). Effects of spaceflight on astronaut brain structure as indicated on MRI. *N. Engl. J. Med.*, *377*(18), 1746–1753.

Robertson, D., Convertino, V. A., & Vernikos, J. (1994). The sympathetic nervous system and the physiologic consequences of spaceflight: a hypothesis. *Am.J. Med. Sci.*, *308*, 126–132.

Rössler, A., Noskov, V., László, Z., Polyakow, V. V., & Hinghofer-Szalkay, H. G. (2001). Permanent depression of plasma cGMP during long-term space flight. *Physiol. Res./Acad. Sci. Bohemosl.*, *50*(1), 83–90.

Scott, J. P. R., Weber, T., & Green, D. A. (2019). Editorial: Optimization of exercise countermeasures for human space flight-lessons from terrestrial physiology and operational implementation. *Front. Physiol.*, *10*, 1567.

Simanonok, K. E., & Charles, J. B. (1994). Space sickness and fluid shifts: a hypothesis. *J. Clin. Pharmacol.*, *34*(6), 652–663.

Stahn, A. C., Gunga, H. C., Kohlberg, E., Gallinat, J., Dinges, D. F., & Kühn, S. (2019). Brain changes in response to long Antarctic expeditions. *N. Engl. J. Med.*, *381*(23), 2273–2275.

Stampi, C. (1994). Sleep and circadian rhythms in space. *J. Clin. Pharmacol.*, *34*(5), 518–534.

Stout, M. S., Watenpaugh, D. E., Breit, G. A., & Hargens, A. R. (1995). Simulated microgravity increases cutaneous blood flow in the head and leg of humans. *Aviat. Space Environ. Med.*, *66*(9), 872–875.

Takenaka, K., Suzuki, Y., Kawakubo, K., Haruna, Y., Yanagibori, R., Kashihara, H., et al. (1994). Cardiovascular effects of 20 days bed rest in healthy young subjects. *Acta Physiol. Scand. Suppl.*, *616*, 59–63.

Taylor, H. L., Erickson, L., Henschel, A., & Keys, A. (1945). The effect of bed rest on blood volume of normal young men. *Am. J. Physiol.*, *144*, 227–232.

Thornton, W. E., Hoffler, W. G., & Rummel, J. A. (1977). Athropometric changes and fluid shifts. In R. S. Johnston, & L. F. Dietlein (Eds.), *Biomedical results from Skylab. (NASA SP 377)* (pp. 330–338). Washington, DC: U.S. Government Printing Office.

Thornton, W. E., Moore, T. P., Pool, S. L., & Vanderploeg, J. (1987). Clinical characterization and etiology of space motion sickness. *Aviat. Space Environ. Med.*, *58*(9 Pt 2), A1–8.

Udden, M. M., Driscoll, T. B., Pickett, M. H., Leach-Huntoon, C. S., & Alfrey, C. P. (1995). Decreased production of red blood cells in human subjects exposed to microgravity. *J. Lab. Clin. Med.*, *125*(4), 442–449.

Van Ombergen, A., Jillings, S., Jeurissen, B., Tomilovskaya, E., Rühl, R. M., Rumshiskaya, A., ... Wuyts, F. L. (2018). Brain tissue-volume changes in cosmonauts. *N. Engl. J. Med.*, *379*(17), 1678–1680.

Verheyden, B., Liu, J., Beckers, F., & Aubert, A. E. (2009). Adaptation of heart rate and blood pressure to short and long duration space missions. *Respir. Physiol. Neurobiol.*, *169*(Suppl 1), S13–16.

Verheyden, B., Liu, J., Beckers, F., & Aubert, A. E. (2010). Operational point of neural cardiovascular regulation in humans up to 6 months in space. *J. Appl. Physiol. (Bethesda, Md.: 1985)*, *108*(3), 646–654.

Vernikos, J., Dallman, M. F., Keil, L. C., O'Hara, D., & Convertino, V. A. (1993). Gender differences in endocrine responses to posture and 7 days of -6 degrees head-down bed rest. *Am. J. Physiol.*, *265*(1 Pt 1), E153–E161.

Volicer, L., Jean-Charles, R., & Chobanian, A. V. (1976). Effects of head-down tilt on fluid and electrolyte balance. *Aviat. Space Environ. Med.*, *47*(10), 1065–1068.

Waters, W. W., Ziegler, M. G., & Meck, J. V. (2002). Postspaceflight orthostatic hypotension occurs mostly in women and is predicted by low vascular resistance. *J. Appl. Physiol. (Bethesda, Md.: 1985)*, *92*(2), 586–594.

Whitson, P. A., Cintron, N. M., Pietrzyk, R. A., Scotto, P., & Loeppky, J. A. (1994). Acute effects of head-down tilt and hypoxia on modulators of fluid homeostasis. *J. Clin. Pharmacol.*, *34*(5), 427–433.

Zhang, L. F. (2001). Vascular adaptation to microgravity: what have we learned? *J. Appl. Physiol. (Bethesda, Md.: 1985)*, *91*(6), 2415–2430.

Zu Eulenburg, P., Buchheim, J. I., Ashton, N. J., Vassilieva, G., Blennow, K., Zetterberg, H., & Choukér, A. (2021). Changes in blood biomarkers of brain injury and degeneration following long-duration spaceflight. *JAMA Neurol.*, *78*, 1525–1527.

CHAPTER 13

Testing autonomic function

The regulatory functions of the autonomic nervous system depend on the actions of a multitude of autonomic reflexes. Autonomic reflex arcs consist of afferent pathways, central processing systems, and efferent pathways. An indication of the overall integrity and the function of afferent, central, and efferent limbs of an autonomic reflex arc can be obtained by autonomic testing. To test an autonomic reflex, efferent neural activity can be recorded directly from nerves, or measured indirectly by recording responses from end organs, part of the efferent pathway. An ideal autonomic test should be simple, safe, noninvasive, reproducible, sensitive, specific, and be suitable for longitudinal studies (Cheshire et al., 2021).

The information derived from autonomic testing is most useful when the findings from a comprehensive history and physical examination are used to guide the selection of tests. In a clinical setting, autonomic tests are usually performed to confirm the presence of autonomic dysfunction, assess the extent of autonomic involvement, and localize the site of the lesion. While most tests provide an indication of the overall integrity of a reflex pathway (Table 13.1), more sophisticated tests permit investigation of afferent, efferent, and central pathways in isolation (Table 13.2). Although many of the tests are easily performed, interpretation of results is complicated by the complexity of the pathways involved and the multiplicity of factors that influence the measured responses (Eckberg, 1980). Because of this, for diagnostic purposes, it is preferable to perform a battery of tests rather than to rely on a single test (Cheshire et al., 2021; Ewing & Clarke, 1982; O'Brien, O'Hare, & Corrall, 1986).

Several abnormal test results are usually required to reach a diagnosis of autonomic failure. Various scoring systems (autonomic scores) have been proposed to help determine the significance of abnormal test results, (Ewing, Martyn, Young, & Clarke, 1985; Low, 1993). Correct use of an autonomic test depends on an understanding of the principles involved and the limitations of the tests' technique. Abnormal test results do not always imply the presence of autonomic symptoms nor do autonomic symptoms imply an abnormal test result. As with any diagnostic test, the predictive value of an autonomic reflex test depends on the prevalence of

Introduction to Clinical Aspects of the Autonomic Nervous System
DOI: https://doi.org/10.1016/B978-0-323-96011-3.00012-7

Table 13.1 Useful and simple noninvasive tests of cardiovascular autonomic function.

Test	Methods	Response Normal[a]	Abnormal	Cause[b]
Active standing	BP and HR supine after 20 min and standing after 2 min HR at beat 30/beat 15 after active standing	↑HR < 30 beats/min Stable BP or ↓SBP < 30 mmHg, ↓DBP < 15 mmHg > 1.03	Neurogenic OH: ↓BP with ΔHR/ΔSBP <0.5 Hypoadrenergic: ↓BP with ↑HR ↓BP with stable HR Hyperadrenergic: ↑HR > 30 beats/min < 1.00	↑sympathetic, ↓ para sympathetic (? Due to exaggerated afferent discharge) Sympathetic failure with intact vagal withdrawal Sympathetic and vagal failure Hypovolemia, deconditio ning, postural tachycardia, after microgravity Vagal failure
Deep breathing (6 breaths/min)	Maximum − minimum heart rate	>10 beats/min (age dependent)	<7 beats/min	Cardiovagal failure
E/I ratio	Maximum R-R during expiration/minimum R-R during inspiration	>1.10	<1.05	Cardiovagal failure

Valsalva maneuver	Longest R-R after maneuver/ shortest R-R	>1.15	<1.10	No rebound bradycardia during Phase IV: vagal failure No tachycardia during Phase II: sympathetic failure
Sustained handgrip (isometric exercise)	BP before and 90 s into handgrip at 30% of maximum voluntary contraction	↑SBP >15 mmHg	↑SBP <10 mmHg	Sympathetic failure

BP, blood pressure; SBP, DBP, systolic, diastolic BP; HR, heart rate; R-R, electrocardiographic R wave to R wave intervals;

[a]Values are for healthy adult men and women <60 years of age.

[b]Cause if afferent and central baroreflex pathways are assumed to be intact.

Source: Modified from Oribe E, Appenzeller O. Evaluation of autonomic reflexes. In: J.P. Mohr & J.G. Gautier (Eds.). Guide to clinical neurology. New York: Churchill Livingstone; 1995a. pp. 213–218; Oribe, E., O. Appenzeller: Long duration running at altitude decreases heart rate variability: a sign of transient aging? Clin. Autonom. Res., 5 (1995b) 327.

Table 13.2 Selected tests for the clinical assessment of sweating.

Test	Methods	Normal response	Mechanism tested	Clinical signs or symptoms	Abnormal responses
Thermoregulatory sweat test	Subject lies in heat box until core temp rises by 1°C or to 38°C. Anterior body surface is powdered with quinizarin red powder, which changes to purple when wet	Sweating changes quinizarin powder to purple	Central, postganglionic, skin and sweat glands	Hypo- or anhidrosis, heat intolerance	Areas of anhidrosis categorized by type (global, regional, focal, length-dependent, etc.)
Axon reflex	Activation of axon terminal by iontophoresis of acetylcholine with quantification of sweat volume	Sweating produced	Sympathetic postganglionic sudomotor axon, skin or sweat glands	Hypo- or anhidrosis	Reduced or absent sweating
Silicone imprint	Sweat droplets from an imprint in plastic dissolved in a solvent. Sweating induced by acetylcholine[a] iontophoresis.	Sweating produced	Sympathetic postganglionic sudomotor axon, skin, or sweat glands	Hypo- or anhidrosis	Reduced or absent sweating
Electro-chemical skin conductance	Electrochemical skin conductance of hands and feet through reverse iontophoresis	Change in voltage or current flow	Central, postganglionic, skin and sweat glands	Hypo- or anhidrosis	No change in voltage or current flow
Skin biopsy	Skin punch biopsy	Normal histologic and histochemical appearance of skin and sweat glands	Skin and sweat glands	Hypo- or anhidrosis	Abnormal sweat gland composition, morphologic changes

[a]Or other cholinergic agonist such as pilocarpine or methacholine.

autonomic dysfunction in the population being tested. The higher the prevalence of a disease, the higher the probability of a patient having the disease if the test is positive (i.e., positive predictive value). Opinions about the clinical value of autonomic tests in the management of patients vary. One opinion is that a good history and clinical examination together with a sphygmomanometer is more valuable to the patient than extensive unguided testing in a laboratory is well-founded (Landau & Nelson, 1996). However, noninvasive autonomic testing has the ability to aid in the diagnosis and management of patients with autonomic disorders along with good history and clinical examination (Cheshire, 2020).

13.1 Key concepts

Approaches to autonomic testing should have several attributes. The test should be sensitive and specific as well as reproducible. Testing should be clinically and physiologically relevant and should be noninvasive. The test stimulus should be standardized with confounding variables identifiable. The technology should be readily available and the equipment should be affordable (Cheshire et al., 2021).

When testing an autonomic reflex, the stimulus applied should be sufficient to activate the afferent limb of the reflex and the end organs involved in the response should be functional. The effects of overlapping compensatory mechanisms (physiological redundancy) should always be considered when interpreting test results, as autonomic dysfunction not apparent under normal testing circumstances may be uncovered by autonomic stressors such as increased temperature, exercise, and after meals. For example, in subjects with mildly deficient sympathetic vasoconstriction, a fall in upright blood pressure may be present only after exercise, meals, or a hot bath.

As autonomic responses are clearly age-dependent (Murata, Landrigan, & Araki, 1992; Piha & Voipio-Pulkki, 1993; Taylor, Hand, Johnson, & Seals, 1992; Wieling, van Brederode, de Rijk, Borst, & Dunning, 1982) and, in some instances, depend on gender and body habitus (Freeman, Weiss, Roberts, Zbikowski, & Sparrow, 1995; Murata et al., 1992; Piha & Voipio-Pulkki, 1993; Schondorf & Low, 1992), a range of normal test values must be determined for each laboratory to permit optimal interpretation. As autonomic test results are very sensitive to testing conditions, protocols should be carefully standardized. Autonomic responses may differ depending on the subjects emotional state, posture during testing (Navarro & Kennedy, 1992; Parsons, Scott, & Macdonald, 1992), instrumentation (i.e., intravenous and arterial catheters), degree of hydration (Tohmeh, Shah, & Cryer, 1979), physical fitness (Gallagher, Terenzi, & de Meersman, 1992) and various other factors such as ingestion of stimulant containing beverages, medications, and smoking. In addition, the timing of testing in relation to meals (Sidery & Macdonald, 1992) and exercise (Furlan et al., 1990; Oribe & Appenzeller, 1995a, 1995b) may modify autonomic responses. Both hypo- and hyperglycemia have the potential to alter testing results in diabetics (Achem-Karam, Funakoshi, Vinik, & Owyang, 1985; Mitrakou et al., 1991; Schwartz, Clutter, Shah, & Cryer, 1987). Autonomic responses are also influenced by the ambient temperature and the presence of background sounds (temperature control and soundproofing are recommended). With the substantial increase in tests and laboratories evaluating autonomic function, a consensus has formed on the appropriate methodology of autonomic testing (Cheshire et al., 2021). Several societies have partnered to facilitate international collaboration for the development and endorsement of current consensus guidelines (Cheshire et al., 2021; Freeman et al., 2011; Gibbons et al., 2017).

13.2 Tests of cardiovascular autonomic function

Autonomic function is frequently tested by measuring heart rate and blood pressure responses to maneuvers such as deep breathing, the Valsalva maneuver, and standing. Because resting and stimulated changes in heart rate and blood pressure can be easily measured, and abnormal findings usually indicate diffuse autonomic damage, autonomic cardiovascular reflexes are the most commonly used tests of autonomic function.

The afferent cardiovascular autonomic reflex limb consists of receptors (mechano- and chemoreceptors) connected to the brainstem and central autonomic pathways via afferent nerves (i.e., glossopharyngeal, vagus, and carotid sinus nerves). Mechanoreceptors that are sensitive to stretch, pressure, and flow are located in the carotids and aortic arch (arterial baroreceptors), large veins, right atrium, lungs, and pleura (cardiopulmonary), and ventricles (cardiac baroreceptors). Chemoreceptors sensitive to changes in blood chemistry (PO_2, PCO_2, and prostaglandins) are located in the carotids (PO_2, peripheral chemoreceptors), the heart (prostaglandins), and in the central nervous system (PCO2, primary central chemoreceptors). Catecholamines (dopamine, noradrenaline, adrenaline, and 5-hydroxytryptamine) and nitric oxide found in the carotid bodies appear to play an important role in the cardiovascular response to chemosensory changes. Additional influences are provided by kidney mechano- and chemoreceptors which produce reflex cardiovascular responses when activated (Ammons, 1992; Stella, 1992).

Within the central nervous system, complex central autonomic network interactions integrate afferent inputs from baro- and chemoreceptors and coordinate autonomic responses. Some of the physiological aspects of the central reflex pathways involved in cardiovascular autonomic control in animals have been reviewed (Benarroch, 1993; Gilbey & Spyer, 1993). Afferent baroreceptor pathways terminate in the nucleus of the tractus solitarius, from where second-order neurons appear to project to the ventrolateral medulla, a sympathoexcitatory region. In addition to receiving baroreceptor input, the rostral ventrolateral medulla also receives projections from vasodepressor areas located in the caudal ventrolateral medulla. Microinjection experiments have indicated that caudal ventrolateral medulla inhibitory interneurons are activated by the release of excitatory amino acids from afferent neurons and that they inhibit sympathoexcitatory neurons via GABAergic projections. Central projections from the ventrolateral medulla may mediate some of the cardiovascular changes elicited by the activity of midbrain and pontine reticular neurons. The A5 group of neurons located in the ventrolateral pons, which are noradrenergic, project to the intermediolateral cell columns of the spinal cord and have important effects on sympathetic nerve discharges. Vagal cardioinhibitory neurons located in the ventrolateral nucleus ambiguus receive afferents from numerous areas.

The changes in cardiac output and ventilation that follow changes in the level of activity and metabolic demand of an individual rely on the interactions between brainstem neurons regulating respiratory rhythm and cardiac vagal neurons regulating heart rhythm (Daly, 1985). During inspiration, respiratory center neurons inhibit cardiac vagal motoneurons producing acceleration of the heart rate (i.e., sinus arrhythmia). Thus, sinus arrhythmia, the physiological oscillation of the cardiac rhythm, results from the respiratory modulation of cardiac vagal outflow. The modulation of the heart rate by respirations depends not only on the actions of a central respiratory control mechanism, but also on its interactions with reflexes evoked during respiratory movements. Indeed, experiments in animals have shown that afferents originating from pulmonary receptors regulate the sensitivity of both vagal and sympathetic neurons. In addition to regulating heart rate, pulmonary afferents are also able to regulate respiratory rate and inhibit somatic muscle activity under certain conditions. It appears that reflex tachypnea and reflex inhibition of somatic muscles occurs when juxtapulmonary capillary receptors located near the alveoli (J receptors or pulmonary C fibers) are stimulated by an increase in interstitial volume (Paintal, 1995). In normal individuals, the usually dormant J receptors are stimulated when the pulmonary capillary pressure rises during exercise (and in other conditions where pulmonary capillary pressure increases such as high-altitude pulmonary edema). Under these circumstances, enhanced stimulation of J receptors may trigger reflex inhibition of all somatic muscles (the J reflex) leading to the termination of exercise. Stimulation of the J receptors also appears to produce choking and pressure sensations in the throat and upper chest and dry cough.

Areas within the hypothalamus, the insular cortex, and the cerebellar uvula have significant effects on cardiovascular function. Within the hypothalamus, activation of the paraventricular nucleus neurons by microinjection of excitatory amino acids and by emotional behavior produces increases in blood pressure, heart rate, and circulatory catecholamines. These and other areas, part of the central autonomic network, project to the autonomic effector organs. Efferents from sympathetic and parasympathetic neurons located in the brainstem synapse with second-order neurons within ganglia, and postganglionic neurons innervate blood vessels, the heart, kidneys, adrenals, sweat glands, and other organs (Fig. 13.1).

In general, variations in heart rate from one heartbeat to the next (i.e., heart rate variability) are measured to assess vagal function, and changes in blood pressure in response to loading or unloading of baroreceptors are used as an indirect measure of sympathetic vasomotor tone. Although the sinus node cardiac pacemaker receives parasympathetic and sympathetic innervation (and is influenced by activation of stretch receptors and neurohormonal factors), under physiological conditions heart rate variability is modulated mainly by respirations through the effects of vagal innervation. While valuable information on the integrity of autonomic cardiovascular reflexes can be derived from simple tests performed by measuring the pulse and blood pressure manually, more sophisticated analyses of heart rate and blood pressure variability require beat to beat heart rate and blood pressure measurements. These methods can record the almost instantaneous changes in heart rate that reflect changes in autonomic tone (changes in heart rate mediated by the vagus have latencies <0.5 seconds while those mediated by the sympathetic have latencies >1 second). Analysis of oscillations in heart rate (or other biological signals such as blood pressure) over time (i.e., analysis in the time domain) can be complemented by frequency analysis (i.e., frequency domain analysis) where

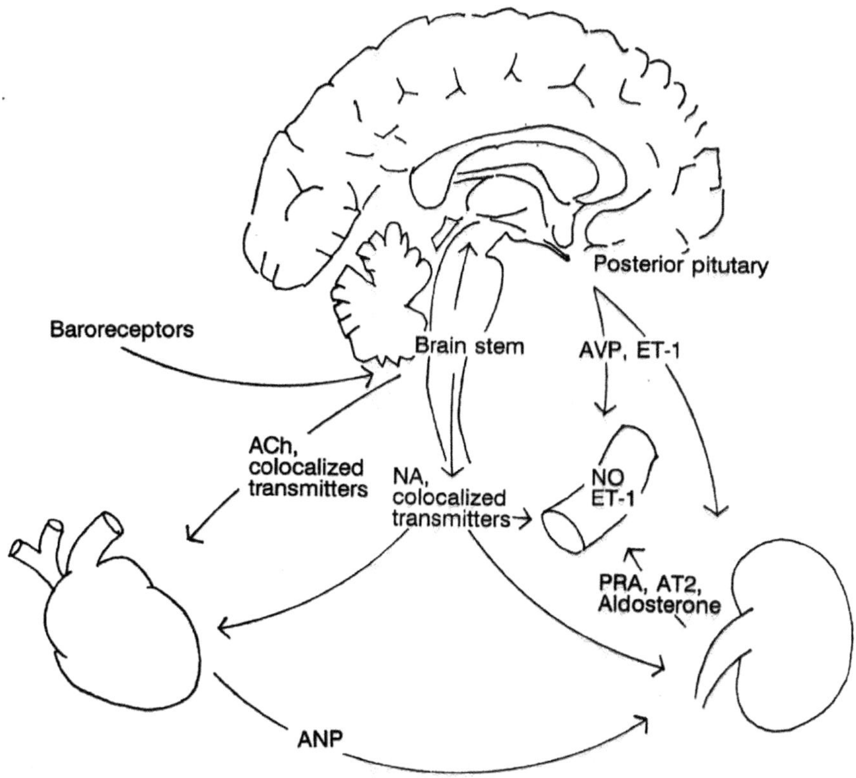

Figure 13.1 Proposed neural, neuroendocrine, and endothelial responses to baroreflex unloading. Decreases in blood pressure (BP) and volume are sensed by high (carotid) and low pressure (cardiopulmonary) baroreceptors and information is transmitted via afferent nerves (i.e., glossopharyngeal, vagus, and carotid sinus nerves) to the cardiovascular centers in the brainstem and their connections. Efferent pre- and postganglionic sympathetic and parasympathetic signals travel from brainstem nuclei and the spinal cord through different nerves to innervate blood vessels and organs. The posterior pituitary releases vasopressin (AVP) and endothelin-1 (ET-1), which have effects on the kidney and produce constriction of blood vessels. The kidneys in turn release renin (PRA) which increases the production of angiotensin (AT2) and aldosterone resulting in vasoconstriction and volume expansion. Decreased venous return also results in decreased production of atrial natriuretic peptide (ANP), which promotes volume expansion. The influence of circulating neuroendocrine factors and mechanical effects on blood vessel walls (shear-stress) modulate the release of potent endothelial vasoactive substances that include nitric oxide (NO) and ET-1. BP and volume rise because of these complex autonomic responses.

oscillations are described as a function of frequency rather than time. Frequency domain analysis techniques such as power spectral analysis are used to describe periodic components within a time series that would otherwise go unnoticed with time-domain analysis methods. As selective autonomic blockade experiments have indicated that sympathetic, parasympathetic, neurohormonal, and other influences on heart rate variability are conveniently segregated around specific frequency bandwidths, frequency domain analysis can provide simultaneous indices of both sympathetic and parasympathetic outflow to the heart by measuring the amplitude of oscillations at different frequencies. Measurement of catecholamines and other biochemical markers can be used to investigate the integrity of certain autonomic pathways. Direct recordings of sympathetic traffic to muscle and skin blood vessels, done in a research setting, provide information on sympathetic outflow to blood vessels.

13.2.1 Heart rate and blood pressure measurements

Most tests of autonomic cardiovascular function can be done without the need for sophisticated equipment (Table 13.1). In the autonomic laboratory, more complex tests require instruments that can record rapid changes in heart rate, blood pressure, and blood flow (Table 13.2). Electrocardiography devices (and Holter prolonged electrocardiography recorders) provide the means to measure beat-to-beat heart rate and R-R intervals, sphygmomanometers (manual or automatically triggered) provide intermittent blood pressure recordings, and sophisticated pressure transducers provide beat to beat arterial pressure data. While arterial pressure recordings obtained directly through arterial cannulation remain the gold standard, beat-to-beat arterial pressure can now be measured noninvasively with reasonable accuracy and reliability employing commercially available tonometric and photoplethysmographic devices. Tonometric devices permit arterial pressure measurements through a pressure sensor applied to the skin overlying an adequately sized and supported artery (such as the radial and dorsalis pedis arteries supported by underlying bone). The sensor records arterial wall displacement, continuously generating a waveform that, once calibrated against oscillometric cuff blood pressure, permits pressure measurements throughout the entire cardiac cycle. The photoplethysmographic devices rely on the measurement of changes in finger volume that follow arterial pressure changes during the cardiac cycle. A small finger cuff and infrared photoplethysmograph linked to a fast response servo control pump device continuously regulate pressure within the finger cuff bladder. Beat-to-beat blood pressure measurements are derived from the pressures required to maintain original finger volume (and arterial size) in a volume clamp system. The waveforms and pressure values generated by both types of devices can be conveniently analyzed in greater detail with the aid of a computer and appropriate software. Recent developments include the ability to estimate cardiac output from analysis of blood pressure pulse contour (Wieling, Veerman, Dambrink, & Imholz, 1992). While absolute beat-to-beat blood pressure values obtained by tonometry and photoplethysmography are usually quite accurate when compared to intra-arterial recordings, the benefit of their use lies mainly in their ability to record relative changes in blood pressure noninvasively.

13.2.2 Blood flow measurements

Indirect measurements of vascular tone (other than by measuring blood pressure) rely on recording peripheral blood flow. Measuring vascular flow can provide an estimate of vascular tone, as flow through a vessel is proportional to the fourth power of its radius (Poiseuille's law). Measurement of tissue blood flow can be obtained indirectly by measuring changes in tissue temperature, volume, and Doppler shifts produced by circulating blood. Doppler techniques rely on measuring frequency shifts in a carrier signal (i.e., ultrasound or laser light) as it is scattered by blood flowing through a vessel. Here, the frequency shift is proportional to the velocity of the flowing blood. In an autonomic laboratory setting, blood flow is most frequently estimated by skin thermometry, Doppler ultrasound, laser Doppler, strain gage plethysmography, and optical methods such as photoplethysmography and transcutaneous oximetry.

Peripheral blood flow through a limb can be measured by plethysmography. Plethysmography with water-filled plethysmographs and with mercury in rubber strain gage plethysmography devices (Greenfield, Whitney, & Mowbray, 1963) allow quantitative assessment of blood flow through limb tissues. Total blood flow can be reliably measured by venous occlusion plethysmography (Brodie & Russell, 1905) which relies on measuring the changes in the volume of a limb once the venous return is arrested for a few seconds without interfering with the arterial inflow. The initial rate of swelling of the limb beyond the venous occlusion is a measure of the rate of arterial inflow and it is assumed that this rate is the same as that immediately preceding venous occlusion. The method is of value as a measure of skin blood flow. Mercury in rubber strain gages (Whitney, 1953), that rely on measuring changes in limb circumference rather than volume (changes in volume and circumference of a cylinder or subcylindrical object follow a linear equation) are mostly used for the determination of muscle blood flow (i.e., forearm and calf blood flows). By recording from several instruments on the forearm, separate blood flows for skin and muscles may be obtained (Clarke, Ginsburg, & Hellon, 1958; Hellon & Clarke, 1959). A simple strain gage consists of a fine bore silicone rubber tube filled with mercury and sealed at the ends by copper plugs in electrical continuity with the mercury. Changes in electrical resistance between the copper plugs are directly proportional to changes in limb circumference (Whitney, 1953).

The measurement of skin temperature can give information about alterations in local skin blood flow. It is, however, influenced by environmental conditions and blood flow to regions other than the one being examined. Tissue temperature can be measured by thermoelectric thermometers (thermocouples and thermistors) and by infrared thermography. Local calorimetry can be done by using a Hatfield Turner copper tellurium heat flow disk, which is useful for measuring heat transfer between a small area of skin and surrounding water below that of central body temperature (Hatfield, 1950; Roddie & Shepherd, 1956). Readings are closely related to blood flow in the skin immediately under the disk but are also influenced by blood flowing to adjacent parts. The principle of internal calorimetry depends on the measurement of the temperature of a thermocouple heated electrically to an equilibrium temperature above that of the surrounding tissue. The temperature measured depends on the tissue thermal conductivity and blood flow which carries heat away and later alters apparent tissue conductivity. A useful needle calorimeter described by Mowbray (1959) provides satisfactory measurements of thyroid, muscle, and bone blood flow (McPherson, Scales, & Gordon, 1961). While the needle calorimeter is used in tissues not easily accessible with other methods of blood flow measurement, it gives values for only a unit mass of tissue and cannot be used for representative measurements of the tissue as a whole.

Thermography relies on the measurement of the natural infrared emission of the human body that can be translated into a visible representation. Liquid crystal thermography (or an electronic equivalent) obtains quantitative data on the temperature of the skin, giving a geographic distribution of skin thermal gradients which can be helpful in static studies of changes in skin temperature. This method has been shown not to reflect any changes in temperature deeper than 6 mm and therefore does not give an indication of changes in blood flow (temperature) of muscles or other tissues underlying the surface from which a recording has been made. Different body areas of interest are compared in search for asymmetry reflecting changes in cutaneous blood flow. Computer-assisted image acquisition and analysis techniques allow more detailed studies.

Infrared photoplethysmography provides a sensitive means of detecting changes in blood volume in the microcirculation near the skin surface. A light-emitting diode housed in a small head of the photo pulse sensor transmits infrared light which, when reflected from the blood in the small superficial vessels, is detected by the phototransistor and converted to an electrical signal. This signal is modulated depending on the changing volume of the blood in the microcirculation. As different wavelengths are reflected depending on the amount of oxygenated blood, some photoplethysmographic devices can distinguish between arterial and venous flow and determine the percentage of hemoglobin oxygen saturation (transcutaneous pulse oximetry).

Laser Doppler flowmetry (LDF) has been used to study cutaneous blood flow (Nilsson, Tenland, & Oberg, 1980). With this technique, a laser light (usually 2 mW) at a wavelength of 632.8 nm reaches the skin through a small fiberoptic probe, penetrating to a depth of about 1 mm. The frequency shift of laser light produced by red blood cells as they circulate through arteriolar capillary beds at depths of up to 0.5 to 0.7 nm is reflected back to the probe and measured in arbitrary units by photodetectors. Under proper conditions, LDF appears to provide sensitive indices of cutaneous vascular caliber and is sufficiently sensitive to permit evaluation of the normal sinusoidal oscillations that occur in vasomotor tone (the physiological sinusoidal changes in vessel diameter result in smaller resistances to blood flow than that of a vessel of the same average but constant diameter). Spectral analysis cross-correlation between skin blood flow and heart rate at certain frequencies (10-second period fluctuations) has been used as an index of the influence of the baroreceptors on autonomic tone and skin blood circulation (Bernardi et al., 1989). LDF has been adapted to measure superficial blood flow through various tissues including the cerebral cortex, spinal cord, bladder, gastric, colonic, and uterine tissues.

Ultrasound techniques can assess vascular flow at penetration depths greater than optical techniques. Duplex scanning (a combination of b mode and bidirectional Doppler ultrasound) can provide quantitative measurements of blood flow through larger and deeper vessels. Doppler ultrasound techniques have been used successfully to obtain measures of blood flow through aortic, mesenteric, celiac, and other deep arteries (Chaudhuri, Thomaides, Hernandez, Alam, & Mathias, 1991). Transcranial Doppler ultrasound permits measurement of flow through cerebral arteries and can be used to investigate cerebral autoregulation. In research settings, Doppler ultrasound probes can be implanted in the proximity of blood vessels to better measure blood flow.

Cardiac output measurements can be used to calculate total peripheral resistance (TPR), a function of vascular tone. Cardiac output can be measured indirectly by Doppler echocardiography and impedance cardiography. Impedance cardiography provides estimates of cardiac stroke volume by measuring changes in thoracic impedance present during the cardiac cycle. While stroke volumes measured by impedance cardiography correlate well with ventriculogram measurements, the relative changes generated by this application are of most value for autonomic testing purposes (Schondorf, 1993).

13.2.3 Baroreflex pathway testing

Baroreflex pathway function can be tested by measuring the changes in heart rate, blood pressure, and plasma concentration of neurotransmitters and hormones produced in response to maneuvers that activate baroreceptors. Sympathetic baroreflex function can be investigated by measuring the changes in heart rate, blood pressure, and plasma norepinephrine concentration, triggered by unloading baroreceptors (i.e., upright posture, lower body negative pressure, and administration of vasodilator agents). Sympathetic efferent activity can be assessed independently by bypassing baroreflex afferent pathways by cortical arousal (i.e., mental stress) or by direct stimulation of cholinergic postganglionic sympathetic neurons (i.e., administration of the cholinesterase inhibitor edrophonium). Efferent sympathetic nerve traffic to muscle and skin blood vessels can be measured directly by intraneural microneurography (muscle and skin sympathetic nerve activity).

The integrity of parasympathetic cardiac control can be estimated by measuring heart rate variability responses to maneuvers that either inhibit or stimulate parasympathetic outflow to the heart. Maneuvers that inhibit parasympathetic efferent activity include the early responses to standing or head-up tilt and the initial response to isometric exercise. Maneuvers that increase parasympathetic outflow include deep breathing and increases in blood pressure produced during the Valsalva maneuver and by the administration of pressor drugs. Parasympathetic efferent pathway integrity can be evaluated in isolation (i.e., independently of afferent baroreflex influences) during maneuvers such as carotid sinus massage, eyeball pressure, during the diving or simulated diving tests, and during administration of atropine. Central autonomic pathways can be assessed by measuring the release of neurohormones in response to baroreflex activation, drugs, or hypoglycemia.

13.2.4 Active standing and passive head-up tilting

Changes in blood pressure and heart rate measured intermittently or manually in response to active standing provide important information regarding the integrity of autonomic responses. The heart rate and blood pressure responses to standing are complex, relying on functional arterial (high pressure) and cardiopulmonary (low pressure) baroreflexes. Additional less defined roles are played by the exercise reflex (Coote, Hilton, & Perez-Gonzalez, 1971), the venoarteriolar reflex (Henriksen, 1977), and through the effects of endothelial and neurohormonal factors that modulate baroreflexes or act directly on blood vessels.

The cardiovascular adaptations to active standing from a supine posture have been grouped into initial ($<$ 0.5 minutes), early steady-state (1–2 minutes), delayed (2–5 minutes), and prolonged ($>$ 5 minutes) responses based on their typical beat to beat heart rate and blood pressure patterns (Imholz, Dambrink, Karemaker, & Wieling, 1990; Wieling & Shepherd, 1992). The changes in heart rate and blood pressure produced by standing reflect autonomic reflex responses activated to compensate for the sudden and marked fluid shifts due to gravitational pooling and muscle contraction. The immediate redistribution of 500–1000 mL of circulatory blood to the lower body on standing produces a fall in venous return, stroke volume, cardiac output, and blood pressure which unloads baroreceptors and activates baroreflexes (Borst et al., 1982). Baroreflex unloading results in a reflex withdrawal of vagal tone and an increase in heart rate that is most prominent at about the 15th heartbeat. Within seconds from this initial response, venous return rises due to the pumping effects of limb and abdominal muscles contractions during the effort of standing, and blood pressure overshoots transiently. The rise in blood pressure is detected by baroreceptors (baroreceptors are loaded) leading to reflex parasympathetic activation with slowing of the heart rate that reaches a nadir at about the 30th heartbeat (Fig. 13.2). The initial

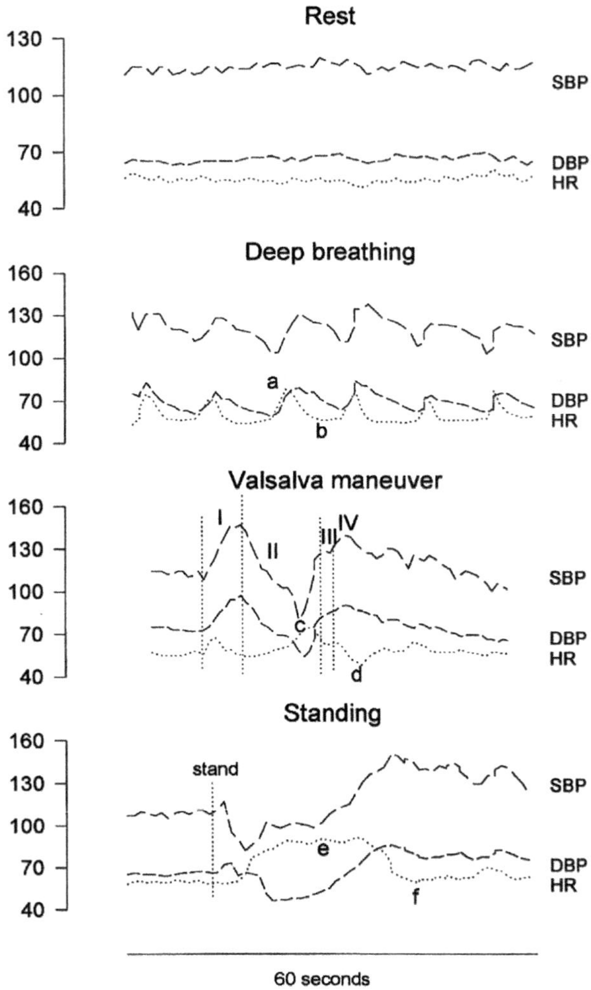

Figure 13.2 Beat-to-beat heart rate (HR) and blood pressure (BP) recorded noninvasively during rest, deep breathing, Valsalva maneuver and active standing in a healthy 28-year-old man. At rest, HR normally increases with inspiration and decreases with expiration. This effect is modulated by the vagus nerve. Deep breathing at a rate of 6 breaths/minute produces the large HR variation (a) and (b) used as a measure of vagal integrity. The response to the Valsalva maneuver results from circulatory adjustments to changes in BP that are mainly dependent on intrathoracic pressure changes produced by the 20-s strain. The HR increase during Phase II (c) and the decrease during Phase IV (d) of the response rely on intact vagal pathways. Active standing from a supine posture results in an acute fall in BP produced by redistribution of part of the circulatory volume to the lower body; this is followed by a BP rebound due to muscle contraction during the effort of standing. An initial increase in HR at about the 15th beat (e) and subsequent slowing at about the 30th beat (f) are mediated by the vagus. For correct interpretation, vagally mediated HR responses are reviewed together with simultaneous BP recordings to exclude sympathetic vasoconstrictor defects that may decrease vagal indices. SBP, DBP = systolic, diastolic pressure (mmHg); HR, heart rate (beats/minute); I, II, III, IV = different phases of the response to Valsalva maneuver.

cardiovascular adjustments to standing are followed by a gradual stabilization of blood pressure and heart rate within the next 1–2 minutes. During the later stages, increases in heart rate are due to the combined effects of vagal withdrawal and sympathetic activation (Wieling, 1992). Through the effects of enhanced sympathetic outflow to the blood vessels and the heart, peripheral resistance, cardiac contractility, and heart rate increase. Most of the autonomic adjustments occurring during the delayed response are complete by 5 minutes. The compensatory responses to prolonged standing also rely on the effects of neurohormones that regulate vascular tone and fluid volume.

In patients with orthostatic symptoms, analysis of blood pressure and heart rate changes during the early steady-state response to standing (at 2 minutes after standing) provides valuable diagnostic information (Wieling, 1992; Wieling, ten Harkel, & van Lieshout, 1991). Typical patterns of response suggest different causes of autonomic dysfunction (Table 13.1 and Fig. 13.3). During the early steady-state response to standing, systolic pressure normally decreases minimally or is unchanged, diastolic pressure rises slightly, and heart rate increases less than 30 beats/minute. While the definition of orthostatic hypotension based on absolute pressure values varies (there is a lack of a consensus as to the definition of orthostatic hypotension), we consider a continued fall in systolic pressure >30 mmHg accompanied by symptoms to be abnormal. Abnormal falls in blood pressure usually indicate defective efferent sympathetic vasoconstrictor tone; lack of heart rate increases in response to hypotension usually indicate defective vagal and sympathetic cardiac innervation. Possible vasoactive or cardiac medication effects should be considered when interpreting abnormal responses.

Patients with neurogenic orthostatic hypotension have a hypoadrenergic response to standing (Fig. 13.3). Blood pressure falls and heart rate either remains unchanged or only minimally rises depending on the relative integrity of vagal and sympathetic cardiac innervation (Table 13.1). With intact cardiovagal function, the rise in heart rate does not usually reach beyond 100 beats/minute unless there is relative preservation of sympathetic innervation. Using the change in heart rate/ change in systolic blood pressure ratio of 0.5 has excellent sensitivity and specificity to distinguish between patients with neurogenic and nonneurogenic orthostatic hypotension (Norcliffe-Kaufmann et al., 2018). The heart rate varies only slightly with severe autonomic dysfunction and in disconnected (transplanted) hearts. In rare instances of profound autonomic failure, the

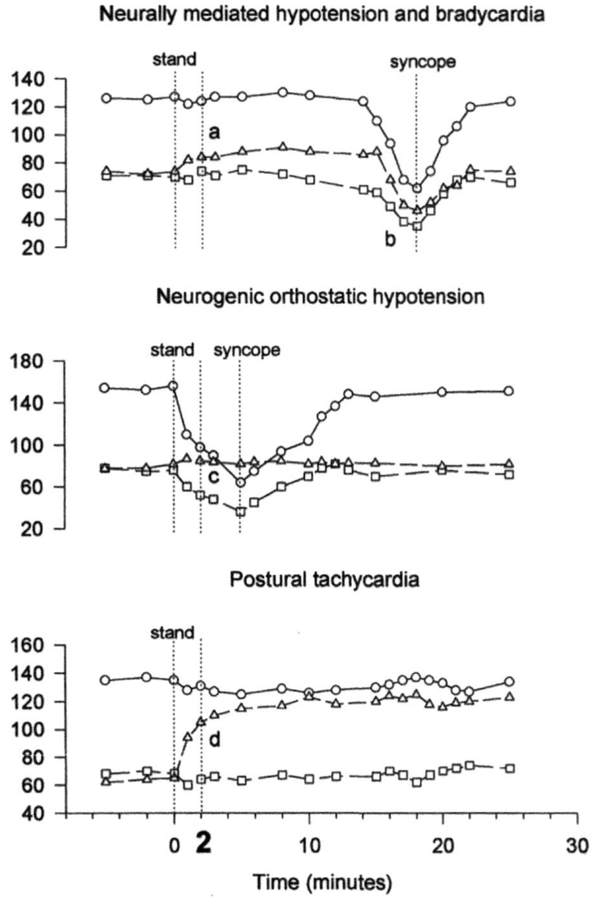

Figure 13.3 Blood pressure (BP) and heart rate (HR) recorded intermittently while supine and after active standing (stand) in three patients referred for evaluation of orthostatic symptoms. In many instances, abnormal response patterns can be identified by measuring BP and HR changes after 2 min of active standing (2). In neurally mediated hypotension and bradycardia (NMS), a normal initial response to standing (a) is followed, after variable lengths of time (19 min in this subject), by acute hypotension and bradycardia typical of the ictal phase of NMS (b). In contrast, with autonomic failure (neurogenic orthostatic hypotension), hypotension occurs immediately upon standing (c). In this patient, a fall in BP is accompanied by little change in HR indicating both vasomotor and cardiomotor failure. With postural tachycardia, there is an exaggerated increase in heart rate (>30 beats/minute) upon standing (d). The subjects with NMS and orthostatic hypotension suffered brief syncope or near-syncope while upright (syncope). □: systolic, diastolic pressure (mmHG); Δ: HR (beats/min).

heart rate has been reported to increase in parallel with maneuvers that modify venous return (i.e., increase with head-down tilt and decrease with head-up tilt) suggesting that large vessel and atrial stretching may modulate heart rate in the absence of autonomic influences (Perera, Oribe, & Kaufmann, 1993). In patients with severe orthostatic blood pressure falls that preclude standing for more than a few seconds, recording the time to orthostatic symptoms after standing (standing time) can be helpful as a measure of orthostatic tolerance. In patients with neurogenic orthostatic hypotension, measurement of plasma concentrations of vasopressin in response to baroreceptor unloading during upright tilt may be used to explore the integrity of afferent and central baroreflex pathways. The expected increase in vasopressin in response to hypotension is blunted in patients with lesions of the central baroreflex pathways (i.e., MSA) whereas it is present in patients with peripheral efferent lesions (i.e., PAF) (Kaufmann et al., 1992).

The neurally mediated (i.e., vasovagal or vasodepressor, or neurocardiogenic) response consists of hypotension accompanied by an inappropriately slow or unchanged heart rate (Fig. 13.3). Compared with the hypoadrenergic response, with the neurocardiogenic response autonomic cardiovascular innervation appears to be intact and dysfunction is transient and limited to the ictal period. Most patients have a normal initial hemodynamic response to standing and develop typical neurocardiogenic blood pressure and heart rate changes later (i.e., during prolonged standing). While the mechanisms involved in the response are for the most part hypothetical, it appears that exaggerated afferent stimuli originating either in the cardiac ventricles, the atria, or at other sites act as triggers. Hypotension is due to vasodilatation resulting from sympathetic withdrawal (passive vasodilatation) and probably through parasympathetic activation with release of endothelial nitric oxide (a powerful vasodilator) (Kaufmann, Berman, Oribe, & Oliver, 1993). The slow heart rate present with the neurocardiogenic response is vagally mediated and can be abolished with atropine (Lewis, 1932). While manifestations of autonomic activation (i.e., apprehension, nausea, vomiting, sweating, pallor, mydriasis) are usually present with neurally mediated responses, they are frequently absent in patients with hypotension due to autonomic failure.

A few subjects who complain of orthostatic symptoms have delayed orthostatic hypotension, a condition where a normal initial hemodynamic response to the upright posture is followed by a fall in blood pressure that usually occurs after 10–15 minutes. The fall in blood pressure can be reversed by the application of a pressure suit, suggesting the presence of a defective venous innervation in the legs (Streeten & Anderson, 1992). The presence of delayed orthostatic hypotension has clinical relevance with an association with developing an α-synucleinopathy, baseline diabetes, and later abnormalities on autonomic function testing. Additionally, subjects with delayed orthostatic hypotension had a 10-year mortality rate near 30% (Gibbons & Freeman, 2015).

Some subjects respond to orthostatic stress produced by standing with an exaggerated increase in heart rate (i.e., consistently increased >30 beats/minute or beyond >120 beats/minute), which may be accompanied by falls in systolic, large rises in diastolic, or abnormal oscillations in blood pressure (Fig. 13.3). The response is a frequent occurrence in hospitalized patients who are allowed to stand after a prolonged period of bed rest. While similar responses may be present with hypovolemia (such as with dehydration and hemorrhage), idiopathic hypovolemia (Fouad, Tadena Thome, Bravo, & Tarazi, 1984), in deconditioned subjects and after exposure to microgravity during spaceflight (Chapter 22), they may be also present in some active individuals despite their lack of prolonged immobility or bed rest. Some of these subjects have evidence of a distal autonomic neuropathy (sudomotor tests are abnormal) with excessive dependent venous pooling while in the upright posture; others have mild and unexplained hypovolemia, and in others, there are no apparent causes. The tachycardic (or hyperadrenergic) response to standing has been considered a cardinal manifestation of the postural tachycardia syndrome (also known as postural orthostatic tachycardia syndrome, POTS). The terms orthostatic tachycardia (Mac Lean, Allen, & Magath, 1944), postural tachycardia syndrome (Rosen & Cryer, 1982), sympathotonic orthostatic hypotension (Hoeldtke, Dworkin, Gaspar, & Israel, 1989), hyperdynamic beta-adrenergic state, idiopathic hypovolemia, sympathicotonic orthostatic hypotension, orthostatic tachycardia, and idiopathic postural orthostatic tachycardia syndromes have been used for what appears to be the same entity (see Chapter 24) (Khurana, 1995; Low, Opfer-Gehrking, Textor, et al., 1995b; Schondorf & Low, 1993).

The heart rate and blood pressure responses to passive changes in posture (i.e., head up tilting), are different from those described during active standing. For passive tilt testing, subjects are secured with chest and knee straps into position on a tilt table with foot support. The knee straps and foot support minimize active muscle contractions that are usually present during active standing. After at least 5–20 minutes in the supine position, and once blood pressure and heart rate are stable, the table is tilted to a 60-degree head-up position (other common protocols use tilt angles of 70–90 degree). On assuming the head-up position, there is a gradual rise in diastolic pressure and little or no change in systolic pressure. With passive tilt, the initial changes in heart rate and blood pressure described with active standing are absent (Wieling, 1992). During head-up tilt, blood pressure may transiently decrease but should be rapidly restored to baseline values within 1–2 minutes in normal subjects (Imholz et al., 1990; van Lieshout, 1989). As occurs with active standing, abnormal blood pressure and heart rate responses during tilt provide valuable diagnostic information in patients with orthostatic symptoms. In addition to upright hypotension (systolic pressure fall >30 mmHg), excessive oscillations in blood pressure (mean arterial pressure oscillations >20 mmHg), >50% reductions in pulse pressure, increases in diastolic pressure (> 20 mmHg), and exaggerated increases in heart rate (> 30 beats/minute above baseline or a heart rate of 120 beats/minute) present >50% of the time during a 5-minute upright tilt are considered abnormal (Low, Opfer-Gehrking, McPhee, et al., 1995a). As with active standing, exaggerated increases in heart rate on tilting (> 30 beats/minute) can result from hypovolemia and deconditioning produced by bed rest (and microgravity) and as a manifestation of the postural tachycardia syndrome (see above).

Prolonging the duration of head-up tilt to 20–60 minutes is widely used as a diagnostic test to determine the cause of unexplained syncope, as the procedure can trigger neurally mediated syncope (see above) (Almquist et al., 1989; Fitzpatrick & Sutton, 1989; Grubb, Temesy-Armos, Hahn, & Elliott, 1991; Kenny, Ingram, Bayliss, & Sutton, 1986; Raviele et al., 1990) and delayed orthostatic hypotension (Gibbons & Freeman, 2015; Streeten & Anderson, 1992), in those who are susceptible. When used as a diagnostic tool for neurally mediated syncope, head-up tilt is defined as positive if hypotension, accompanied by slowing or no changes in heart rate, and symptoms of decreased cerebral perfusion develop (i.e., syncope) reproducing symptoms identical or very similar to those suffered by the patient during a spontaneously suffered syncope. It is assumed that neurally mediated syncope is the actual cause for spontaneous syncope when patients recognize the symptoms induced during tilt to be identical or very similar to those suffered during a spontaneous syncope (the clinical features of a spontaneously occurring syncopal attack are the tests gold standard). The value of the 40 minutes 60-degree head-up tilt test in the diagnosis of neurally mediated syncope was prospectively examined in 201 patients with a history of syncope of unknown cause and 102 age and gender-matched control subjects (Oribe et al., 1997). Head-up tilt elicited syncope (i.e., was positive) in 74 of the 201 patients (37%) with a history of unexplained syncope and in only 6 of the 102 controls (6%). The specificity of the test was highest at 100% in patients 60 years of age and older. In this study, the positive predictive value of a positive response was 93% and the negative predictive value was 43%. The results indicate that the prolonged head-up tilt test is a very specific procedure of high diagnostic value in patients with a history of unexplained syncope. Because of its noninvasive nature and relative safety, the test is particularly useful in the elderly age groups who have a high incidence of syncope. Combining the head-up tilt test with infusion of isoproterenol may reduce the time to syncope during tilt and may increase the sensitivity of the test to above 80% (Almquist et al., 1989). While this combination may trigger a higher rate of syncope in controls and reduce the specificity of the test (Kapoor & Brant, 1992), another study employing low doses of isoproterenol has reported increased sensitivity (61%) without any detriment in specificity (92%) (Morillo, Klein, Zandri, & Yee, 1995). The use of nitroglycerin is also used to induce tilt which provides for sublingual administration (Teodorovich & Swissa, 2016). Implantable loop recorders have also been used to document cardioinhibitory reflex syncope (Teodorovich & Swissa, 2016).

13.2.5 Lower body negative pressure

Lower body negative pressure (LBNP) can be used to stimulate baroreflex activity because it can mimic some of the effects of gravitational stress and hypovolemia on the circulation. The method is simple and consists of placing the lower part of the body into an airtight chamber (usually sealed around or below the umbilicus) connected to a vacuum source (Fig. 12.2). Different subatmospheric pressures are maintained within the chamber by the use of a valve. While LBNP results in volume shifts that are in many ways similar to those produced by the upright posture, the cardiovascular responses may differ to some degree. This may be explained in part by physical differences in the direction and nature of the pressure changes (with LBNP, negative pressure is applied evenly over the whole of the lower body, while the gravitational pull during standing is graded, being largest in the feet and progressively less near to the heart). Also, with LBNP diastolic pressure remains either unchanged or decreases, whereas it invariably increases in the head-up posture.

Normally LBNP is used as a research tool in subjects who are unable to stand or who have other measurements made which make tilting or Valsalva maneuver inconvenient. The procedure has been used to selectively test low-pressure cardiopulmonary baroreflex function as central venous pressure can be decreased at low suction pressures without affecting arterial pressure. Interpretation of the responses has been complicated by the finding that intact arterial baroreceptors are required to maintain arterial pressure during low LBNP suction pressures (Aksamit, Floras, Victor, & Aylward, 1987) indicating that arterial and cardiopulmonary baroreceptors act in concert under these conditions. A combination of upright tilt and LBNP has been reported to be useful as a diagnostic procedure in cases of unexplained syncope (Hainsworth & el-Bedawi, 1994). Because it can simulate the effects of standing and gravity in a microgravity environment, LBNP has been used extensively to study cardiovascular adaptations during spaceflight (Chapter 22).

13.2.6 The Valsalva maneuver

Cardiovascular changes triggered by the Valsalva maneuver rely on the integrity of complex afferent, central and efferent baroreflex pathways. Measurements of the blood pressure and heart rate responses elicited by the maneuver have been used as an index of sympathetic and vagal function. Subjects are asked to take a deep breath and blow through a mouthpiece to generate a pressure of 40 mmHg (5.3 kPa) for at least 15 seconds (i.e., Valsalva's maneuver) (Levin, 1966). The maneuver results in tachycardia and peripheral vasoconstriction during the strain after which there is an overshoot rise in blood pressure accompanied by reflex bradycardia (Fig. 13.2). Blood pressure and heart rate changes induced by the Valsalva maneuver are divided into four phases and may persist for as much as 1–2 minutes after the maneuver.

There are four phases of the Valsalva maneuver. During the initial phase (Phase I), at the onset of straining, a rise in arterial pressure due to compression of the aorta is accompanied by a reflex reduction in heart rate. During Phase II, while the subject forcefully blows through the mouthpiece against resistance, intrathoracic pressure increases leading to a fall in venous return, cardiac output, and blood pressure, accompanied by tachycardia due to parasympathetic withdrawal and increased sympathetic outflow. During Phase III, once the strain ends, intrathoracic pressure falls, and pulmonary venous capacitance rises resulting in further falls in systemic blood pressure and a reflex increase in heart rate. During Phase IV, blood pressure rises transiently beyond baseline (i.e., overshoots) due to continued vasoconstriction and increased cardiac output. During this phase, there is reflex bradycardia followed by a gradual return of heart rate to baseline as blood pressure normalizes. Hence, all phases are triggered by an adequate stimulus (a fall in blood pressure) secondary to the maneuver.

The changes in heart rates can be measured from an electrocardiogram and the Valsalva ratio can be calculated (Baldwa & Ewing, 1977; Levin, 1966). This is the longest beat-to-beat (R-R) interval after the maneuver (Phase IV, the overshoot bradycardia) to the shortest R-R interval during the strain (Phase II). A ratio of 1.2 or greater is thought to be normal, 1.11−1.20 borderline, and less than 1.10 abnormal although there are age-related normative values (Ewing, Campbell, & Clarke, 1980) (Table 13.1). In addition, when beat-to-beat blood pressure is recorded, an index of the baroreflex sensitivity can be determined by dividing the difference between R-R during Phases IV and II by the difference of systolic pressure during Phases IV and II. In autonomic failure, defective peripheral vasoconstriction leads to marked decreases in blood pressure during Phase II (>30 mmHg systolic) and lack of the normal blood pressure overshoot and reflex bradycardia during Phase IV of the Valsalva maneuver. While lack of rebound bradycardia during Phase IV may indicate vagal dysfunction (Fig. 13.4), it may also reflect the absence of the blood pressure overshoot due to deficient sympathetic vasoconstriction (Wieling, 1992).

Detailed analysis of the heart rate and blood pressure responses during the different phases of the maneuver by using beat-to-beat recording techniques may permit recognition of different patterns and degrees of autonomic dysfunction (Sandroni, Benarroch, & Low, 1991). Pharmacologic dissection studies have shown that late Phase II is largely mediated by α receptors, which can be blocked by phentolamine, while Phase IV overshoot is mediated by ß receptors, which can be blocked by phentolamine (Sandroni et al., 1991). Abnormal response patterns with lack of tachycardia and increased blood pressure during Phase II and lack of blood pressure overshoot during Phase IV have been reported with severe congestive heart failure (Nishimura & Tajik, 1986; Zema, 1999). The heart rate responses to the Valsalva maneuver clearly decrease with advancing age (Fig. 13.5).

The heart rate and blood pressure responses to squatting can also be recorded to study autonomic cardiovascular reflexes. The heart rate and

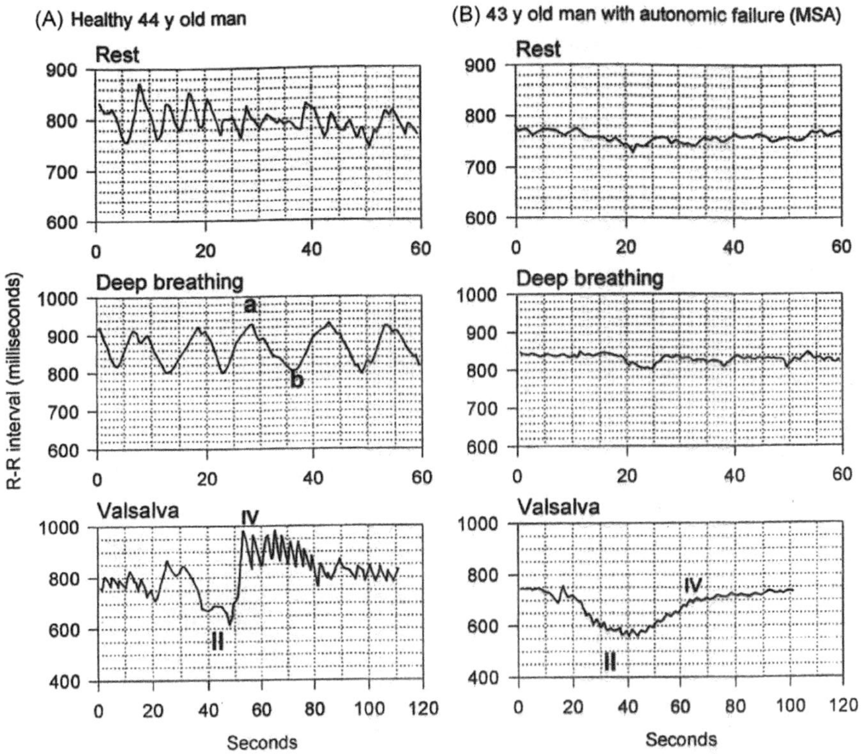

Figure 13.4 Electrocardiographic R-R interval (R-R) time series recorded during quiet rest, deep breathing, and Valsalva maneuver in (A) a healthy 44-year-old man, and (B) a 43-year-old man with autonomic failure (multiple system atrophy, MSA). The healthy subject has normal R-R variability during quiet rest, R-R changes in response to deep breathing (a, expiration; b, inspiration), and during Phases II and IV of the Valsalva maneuver. R-R variability and R-R changes elicited by the maneuvers are decreased in the patient with autonomic failure.

blood pressure responses after squatting for 1 minute and standing during inspiration have been used to assess the integrity of parasympathetic and sympathetic cardiac and vasomotor innervation in diabetic patients (Marfella et al., 1994).

13.2.7 Baroreflex sensitivity

Baroreflex function can be characterized by defining the relationship between changes in heart rate and arterial pressure. Negative or positive pressure applied intermittently at the level of the carotid baroreceptors with airtight neck chamber devices allows manipulation of the arterial

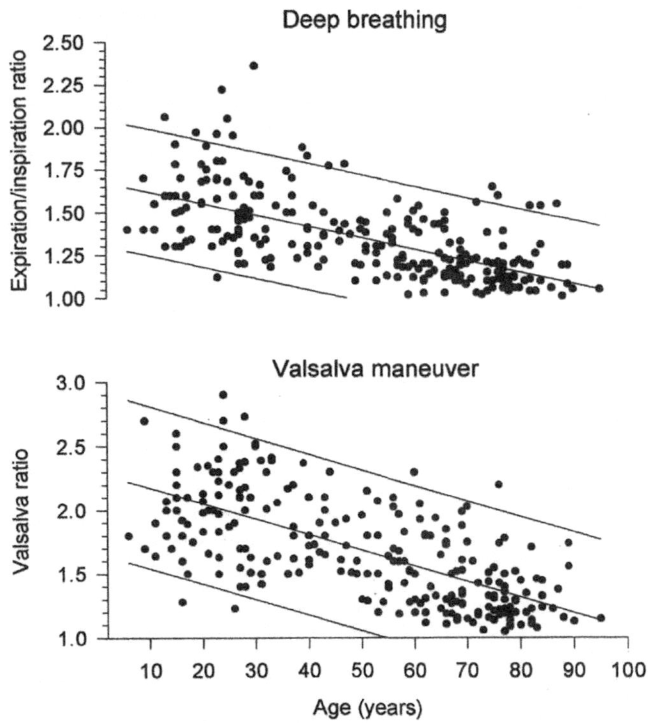

Figure 13.5 Variability of responses and the effect of age on expiration/inspiration (E/I) and Valsalva ratio (VR) in healthy men and women. Regression lines of E/I (230 subjects) and VR (237 subjects) on age, with 95% prediction intervals.

baroreflex environment without affecting low-pressure baroreceptors (Ernsting & Parry, 1957). Changes in neck pressure can be timed to coincide with the cardiac cycle by linking the pressure valve with an electrocardiograph device. A curve that describes the relationship between pressure and heart rate can be obtained by plotting changes in electrocardiographic R-R intervals against carotid distending pressure within the neck chamber (Eckberg & Fritsch, 1993). Sinusoidal neck suction at different frequencies has been used to activate baroreceptors independently of the effects of the cardiac cycle and respiration to study autonomic reinnervation following cardiac transplantation (Bernardi et al., 1995).

Drugs that alter blood pressure can also be used to test the afferent limb of the baroreflex and to determine baroreflex sensitivity. Agents such as phenylephrine and angiotensin and nitrites are given to induce changes in blood pressure while the corresponding changes in heart rate are measured. Beat to beat changes in systolic blood pressure are plotted against the changes in the subsequent R-R interval (Smyth, Sleight, & Pickering, 1969). As there is a linear relationship between the changes in systolic pressure and the changes in heart rate, the slope of the correlation can provide an index of baroreflex sensitivity. Lower baroreflex sensitivity has been reported in patients with hypertension (Sleight, 1988) and its presence may identify patients that have a higher risk of cardiac arrhythmia during cardiac ischemia. Infusions of the ß adrenoceptor agonist isoproterenol can be used to assess the integrity of the baroreflex as it produces vasodilatation and a fall in blood pressure which normally elicits tachycardia due to vagal withdrawal (Arnold & McDevitt, 1983). The heart rate and blood pressure responses to the Valsalva maneuver can also provide an index of baroreflex sensitivity (see below). Algorithms to assess baroreflex sensitivity include time-domain methods and frequency domain methods (Persson et al., 2001). Most time-domain methods use sequences during which heart rate and blood pressure increase or decrease, measure slopes in these sequences, average the slopes, eventually separating positive and negative slopes. Algorithms in the frequency domain make it possible to focus on specific frequency bands (Swenne, 2013).

13.3 Efferent sympathetic pathway testing

13.3.1 Microneurography

Direct neural recordings (microneurography) of sympathetic traffic to muscle and skin blood vessels have been used to measure sympathetic vascular efferent outflow in a research setting. Microneurography has proven to be an effective tool to study the sympathetic responses to standing, lower body negative pressure, and various maneuvers commonly used as stimuli during autonomic testing. A microelectrode is manipulated within the substance of a peripheral nerve (usually the peroneal nerve) until a burst pattern characteristic of sympathetic nerve activity is found (Vallbo, Hagbarth, Torebjork, & Wallin, 1979; Wallin & Fagius, 1988). Muscle nerve sympathetic activity (MNSA) carries vasoconstrictor impulses to muscle blood vessels, while skin sympathetic nerve activity (SSNA) carries vasoconstrictor and sudomotor traffic to the skin. MNSA occurs in bursts that are synchronous with heartbeats through modulation by high-pressure baroreceptors (Fagius, Wallin, Sundlof, Nerhed, & Englesson, 1985), and SSNA occurs in irregular bursts unrelated to the heart rhythm (Hagbarth, Hallin, Hongell, Torebjork, & Wallin, 1972). Changes in temperature can result in separate sudomotor and vasoconstrictor sympathetic nerve activity. Microneurography has also been used to detect impairments in central neurological disorders, including MSNA, SSNA, or abnormal responses to arousal (Donadio & Liguori, 2015).

13.3.2 Cold pressor test

The cold pressor test was developed by Edgar J. Hines, Jr. as a tool to study blood pressure variability (Lamotte, Boes, Low, & Coon, 2021). The normal response to the immersion of a hand (or foot) in ice water (the cold pressor test) for 2 minutes involves reflex arteriolar vasoconstriction producing an increase in blood pressure and cardiac output (Hines & Brown, 1936) probably triggered by cutaneous pain and cold receptors. Blood pressure is raised mainly through increased vascular resistance due to enhanced sympathetic activity (Victor, Leimbach, Seals, Wallin, & Mark, 1987). While the cold pressor test has been used to assess efferent sympathetic outflow, the response involves a reflex arc that includes afferent sensory nerves (pain), spinothalamic tracts, suprapontine, and intrathalamic relays in addition to efferent sympathetic pathways, peripheral sympathetic nerves, and vascular receptors (Quispe & Novak, 2021).

13.3.3 Isometric exercise (sustained handgrip)

Sustained isometric exercise causes increases in blood pressure and heart rate that have been used as an indication of sympathetic and parasympathetic cardiovascular control. Heart rate increases due to parasympathetic withdrawal and cardiac output increases due to muscle vasoconstriction producing a rise in blood pressure. Part of the responses may be related to a central command and an exercise reflex (Borst, Hollander, & Bouman, 1972). To perform the test, a subject uses a handgrip dynamometer (or similar device) which is calibrated to demand 30% of the individual's maximum voluntary contraction. Blood pressure is measured during a >1.5- to 3-minute handgrip (if possible). Patients with autonomic neuropathy have an abnormally low diastolic blood pressure response. A rise in diastolic pressure of 16 mmHg or more is normal, 11–15 mmHg borderline, and less than 10 mmHg evidence of autonomic neuropathy (Ewing, Irving, Kerr, Wildsmith, & Clarke, 1974).

13.3.4 Mental stress test

The integrity of central and efferent sympathetic pathways can be tested by measuring the rise in heart rate and blood pressure in response to the emotional stress that accompanies certain mental tasks (mental stress test) performed under pressure. Stressful mental tasks such as mental arithmetic (i.e., successive subtraction of 7 from 100, maze tests) or tests during which the subject is pressed to make choices result in a reflex increase in heart rate and blood pressure. Heart rate changes are inhibited by ß blockers and decreased in diabetes with autonomic neuropathy. The lack of a blood pressure rise indicates a lesion of efferent vasoconstrictor fibers. Standards for these tests are difficult to establish as there are large intra- and inter-individual variations in responsiveness (a particular mental task may have different stressor effects on different subjects).

13.3.5 Inspiratory gasping

A deep inspiratory gasp produces skin vasoconstriction (which can be measured by laser Doppler flowmetry). There is a high degree of variability of results, such as the influence of vasoconstrictor episodes prior to testing, which limits the clinical applicability of this test (Mueck-Weymann & Rauh, 2002).

13.3.6 Plasma catecholamines and pharmacologic testing

Plasma norepinephrine concentrations reflect the balance from the norepinephrine spillover from postganglionic sympathetic neurons and its clearance. Plasma norepinephrine concentrations change rapidly in response to stimuli that increase the sympathetic outflow (i.e., mental stress, head-up tilt, etc.) providing an indirect estimate of sympathetic nervous system activity. At rest, forearm venous plasma norepinephrine concentrations correlate with muscle sympathetic nerve activity (Wallin et al., 1981) probably reflecting the significant contribution of skeletal muscle to norepinephrine spillover (about 20%). Measurement of ratios of plasma norepinephrine to its metabolites provides an estimate of the activity of sympathetic neurons as norepinephrine released by postganglionic sympathetic terminals is eliminated mainly by reuptake into neurons where it is metabolized to vanillylmandelic acid (VMA) and 3 methoxy 4 hydroxyphenylglycol (MHPG). Changes in plasma norepinephrine in response to baroreflex unloading during active standing or upright tilt can be used as a measure of the capacity of baroreceptor-mediated stimuli to activate efferent sympathetic neurons. Blood should be sampled while the subject is supine and again after at least 10 minutes in the upright position. While norepinephrine normally increases over 50%–100% of supine values during the upright posture, increases are subnormal in patients with autonomic failure. Norepinephrine may increase unexpectedly in some patients with mild autonomic failure who have marked intravascular volume depletion. The diagnosis of dopamine ß hydroxylase deficiency, a rare condition, relies on the detection of low norepinephrine and high dopamine levels (Robertson et al., 1986). Supine plasma norepinephrine concentrations sometimes help to distinguish autonomic failure due to central lesions (i.e., MSA) from that caused by peripheral lesions (i.e., PAF). MSA patients typically have normal supine norepinephrine levels, while PAF patients typically have low norepinephrine levels (Ziegler, Lake, & Kopin, 1977). The normal rise in plasma norepinephrine levels in response to the upright posture is usually absent in both MSA and PAF patients (Polinsky, Kopin, Ebert, & Weise, 1981).

The integrity of sympathetic postganglionic neurons can be tested independently of baroreceptor afferent pathways by measuring changes in plasma concentrations of norepinephrine following injection of the cholinesterase inhibitor edrophonium. Normally, norepinephrine concentrations should rise within the first few minutes following preganglionic stimulation with 10 mg of edrophonium given intravenously. This test has been used to differentiate between central (i.e., preganglionic) and peripheral (i.e., postganglionic) sympathetic neuronal damage (Gemmill, Venables, & Ewing, 1988).

The release of endogenous adrenaline can be assessed during the insulin tolerance test. In normal fasting subjects, the administration of 0.1 U/kg of body weight will cause a 50% reduction of the blood sugar within half an hour and by two hours the blood sugar level is back to normal if the release of adrenaline (and other counterregulatory neurohormones such as glucagon, growth hormone, and cortisol) is normal. A dramatic rise in adrenaline coinciding with insulin-induced hypoglycemia results in a rapid increase in glucose. This initial response is absent in patients with dysfunction of adrenal medullary innervation (Polinsky, 1992).

Some pharmacologic tests can be used to determine if a defect is pre- or postganglionic. For example, low doses of the a1 adrenoceptor agonist phenylephrine produce an increase in blood pressure in patients with autonomic failure who have postganglionic lesions (i.e., diabetic autonomic neuropathy and PAF) and denervation supersensitivity. Denervation supersensitivity refers to the exaggerated responsiveness exhibited by a denervated organ when it is stimulated (Canon's law). The response is thought to result in part from increased quantity and sensitivity of denervated end-organ receptors (upregulation). The increase in blood pressure is slight or absent in patients with preganglionic sympathetic lesions (i.e., MSA) and absent in normal subjects. Measurement of norepinephrine concentrations in plasma following graded boluses of tyramine, which increases the release of norepinephrine from postganglionic sympathetic terminals, may indicate if a lesion is pre- or postganglionic. In the latter case, administration of tyramine is not followed by increased norepinephrine levels (Polinsky et al., 1981). With postganglionic lesions, there is a subnormal increase in norepinephrine concentrations. Furthermore, a lack of a rise in blood pressure with tyramine administration suggests severe postganglionic sympathetic lesions, as the release of even small amounts of norepinephrine should be sufficient to raise blood pressure if denervation supersensitivity is present. In normal subjects, increases in blood pressure triggered by tyramine are proportional to the amount of norepinephrine released and can be plotted against the tyramine doses to provide an estimate of postganglionic norepinephrine stores. In patients with hyperadrenergic states, the blood pressure responses to tyramine are normal due to adrenoceptor downregulation. While these tests are cumbersome and invasive, the presence of denervation supersensitivity in a patient with autonomic failure can be easily determined by testing pupillary responses to the instillation of dilute solutions of pharmacologic agents into the conjunctival sac (see below).

13.3.7 Tests of cardiovagal function

Normally during rest, the heart rate increases during inspiration and decreases during expiration (Fig. 13.2). This variation in heart rate (respiratory sinus arrhythmia) is modulated by afferents from cardiac and aortic baroreceptors and stretch receptors in the lungs and chest wall and depends on the integrity of efferent vagal neurons. Respiratory sinus arrhythmia is abolished after atropinization and cooling of the vagus nerves (Fouad, Tadena-Thome, Bravo, & Tarazi, 1986; Katona & Jih, 1975). Patients with autonomic failure and vagal dysfunction have decreased heart rate variability (Fig. 13.4). While heart rate variability is frequently measured to study efferent vagal pathways, the reflex arcs involved are complex and include vagal afferents, the nucleus ambiguus, and dorsomotor vagal nucleus which, in turn, are subject to other afferent influences. The variations in beat-to-beat heart rate or in heartbeat intervals (electrocardiographic R-wave to R-wave intervals, R-R) are most noticeable during slow heart rates and deep breathing. Faster heart rates, older subjects, cardiac failure, and some intracranial lesions influence beat-to-beat variations. Heart rate variability is frequently studied during rest and provocative maneuvers such as deep breathing, the Valsalva maneuver, active standing, isometric exercise, and a variety of stimuli such as pain, temperature, and administration of various pharmacologic agents. Measures of heart rate variability are affected by the presence of ectopic heartbeats and artifacts, and in many instances, by heart rate (i.e., lower variability with higher heart rates). Accurate assessment of cardiac vagal tone is limited with resting heart rates above 100 beats/minute.

The integrity of vagal cholinergic pathways can be tested independently by measuring changes in plasma concentrations of pancreatic polypeptide (PP) in response to insulin-induced hypoglycemia (Polinsky, 1992). Normally, PP concentrations rise dramatically. This response is absent in patients with PAF and in most patients with MSA.

13.3.8 Resting heart rate variability

During quiet supine rest, the standard deviation of the mean R-R interval (Murray, Ewing, Campbell, Neilson, & Clarke, 1975) and the coefficient of variation are frequently used as a measure of parasympathetic cardiac outflow (Bennett, 1978). Other methods such as calculating the mean circular resultant (Genovely & Pfeifer, 1988), the mean square of the successive differences between R-R intervals (MSSD) (Gundersen & Neubauer, 1977), its square root (rMSSD), and the pNN50 (the amount of times a difference of >50 milliseconds occurs between successive R-R intervals) are used as a measure of vagally mediated beat to beat variation between successive heartbeats (Ewing, Neilson, & Travis, 1984). Analysis of heart rate variability from Holter monitor ECG recordings has been used in research applications. Pharmacological blockade experiments have shown that the resting heart rate depends mainly on parasympathetic influences. A heart rate of about 80–90 beats/minute that does not vary with respiration, exercise, stress, or sleep (i.e., that is fixed) indicates severe cardiac denervation.

13.3.9 Standing

The heart rate changes during the initial 30 seconds after active standing can provide estimates of cardiac parasympathetic control as they are mediated by parasympathetic withdrawal and activation and reflect the changes in pressure and volume receptor afferent traffic. Beat to beat recordings during the immediate response to standing (Fig. 13.2) have demonstrated heart rate peaks at about 3 and 12 seconds (reaching a maximum after about the 15th beat), followed by slowing at about 20 seconds (reaching a minimum at about the 30th beat) (Borst et al., 1972). A reproducible ratio calculated from the maximum and minimum R-R intervals at about the 15th and 30th beats after standing has been used as a measure of vagal function (30:15 index) (Ewing & Clarke, 1982). Normally, the ratio is greater than 1.03 but in diabetic patients with neuropathy, the values are usually less than 1. Calculating the ratio of the maximum and minimum heart rates during the first 30 seconds of standing (HR_{max}/HR_{min}) may be preferable as a test of vagal function as it allows for the considerable normal variation in the timing of the initial heart rate peaks (Wieling, 1992).

13.3.10 Deep breathing

During the deep breathing test, the subject is asked to take 6 breaths/minute (0.1 Hz), a rate found to produce maximum heart rate variations (Fig. 13.2) (Angelone & Coulter, 1964; Hirsch & Bishop, 1981; Wheeler & Watkins, 1973). The differences and the ratios between the maximum and minimum heart rates during deep breathing (inspiration expiration difference and ratio), and the ratio between the longest R-R interval during expiration and the shortest R-R during expiration (expiration/inspiration ratio) provide easily quantifiable indices of respiratory sinus arrhythmia. During deep breathing, a difference between the maximum and minimum heart rate of 15 beats/minute or more is normal; 10 beats/minute or less is abnormal (Table 13.1). As with many provocative tests, the response is related to the subject's effort during the maneuver. Larger tidal volumes during deep breathing result in greater heart rate variability (Angelone & Coulter, 1964; Wheeler & Watkins, 1973). Similar to most autonomic responses, the heart rate variability responses to deep breathing decrease with age (Fig. 13.5) (Ewing et al., 1985; Smith & Smith, 1983).

13.3.11 Diving response (cold face test)

Immersion of the face in cold water stimulates trigeminal nerve receptors on the face and produces reflex apnea, sinus bradycardia, a negative inotropic effect on the ventricles, and vasoconstriction (the diving response). The diving response can be reproduced by performing the cold face test (Heath & Downey, 1990; Khurana, 1995) which provides an indication of trigeminal brainstem cardiovagal pathways independently from vagal and thoracic mechanoreceptor afferents. Either the subject lies face down and is asked to submerge the face into a tub of cold water or a cold compress is applied to the face for 30–90 seconds while avoiding a Valsalva maneuver. While cardiac output falls (due to bradycardia), blood pressure changes little (due to increased vasoconstriction). The pathways involved in the response are complex. The powerful inhibition of the respiratory drive produced by trigeminal stimulation results in apnea with arterial hypoxemia and hypercapnia. Peripheral arterial chemoreceptors in the carotid and aortic bodies are then stimulated and act by producing vagally mediated reflex bradycardia and sympathetically mediated vasoconstriction. Stimulation of other areas of the body (upper airways) can give rise to similar reflex responses. The physical training status of the subjects should be considered when interpreting test results (Chapter 17).

13.3.12 Carotid sinus massage

Carotid sinus massage, by directly stimulating carotid arterial baroreceptors, can be used to test carotid baroreceptor sensitivity and as an index of parasympathetic efferent activity. Carotid sinus massage results in a moderate slowing of the heart rate and minor decreases in blood pressure in normal subjects but can produce prominent bradycardia and vasodilatation in patients with carotid sinus sensitivity (asystole > 3 seconds and a systolic pressure fall > 50 mmHg is considered diagnostic). In the elderly, carotid sinus massage is not recommended as a test of autonomic function as it may produce prolonged asystole and hypotension and because of the risk of embolization from carotid plaques.

13.3.13 Swallow induced tachycardia

The immediate heart rate increase in response to swallowing (swallow-induced tachycardia) appears to be a result of an autonomic reflex that originates in the brain stem and produces tachycardia by overriding cardiac vagal control. For this test, subjects swallow a bolus of 5–10 cc of water while beat to beat heart rate is recorded. In healthy individuals, there is an immediate slowing of the heart rate at the time of swallowing which is followed by an increase within the first 2–5 heartbeats (i.e., swallow-induced tachycardia; Fig. 13.6). The response is blocked by atropine and the increase in heart rate is inversely related to the baseline heart rate and decreases with advancing age, hypoxia, pregnancy, and diabetes (Appenzeller, Qualls, & Martignoni, 1993; Gandevia, McCloskey, & Potter, 1978).

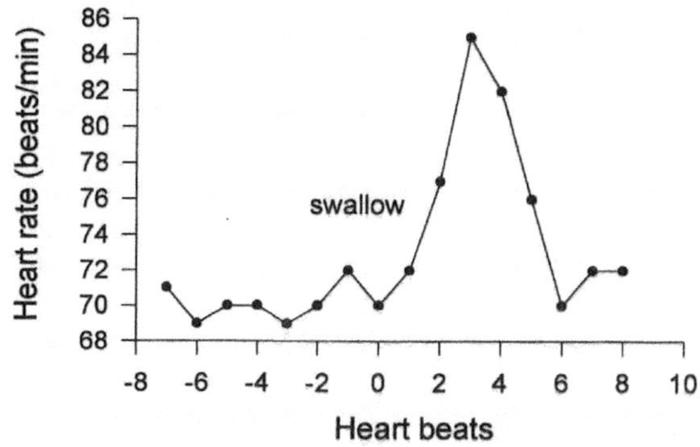

Figure 13.6 Swallow-induced tachycardia in a healthy 28-year-old woman. Swallowing (swallow) a bolus of 3 mL of water on command elicited tachycardia that peaked at the 3rd heartbeat after the swallow. The tachycardia is reproducible and decreases with increasing baseline heart rates and increasing age. It is diminished in diabetic patients with autonomic neuropathy.

13.3.14 Coughing

Changes in intrathoracic pressure produced by a brief voluntary cough result in a prompt (within the first 2–4 seconds) increase in heart rate followed by a gradual return to baseline by 20 seconds. The increase in heart rate is used as a measure of parasympathetic cardiac function as the response is blocked by atropine and unaffected by propranolol (Fig. 13.7). The magnitude of the response decreases with aging and in diabetics with autonomic neuropathy.

13.3.15 Frequency domain analysis of heart rate variability

Analysis of the frequencies (frequency domain analysis) within beat to beat heart rate fluctuations are used to provide a simultaneous measure of sympathetic and parasympathetic influences on the cardiac sinus node. Analysis of heart rate variability in the frequency domain is usually done by applying power spectral analysis techniques to break down heart rate (or R-R interval) time series signals into their constituent frequencies and measure their contributions to overall heart rate variability. Power spectral analysis of heart rate variability has demonstrated the presence of 2 main frequency bands. High-frequency oscillations (0.15 to 0.4 Hz) reflect mainly parasympathetic outflow to the heart as they correlate with respiration, are reduced by atropine, and are abolished by vagotomy. The high-frequency heart rate oscillation peak is most prominent coinciding with

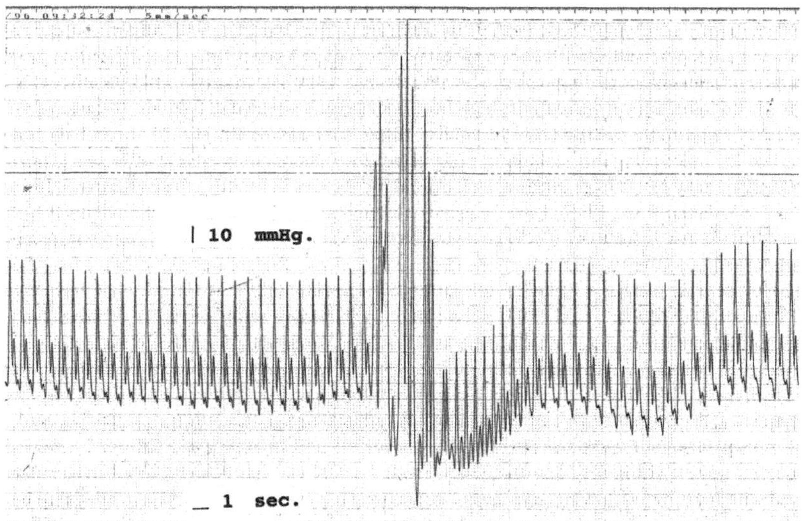

Figure 13.7 Arterial pressure and heart rate changes triggered by voluntary coughing in a normal subject in the seated position. A fall in mean arterial pressure (101–62 mmHg) accompanied by tachycardia (75–100 beats/min.) occurred immediately after coughing. Arterial pressure recovered and systolic pressure overshot the baseline at 7 s by 14 mmHg while heart rate slowed to 60 beats/min. A subsequent systolic overshoot occurred at 21 s. The heart rate response is reproducible and decreases with increasing age. It is reduced in patients with diabetes (Cardone et al., 1990).

the breath rate frequency. For example, at 15 breaths a minute (1/4 of a breath/second) the high-frequency peak is situated at 0.25 Hz. Because of the effect of respiration on heart rate variability, respiratory parameters should be carefully controlled when using this technique. High-frequency fluctuations present after heart transplantation may result from atrial stretch caused by changes in venous return (Bernardi et al., 1995). Low-frequency oscillations (0.03–0.15 Hz) are increased with postural tilt and reduced by ß adrenoceptor blockade and may be generated from baroreceptor sensed blood pressure fluctuations (Bernardi et al., 1994). Many frequency domain measures of heart rate variability correlate well with time-domain measures (Freeman et al., 1991). High-frequency spectral components are decreased suggesting the presence of a diminished parasympathetic outflow to the heart in various conditions such as aging, autonomic neuropathies, primary autonomic failure, ischemic heart disease, hypertension, and fetal distress. Spectral analysis has also shown a total (high and low frequency) decrease in heart rate variability with advancing age, heart failure, following cardiac transplantation (where the heart is denervated), and in autonomic failure. Decreased spectral power within very low-frequency bands (<0.04 Hz) has been reported to correlate well with mortality following myocardial infarction (Bigger et al., 1992). The relationship between oscillations that are common to heart rate, respiration, blood pressure, and blood flow (and other variables) can be quantified by spectral analysis cross-correlation techniques. Beat-to-beat oscillations of the heart rate that are synchronous with respiration and blood pressure waves, are considered to be valid markers of sympathetic vagal interactions in man and animals (Pagani et al., 1986).

13.3.16 Central autonomic pathway testing

13.3.16.1 Neuroendocrine function

Baroreflex-mediated vasopressin release may distinguish between patients with MSA and a central autonomic pathway lesion from those with PAF with a peripheral autonomic pathway lesion. In peripheral lesions, the release of vasopressin is normal (Kaufmann et al., 1992; Puritz, Lightman, Wilcox, Forsling, & Bannister, 1983). In MSA, there is also a subnormal release of vasopressin in response to metoclopramide (Norbiato et al., 1992). Similarly, clonidine stimulated growth hormone release (Thomaides et al., 1992) and hypoglycemia-induced increases in ACTH/β endorphin (Polinsky, 1992) are absent in patients with MSA but not PAF.

13.4 Chemoreceptor testing (hypoxic ventilatory drive)

Hypoxic ventilatory drive can be tested to assess the functional integrity of chemosensitivity to increasing levels of hypoxia. For this, subjects are tested in the supine and seated positions separately while breathing into a spirometer in which the oxygen concentration is lowered over time. Since ventilation reflexively increases in relation to the severity of hypoxia, the ventilatory responses measured during the test can be plotted against the oxygen concentration in the inspired gas.

13.5 Sudomotor testing

Sympathetic innervation of eccrine sweat glands can be evaluated through sudomotor testing (Table 13.3). Central and peripheral sympathetic pathways to sweat glands can be assessed by studying the sweat response to raising the body temperature by heating and to iontophoresis or

Table 13.3 Tests of autonomic function.

Reflex pathway tested		Autonomic test	Abnormal response
Sympathetic	Vasomotor	Standing or head-up tilt Valsalva maneuver Isometric exercise/hand grip, mental stress, cold pressor	Orthostatic hypotension Lack of late Phases II and IV with prolonged blood pressure recovery time Lack of ↑BP
	Cardiomotor	Standing or head-up tilt Valsalva maneuver Isometric exercise/hand grip, mental stress, cold pressor	Lack of ↑HR (after 2 min) Lack of ↑HR during Phase IV Lack of ↑HR
	Sudomotor	Thermal sweating	Lack of sweating
	Postganglionic sympathetic	Pupillary response to dilute pheny lephrine Norepinephrine supine and upright at 10 min Sweating response to acetylcholine/ QSART	Dilatation (indicates denervation supersensitivity) Lack of rise in norepinephrine Sweating absent/ reduced
Parasympathetic	Cardiovagal	Rest Standing Deep breathing Valsalva maneuver Carotid sinus massage Atropine	Lack of HR variation (sinus arrhythmia) Lack of ↑HR, subnormal HR_{max}/HR_{min} or 15:30 ratio Lack of HR variation, subnormal E/I or $HR_{max}—HR_{min}$

(*Continued*)

Table 13.3 (Continued)

Reflex pathway tested		Autonomic test	Abnormal response
			Lack of ↑HR despite presence of BP overshoot during Phase IV of maneuver Lack of ↑HR Lack of ↑HR
	Vagus	Pancreatic polypeptide in response to hypoglycemia	Lack of ↑ pancreatic polypeptide
Sensory afferents	Nociceptive C-fibers	Skin flare to histamine or capsaicin	Flare absent
Central autonomic pathways		AVP supine and upright	Lack of upright ↑ AVP despite upright BP fall
		ACTH/ β-endorphin response to hypoglycemia	Lack of ↑ ACTH/ β-endorphin

AVP, arginine vasopressin; BP, blood pressure; E/I, expiration/inspiration ratio; HR, heart rate; QSART, quantitative sudomotor axon reflex test.

intradermal injections of cholinergic drugs. Sudomotor testing is very useful for detecting evidence of distal small fiber neuropathies.

Cholinergic agents (i.e., acetylcholine or pilocarpine) delivered by iontophoresis or injection can be used to trigger sweat production by stimulating postganglionic fibers and sweat glands directly (Kennedy & Navarro, 1989; Macmillan & Spalding, 1969). A preganglionic lesion is suspected as a cause of reduced sweating in the presence of a normal sweat response to pilocarpine given by iontophoresis or intradermally.

The thermoregulatory sweat test (TST) assesses the function of the entire thermoregulatory pathway: somatic and sympathetic afferents, hypothalamic structures, and the pre- and postganglionic sympathetic pathways. Indicator substances that change in color when in contact with sweat (i.e., iodine with starch; 25% alizarin red and 25% calcium carbonate in corn starch) are spread on the body to reveal the area of the body surface which sweats in response to raising the central body temperature by 1°C or to a core temperature of 38°C. To raise body temperature, subjects are placed in a heating chamber (usually a cabinet with well-controlled humidity and heating elements). The distribution and percentage of body surface that sweats are recorded as assessed over the anterior body (Fealey, Low, & Thomas, 1989). The TST is sensitive but has a low specificity. The advantage of the TST relies on its ability to simultaneously display sweat patterns over all the anterior body surface. Typical patterns of sweat loss (Fealey, 1996) help support a diagnosis of peripheral neuropathy (distal sweat loss), thoracoabdominal neuropathy (loss of sweat over the anterior trunk and distal limbs), truncal radiculopathy (asymmetric sweat loss over the anterior distribution of thoracic dermatomes), lepromatous neuropathy (multifocal or patchy sweat loss), and autonomic failure (usually a global loss of sweating) (Fig. 13.8).

Imprint methods that test postganglionic sweat gland function rely on studying sweating induced by iontophoresis of acetylcholinesterase inhibitors. With the imprint methods, sweat droplets imprinted into silastic or other matrices applied to the skin are counted to reveal sweat gland density, sweat volume, and droplet size (Kennedy & Navarro, 1989; Kennedy, Sakuta, Sutherland, & Goetz, 1984).

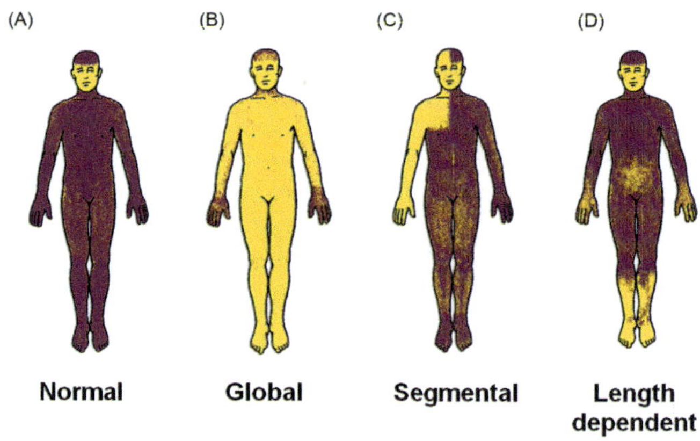

Figure 13.8 Thermoregulatory sweat test patterns are shown. Normal sweating recruitment and distribution is shown (A) with yellow depicting the areas that do not sweat and purples as areas that sweat (the face is not powdered). Global anhidrosis is shown in (B) with some areas of preserved sweating over the hands and forehead, such as can be seen in multiple system atrophy. Segmental anhidrosis is shown in (C) which may be due to disruption of the right sympathetic chain that innervates the upper limb and face. A neuropathic length-dependent pattern is shown (D) with hypo/anhidrosis in hands, medial abdomen, and lower legs.

The quantitative sudomotor axon reflex test (QSART) assesses the integrity of an axon reflex pathway involving postganglionic sympathetic sudomotor axons. A sudorometer linked to a computer quantifies the amount and the latency of sweat induced by iontophoresis of acetylcholine. The pattern of distribution of abnormal responses can be helpful as it may follow that of different neuropathies (i.e., distal in the case of a diabetic polyneuropathy). The test has been standardized and it is sensitive, specific, and reproducible. The site of a sudomotor pathway lesion can be inferred by combining QSART and TST. For example, QSART and TST responses are absent with postganglionic sympathetic sudomotor lesions, whereas with preganglionic lesions QSART responses are maintained.

The sympathetic skin response is a change in skin potentials produced in response to a variety of stimuli such as electric shocks, sudden loud noise, and inspiratory gasps. The response results from synchronized activation of sweat glands and sweat release (additional contributions of the skin itself to the response have been proposed). In general, a square wave pulse of 200 milliseconds duration delivered to the skin of either the hand or foot is used as a stimulus, and the response consists of a biphasic potential with an initial negative peak (Shahani, Halperin, Boulu, & Cohen, 1984). The response habituates on repeated stimulation. Because sweat glands are an integral part of autonomic responsiveness to various stresses, the sympathetic skin response is thought to give a good measure of postganglionic sudomotor fiber activity. The sympathetic responses have also been correlated with microneurographic techniques for recording from single and small numbers of myelinated and unmyelinated fibers in conscious humans (Satchell & Seers, 1987). When the contralateral nerve trunk was stimulated or sweat glands were activated by tone bursts randomly delivered, it was found that the evoked sympathetic skin response to both stimuli was identical in latency and duration. With this method, an average evoked response was about 1500 milliseconds and an average mean latency was about 170 milliseconds. The drawback of this test is the great variability in the amplitude of the response and reproducibility. Due to its variability, the sympathetic skin response is of value only when absent.

13.6 The flare component of the triple response of Lewis

The integrity of sensory afferents can be tested by checking for the presence of a skin axon reflex vasodilatation (or skin flare) response to the injection, iontophoresis, or application of substances to the skin that trigger the release of vasoactive agents from afferent C fiber terminals and of histamine from mast cells (Anand, 1992). The reflex arc includes afferent sensory neurons and their collaterals and their terminal branches from where the vasoactive substances are released antidromically. For example, intradermal injection of histamine (0.1 mL of a 1:1000 solution) produces a response that includes local cutaneous vasodilatation (a flare) surrounding the injection site. The presence of the skin flare is used to indicate the integrity of dorsal root ganglia and their peripheral axons which mediate the response through the release of vasoactive substances. Skin flare responses may be absent with peripheral small myelinated sensory fiber dysfunction and plexus lesions distal to dorsal root ganglia. Similar flare responses are elicited by the cutaneous application of capsaicin (red chili pepper extract). A drop of 1% capsaicin in olive oil is applied to the skin and the increase in skin temperature (or the size of the flare or changes in blood flow measured by LDF) is recorded to provide data on the degree and latency of the vasodilatation produced by the release of neuropeptides (such as substance P and calcitonin gene-related peptide, CGRP) or ATP from terminal branches of sensory afferents.

13.7 Testing pupillomotor function

The autonomic innervation to the pupil can be assessed by pupillomotor testing where the pupils' responses to illumination or to drugs are measured. The pupillary diameter at rest in darkness (measured as the % of the iris diameter taken by the dilated pupil) provides a good indication of pupillary function as is highly reproducible. The diameter depends on the balance between parasympathetic inhibition and sympathetic activation. The pupil dilatates due to relaxation of the iris sphincter that results from the inhibition of preganglionic parasympathetic neurons of the Edinger Westphal nucleus, and from activation of the sympathetic efferents to the dilator pupillae. Conversely, the pupil constricts due to stimulation of parasympathetic and inhibition of sympathetic efferents (see Chapter 13).

The pupil cycle time (PCT) is a simple method of testing pupillary autonomic function that can be done at bedside. A narrow beam of light is focused on the pupillary margin. As the pupil constricts in response to light falling on the retina, the light is interrupted, and the pupil again dilates. A stopwatch is used to record the time taken to complete 100 cycles and the PCT is calculated as the average length of a cycle. PCT above 1150 milliseconds are considered prolonged. More sophisticated analysis employing infrared photography or television can be done in an autonomic laboratory. The amplitude of change in a pupils' diameter produced by illumination can be used as a measure of parasympathetic responses. Recording a pupil's latency to reach 3/4 of maximum dilatation provides a good measure of sympathetic activity.

Conjunctival instillation of drugs can be used to investigate whether sympathetic and parasympathetic pupillary innervation dysfunction are present and can also help determine if a lesion is pre- or postganglionic. Instillation of a 2%–10% cocaine solution will dilate a constricted pupil if sympathetic innervation is intact (as cocaine interferes with the elimination of norepinephrine by reuptake). Instillation of phenylephrine (1%–2% solution) will dilate a constricted pupil if denervation supersensitivity is present due to a defect in sympathetic innervation. A constricted pupil will not dilate with instillation of drugs such as hydroxyamphetamine (0.5%–1% solution) that release norepinephrine but do not act directly on norepinephrine receptors if the defect is postganglionic. A pupil that is dilated due to a parasympathetic lesion (where denervation supersensitivity is present) will constrict with dilute pilocarpine solutions (0.125%) whereas constriction will be absent if the pupil is damaged or under the effects of anticholinergic agents.

13.8 Imaging techniques and autonomic disorders

Imaging techniques such as positron emission tomography (PET) scanning can be used in the diagnosis and assessment of the clinical course and treatment of autonomic disorders. Cranial PET scanning using 6 [18 F] fluorodopa or [11 C] nomifensine has shown decreased radioactivity in the striatum consistent with decreased dopaminergic neuronal function in patients with MSA. The integrity of sympathetic cardiac innervation can be assessed using the guanethidine analog [123I] metaiodobenzylguanidine (MIBG) as a tracer. The tracer's uptake and washout dynamics have been used to provide an indirect measure of myocardial noradrenergic activity and innervation. Cardiac MIBG PET has shown decreased tracer activity in patients with PAF and other peripheral autonomic lesions (Goldstein, Holmes, Cannon, Eisenhofer, & Kopin, 1997). PET scans using tracers that visualize sympathetic nerve terminals (such as 6 [18 F] fluorodopa) offer superior imaging resolution (Chapter 19). Use of 18F-dopamine PET neuroimaging along with 11C-methylreboxetine, a selective ligand for the cell membrane norepinephrine transporter, has been used to assess intraneuronal vesicular storage. The findings showed that PAF entailed moderately decreased cardiac sympathetic innervation and a substantial vesicular storage defect in residual nerves (Goldstein, Isonaka, Holmes, Ding, & Sharabi, 2020). The results of this study suggest that the majority of residual cardiac postganglionic sympathetic nerves are dysfunctional in PAF, in other words, "sick-but-not-dead" (Goldstein et al., 2020; Goldstein, 2020).

13.9 Testing genitourinary function

13.9.1 Bladder dysfunction

The urological evaluation of patients with suspected autonomic bladder dysfunction should include measurement of the urinary postvoid residual volume. Increased volumes (measured by catheterization or by ultrasound), in the absence of bladder outlet obstruction, indicate bladder motor dysfunction. Significant bladder outflow obstruction can be screened for by measuring urinary flow rates (uroflowmetry) and measuring postvoid bladder residual volume by ultrasonography.

Urodynamic tests such as cystometry investigate the pressure-volume responses of the bladder. For cystometry, the bladder is catheterized and the intravesical and intraabdominal (rectal) pressures are recorded at different bladder volumes while it is being filled with water and during micturition attempts. Cystometry can measure bladder filling and emptying volumes, pressures, and times, giving an indication of detrusor function. Lack of bladder contraction as it fills (areflexic bladder) can be due to lack of voluntary release of detrusor inhibition (the cause may be psychogenic), and due to afferent sensory or efferent motor pathway dysfunction. Administration of the muscarinic agonist bethanechol during cystometry can help to investigate the integrity of efferent motor pathways. The presence of motor lesions is suggested by an exaggerated rise in bladder pressure (i.e., there is evidence of denervation supersensitivity) which is absent with afferent sensory lesions and psychogenic causes.

Bladder sensation can be measured by a variety of tests (Medina Lucena & Tincello, 2019). Proprioception can be assessed while the bladder is being filled by recording when the patient first senses a desire to void and when bladder distention becomes painful. Temperature sense thresholds can be determined by instilling water at different temperatures. The electrical (pain) perception threshold can be measured by delivering graded electric shocks at the level of the vesicoureteral junction. When raised, it suggests afferent sensory pathway dysfunction. Electrical stimulation at the urethra or vesicoureteral junction, while anal sphincter contractions are recorded, can provide an indication of the integrity of a reflex pathway that includes afferent sensory autonomic fibers within the pelvic nerves and efferent somatic motor fibers within the pudendal nerves. If the evoked response is abnormal, testing the bulbocavernosus reflex (see below) can provide an indication if the efferent motor pathway is intact. A normal bulbocavernosus latency pinpoints the lesion to the sensory limb of the reflex arc. Striated periurethral and anal sphincter EMG can be used to demonstrate denervation/reinnervation changes which may be helpful when investigating the causes of bladder dysfunction (methods to study detrusor smooth muscle function are lacking). Sphincter EMG of the periurethral striated muscle (or anal sphincter activity which correlates closely with the external urinary sphincter activity) has been used to assist in differentiating MSA and Parkinson's disease with bladder involvement from other causes of reinnervation changes (Pramstaller et al., 1995). While reinnervation changes are marked with MSA, they are less prominent in multiparous women and after pelvic surgery.

13.9.2 Sexual dysfunction

Multiple factors can result in sexual dysfunction. Investigation of the causes of male erectile dysfunction requires a careful history and physical examination to assess the relative contribution of psychologic, endocrine, and metabolic, neurologic, and vascular elements. Once endocrine and metabolic factors are excluded (usual screening tests include blood levels of testosterone, prolactin, luteinizing, follicle-stimulating hormone, and thyroid function tests), psychogenic, neurogenic, or vascular causes of male erectile dysfunction should be investigated. Diagnostic tests should be aimed at establishing the degree of penile erectile dysfunction and determining possible causes (Nehra et al., 2012). As autonomic innervation of the penis cannot be studied directly, other causes of erectile dysfunction should be excluded before attributing erectile impotence to autonomic dysfunction.

Nocturnal penile tumescence (NPT) tests are used to measure erectile function. Postage stamps and other materials are placed around a flaccid penis at bedtime to screen for the presence of nocturnal erections. If found broken in the morning, erections may have occurred. These tests include the Poten (or postage stamp) and snap gage tests. While NPT tests assess changes in penile circumference, they cannot measure rigidity or the number of erections. More sophisticated tests include the use of strain gages that measure changes in penile size. This can be complemented by direct visualization of erections and EEG monitoring to correlate with REM sleep. The use of a rigidometer device allows measurement of both penile circumference and rigidity outside a sleep laboratory setting. The presence of normal erections during sleep helps differentiate psychogenic from organic causes of impotence.

The presence of an erection in a patient with erectile dysfunction following the intracorporeal injection of papaverine discriminates between neurogenic or psychogenic and vascular causes (an erection is absent with vascular dysfunction). Similar responses can be obtained with intracavernosal injection of prostaglandin E1 and a combination of papaverine and phentolamine. Self-stimulation may increase the sensitivity of the test. The presence of an erection after intracavernosal injection of vasoactive agents has been reported to be 90% effective in ruling out venous insufficiency as a cause of erectile impotence. Vascular dysfunction is likely if there is no erectile response after up to five intracavernosal injections. In cases where vascular insufficiency is suspected, the blood supply to the penis can be investigated by duplex ultrasonography.

While measurements of somatic nerve function may be helpful in defining some lesions that can account for impotence, the usefulness of the tests has not been established. Pudendal nerve function is most relevant to penile sensation and ejaculation. Bulbocavernous reflex responses can be measured to assess pudendal nerve conduction (Ertekin & Reel, 1976). The dorsal nerve of the penis is stimulated, and the response is recorded at the rectal sphincter or over the perineum. Penile nerve conduction velocity may be measured by stimulating the distal and proximal segments of the penile shaft. Somatosensory evoked potential (SSEP) responses can be used to evaluate sensory pathways. When used in combination, bulbocavernous reflex and pudendal SSEP testing with cystometry and NPT, can distinguish between upper and lower motor neuron lesions. When used alone, pudendal SSEP has less value as it may be normal with unilateral pudendal lesions. The vibratory perception threshold of the penis appears to correlate well with electrophysiological testing and once validated may have promise as a helpful screening test.

Investigation of female sexual dysfunction involves history and physical examination and may include sexuality questionnaires (Faubion & Rullo, 2015; Kingsberg & Althof, 2009). Female sexual dysfunction due to autonomic deficits has not been well defined. Nonstandardized tests such as vaginal plethysmography may be performed to measure lubrication and vaginal flushing. SSEP recorded from the rectal sphincter or over the perineum while stimulating the clitoris may offer an indication of the integrity of sensory pathways (Krakowsky & Grober, 2018).

13.10 Testing gastrointestinal function

Plain and barium contrasted X-rays may be helpful if they reveal small bowel dilatation suggestive of chronic intestinal pseudo-obstruction (and by excluding mechanical obstruction). Gastric emptying and intestinal transit can be assessed by measuring the transit of radiopaque pellets and meals containing radioactive tracers. Overall transit time (gastric, small bowel, and colonic) can be measured by counting the number of radiopaque pellets found on an abdominal flat plate taken at day 5 after ingestion of 20 pellets (Hinton, Lennard-Jones, & Young, 1969). Normally 80% should have been eliminated by then. More specific information regarding transit through different parts of the gastrointestinal tract can be obtained by radioscintigraphy. Here, a delayed-release capsule containing radioactive particles is ingested and the passage of radioactivity is followed by scintigraphy to determine transit times (Camilleri & Zinsmeister, 1992). While transit studies are helpful as screening tools, they do not provide detailed information regarding motility. Measuring intraluminal pressures at different sites of the gastrointestinal tract by manometry can be helpful to differentiate neuropathic from myopathic causes of constipation and pseudo-obstruction. Neuropathic lesions produce normal amplitude but dysrhythmic (uncoordinated) phasic pressure peaks, whereas myopathic lesions (as with amyloidotic infiltration) result in low amplitude phasic pressure peaks.

Constipation can result from defecation disturbances. A history of the need to perform digital extraction of stool from the rectum, to apply pressure on the posterior wall of the vagina, or to support the perineum during straining suggests the presence of defecation disturbances. Involvement of the pelvic floor is suggested when constipation is associated with the inability to evacuate the rectum, feelings of persistent rectal fullness, rectal pain, pelvic floor descent, and straining (Pemberton, 1990). Once the determination that a disturbance of defecation exists is reached, patients should undergo a psychological evaluation in search of psychiatric causes of defecation disturbances such as need to control, depression, and eating disorders (Fleshman, Dreznik, Cohen, Fry, & Kodner, 1992). Further testing should be done if evaluations are unrevealing and the symptoms are persistent (Camilleri, Thompson, Fleshman, & Pemberton, 1994).

Motion of the perineal muscles during descent can be assessed by asking the subject to strain to defecate. Observation with the subject in a lateral decubitus position can reveal ballooning of the pelvic floor or lack of motion of the perineal muscles when pelvic floor dysfunction is present. A normally functioning puborectalis muscle moves posteriorly during the strain to expel a finger placed in the rectum during rectal examination. Studying the expulsion of a Foley catheter placed in the rectum and inflated to 50 mL is a simple screening test that enables the investigation of the defecatory process and the pelvic floor function (the balloon expulsion test) (Fleshman et al., 1992). Inability to expel the inflated balloon suggests pelvic floor dysfunction, anatomical abnormalities of the rectum or anismus. Rectal sensation can be tested by inflating a balloon in the rectum to determine the minimal volume needed to produce a sensation of filling or the maximum volume tolerated before the urge to defecate is overwhelming. While defecation proctography frequently detects abnormalities that can explain a patient's defecation disturbance, the specificity of the test is low and requires an experienced radiologist for interpretation. In some situations, anorectal manometry is helpful in supporting the diagnosis of anismus (where there is a high resting and squeeze anal sphincter tone). External anal sphincter or puborectalis electromyography and pudendal nerve motor latency are more useful to study rectal incontinence rather than defecation dysfunction (Camilleri et al., 1994).

13.11 The autonomic laboratory

An autonomic laboratory can be assembled using standard medical equipment. The following is intended as a guideline on how to set up a basic laboratory to perform some of the most used standard tests. These include tests of cardiovagal, adrenergic vasomotor, and sudomotor function. Additional investigations require more sophisticated equipment.

13.11.1 Testing environment and equipment

A small autonomic laboratory requires a surface of approximately 200 square feet to allow for comfortable testing conditions. The ambient temperature should be kept between 21°C and 23°C, and the premises should be situated in a quiet area of the building or be soundproofed.

An electrocardiographic (ECG) monitor with a chart recorder that can generate a rhythm strip is used to calculate the heart rate or measure ECG R-R intervals manually during the different testing procedures. More sophisticated equipment options include ECG monitors linked to electronic chart recorders with software applications that permit on or off-line analysis of the data.

A sphygmomanometer is required to measure blood pressure. Automatic intermittent blood pressure monitors/recorders can be helpful by freeing the operator for other activities during testing. Continuous beat-to-beat blood pressure monitor/recorders (using finger plethysmography or the arterial tonometry principles) are required to determine rapid blood pressure responses. As the sensors used by the beat-to-beat blood pressure devices are quite sensitive to limb movements and changes in limb position, the limb to which they are attached should be immobilized either with a sling or an arm board and kept at the level of the heart during testing. The signals and waveforms generated by monitors can be digitized, acquired, and analyzed with the aid of a computer.

A tilt table (manual or electric) with foot support and chest, hip, and knee straps is used for prolonged tilt testing and to evaluate the early cardiovascular responses to the head-up position. The knee straps and foot support are necessary to minimize muscle contractions and thus favor dependent blood and fluid pooling which increases orthostatic stress.

A manometer should be used to measure the expiration pressure generated during the strain phase of the Valsalva maneuver (40 mmHg). A sphygmomanometer can be easily modified for this use by attaching a mouthpiece with a small air leak. Similarly, an inflated sphygmomanometer cuff can be used to measure the force exerted during handgrip when testing the responses to sustained handgrip (isometric exercise).

To test sweating, electric lamps (i.e., infrared) can be used to heat the trunk. More sophisticated sweat testing requires a heating chamber, thermometers, and devices to measure sweat production (i.e., sudorometers, QSART). An iontophoresis device can be useful to deliver cholinergic agents painlessly when testing postganglionic sweat fibers and glands directly.

13.11.2 Testing conditions

Most laboratories advise studying subjects in the mornings, after an 8-hour overnight fast as this may increase the yield of certain tests (i.e., the sensitivity of the prolonged head-up tilt test used to investigate the cause of unexplained syncope appears to be higher when testing is done after an overnight fast). The testing environment should be quiet and comfortable, as an environment that is a source of stress may alter results. Subjects should also be given time to practice the procedure prior to testing to assure optimal performance and to reduce the stresses associated with new tasks performed under pressure.

Subjects should be asked to avoid activities and medications that have the potential to modify autonomic responses confounding the interpretation of abnormal test results. For example, instructions should be given to refrain from vigorous exercise for 24–36 hours, and from taking drugs with significant autonomic effects (i.e., drugs with anticholinergic effects, antihistamines, fludrocortisone, diuretics, sympathomimetic and parasympathomimetic medications, cough and cold medications, and aspirin) for 18 hours. Meals, caffeine, tobacco products, or alcohol should be avoided for at least 8 hours. Compressive clothing or Jobst stockings should be removed prior to testing. Testing should be deferred in subjects with an acute illness <48 hours prior and in diabetic patients with recent (<12 hours) hypoglycemic episodes. Also, in patients with diabetes, blood glucose should be stable, and the last dose of insulin should have been given >8 hours prior to testing. In general, if the results of testing are normal despite the presence of confounding factors, there is no need for retesting. On the other hand, if testing results are abnormal, the testing should be repeated in the absence of confounding factors.

13.12 Autonomic tests

13.12.1 Standing and upright tilt

To study the cardiovascular response to standing, subjects are asked to stand up abruptly after 10 minutes of quiet, supine rest while heart rate (and blood pressure) are recorded. Blood pressure should be recorded several times during the first five minutes of standing (usually blood pressure and heart rate values at 3 minutes are sufficient to detect early orthostatic hypotension and determine if the heart rate response is normal). Falls in upright blood pressure suggest sympathetic vasomotor failure (or sympathetic vasomotor tone withdrawal in instances of neurally mediated hypotension). The ratio calculated from the maximum and minimum R-R intervals at about the 15th and 30th beats after standing (the 30:15 index) and the ratio of the maximum and minimum heart rates during the first 30 seconds of standing (HRmax/HRmin) are used as a measure of cardiovagal function (Fig. 13.1).

For passive upright tilt testing, subjects are secured into position onto the tilt table with chest and knee straps in place and are tilted to a 60- to 70-degree head-up position. A period of supine rest (10–20 minutes) and stable blood pressure and heart rate are necessary prior to upright tilt. Prolonged head-up tilt lasting 20–60 minutes is used as a diagnostic test to investigate the cause of unexplained syncope.

Supine and upright norepinephrine and vasopressin concentrations can be measured in venous blood sampled from indwelling catheters. Baseline supine blood samples should be drawn during quiet rest and at least 20 minutes after cannulation to avoid elevations due to venipuncture that could be a confounding source. Norepinephrine should be measured after at least 10 minutes in the upright position and vasopressin after 20 minutes.

13.12.2 Valsalva maneuver

Subjects, connected to an ECG device (and if available, a beat-to-beat blood pressure device) lay flat resting for 5 minutes and are then asked to take a deep breath and immediately blow into a mouthpiece (with a small leak) connected to a manometer and maintain a pressure of 40 mmHg for 15–20 seconds (duration dependent on subjects' clinical status). Recordings are continued for at least a minute after releasing the strain. The normal response is shown in Fig. 13.1 and compared with an abnormal response in Fig. 13.4. The Valsalva ratio is calculated as the maximum R-R interval during Phase IV divided by the minimum R-R during Phase II of the maneuver.

13.12.3 Deep breathing

For this test, subjects are instructed to take a series of deep breaths at a rate of 6/minute (5 seconds of inspiration followed by 5 seconds of expiration) while the heart rate is recorded continuously (Figs. 13.1 and 13.4). While a metronome is desirable, a combination of visual and verbal cues can be sufficient to help a subject pace deep breaths. The test can be performed either in a seated or supine posture. Although the response has been reported to be enhanced by testing in the supine position, this test can be performed in the seated position, for instance for patients who suffer from illnesses that result in significant ventilatory restriction when flat (i.e., patients with parkinsonism). The maximum and minimum heart rates during deep breathing and the expiration/inspiration ratio (the maximum R-R during expiration/minimum R-R during inspiration) are calculated as a measure of cardiovagal function.

13.12.4 Sweating

During thermoregulatory sweat testing, the patterns of sweating are recorded after oral or rectal temperature has risen by 1°C by heating the body under infrared lights in a special cabinet or chamber for 60–90 minutes. Sweating induced by heating involves complex thermoregulatory pathways that include temperature afferents, hypothalamic centers, spinal tracts, sympathetic efferents, and sweat glands. Iontophoresis or dermal injections of cholinergic agents (i.e., pilocarpine, acetylcholine, or methacholine) can be used to test postganglionic sweat fibers and sweat glands.

Sweating can also be assessed by heating the anterior trunk with a lamp for 1–2 minutes, a procedure that normally triggers sweating. Indicators that change in color in the presence of sweat can be spread on the body to map sweat patterns. Iodine starch combinations (i.e., 2 g of iodine in 10 mL of castor oil diluted with alcohol to 100 mL painted on the body followed by a thin coat of starch which turn dark purple in contact with sweat) and powders (i.e., 25% alizarin red and 25% calcium carbonate in corn starch) are used frequently. A simple and useful one-step method for visualizing sweating that requires only soluble starch previously treated with iodine has been developed (Sato, Richardson, Timm, & Sato, 1988). For this, 0.5–1 g of iodine crystals are added to 500 g of soluble starch and kept in an amber bottle at room temperature for several days until the starch turns light yellow/brown. When the powder is spread on the skin, sweat droplets are visualized as discrete dark purple dots. As they are easily wiped off without staining the skin, the same skin site can be restudied consecutively.

13.12.5 Interpretation of results, reference values, and test reproducibility

Determination if an autonomic test result is normal requires comparison with test results obtained from a healthy population. As numerous factors can affect autonomic responses, each laboratory should standardize test procedures and testing conditions. Because age is an important factor, a range of normal values should be determined for different ages. The technical aspects of each test procedure should be refined to achieve high reproducibility (the coefficient of variation should be low, preferably <15%–20%).

Abnormal test results should be interpreted in the context of a patients' symptoms. Additional confirmatory tests, part of a battery of tests, are usually required for diagnostic purposes. Rather than relying on a single abnormal test result, several abnormal results are necessary to reach a diagnosis of autonomic failure. As mentioned elsewhere, abnormal test results do not always imply the presence of autonomic symptoms or autonomic failure.

Performing selected tests can help infer the site of a lesion in an autonomic reflex arc. For example, efferent pathway integrity can be assessed by combining tests that share the same efferent pathways but have different afferent or central pathways. A patient with orthostatic hypotension and an abnormal response to the Valsalva maneuver (indicating a sympathetic vasomotor defect) should undergo handgrip, mental stress, cold pressor tests, or sweat tests to confirm the presence of an efferent lesion (in this situation the responses are abnormal). In this case, normal responses suggest an afferent lesion.

13.12.6 Safety of autonomic reflex testing

Autonomic reflex testing procedures are usually safe. During cardiovascular autonomic testing, the ECG should be continuously monitored, and resuscitation measures should be available as there is a minimal risk of arrhythmias, and some tests can induce marked hypotension. An extensive review of ECG tracings of 925 consecutive subjects taken during a battery of cardiovascular autonomic reflex tests (Valsalva maneuver, deep breathing, orthostatic challenge, and isometric handgrip) showed an increase in the frequency of ventricular extrasystoles during or after the tests in 11% of healthy subjects, in 11% of diabetic subjects and in 23% of subjects with a previous myocardial infarction ($P = .001$ vs healthy subjects). In patients with previous myocardial infarction, the most dysrhythmogenic individual tests were orthostatic and isometric handgrip. While 9 subjects had other cardiac rhythm disturbances (including nonsustained ventricular tachycardia, conduction block, and atrial fibrillation), in all cases, the dysrhythmias were asymptomatic and resolved without medical intervention (Piha & Voipio-Pulkki, 1993).

Subjects should be observed closely during testing as the head-up tilt and the Valsalva maneuver may precipitate abrupt hypotension and syncope (sometimes with motor phenomena that can be confused with epileptic seizures). Subjects should be strapped to the tilt table for the duration of head-up tilt testing to prevent falls. Once the diagnostic criteria sought for have been met, subjects should promptly be returned to the supine posture to minimize hypotension and the risk of severe bradycardia and sinus arrest, and other arrhythmias that may occur during syncope. In the event of symptomatic hypotension, subjects should be returned promptly to the horizontal position and their lower limbs raised to increase venous return to prevent syncope. While life-threatening arrhythmias (including some requiring cardiopulmonary resuscitation) have been reported during head-up tilt, they are rare and usually occur during the ictal phase of tilt-induced syncope. In a series of 303 unmedicated head-up tilt tests, there were no complications (Oribe et al., 1997). The use of isoproterenol and other drugs, however, appear to be riskier than the unmedicated tilt.

The diving response should not be performed in patients with autonomic failure. Carotid sinus massage and ocular pressure are dangerous and should not be performed in the elderly. When testing sweating, care should be given to prevent excessive heating and burns, and humidity within heating cabinets should be carefully controlled.

13.12.7 Lie detector testing

This test uses a polygraph to record physiological measures of autonomic function such as blood pressure, pulse rate, respiration rate, and skin conductivity during an interview. During the test, the subject is asked several questions. The belief is that responses to untruthful answers can be differentiated from those that are true using the recording produced by the machine. However, there is no physiological reaction that is characteristic of lying. Moreover, many testers have a scoring system specific to their detectors, and computerized results can also be made available in legal challenges to testimony. Additionally, results are subject to inaccuracies in patients with autonomic disorders and may be confounded by multiple medications (Cook & Mitschow, 2019).

References

Achem-Karam, S. R., Funakoshi, A., Vinik, A. I., & Owyang, C. (1985). Plasma motilin concentration and interdigestive migrating motor complex in diabetic gastroparesis: Effect of metoclopramide. *Gastroenterology*, *88*(2), 492–499.

Aksamit, T. R., Floras, J. S., Victor, R. G., & Aylward, P. E. (1987). Paroxysmal hypertension due to sinoaortic baroreceptor denervation in humans. *Hypertension*, *9*(3), 309–314.

Almquist, A., Goldenberg, I. F., Milstein, S., Chen, M. Y., Chen, X. C., Hansen, R., ... Benditt, D. G. (1989). Provocation of bradycardia and hypotension by isoproterenol and upright posture in patients with unexplained syncope. *N. Engl. J. Med.*, *320*(6), 346–351.

Ammons, W. S. (1992). Bowditch Lecture. Renal afferent inputs to ascending spinal pathways. *Am. J. Physiol.*, *262*(2 Pt 2), R165–R176.

Anand, P. (1992). Skin axon reflex vasodilatation mechanisms, testing, and implications in autonomic disorders. In: Autonomic failure. A textbook of clinical disorders of the autonomic nervous system, 3rd ed. Bannister, R. and C.J. Mathias (Eds.). Oxford: Oxford University Press, 479–488.

Angelone, A., & Coulter, N. A., Jr. (1964). Respiratory sinus arrhythemia: A frequency dependent phenomenon. *J. Appl. Physiol. (Beth., Md.: 1985)*, *19*, 479–482.

Appenzeller, O., Qualls, C., & Martignoni, E. (1993). Swallow induced tachycardia: Effects of age, altitude, and ethnicity on this brainstem autonomic reflex. *Neurology*, *43*, 757S.

Arnold, J. M., & McDevitt, D. G. (1983). Heart rate and blood pressure responses to intravenous boluses of isoprenaline in the presence of propranolol, practolol and atropine. *Br. J. Clin. Pharmacol.*, *16*(2), 175–184.

Baldwa, V. S., & Ewing, D. J. (1977). Heart rate response to Valsalva manoeuvre. Reproducibility in normals, and relation to variation in resting heart rate in diabetics. *Br. Heart J.*, *39*(6), 641–644.

Benarroch, E. E. (1993). The central autonomic network: Functional organization, dysfunction, and perspective. *Mayo Clinic Proceedings. Mayo Clinic*, *68*(10), 988–1001.

Bennett, D. L. (1978). The adolescent with diabetes mellitus. *Pediatr. Ann.*, 7(9), 626–632.

Bernardi, L., Bianchini, B., Spadacini, G., Leuzzi, S., Valle, F., Marchesi, E., et al. (1995). Demonstrable cardiac reinnervation after human heart transplantation by carotid baroreflex modulation of RR interval. *Circulation*, *92*(10), 2895–2903.

Bernardi, L., Leuzzi, S., Radaelli, A., Passino, C., Johnston, J. A., & Sleight, P. (1994). Low-frequency spontaneous fluctuations of R-R interval and blood pressure in conscious humans: A baroreceptor or central phenomenon? *Clin. Sci. (Lond.)*, *87*(6), 649–654.

Bernardi, L., Rossi, M., Fratino, P., Finardi, G., Mevio, E., & Orlandi, C. (1989). Relationship between phasic changes in human skin blood flow and autonomic tone. *Microvasc. Res.*, *37*(1), 16–27.

Bigger, J. T., Jr., Fleiss, J. L., Steinman, R. C., Rolnitzky, L. M., Kleiger, R. E., & Rottman, J. N. (1992). Correlations among time and frequency domain measures of heart period variability two weeks after acute myocardial infarction. *Am. J. Cardiol.*, *69*(9), 891–8. Available from https://doi.org/10.1016/0002-9149(92)90788-z.

Borst, C., Hollander, A. P., & Bouman, L. N. (1972). Cardiac acceleration elicited by voluntary muscle contractions of minimal duration. *J. Appl. Physiol. (Beth., Md.: 1985)*, *32*(1), 70–77.

Borst, C., Wieling, W., van Brederode, J. F., Hond, A., de Rijk, L. G., & Dunning, A. J. (1982). Mechanisms of initial heart rate response to postural change. *Am. J. Physiol.*, *243*(5), H676–H681.

Brodie, T. G., & Russell, A. E. (1905). On the determination of the rate of blood flow through an organ. *J. Physiol. (Lond.)*, *32*, 47P.

Camilleri, M., Thompson, W. G., Fleshman, J. W., & Pemberton, J. H. (1994). Clinical management of intractable constipation. *Ann. Inter. Med.*, *121*(7), 520–528.

Camilleri, M., & Zinsmeister, A. R. (1992). Towards a relatively inexpensive, noninvasive, accurate test for colonic motility disorders. *Gastroenterology*, *103*(1), 36–42.

Cardone, C., Paiusco, P., Marchetti, G., Burelli, F., Feruglio, M., & Fedele, D. (1990). Cough test to assess cardiovascular autonomic reflexes in diabetes. *Diab. Care*, *13*(7), 719–724.

Chaudhuri, K. R., Thomaides, T., Hernandez, P., Alam, M., & Mathias, C. J. (1991). Noninvasive quantification of superior mesenteric artery blood flow during sympathoneural activation in normal subjects. *Clin. Autonom. Res.*, *1*(1), 37–42.

Cheshire, W. P., Jr. (2020). Autonomic history, examination, and laboratory evaluation. *Continuum (Minneap. Minn.)*, *26*(1), 25–43.

Cheshire, W. P., Freeman, R., Gibbons, C. H., Cortelli, P., Wenning, G. K., Hilz, M. J., ... Singer, W. (2021). Electrodiagnostic assessment of the autonomic nervous system: A consensus statement endorsed by the American Autonomic Society, American Academy of Neurology, and the International Federation of Clinical Neurophysiology. *Clin. Neurophysiol.*, *132*(2), 666–682.

Clarke, R. S., Ginsburg, J., & Hellon, R. F. (1958). Use of the strain gauge plethysmograph in assessing the effect of certain drugs on the blood flow through the skin and muscle of the human forearm. *J. Physiol.*, *140*(2), 318–326.

Cook, L. G., & Mitschow, L. C. (2019). Beyond the polygraph: Deception detection and the autonomic nervous system. *Fed. Pract.*, *36*(7), 316–321.

Coote, J. H., Hilton, S. M., & Perez-Gonzalez, J. F. (1971). The reflex nature of the pressor response to muscular exercise. *J. Physiol.*, *215*(3), 789–804.

Daly, M. (1985). Interactions between respiration and circulation. *In: Handbook of physiology. The respiratory system II. Bethesda, maryland: American Physiological Society*, 529–594.

Donadio, V., & Liguori, R. (2015). Microneurographic recording from unmyelinated nerve fibers in neurological disorders: An update. *Clin. Neurophysiol.*, *126*(3), 437–445.

Eckberg, D. L. (1980). Parasympathetic cardiovascular control in human disease: a critical review of methods and results. *Am. J. Physiol.*, *239*(5), H581–93. Available from https://doi.org/10.1152/ajpheart.1980.239.5.H581.

Eckberg, D. L., & Fritsch, J. M. (1993). How should human baroreflexes be tested? *News Physiol. Sci.*, *8*, 7–12.

Ernsting, J., & Parry, D. J. (1957). Some observations on the effects of of stimulating the stretch receptors in the carotid artery of man. *J. Physiol. (Lond.)*, *137*, 45P–46P.

Ertekin, C., & Reel, F. (1976). Bulbocavernosus reflex in normal men and in patients with neurogenic bladder and/or impotence. *J. Neurol. Sci.*, *28*(1), 1–15.

Ewing, D. J., Campbell, I. W., & Clarke, B. F. (1980). Assessment of cardiovascular effects in diabetic autonomic neuropathy and prognostic implications. *Ann. Intern. Med.*, *92*(2 Pt 2), 308–11. Available from https://doi.org/10.7326/0003-4819-92-2-308.

Ewing, D. J., & Clarke, B. F. (1982). Diagnosis and management of diabetic autonomic neuropathy. *Br. Med. J. (Clin. Res. Ed.)*, *285*(6346), 916–918.

Ewing, D. J., Irving, J. B., Kerr, F., Wildsmith, J. A., & Clarke, B. F. (1974). Cardiovascular responses to sustained handgrip in normal subjects and in patients with diabetes mellitus: A test of autonomic function. *Clin. Sci. Mol. Med.*, *46*(3), 295–306.

Ewing, D. J., Martyn, C. N., Young, R. J., & Clarke, B. F. (1985). The value of cardiovascular autonomic function tests: 10 years experience in diabetes. *Diab. Care*, *8*(5), 491–498.

Ewing, D. J., Neilson, J. M. M., & Travis, P. (1984). New method for assessing cardiac parasympathetic activity using 24 hour electrocardiograms. *Br. Heart J.*, *52*, 396–402.

Fagius, J., Wallin, B. G., Sundlof, G., Nerhed, C., & Englesson, S. (1985). Sympathetic outflow in man after anaesthesia of the glossopharyngeal and vagus nerves. *Brain*, *108* (Pt 2), 423–438.

Faubion, S. S., & Rullo, J. E. (2015). Sexual dysfunction in women: A practical approach. *Am. Fam. Phys.*, *92*(4), 281–288.

Fealey, R. D. (1996). Thermoregulatory sweat test. *ANS News*, *3*, 6–7.

Fealey, R. D., Low, P. A., & Thomas, J. E. (1989). Thermoregulatory sweating abnormalities in diabetes mellitus. *Mayo Clinic Proceedings. Mayo Clinic*, *64*(6), 617–628.

Fitzpatrick, A., & Sutton, R. (1989). Tilting towards a diagnosis in recurrent unexplained syncope. *Lancet*, *1*(8639), 658–660.

Fleshman, J. W., Dreznik, Z., Cohen, E., Fry, R. D., & Kodner, I. J. (1992). Balloon expulsion test facilitates diagnosis of pelvic floor outlet obstruction due to nonrelaxing puborectalis muscle. *Dis. Colon Rect.*, *35*(11), 1019–1025.

Fouad, F. M., Tadena Thome, L., Bravo, E. M., & Tarazi, R. C. (1984). Idiopathic hypovolemia. *Ann. Int. Med.*, *104*, 298–303.

Fouad, F. M., Tadena-Thome, L., Bravo, E. L., & Tarazi, R. C. (1986). Idiopathic hypovolemia. *Ann. Intern. Med.*, *104*(3), 298–303. Available from https://doi.org/10.7326/0003-4819-104-3-298.

Freeman, R., Saul, J. P., Roberts, M. S., Berger, R. D., Broadbridge, C., & Cohen, R. J. (1991). Spectral analysis of heart rate in diabetic autonomic neuropathy. A comparison with standard tests of autonomic function. *Arch. Neurol.*, *48*(2), 185–190.

Freeman, R., Weiss, S. T., Roberts, M., Zbikowski, S. M., & Sparrow, D. (1995). The relationship between heart rate variability and measures of body habitus. *Clin. Autonom. Res.*, *5*(5), 261–266.

Freeman, R., Wieling, W., Axelrod, F. B., Benditt, D. G., Benarroch, E., Biaggioni, I., ... van Dijk, J. G. (2011). Consensus statement on the definition of orthostatic hypotension, neurally mediated syncope and the postural tachycardia syndrome. *Clin. Autonom. Res.*, *21*(2), 69–72.

Furlan, R., Guzzetti, S., Crivellaro, W., Dassi, S., Tinelli, M., Baselli, G., ... Malliani, A. (1990). Continuous 24-hour assessment of the neural regulation of systemic arterial pressure and RR variabilities in ambulant subjects. *Circulation*, *81*(2), 537–547.

Gallagher, D., Terenzi, T., & de Meersman, R. (1992). Heart rate variability in smokers, sedentary and aerobically fit individuals. *Clin. Autonom. Res.*, *2*(6), 383–387.

Gandevia, S. C., McCloskey, D. I., & Potter, E. K. (1978). Reflex bradycardia occurring in response to diving, nasopharyngeal stimulation and ocular pressure, and its modification by respiration and swallowing. *J. Physiol.*, *276*, 383–394.

Gemmill, J. D., Venables, G. S., & Ewing, D. J. (1988). Noradrenaline response to edrophonium in primary autonomic failure: Distinction between central and peripheral damage. *Lancet*, *1*(8593), 1018–1021.

Genovely, H., & Pfeifer, M. A. (1988). RR-variation: the autonomic test of choice in diabetes. *Diab./Metab. Rev.*, *4*(3), 255–271.

Gibbons, C. H., & Freeman, R. (2015). Clinical implications of delayed orthostatic hypotension: A 10-year follow-up study. *Neurology*, *85*(16), 1362–1367.

Gibbons, C. H., Schmidt, P., Biaggioni, I., Frazier-Mills, C., Freeman, R., Isaacson, S., ... Kaufmann, H. (2017). The recommendations of a consensus panel for the screening, diagnosis, and treatment of neurogenic orthostatic hypotension and associated supine hypertension. *J. Neurol.*, *264*(8), 1567–1582.

Gilbey, M. P., & Spyer, K. M. (1993). Physiological aspects of autonomic nervous system function. *Curr. Opin. Neurol. Neurosurg.*, *6*(4), 518–523.

Goldstein, D. S. (2020). The "sick-but-not-dead" phenomenon applied to catecholamine deficiency in neurodegenerative diseases. *Semin. Neurol.*, *40*(5), 502–514.

Goldstein, D. S., Holmes, C., Cannon, R. O., 3rd, Eisenhofer, G., & Kopin, I. J. (1997). Sympathetic cardioneuropathy in dysautonomias. *N. Engl. J. Med.*, *336*(10), 696–702.

Goldstein, D. S., Isonaka, R., Holmes, C., Ding, Y. S., & Sharabi, Y. (2020). Cardiac sympathetic innervation and vesicular storage in pure autonomic failure. *Ann. Clin. Transl. Neurol.*, 7(10), 1908–1918.

Greenfield, A. D., Whitney, R. J., & Mowbray, J. F. (1963). Methods for the investigation of peripheral blood flow. *Br. Med. Bull.*, *19*, 101–109.

Grubb, B. P., Temesy-Armos, P., Hahn, H., & Elliott, L. (1991). Utility of upright tilt-table testing in the evaluation and management of syncope of unknown origin. *Am. J. Med.*, *90*(1), 6–10.

Gundersen, H. J., & Neubauer, B. (1977). A long-term diabetic autonomic nervous abnormality. Reduced variations in resting heart rate measured by a simple and sensitive method. *Diabetologia*, *13*(2), 137–140.

Hagbarth, K. E., Hallin, R. G., Hongell, A., Torebjork, H. E., & Wallin, B. G. (1972). General characteristics of sympathetic activity in human skin nerves. *Acta Physiol. Scand.*, *84*(2), 164–176.

Hainsworth, R., & el-Bedawi, K. M. (1994). Orthostatic tolerance in patients with unexplained syncope. *Clin. Autonom. Res.*, *4*(5), 239–244.

Hatfield, H. S. (1950). A heat flow meter. *J. Physiol. (Lond.)*, *111*, 10P.

Heath, M. E., & Downey, J. A. (1990). The cold face test (diving reflex) in clinical autonomic assessment: methodological considerations and repeatability of responses. *Clin. Sci. (Lond.)*, *78*(2), 139–147.

Hellon, R. F., & Clarke, R. S. (1959). Changes in forearm blood flow with age. *Clin. Sci.*, *18*(1), 1–7.

Henriksen, O. (1977). Local sympathetic reflex mechanism in regulation of blood flow in human subcutaneous adipose tissue. *Acta Physiol. Scand.*, *450*, 1–48.

Hines, E. A., & Brown, G. E. (1936). The cold pressor test for measuring the reactibility of the blood pressure: data concerning 571 normal and hypertensive subjects. *Am. Heart J.*, *11*, 1–9.

Hinton, J. M., Lennard-Jones, J. E., & Young, A. C. (1969). A ne method for studying gut transit times using radioopaque markers. *Gut*, *10*(10), 842–847.

Hirsch, J. A., & Bishop, B. (1981). Volume, flow, and timing of each breath during negative airway pressure in humans. *J. Appl. Physiol. Respir. Environ. Exerc. Physiol.*, *50*(3), 552–560.

Hoeldtke, R. D., Dworkin, G. E., Gaspar, S. R., & Israel, B. C. (1989). Sympathotonic orthostatic hypotension: a report of four cases. *Neurology*, *39*(1), 34–40.

Imholz, B. P., Dambrink, J. H., Karemaker, J. M., & Wieling, W. (1990). Orthostatic circulatory control in the elderly evaluated by non-invasive continuous blood pressure measurement. *Clin. Sci. (Lond.)*, *79*(1), 73–79.

Kapoor, W. N., & Brant, N. (1992). Evaluation of syncope by upright tilt testing with isoproterenol. A nonspecific test. *Ann. Inter. Med.*, *116*(5), 358–363.

Katona, P. G., & Jih, F. (1975). Respiratory sinus arrhythmia: noninvasive measure of parasympathetic cardiac control. *J. Appl. Physiol. (Beth., Md.: 1985)*, *39*(5), 801–805.

Kaufmann, H., Berman, J., Oribe, E., & Oliver, J. (1993). Possible increase in the synthesis of endothelial derived relaxing factor (EDRF) during vasovagal syncope. *Clin. Autonom. Res.*, *3*, 77.

Kaufmann, H., Oribe, E., Miller, M., Knott, P., Wiltshire-Clement, M., & Yahr, M. D. (1992). Hypotension-induced vasopressin release distinguishes between pure autonomic failure and multiple system atrophy with autonomic failure. *Neurology*, *42*(3 Pt 1), 590–593.

Kennedy, W. R., & Navarro, X. (1989). Sympathetic sudomotor function in diabetic neuropathy. *Arch. Neurol.*, *46*(11), 1182–1186.

Kennedy, W. R., Sakuta, M., Sutherland, D., & Goetz, F. C. (1984). Quantitation of the sweating deficiency in diabetes mellitus. *Ann. Neurol.*, *15*(5), 482–488.

Kenny, R. A., Ingram, A., Bayliss, J., & Sutton, R. (1986). Head-up tilt: a useful test for investigating unexplained syncope. *Lancet*, *1*(8494), 1352–1355.

Khurana, R. K. (1995). Orthostatic intolerance and orthostatic tachycardia: a heterogeneous disorder. *Clin. Autonom. Res.*, *5*(1), 12–18.

Kingsberg, S., & Althof, S. E. (2009). Evaluation and treatment of female sexual disorders. *Int. Urogynecol. J. Pelvic Floor Dysfunct.*, *20*(Suppl 1), S33–S43.

Krakowsky, Y., & Grober, E. D. (2018). A practical guide to female sexual dysfunction: An evidence-based review for physicians in Canada. *Canad. Urol. Assoc. J.*, *12*(6), 211–216.

Lamotte, G., Boes, C. J., Low, P. A., & Coon, E. A. (2021). The expanding role of the cold pressor test: a brief history. *Clin. Autonom. Res.*, *31*(2), 153–155.

Landau, W. M., & Nelson, D. A. (1996). Clinical neuromythology XV. Feinting science: Neurocardiogenic syncope and collateral vasovagal confusion. *Neurology*, *46*(3), 609–18.

Levin, A. B. (1966). A simple test of cardiac function based upon the heart rate changes induced by the Valsalva maneuver. *Am. J. Cardiol.*, *18*(1), 90–99.

Lewis, T. (1932). Lecture on vasovagal syncope and carotid sinus mechanism. *Brit. Med. J.*, *1*, 873–876.

Low, P. A. (1993). Composite autonomic scoring scale for laboratory quantification of generalized autonomic failure. *Mayo Clinic Proceedings. Mayo Clinic*, *68*(8), 748–752.

Low, P. A., Opfer-Gehrking, T. L., McPhee, B. R., Fealey, R. D., Benarroch, E. E., Willner, C. L., et al. (1995a). Prospective evaluation of clinical characteristics of orthostatic hypotension. *Mayo Clinic Proceedings. Mayo Clinic*, *70*(7), 617–622.

Low, P. A., Opfer-Gehrking, T. L., Textor, S. C., Benarroch, E. E., Shen, W. K., Schondorf, R., ... Rummans, T. A. (1995b). Postural tachycardia syndrome (POTS). *Neurology*, *45*(4 Suppl 5), S19–S25.

Mac Lean, A. R., Allen, E. V., & Magath, T. B. (1944). Orthostatic tachycardia and orthostatic hypotension: defects in the return of venous blood to the heart. *Am. J. Med.*, *27*, 145–168.

Macmillan, A. L., & Spalding, J. M. (1969). Human sweating response to electrophoresed acetylcholine: A test of postganglionic sympathetic function. *J. Neurol., Neurosurg., Psychiatr.*, *32*(2), 155–160.

Marfella, R., Giugliano, D., di Maro, G., Acampora, R., Giunta, R., & D'Onofrio, F. (1994). The squatting test. A useful tool to assess both parasympathetic and sympathetic involvement of the cardiovascular autonomic neuropathy in diabetes. *Diabetes*, *43*(4), 607–612.

McPherson, A., Scales, J. T., & Gordon, L. H. (1961). A method of estimating qualitative changes of blood flow in bone. *J. Bone Jt. Surg.*, *43B*, 791–799.

Medina Lucena, H., & Tincello, D. G. (2019). Methods of assessing and recording bladder sensation: a review of the literature. *Int. Urogynecol. J.*, *30*(1), 3–8.

Mitrakou, A., Ryan, C., Veneman, T., Mokan, M., Jenssen, T., Kiss, I., ... Gerich, J. (1991). Hierarchy of glycemic thresholds for counterregulatory hormone secretion, symptoms, and cerebral dysfunction. *Am. J. Physiol.*, *260*(1 Pt 1), E67–E74.

Morillo, C. A., Klein, G. J., Zandri, S., & Yee, R. (1995). Diagnostic accuracy of a low-dose isoproterenol head-up tilt protocol. *Am. Heart J.*, *129*(5), 901–906.

Mowbray, J. F. (1959). Measurement of tissue blood flow using small heated thermocouple needles. *J. Appl. Physiol. (Beth., Md.: 1985)*, *14*, 647–648.

Mueck-Weymann, M., & Rauh, R. (2002). Do preceding vasoconstrictions influence the 'inspiratory gasp test'? *Clin. Physiol. Funct. Imag.*, *22*(3), 206–209.

Murata, K., Landrigan, P. J., & Araki, S. (1992). Effects of age, heart rate, gender, tobacco and alcohol ingestion on R-R interval variability in human ECG. *J. Autonom. Nerv. Syst.*, *37*(3), 199–206.

Murray, A., Ewing, D. J., Campbell, I. W., Neilson, J. M., & Clarke, B. F. (1975). RR interval variations in young male diabetics. *Br. Heart J.*, *37*(8), 882–885.

Navarro, X., & Kennedy, W. R. (1992). Influence of posture on cardiorespiratory reflexes in normal subjects and patients with diabetes mellitus. *Clin. Autonom. Res.*, *2*(6), 393–396.

Nehra, A., Jackson, G., Miner, M., Billups, K. L., Burnett, A. L., Buvat, J., ... Wu, F. C. (2012). The Princeton III Consensus recommendations for the management of erectile dysfunction and cardiovascular disease. *Mayo Clinic Proceedings. Mayo Clinic*, *87*(8), 766–778.

Nilsson, G. E., Tenland, T., & Oberg, P. A. (1980). Evaluation of a laser Doppler flowmeter for measurement of tissue blood flow. *IEEE Transact. Bio-med. Eng.*, *27*(10), 597–604.

Nishimura, R. A., & Tajik, A. J. (1986). The Valsalva maneuver and response revisited. *Mayo Clinic Proc. Mayo Clinic*, *61*(3), 211–217.

Norbiato, G., Bevilacqua, M., Righini, V., Chebat, E., Vago, T., Bertora, P., ... Mangoni, A. (1992). Altered vasopressin response to metoclopramide in multiple system atrophy: evidence of a cholinergic defect in the hypothalamus. *Acta Neurol. Scand.*, *85*(4), 299–303.

Norcliffe-Kaufmann, L., Kaufmann, H., Palma, J. A., Shibao, C. A., Biaggioni, I., Peltier, A. C., ... Autonomic Disorders, C. (2018). Orthostatic heart rate changes in patients with autonomic failure caused by neurodegenerative synucleinopathies. *Ann. Neurol.*, *83*(3), 522–531.

O'Brien, I. A., O'Hare, P., & Corrall, R. J. (1986). Heart rate variability in healthy subjects: effect of age and the derivation of normal ranges for tests of autonomic function. *Br. Heart J.*, *55*(4), 348–354.

Oribe, E., & Appenzeller, O. (1995a). Evaluation of autonomic reflexes. In J. P. Mohr, & J. G. Gautier (Eds.), *Guide to clinical neurology* (pp. 213–218). New York: Churchill Livingstone.

Oribe, E., & Appenzeller, O. (1995b). Long duration running at altitude decreases heart rate variability: a sign of transient aging? *Clin. Autonom. Res.*, *5*, 327.

Oribe, E., Caro, S., Perera, R., Winters, S. L., Gomes, J. A., & Kaufmann, H. (1997). Syncope: the diagnostic value of head-up tilt testing. *Pac. Clin. Electrophysiol. PACE*, *20*(4 Pt 1), 874–879.

Pagani, M., Lombardi, F., Guzzetti, S., Rimoldi, O., Furlan, R., Pizzinelli, P., et al. (1986). Power spectral analysis of heart rate and arterial pressure variabilities as a marker of sympatho-vagal interaction in man and conscious dog. *Circ. Res.*, *59*(2), 178–193.

Paintal, A. S. (1995). Some recent advances in studies on J receptors. *Adv. Exp. Med. Biol.*, *381*, 15–25.

Parsons, S., Scott, A. R., & Macdonald, I. A. (1992). The effect of posture and environmental temperature on cardiovascular reflexes in normal subjects and diabetes mellitus. *Clin. Autonom. Res.*, *2*(3), 147–151.

Pemberton, J. H. (1990). Anorectal and pelvic floor disorders: putting physiology into practice. *J. Gastroenterol. Hepatol.*, *5*(Suppl 1), 127–143.

Perera, R., E. Oribe, & H. Kaufmann (1993). Orthostatic bradycardia in autonomic failure. Abstracts: 1993 Scientific sessions of the American Autonomic Society, Nashville, Tennessee: 23–24 October.

Persson, P. B., DiRienzo, M., Castiglioni, P., Cerutti, C., Pagani, M., Honzikova, N., ... Parati, G. (2001). Time vs frequency domain techniques for assessing baroreflex sensitivity. *J. Hypertens.*, *19*(10), 1699–1705.

Piha, S. J., & Voipio-Pulkki, L. M. (1993). Cardiac dysrhythmias during cardiovascular autonomic reflex tests. *Clin. Autonom. Res.*, *3*(3), 183–187.

Polinsky, R. J. (1992). Neuropharmacological investigation of autonomic failure. In R. Bannister, & C. J. Mathias (Eds.), *Autonomic failure. A textbook of clinical autonomic disorders of the autonomic nervous system* (3rd ed., pp. 334–358). Oxford: Oxford University Press.

Polinsky, R. J., Kopin, I. J., Ebert, M. H., & Weise, V. (1981). Pharmacologic distinction of different orthostatic hypotension syndromes. *Neurology*, *31*(1), 1–7.

Pramstaller, P. P., Wenning, G. K., Smith, S. J., Beck, R. O., Quinn, N. P., & Fowler, C. J. (1995). Nerve conduction studies, skeletal muscle EMG, and sphincter EMG in multiple system atrophy. *J. Neurol. Neurosurg. Psychiatr.*, *58*(5), 618–621.

Puritz, R., Lightman, S. L., Wilcox, C. S., Forsling, M., & Bannister, R. (1983). Blood pressure and vasopressin in progressive autonomic failure. *Resp. Post. Stimul. L-dopa and Naloxone. Brain*, *106*(Pt 2), 503–511.

Quispe, R. C., & Novak, P. (2021). Auxiliary tests of autonomic functions. *J. Clin. Neurophysiol.*, *38*(4), 262–273.

Raviele, A., Gasparini, G., Di Pede, F., Delise, P., Bonso, A., & Piccolo, E. (1990). Usefulness of head-up tilt test in evaluating patients with syncope of unknown origin and negative electrophysiologic study. *Am. J. Cardiol.*, *65*(20), 1322–1327.

Robertson, D., Goldberg, M. R., Onrot, J., Hollister, A. S., Wiley, R., Thompson, J. G., Jr., & Robertson, R. M. (1986). Isolated failure of autonomic noradrenergic neurotransmission. Evidence for impaired beta-hydroxylation of dopamine. *N. Engl. J. Med.*, *314*(23), 1494–1497.

Roddie, I. C., & Shepherd, J. T. (1956). The reflex nervous control of human skeletal muscle blood vessels. *Clin. Sci.*, *15*(3), 433–440.

Rosen, S. G., & Cryer, P. E. (1982). Postural tachycardia syndrome. Reversal of sympathetic hyperresponsiveness and clinical improvement during sodium loading. *Am. J. Med.*, *72*(5), 847–850.

Sandroni, P., Benarroch, E. E., & Low, P. A. (1991). Pharmacological dissection of components of the Valsalva maneuver in adrenergic failure. *J. Appl. Physiol. (1985)*, *71*(4), 1563–1567.

Satchell, P. M., & Seers, C. P. (1987). Evoked skin sympathetic nerve responses in man. *J. Neurol. Neurosurg. Psychiatr.*, *50*(8), 1015–1021.

Sato, K. T., Richardson, A., Timm, D. E., & Sato, K. (1988). One-step iodine starch method for direct visualization of sweating. *Am. J. Med. Sci.*, *295*(6), 528–31. Available from https://doi.org/10.1097/00000441-198806000-00006.

Schondorf, R. (1993). New investigations of autonomic nervous system function. *J. Clin. Neurophysiol.*, *10*(1), 28–38.

Schondorf, R., & Low, P. A. (1992). Gender related differences in the cardiovascular responses to upright tilt in normal subjects. *Clin. Autonom. Res.*, *2*(3), 183–187.

Schondorf, R., & Low, P. A. (1993). Idiopathic postural orthostatic tachycardia syndrome: an attenuated form of acute pandysautonomia? *Neurology*, *43*(1), 132–137.

Schwartz, N. S., Clutter, W. E., Shah, S. D., & Cryer, P. E. (1987). Glycemic thresholds for activation of glucose counterregulatory systems are higher than the threshold for symptoms. *J. Clin. Investig.*, *79*(3), 777–781.

Shahani, B. T., Halperin, J. J., Boulu, P., & Cohen, J. (1984). Sympathetic skin response—a method of assessing unmyelinated axon dysfunction in peripheral neuropathies. *J. Neurol. Neurosurg. Psychiatr.*, *47*(5), 536–542.

Sidery, M. B., & Macdonald, I. A. (1992). The effect of time of day on orthostatic tolerance and the cardiovascular effects of a high carbohydrate meal in healthy young subjects. *Clin. Autonom. Res.*, *2*(4), 271–276.

Sleight, P. (1988). Importance of cardiovascular reflexes in disease. *Am. J. Med.*, *84*(3A), 92–96.

Smith, S. A., & Smith, S. E. (1983). Evidence for a neuropathic aetiology in the small pupil of diabetes mellitus. *Br. J. Ophthalmol.*, *67*(2), 89–93.

Smyth, H. S., Sleight, P., & Pickering, G. W. (1969). Reflex regulation of arterial pressure during sleep in man. A quantitative method of assessing baroreflex sensitivity. *Circ. Res.*, *24*(1), 109–121.

Stella, A. (1992). The kidney as a sensor: Functional evidence. *J. Hypertens.*, *10*(7), S113–S119.

Streeten, D. H., & Anderson, G. H., Jr. (1992). Delayed orthostatic intolerance. *Arch. Inter. Med.*, *152*(5), 1066–1072.

Swenne, C. A. (2013). Baroreflex sensitivity: Mechanisms and measurement. *Netherl. Heart J.*, *21*(2), 58–60.

Taylor, J. A., Hand, G. A., Johnson, D. G., & Seals, D. R. (1992). Sympathoadrenal-circulatory regulation of arterial pressure during orthostatic stress in young and older men. *Am. J. Physiol.*, *263*(5 Pt 2), R1147–R1155.

Teodorovich, N., & Swissa, M. (2016). Tilt table test today—State of the art. *World J. Cardiol.*, *8*(3), 277–282.

Thomaides, T. N., Chaudhuri, K. R., Maule, S., Watson, L., Marsden, C. D., & Mathias, C. J. (1992). Growth hormone response to clonidine in central and peripheral primary autonomic failure. *Lancet*, *340*(8814), 263–266.

Tohmeh, J. F., Shah, S. D., & Cryer, P. E. (1979). The pathogenesis of hyperadrenergic postural hypotension in diabetic patients. *Am. J. Med.*, *67*(5), 772–778.

Vallbo, A. B., Hagbarth, K. E., Torebjork, H. E., & Wallin, B. G. (1979). Somatosensory, proprioceptive, and sympathetic activity in human peripheral nerves. *Physiol. Rev.*, *59* (4), 919–957.

van Lieshout, J. J. (1989). *Cardiovascular reflexes in orthostatic disorders (Academisch proefscchrift. University of Amsterdam)*. Amsterdam: Rodopi.

Victor, R. G., Leimbach, W. N., Jr., Seals, D. R., Wallin, B. G., & Mark, A. L. (1987). Effects of the cold pressor test on muscle sympathetic nerve activity in humans. *Hypertension*, *9*(5), 429–436.

Wallin, B. G., & Fagius, J. (1988). Peripheral sympathetic neural activity in conscious humans. *Ann. Rev. Physiol.*, *50*, 565–576.

Wallin, B. G., Sundlof, G., Eriksson, B. M., Dominiak, P., Grobecker, H., & Lindblad, L. E. (1981). Plasma noradrenaline correlates to sympathetic muscle nerve activity in normotensive man. *Acta Physiol. Scand.*, *111*(1), 69–73.

Wheeler, T., & Watkins, P. J. (1973). Cardiac denervation in diabetes. *Br. Med. J.*, *4* (5892), 584–586.

Whitney, R. J. (1953). Measurement of volume changes in human limbs. *J. Physiol. (Lond.)*, *121*, 27P.

Wieling, W. (1992). Non invasive continuous recording of heart rate and blood pressure in the evaluation of neurocardiovascular control. In R. Bannister, & C. J. Mathias (Eds.), *Autonomic failure. A textbook of clinical disorders of the autonomic nervous system* (3rd (ed.), pp. 291–311). Oxford: Oxford University Press.

Wieling, W., & Shepherd, J. T. (1992). Initial and delayed circulatory responses to orthostatic stress in normal humans and in subjects with orthostatic intolerance. *Int. Angiol.*, *11*(1), 69–82.

Wieling, W., ten Harkel, A. D., & van Lieshout, J. J. (1991). Spectrum of orthostatic disorders: classification based on an analysis of the short-term circulatory response upon standing. *Clin. Sci. (Lond.)*, *81*(2), 241–248.

Wieling, W., van Brederode, J. F., de Rijk, L. G., Borst, C., & Dunning, A. J. (1982). Reflex control of heart rate in normal subjects in relation to age: a data base for cardiac vagal neuropathy. *Diabetologia*, *22*(3), 163–166.

Wieling, W., Veerman, D. P., Dambrink, J. H., & Imholz, B. P. (1992). Disparities in circulatory adjustment to standing between young and elderly subjects explained by pulse contour analysis. *Clin. Sci. (Lond.)*, *83*(2), 149–155.

Zema, M. J. (1999). Diagnosing heart failure by the Valsalva maneuver: Isn't It finally time? *Chest*, *116*(4), 851–853.

Ziegler, M. G., Lake, C. R., & Kopin, I. J. (1977). The sympathetic-nervous-system defect in primary orthostatic hypotension. *N. Engl. J. Med.*, *296*(6), 293–297.

Further reading

Holden, W., M. Mendall, P. Andrews and S. Whitehead: Swallow induced tachycardia: An index of brain stem function in man? Presented at the Clinical Autonomic Research Society Meeting, London, (1992).

Mano, T. (1994). Microneurographic analysis of sympathetic nerve response to environmental stress. In B. N. Mallick, & R. Singh (Eds.), *Environment and physiology* (pp. 91–103). New Delhi: Narosa.

Proceedings of a consensus development conference on standardized measures in diabetic neuropathy. Autonomic nervous system testing. Neurology, (1992) 42(9), 1831–1837.

CHAPTER 14

Other forms of autonomic dysfunction

14.1 The autonomic nervous system and COVID-19

The novel severe acute respiratory syndrome coronavirus type 2 (SARS-CoV-2) and resulting coronavirus disease 2019 (COVID-19) emerged in late 2019. It quickly evolved into a pandemic that disrupted many aspects of daily life across the globe with profound economic, health, and social consequences. The outbreak of this viral disease was, and continues to be, stressful for many individuals. Isolation or quarantine became necessary after close contact (within 6 feet for a cumulative total of 15 minutes or more over 24 hours) with someone who has COVID-19. Changes in daily life functioning to slow the spread of the virus have affected the family structure. For example, homeschooling helps contain COVID-19 and many schools moved to online learning. In addition, many people are being asked to work from home. People over 60 are at higher risk of severe COVID-19 disease and should not be around children who may be ill with mild symptoms. People who are overweight may be at increased risk of severe illness from COVID-19 and hospitalization. Pneumonitis and pulmonary dysfunction often dominate the clinical picture; however, SARS-CoV-2 affects other body organs and systems, including the heart, gut, kidneys, and brain. The SARS-CoV-2 infection has been associated with multiple parainfectious neurological and autonomic abnormalities. Furthermore, people respond to the stress of disease in different ways and it is normal to experience a range of emotions, including fear, anxiety, and grief, all of which activate the autonomic nervous system (ANS). There is concern that postacute sequelae of COVID-19 (also called "long-haul COVID," "long-COVID," or "post-COVID syndrome") include abnormalities of autonomic cardiovascular function, which may be associated with various symptoms such as fatigue, "brain fog," sensory changes, and orthostatic intolerance (Goldstein, 2020b). Numerous new variants have emerged during the pandemic. For example, samples initially collected in Botswana, South Africa, and in Hong Kong, subsequently called *Omicron*, proved to be a new variant of COVI-19 of great concern. This

Introduction to Clinical Aspects of the Autonomic Nervous System
DOI: https://doi.org/10.1016/B978-0-323-96011-3.00007-3

variant has many mutations known to be important in evading immunity. Therefore, the spread of the virus causes great anxiety and affects autonomic nervous system function throughout the World (Kupferschmidt, 2021a,b).

14.2 Historical aspects

On occasions, apocalyptic scenarios, resembling Dante's Inferno have been witnessed in present times such as "multitudes of patients entering a suffering city, where could be read above the gates Abandon every hope, who enter here." Hospitals were overwhelmed and intensive care units were overflowing. Often, we sighted an endless line of ambulances bringing more patients into the emergency room, while a refrigerated truck awaited outside for the deceased. The Divine Comedy narrates Dante's (Fig. 14.1) journey to the afterlife, through hell (Inferno), purgatory (Purgatorio), and finally, heaven (Paradiso), where he meets his beloved Beatrice (Lemos, 2021).

Figure 14.1 Allegorical Portrait of Dante Alighieri, Unknown Master, late 16th century, Florence. Oil on panel. 126 × 120 cm. *From Lemos, C. Evoking Dante amidst the COVID-19 pandemic. Am. J. Med., (2021) 135(1), 126–127. Samuel H. Kress Collection. Courtesy National Gallery of Art, Washington, DC.*

14.3 Long COVID syndrome—"long haulers"

The prevalence of long-haul COVID is not known but it has been reported that more than 50% are left with ongoing symptoms several months after acute infection, many of which appear to be secondary to dysfunction of the ANS (Huang et al., 2021). Long-term activation of the extended autonomic nervous system may happen in long-COVID. The extended autonomic nervous system regroups the sympathetic nervous system, parasympathetic nervous system, and enteric nervous systems as well as neuroendocrine (sympathetic adrenergic system, hypothalamic-pituitary-adrenocortical axis, arginine vasopressin, renin-angiotensin-aldosterone system) and neuroimmune (cholinergic antiinflammatory and sympathetic inflammasomal pathways) systems (Goldstein, 2020a). Internet-based symptom surveys of patients with persistent symptoms after acute illness have indicated a high prevalence of fatigue, orthostatic intolerance, episodic tachycardia, anosmia and dysgeusia, exercise intolerance, sensory disturbances, gastrointestinal complaints, headache, and cognitive dysfunction ("brain fog") (Davis et al., 2020). Other symptoms include shortness of breath or difficulty breathing, cough, joint pain, muscle pain, chest pain, and sleep disturbances. It is widely suspected that these symptoms are associated with abnormalities of the ANS (Dani et al., 2020; Lo, 2020; Su et al., 2020; Umapathi et al., 2020). These symptoms or signs also activate the ANS and, thus, are likely to contribute to the patient's complaints and lack of wellbeing after this viral infection (Carfi, Bernabei, & Landi, 2020).

The pathophysiological mechanisms of postacute cardiovascular dysautonomia in COVID-19 are not well understood. Similar mechanisms may be involved in the pathophysiology of the myalgic encephalomyelitis/chronic fatigue syndrome.

Hypovolemia may contribute to orthostatic intolerance, tachycardia, and fatigue. Patients often report improvement with increased fluid and salt intake or intravenous saline infusion. Dysfunction of the renin-angiotensin-aldosterone system may contribute to hypovolemia and a "paradoxical" decrease in aldosterone level has been demonstrated in patients with postural tachycardia syndrome (Raj et al., 2005). Excess fluid loss in patients (e.g., functional gastrointestinal disorders with diarrhea) may contribute to hypovolemia, and treating the gastrointestinal disorder will usually improve volume status and orthostatic intolerance.

An abnormal orthostatic shift in blood volume with excessive venous pooling in the lower extremities associated with the upright posture may contribute to symptoms of cerebral hypoperfusion. This can be seen with a restricted autonomic neuropathy causing impaired vasomotor and venomotor tone in the lower limbs as seen in some patients with postural tachycardia syndrome. Distal small fiber neuropathy has been documented in long-haul COVID and the timing of symptoms suggests a possible postinfectious autoimmune trigger (Shouman et al., 2021).

The cause of brain fog in long-haul COVID may be related to decreased cerebral blood flow and impaired cerebrovascular autoregulation. A study demonstrated that 90% of patients with chronic fatigue syndrome—a syndrome with possible shared pathophysiological mechanisms—have an abnormal reduction in cerebral blood flow during tilt-table testing (van Campen, Verheugt, Rowe, & Visser, 2020). Reduction in cardiac output may also contribute to the reduction in cerebral blood flow and brain fog. SARS-CoV-2 can infect myocardial cells and thereby interfere with intrinsic cardiac functions, resulting in fatigue and exercise intolerance.

The role of the neuroendocrine and neuroimmune systems in long-haul COVID is a matter of ongoing research. In patients with chronic fatigue syndrome, where a viral trigger is often reported by the patient, neuroendocrine studies have suggested decreased hypothalamic-pituitary-adrenocortical axis activity (Silverman, Heim, Nater, Marques, & Sternberg, 2010), which might promote increased production of proinflammatory cytokines (Gaab et al., 2005), and increased sympathetic outflows (Kristiansen et al., 2019). Angiotensin-converting enzyme-2 are expressed in endothelial cells lining many tissues in the body. Chronic complications from direct injury to endothelial cells inflicted during the acute and parainfectious periods of COVID-19 is possible. An autoimmune etiology has been suggested in some cases of postural tachycardia syndrome (7/16 cases were preceded by a viral illness in the initial report in 1993; Schondorf & Low, 1993) and patients with autoimmune disorders such as Sjogren's syndrome often have autonomic complaints. Antibodies that have been associated with postural tachycardia syndrome include antibodies against different G-coupled protein receptors, ß-adrenergic antibodies, muscarinic receptor antibodies among others; however, there has been no compelling evidence of the pathogenicity of these antibodies. In patients with long-haul COVID, a variety of inflammatory or immune factors could produce inflammation in the vascular endothelium, vasodilation, or increased fluid extravasation, both of which would be expected to increase cardiac sympathetic outflow reflexively. Abnormal mast cell activation may also contribute to the condition and cause various symptoms such as pruritis, urticaria, flushing, angioedema, wheezing, gastrointestinal symptoms, tachycardia, and labile blood pressure (Afrin, Weinstock, & Molderings, 2020).

The central autonomic network plays a critical role in the stress response, which is triggered by challenges on the homeostasis (physiological stressors) or unpleasant social or environmental situations. Areas of the central autonomic network involved in the stress response include the insular and anterior cingulate cortices, extended amygdala, hypothalamus, periaqueductal gray, and locus coeruleus (Lamotte, Shouman, & Benarroch, 2021). Alteration of the central autonomic network might cause deleterious changes in autonomic outflows to the blood vessel and the heart and contribute to the pathophysiology of long-haul COVID. Studies have sought central abnormalities in anatomic loci in the brain that may modulate autonomic responses to stressors in patients with chronic fatigue syndrome (Barnden, Kwiatek, Crouch, Burnet, & Del Fante, 2016; Vuong, Allison, Finkelmeyer, Newton, & Durham, 2020), but the results have not been consistent (Maksoud et al., 2020).

Other possible contributors to long-haul COVID include persisting cardiopulmonary injury (interstitial lung disease, lung, and cardiac tissue fibrosis), chronic kidney injury, and the hypercoagulable state of COVID-19 that may cause chronic pulmonary emboli. Finally, cardiovascular deconditioning as a consequence of prolonged bed rest can lead to resting tachycardia, orthostatic intolerance, and reduced exercise capacity.

Cases of autonomic dysfunction beyond postural tachycardia syndrome have been reported, including phenomena such as small fiber neuropathy with orthostatic cerebral hypoperfusion syndrome (Novak, 2020), neurally mediated syncope, and post-COVID-19 exacerbation of paroxysmal hypothermia and hyperhidrosis (Lamotte, Benarroch, & Coon, 2021). Post-COVID-19 acute inflammatory demyelinating polyneuropathy, known for its potential association with autonomic dysfunction, has been described as well (Mostel, Ayat, Capric, Trimmingham, & McFarlane, 2021). In a retrospective series of 27 patients with a confirmed history of COVID-19 infection referred for autonomic testing for symptoms concerning for para-/postinfectious autonomic dysfunction, autonomic symptoms developed between 0 and 122 days following the acute infection and included lightheadedness (93%), orthostatic headache (22%), syncope (11%), hyperhidrosis (11%), and burning pain (11%). On standardized autonomic testing, sudomotor function was abnormal in 36%, cardiovagal function in 27%, and cardiovascular adrenergic function in 7%. The most common clinical scenario was orthostatic intolerance without tachycardia or hypotension (41%). Twenty-two percent of patients fulfilled the criteria for postural tachycardia syndrome, and 11% had borderline findings to support orthostatic intolerance. One patient each was diagnosed with autoimmune autonomic ganglionopathy, inappropriate sinus tachycardia, vasodepressor syncope, cough/vasovagal syncope, exacerbation of preexisting orthostatic hypotension, exacerbation of sensory and autonomic neuropathy, and exacerbation of small fiber neuropathy. The authors concluded that abnormalities on autonomic testing were seen in the majority of patients with post-COVID syndrome but were mild in most cases. The most common finding was orthostatic intolerance, often without objective hemodynamic abnormalities on testing (Shouman et al., 2021). In a study of 16 young adults who tested positive for SARS-CoV-2, resting heart rate and blood pressure were not different compared to controls ($N = 14$); however, muscle sympathetic nerve activity was higher in COVID subjects compared to controls during head-up tilt with a group-by-position interaction in muscle sympathetic nerve activity burst incidence, as well as heart rate, in response to tilt. The results of this study suggest that resting sympathetic activity may be elevated following SARS-CoV-2 infection (Stute et al., 2021).

Long-haul COVID appears to affect women more than men which is of interest since men appear to have a greater incidence and mortality from COVID-19 (Acheampong et al., 2020). Sex-based differences in autonomic, innate, and adaptive immune responses to pathogens may explain the difference in the prevalence of long-haul COVID between females and males. Further studies are needed to better understand sex differences in long-haul COVID.

14.4 "Other" COVID-associated syndromes

14.4.1 Coronasomnia

Sleep disruptions have manifested in different ways. Stress and anxiety related to the pandemic can lead to sleep loss and feelings of daytime fatigue. Some frontline workers and essential employees also face the added challenge of working longer shifts that force them to alter their sleep schedules. Additionally, social distancing, remote learning, and restrictions on normal activities have left many people struggling with depression and isolation, increasing their risk for sleep disorders. The right amount of sleep is important to maintaining health and wellness.

Spending time in front of a television may be tempting because of social distancing, but this can cause mental stimulation. Also, blue light from the device can negatively impact sleep cycles. Other healthy sleep strategies include a regular daily exercise routine and spending time outdoors during the day for natural light exposure. See also chapter 12.

14.4.2 The sick building syndrome

The sick building syndrome is a complication that can appear to building occupants along with general, mucosal, and skin symptoms such as headache, fatigue, irritability in the upper respiratory tract, throat, eyes, nose, hands, or facial skin (Hosseini, Fouladi-Fard, & Aali, 2020). The occurrence of SBS symptoms is attributed to individual risk factors like anxiety and stress, smoking and lack of communication and seclusion, and the domestic environment risk factors such as sound pollution, the dissatisfaction of lighting, odor pollution, indoor air pollutants such as airborne bacteria, dust, molds, and airborne fungi (Singh, Yu, & Jeong Tai, 2010).

14.4.3 Lack of sunlight and vitamin D deficiency

Consequences of lockdown include lack of sunlight and high prevalence of hypovitaminosis D. Careful identification of factors affecting vitamin status and supplementation if necessary has now become the mantra for pediatric care (Vierucci, Del Pistoia, Fanos, Erba, & Saggese, 2014).

14.4.4 Chilblains (pernio)

The published literature suggests that several unifying cutaneous and systemic mechanisms may explain COVID-19—associated pernio. SARS-CoV-2 cell infection occurs through the cellular receptor angiotensin-converting enzyme 2 mediated by transmembrane protease serine 2, subsequently affecting the renin-angiotensin-aldosterone system with an increase in the vasoconstricting, proinflammatory, and prothrombotic angiotensin II pathway. Additionally, severe acute respiratory syndrome coronavirus 2 cell infection triggers an immune response with robust IFN-I release in patients predisposed to COVID-19—associated pernio, and age and sex discrepancies correlate with COVID-19 severity and manifestations, including pernio as a sign of mild disease (Cappel, Cappel, & Wetter, 2021).

14.4.5 Migraine

Not all effects of COVID-19 are "negative." Migraine for example improved during the pandemic (Verhagen, van Casteren, de Vries Lentsch, & Terwindt, 2021). This is attributed to working from home, scaling down demanding social lives, and freedom to choose how to organize one's time.

14.4.6 Others

New syndromes have emerged during this epidemic such as "*staring out the window*" (see Fig. 14.2). This is said to be a common sight during lockdown times when work from home is often mandated and socializing significantly limited. New words have also surfaced such as Covidtorians (see Fig. 14.3).

14.5 Orthostatic intolerance—focus on postural tachycardia syndrome

14.5.1 Introduction—definition

Postural tachycardia syndrome (POTS) is a form of orthostatic intolerance, defined as the development upon standing of symptoms of cerebral hypoperfusion (e.g., lightheadedness, weakness, and blurred vision) associated with those of sympathetic activation (such as tachycardia, nausea, and tremulousness) and a sustained excessive heart rate increment of 30 beats/min

Figure 14.2 An idle young woman at home with apparently nothing to do is staring out the window. *Free image from the web.*

Figure 14.3 An artist's view of Epidemic times. "Covidtorians." Looking Out: The Airstream," 2020. *By Anne Kornfeld. With permission.*

(sequential 30-second averages) or more within 10 minutes of standing or head-up tilt in the absence of orthostatic hypotension. For individuals 12–19 years of age, the required increment is at least 40 beats/minute. Both blood pressure and heart rate are often less stable in patients with POTS than in controls, which is reflected by signal oscillations on testing. Female:male ratio is about 4–5:1 and most cases occur between the ages of 15 and 50 years. Symptoms should persist for more than 6 months. Reversible causes (e.g., medications, anemia) and untreated underlying medical disorders (e.g., hyperthyroidism) should be excluded.

14.6 Clinical features

POTS is a condition rather than a disease and there is considerable heterogeneity in the clinical manifestations. Common symptoms include lightheadedness or dizziness, blurred vision, weakness, nausea, tremulousness, palpitation, chest pressure, breathlessness, brain fog, and fatigue. An antecedent event such as a viral illness, injury, or stress is common. A common comorbidity is joint hypermobility and some patients fulfill the criteria for Ehlers Danlos type II.

Apart from symptoms of orthostatic intolerance, patients with chronic POTS often have significant issues with comorbid conditions. Central sensitization is common so that symptoms can far exceed cardiovascular changes. Mast cell activation syndrome has also been described in a subset of patients. Manifestations of mast cell activation syndrome include food intolerance, orthostatic intolerance, gastrointestinal complaints of nausea, abdominal cramps and diarrhea, skin manifestations of flushing, and urticarial. Mast cell degranulation is associated with a high tryptase level; however, the tryptase levels can be normal in between the episodes of mast cell degranulation. There is some uncertain association with autoimmunity and some patients have low levels of autoimmune antibodies. Another subset is patients with known autoimmune disorders (e.g., autoimmune autonomic ganglionopathy) who develop orthostatic tachycardia as part of the autonomic neuropathy.

14.7 Phenotypes of postural tachycardia syndrome

Classifying patients in different phenotypes can be challenging and it should be emphasized that these phenotypes are not mutually exclusive. POTS is not a psychogenic disorder and sham lower body negative pressure is not associated with excessive increase in heart rate in patients; however, psychological testing confirmed that patients with POTS have somatic hypervigilance and perceive and report sensory information in a more dramatic way than control subjects (Owens, Low, Iodice, Critchley, & Mathias, 2017).

14.7.1 Postural tachycardia syndrome and deconditioning—the same or different?

It has been argued, with significant evidence, that deconditioning to some degree is present in virtually all patients with POTS although its presence as either a cause or effect is unclear. Poor conditioning is evident in patients who undergo a period of prolonged bed rest and in patients who have life-long orthostatic intolerance with chronic fatigue. The normal relationships between afferent signals in the cardiovascular system and efferent responses that should be considered reasonable for individuals with small hearts and small plasma volumes argue for the presence of deconditioning in POTS (Joyner & Masuki, 2008; Parsaik et al., 2012).

14.7.2 Neuropathic postural tachycardia syndrome

This entity is best understood when neuropathic POTS is considered within the spectrum of peripheral adrenergic denervation. Partial adrenergic denervation can result in hyperadrenergic OH (or OH with orthostatic tachycardia). More modest adrenergic denervation results in orthostatic tachycardia without OH (i.e., neuropathic POTS). These patients sometimes have postviral and peripheral autonomic denervation, manifest as distal loss of sweating on QSART and thermoregulatory sweat test reflecting involvement of distal sympathetic cholinergic fibers. Abnormalities in intraepidermal nerve fiber density on skin biopsy can also be found in a length-dependent fashion. Impaired vasomotor and venomotor tone in the lower limbs is due to the distal neuropathy and considered to result in venous pooling with orthostatic stress. A possible postinfectious autoimmune trigger is suggested by a frequently reported antecedent viral illness. A study by Jacob et al. (2000) demonstrated that plasma norepinephrine increment to sympathetic activation was intact in blood drawn from arm veins but reduced from corresponding leg veins (Jacob et al., 2000). Adrenergic denervation can be evident from loss of vasoconstrictor reflexes during the Valsalva maneuver with reduced or absent late Phase II.

14.7.3 Hyperadrenergic postural tachycardia syndrome

Three features characterize hyperadrenergic POTS. First, patients have an excessive increment in plasma norepinephrine (NE) with standing values ≥ 600 pg/mL. Second, there is sustained or fluctuating pressor response to head-up tilt (HUT) so that systolic BP increases during HUT. Third, these patients have symptoms of sympathetic overactivity, which can be episodic or sustained. These are characterized by episodes of tachycardia, hypertension, and hyperhidrosis. When these are prominent, they comprise autonomic storms. The episodes can be spontaneous or provoked and episodes may even occur during sleep. Stimuli include stress, activity, and orthostatic stress. Hyperdynamic responses can be observed during the Valsalva maneuver. It is important to look for the possible contribution of medications associated with an increase in circulating norepinephrine (e.g., norepinephrine transporter blockade by tricyclic antidepressants, norepinephrine transporter inhibitors, stimulants). Augmented muscle sympathetic nerve activity (MSNA) response has been documented in patients with POTS; however, MSNA responses for a given change in heart volume are normal in POTS, and the vasoconstriction evoked by these changes in MSNA are also normal (Fu et al., 2010).

14.7.4 Role of hypovolemia, genetic mutations, connective tissue disorders, and autoimmunity

There is evidence for low blood volumes in many, but not all, patients with POTS. Patients often report improvement with volume expansion. Patients with orthostatic tachycardia with hypovolemia have been found to have inappropriately low levels of standing plasma renin activity and aldosterone compared to normovolemic subjects. Postural tachycardia can also be secondary to poor oral intake or excess fluid loss in patients (e.g., functional gastrointestinal disorders with diarrhea).

A number of patients with POTS have hypermobile joints consistent with an underlying disorder of the connective tissue matrix, most commonly the hypermobile type of Ehlers–Danlos syndrome. Though increasingly recognized, autonomic dysfunction is likely undiagnosed in many subjects with Ehlers–Danlos. The mechanisms that cause or contribute to POTS in Ehlers–Danlos syndrome are still poorly understood. Relationships among joint hypermobility, anxiety, and autonomic dysfunction (primarily tachycardia) and elucidated brain–body mechanisms including aberrant engagement of the amygdala and insula have been demonstrated in individuals with joint hypermobility (Baur, Hänggi, Langer, & Jäncke, 2013; Eccles, Owens, Harrison, Grahame, & Critchley, 2016). Rowe et al. suggested that in some cases, the association of orthostatic intolerance with hypermobile Ehlers–Danlos syndrome could be attributed to abnormal connective tissue in dependent blood vessels with veins distending excessively in response to ordinary hydrostatic pressures. This in turn leads to increased venous pooling and its hemodynamic and symptomatic consequences (Rowe et al., 1999). Neuropathy, mast cell activation leading to high circulating levels of histamine, alpha-adrenergic and beta-adrenergic hyperresponsiveness, connective tissue laxity, and vasoactive medication may also play a role in the development of cardiovascular dysfunction in hypermobile Ehlers–Danlos syndrome (De Wandele et al., 2014).

It is not uncommon to encounter a patient with POTS who has a family member with POTS or a similar form of dysautonomia suggesting a possible genetic component. Possible genetic contributions in POTS include genotypic variability in endothelial nitric oxide synthase, ß-2 adrenoreceptor polymorphisms, variability in dopamine ß-hydroxylase activity, and genetic disorders of the connective tissue matrix among others.

At this time, there is no compelling evidence for an active autoimmune process in most patients with POTS, except for the rare cases of postural tachycardia seen as part of an autoimmune autonomic ganglionopathy.

There is agreement that positivity for ganglionic nicotinic antibodies at low titer (i.e., below 0.2 nmol/L) has no clinical relevance for POTS. Additional antibodies that have been associated with POTS include antibodies against different G-coupled protein receptors, beta-adrenergic antibodies, muscarinic receptor antibodies among others. There has been no compelling evidence of the pathogenicity of these antibodies and further research is needed to assess the relevance of an autoimmune basis in the development of POTS.

A study reported an association between postural tachycardia with neuropathic features and cutaneous alpha-synuclein deposition in a very small group of patients (Levine, Bellaire, Gibbons, & Freeman, 2021). Further studies are needed to confirm whether POTS can be seen in the early stage in synucleinopathies. Shibao and colleagues reported that individuals with POTS may have some degree of insulin resistance, that they release greater amounts of glucose-dependent insulinotropic polypeptide after glucose ingestion, and that this greater glucose-dependent insulinotropic polypeptide release may contribute to the exacerbation of orthostatic symptoms via splanchnic vasodilation (Breier et al., 2022). glucose-dependent insulinotropic polypeptide receptor antagonists may be a potential therapeutic target in POTS.

14.8 Management

Treatment of POTS includes volume expansion with fluid intake, water bolus, and a high salt diet. Patients should undertake a program of graded exercise sufficient to treat deconditioning. Other measures used include compression garments with waist-high compression stockings, abdominal binder, and physical counter maneuvers. Drugs that have demonstrated benefit in patients with POTS include propranolol, pyridostigmine, midodrine in low dose, ivabradine, and fludrocortisone. Ideally, medications should be considered short-term support and should be tapered off after 6–12 months, if possible. Comorbidities should be treated appropriately (e.g., mast cell activation syndrome in a subset of patients).

Volume expansion with intravenous saline provides temporary benefit but is poorly sustained. Patients who have a port inserted and receive saline therapy are at risk of complications including sepsis. The use of stimulants to treat brain fog can compound tachycardia and the complexity of management in some patients. Many patients have central sensitization and a multidisciplinary approach (neurology, psychotherapy, physical therapy, pain management) may be very helpful.

The indication for immunotherapy should obviously be dictated by the primary condition. Most patients with POTS who have nonspecific immune markers including low titer antibodies do not need immunotherapy.

References

Acheampong, D. O., Barffour, I. K., Boye, A., Aninagyei, E., Ocansey, S., & Morna, M. T. (2020). Male predisposition to severe COVID-19: Review of evidence and potential therapeutic prospects. *Biomed. Pharmacother.*, *131*, 110748.

Afrin, L. B., Weinstock, L. B., & Molderings, G. J. (2020). Covid-19 hyperinflammation and post-Covid-19 illness may be rooted in mast cell activation syndrome. *Int. J. Infect. Dis.*, *100*, 327–332.

Barnden, L. R., Kwiatek, R., Crouch, B., Burnet, R., & Del Fante, P. (2016). Autonomic correlations with MRI are abnormal in the brainstem vasomotor centre in chronic fatigue syndrome. *Neuroimage Clin.*, *11*, 530–537.

Baur, V., Hänggi, J., Langer, N., & Jäncke, L. (2013). Resting-state functional and structural connectivity within an insula-amygdala route specifically index state and trait anxiety. *Biol. Psychiatr.*, *73*(1), 85–92. (2013).

Breier, Nicholas C, Paranjape, Sachin Y, Scudder, Shea, Mehr, Shahram E, Diedrich, Andre', Flynn, Charles R, ... Shibao, Cyndya A (2022). Worsening Postural Tachycardia Syndrome Is Associated With Increased Glucose-Dependent Insulinotropic Polypeptide Secretion. *Hypertension*, *79*(5), e89–e99. Available from https://doi.org/10.1161/HYPERTENSIONAHA.121.17852.

Cappel, M. A., Cappel, J. A., & Wetter, D. A. (2021). Pernio (chilblains), SARS-CoV-2, and COVID toes unified through cutaneous and systemic mechanisms. *Mayo Clinic Proc. Mayo Clinic*, *96*(4), 989–1005.

Carfi, A., Bernabei, R., & Landi, F. (2020). Persistent symptoms in patients after acute COVID-19. *JAMA*, *324*(6), 603–605.

Dani, M., Dirksen, A., Taraborrelli, P., Torocastro, M., Panagopoulos, D., Sutton, R., & Lim, P. B. (2020). Autonomic dysfunction in 'long COVID': Rationale, physiology and management strategies. *Clin. Med. (Lond., Engl.)*.

Davis, H.E., Assaf, G.S., McCorkell, L., Weid, H., Low, R.J., Re'em, Y., ... Akrami, A. (2020). Characterizing long COVID in an international cohort: 7 months of symptoms and their impact. medRxiv.

De Wandele, I., Rombaut, L., Leybaert, L., Van de Borne, P., De Backer, T., Malfait, F., ... Calders, P. (2014). Dysautonomia and its underlying mechanisms in the hypermobility type of Ehlers-Danlos syndrome. *Semin. Arthritis Rheum.*, *44*, 93–100.

Eccles, J., Owens, A., Harrison, N., Grahame, R., & Critchley, H. (2016). Joint hypermobility and autonomic hyperactivity: An autonomic and functional neuroimaging study. *Lancet*, *387*(2), S40.

Fu, Q., Vangundy, T. B., Galbreath, M. M., Shibata, S., Jain, M., Hastings, J. L., ... Levine, B. D. (2010). Cardiac origins of the postural orthostatic tachycardia syndrome. *J. Am. Coll. Cardiol.*, *55*(25), 2858–2868.

Gaab, J., Rohleder, N., Heitz, V., Engert, V., Schad, T., Schurmeyer, T. H., & Ehlert, U. (2005). Stress-induced changes in LPS-induced pro-inflammatory cytokine production in chronic fatigue syndrome. *Psychoneuroendocrinology*, *30*(2), 188–198.

Goldstein, D. S. (2020a). The extended autonomic system, dyshomeostasis, and COVID-19. *Clin. Autonom. Res.*, *30*(4), 299–315.

Goldstein, D.S. (2020b). Principles of Autonomic Medicine. <https://neuroscience.nih.gov/publications/Principles%20of%20Autonomic%20Medicine%20v.%202.1.pdf>.

Hosseini, M. R., Fouladi-Fard, R., & Aali, R. (2020). COVID-19 pandemic and sick building syndrome. *Indoor Built Environ.*, *29*(8), 1181–1183.

Huang, C., Huang, L., Wang, Y., Li, X., Ren, L., Gu, X., ... Cao, B. (2021). 6-month consequences of COVID-19 in patients discharged from hospital: A cohort study. *Lancet*, *397*(10270), 220–232.

Jacob, G., Costa, F., Shannon, J. R., Robertson, R. M., Wathen, M., Stein, M., ... Robertson, D. (2000). The neuropathic postural tachycardia syndrome. *N Engl. J. Med.*, *343*, 1008–1014.

Joyner, M. J., & Masuki, S. (2008). POTS vs deconditioning: The same or different? *Clin. Autonom. Res.*, *18*(6), 300–307.

Kristiansen, M. S., Stabursvik, J., O'Leary, E. C., Pedersen, M., Asprusten, T. T., Leegaard, T., ... Wyller, V. B. B. (2019). Clinical symptoms and markers of disease mechanisms in adolescent chronic fatigue following Epstein-Barr virus infection: An exploratory cross-sectional study. *Brain, Behav. Immun.*, *80*, 551–563.

Kupferschmidt, K. (2021a). Startling new variant raises urgent questions. *Science (New York, N.Y.)*, *374*(6572), 1178–1180.

Kupferschmidt, K. (2021b). Where did 'weird' Omicron come from? *Science*, *374*(6572), 1179.

Lamotte, G., Benarroch, E. E., & Coon, E. A. (2021). Paroxysmal hypothermia and hyperhidrosis with exacerbation after COVID-19 Infection. *Clin. Autonom. Res.*, *31* (2), 327–329.

Lamotte, G., Shouman, K., & Benarroch, E. E. (2021). Stress and central autonomic network. *Auton. Neurosci. Basic Clin.*, *235*, 102870.

Lemos, C. (2021). Evoking Dante amidst the COVID-19 pandemic. *Am. J. Med.*, *135*, 126–127.

Levine, T. D., Bellaire, B., Gibbons, C., & Freeman, R. (2021). Cutaneous alpha-synuclein deposition in postural tachycardia patients. *Ann. Clin. Transl. Neurol.*, *8*(4), 908–917.

Lo, Y. L. (2020). COVID-19, fatigue, and dysautonomia. *J. Med. Virol.*, *93*, 1213.

Maksoud, R., du Preez, S., Eaton-Fitch, N., Thapaliya, K., Barnden, L., Cabanas, H., ... Marshall-Gradisnik, S. (2020). A systematic review of neurological impairments in myalgic encephalomyelitis/ chronic fatigue syndrome using neuroimaging techniques. *PLoS One*, *15*(4), e0232475.

Mostel, Z., Ayat, P., Capric, V., Trimmingham, A., & McFarlane, S. I. (2021). Guillain-Barré syndrome in a COVID-19 patient: A case report and review of management strategies. *Am. J. Med. Case. Rep.*, *9*(3), 198–200.

Novak, P. (2020). Post COVID-19 syndrome associated with orthostatic cerebral hypoperfusion syndrome, small fiber neuropathy and benefit of immunotherapy: A case report. *eNeurologicalSci*, *21*, 100276.

Owens, A. P., Low, D. A., Iodice, V., Critchley, H. D., & Mathias, C. J. (2017). The genesis and presentation of anxiety in disorders of autonomic overexcitation. *Auton. Neurosci. Basic Clin.*, *203*, 81–87.

Parsaik, A., Allison, T. G., Singer, W., Sletten, D. M., Joyner, M. J., Benarroch, E. E., ... Sandroni, P. (2012). Deconditioning in patients with orthostatic intolerance. *Neurology*, *79*(14), 1435–1439.

Raj, S. R., Biaggioni, I., Yamhure, P. C., Black, B. K., Paranjape, S. Y., Byrne, D. W., & Robertson, D. (2005). Renin-aldosterone paradox and perturbed blood volume regulation underlying postural tachycardia syndrome. *Circulation*, *111*(13), 1574–1582.

Rowe, P. C., Barron, D. F., Calkins, H., Maumenee, I. H., Tong, P. Y., & Geraghty, M. T. (1999). Orthostatic intolerance and chronic fatigue syndrome associated with Ehlers Danlos syndrome. *J. Pediatr.*, *135*, 494–499.

Schondorf, R., & Low, P. A. (1993). Idiopathic postural orthostatic tachycardia syndrome: An attenuated form of acute pandysautonomia? *Neurology*, *43*(1), 132–137.

Shouman, K., Vanichkachorn, G., Cheshire, W. P., Suarez, M. D., Shelly, S., Lamotte, G. J., ... Singer, W. (2021). Autonomic dysfunction following COVID-19 infection: An early experience. *Clin. Auton. Res.*, *31*(3), 385–394.

Silverman, M. N., Heim, C. M., Nater, U. M., Marques, A. H., & Sternberg, E. M. (2010). Neuroendocrine and immune contributors to fatigue. *Phys. Med. Rehab.*, *2*(5), 338–346.

Singh, J., Yu, C. W. F., & Jeong Tai, K. (2010). Building pathology, investigation of sick buildings —toxic moulds. *Indoor Built Environ.*, *19*(1), 40–47.

Stute, N. L., Stickford, J. L., Province, V. M., Augenreich, M. A., Ratchford, S. M., & Stickford, A. S. L. (2021). COVID-19 is getting on our nerves: Sympathetic neural activity and haemodynamics in young adults recovering from SARS-CoV-2. *J. Physiol.*, *599*(18), 4269–4285.

Su, X. W., Palka, S. V., Rao, R. R., Chen, F. S., Brackney, C. R., & Cambi, F. (2020). SARS-CoV-2-associated Guillain-Barre syndrome with dysautonomia. *Muscle Nerve*, *62*(2), E48–E49.

Umapathi, T., Poh, M. Q. W., Fan, B. E., Li, K. F. C., George, J., & Tan, J. Y. L. (2020). Acute hyperhidrosis and postural tachycardia in a COVID-19 patient. *Clin. Auton. Res. Offic. J. Clin. Auton. Res. Soc.*, *30*(6), 571–573.

van Campen, C., Verheugt, F. W. A., Rowe, P. C., & Visser, F. C. (2020). Cerebral blood flow is reduced in ME/CFS during head-up tilt testing even in the absence of hypotension or tachycardia: A quantitative, controlled study using Doppler echography. *Clin. Neurophysiol. Pract.*, *5*, 50–58.

Verhagen, I. E., van Casteren, D. S., de Vries Lentsch, S., & Terwindt, G. M. (2021). Effect of lockdown during COVID-19 on migraine: A longitudinal cohort study. *Cepha.: Int. J. Headach.*, *41*(7), 865–870.

Vierucci, F., Del Pistoia, M., Fanos, M., Erba, P., & Saggese, G. (2014). Prevalence of hypovitaminosis D and predictors of vitamin D status in Italian healthy adolescents. *Ital. J. Pediatr.*, *40*, 54.

Vuong, Q. C., Allison, J. R., Finkelmeyer, A., Newton, J., & Durham, J. (2020). Brain responses in CFS and TMD to autonomic challenges: An exploratory fMRI study. *JDR Clin. Trans. Res.*, *5*(3), 224–232.

Index

Note: Page numbers followed by "*f*" and "*t*" refer to figures and tables, respectively.

B

C

D

E

F

G

I

J

K

L

M

N

Q

R

S

T

U

V

W

Y

Z

9780323960113